Lecture Notes in Civil Engineering

Volume 450

Lecture Notes in Civil Engineering (LNCE) publishes the latest developments in Civil Engineering—quickly, informally and in top quality. Though original research reported in proceedings and post-proceedings represents the core of LNCE, edited volumes of exceptionally high quality and interest may also be considered for publication. Volumes published in LNCE embrace all aspects and subfields of, as well as new challenges in, Civil Engineering. Topics in the series include:

- Construction and Structural Mechanics
- Building Materials
- Concrete, Steel and Timber Structures
- Geotechnical Engineering
- Earthquake Engineering
- Coastal Engineering
- Ocean and Offshore Engineering; Ships and Floating Structures
- Hydraulics, Hydrology and Water Resources Engineering
- Environmental Engineering and Sustainability
- Structural Health and Monitoring
- Surveying and Geographical Information Systems
- Indoor Environments
- Transportation and Traffic
- Risk Analysis
- Safety and Security

To submit a proposal or request further information, please contact the appropriate Springer Editor:

- Pierpaolo Riva at pierpaolo.riva@springer.com (Europe and Americas);
- Swati Meherishi at swati.meherishi@springer.com (Asia—except China, Australia, and New Zealand);
- Wayne Hu at wayne.hu@springer.com (China).

All books in the series now indexed by Scopus and EI Compendex database!

Shashi Mesapam · Anurag Ohri ·
Venkataramana Sridhar · Nitin Kumar Tripathi
Editors

Developments and Applications of Geomatics

Proceedings of DEVA 2022

Springer

Editors
Shashi Mesapam
Department of Civil Engineering
National Institute of Technology Warangal
Hanamkonda, Telangana, India

Anurag Ohri
Department of Civil Engineering
Indian Institute of Technology (Banaras Hindu University)
Varanasi, India

Venkataramana Sridhar
Department of Biological Systems Engineering
Virginia Tech
Blacksburg, VA, USA

Nitin Kumar Tripathi
School of Engineering and Technology
Asian Institute of Technology
Bangkok, Thailand

ISSN 2366-2557 ISSN 2366-2565 (electronic)
Lecture Notes in Civil Engineering
ISBN 978-981-99-8567-8 ISBN 978-981-99-8568-5 (eBook)
https://doi.org/10.1007/978-981-99-8568-5

This Springer imprint is published by the registered company Springer Nature Singapore Pte Ltd.
The registered company address is: 152 Beach Road, #21-01/04 Gateway East, Singapore 189721, Singapore

Foreword

It gives me immense pleasure to write this foreword to the "Lecture notes in Civil Engineering" containing the selected papers presented during the International Virtual Conference on Developments and Applications of Geomatics (DEVA—2022) held during 29th–31st August, 2022. The conference was held to coincide with the superannuation of Prof. Deva Pratap.

NITW has been running an M.Tech. course in Remote Sensing and GIS, in the Department of Civil Engineering, since 2000 and has established itself as a pioneering institute in the field with its alumni occupying key positions. Many Ph.D. level research work and sponsored R&D works are in progress.

I am happy to note that 38 peer-reviewed papers are part of this volume covering various topics where Remote Sensing and GIS can be applied. The National and International Advisory Committee consisted of experts who have been recognised for their contribution.

I am sure the wealth of information contained in this volume will be useful for the Geomatics engineers who use a wide range of technologies for application in various fields like Civil Engineering, Computer Engineering and Software engineering covering a wide range of topics like land development and planning, satellite and information technologies, advanced surveying etc. The topics in Geoinformatics are interdisciplinary in nature. I am confident that this volume will be very useful and add to the knowledge in the interdisciplinary area of Geoinformatics. I convey my appreciation to the Organizing Secretaries of the Conference for the excellent work done.

Warangal, India

Prof. Bidyadhar Subudhi
Director

Preface

National Institute of Technology Warangal, the first among the chain of erstwhile Regional Engineering Colleges has a well established Department of Civil Engineering which runs an undergraduate and 8 Master degree programs besides Ph.D. One of the specializations is the Remote Sensing and GIS, started in the year 2000. Over the years, this programme has grown in stature and attracts many research projects.

A conference was organised to coincide with the superannuation of Prof. Deva Pratap, who initiated the starting of the Master degree program at NIT Warangal, and Shashi Mesapam, Keesara Venkatareddy and Manali Pal being the Organizing Secretaries. The conference attracted researchers from all parts of India as well as other countries. The conference had to be conducted in a virtual mode due to Covid. The papers covered various areas of application of Geomatics like water resources, environmental engineering, ecosystem management, structural health monitoring, transportation engineering, web GIS etc. The selected papers presented in the conference form the contents of this Lecture Note Series.

We are sure the contents of this volume will be useful for researchers working in the area of Remote Sensing and GIS.

Hanamkonda, India — Shashi Mesapam
Varanasi, India — Anurag Ohri
Blacksburg, USA — Venkataramana Sridhar
Bangkok, Thailand — Nitin Kumar Tripathi

Acknowledgements

A conference of this nature covering a wide area cannot be organized without the involvement of many individuals and organizations. We thank the Director and administration of NIT Warangal for all the infrastructure support. The conference had excellent national and international advisory committee for critically reviewing the papers and offering constructive suggestions for improving the quality of papers selected for the conference. Special mention must be made of IITT Navishkar I-Hub Foundation (IITTNiF) and GIS Monk whose financial support made the conference very successful.

Organizing Secretaries
Shashi Mesapam
Keesara Venkatareddy
Manali Pal

Contents

About the Editors

Prof. Shashi Mesapam is currently an Associate Professor at the Department of Civil Engineering, National Institute of Technology Warangal, India. He obtained his B.Tech. (Civil) from Osmania University, Hyderabad, and M.Tech. (Remote Sensing and Photogrammetric Engineering) and Ph.D. from the Indian Institute of Technology, Roorkee. His major areas of research interests include Photogrammetry, Unmanned Aerial Vehicles (UAV) and its Applications, Digital Image Processing, Remote Sensing, Advanced Surveying. He has published 26 papers in reputed international journals and 4 book chapters in reputed publications.

Prof. Anurag Ohri is currently an Associate Professor at the Department of Civil Engineering, Indian Institute of Technology (Banaras Hindu University), Varanasi, India. He obtained his B.Tech. (Civil) from Regional Engineering College, Kurukshetra, and M.Tech. (Remote Sensing and Photogrammetric Engineering) from the Indian Institute of Technology, Roorkee and Ph.D. from the Indian Institute of Technology (Banaras Hindu University). His major areas of research interests include Mapping, Development of GIS based Decision support systems, GIS application in Smart Cities, Land use planning. He has published 34 papers in reputed international journals and 3 book chapters in reputed publications.

Prof. Venkataramana Sridhar is currently an Associate Professor at the Biological Systems Engineering in Virginia Tech, Blacksburg, USA. He obtained his B.E. in Agricultural Engineering from Tamil Nadu Agricultural University, India and M.Eng. in Irrigation Engineering & Management from Asian Institute of Technology, Thailand and Ph.D. from Oklahoma State University, Stillwater, Oklahoma, USA. His major areas of research interests include Global change; urban systems and sustainability; agricultural water management; water resources systems; modeling; extreme events (floods and droughts); food-energy-water nexus; developing, implementing and evaluating earth system models; new hydrologic instrumentation, sensors, drones, databases, and model applications for use by federal, state, and private stakeholders for real-time operational applications; international activities. He has published 90 papers in reputed international journals and 6 book chapters in reputed publications.

Prof. Nitin Kumar Tripathi is currently a Professor at the Asian Institute of Technology, Thailand. He obtained his B.Tech. (Civil) from National Institute of Technology, Warangal, India, and M.Tech. (Remote Sensing) and Ph.D. from the Indian Institute of Technology, Kanpur. He has published 211 papers in reputed international journals, two books and 12 book chapters in reputed publications. He is currently Editor-in-chief of International Journal of Geoinformatics and Editorial Board Member of Int. Jour. of Applied earth Observation and Geoinformation (ITC Journal), Asian J. of Geoinformatics, Indian J. of Remote Sensing and Photogrammetry, Journal of Prince of Songkhla University, West Indian Journal of Civil Engineering, International Journal of Geoinformatics, Arabian Journal for Science and Engineering, International Journal of Remote Sensing—Taylor and Francis, International Journal of Imaging, J. of Chemistry and Environment, Geojournal—Springer, Disaster Advances, Natural Hazards, International Journal of Geographical Information Science.

Abbreviations

σ	Standard Deviation
%	Percent
°C	Degree Celsius
μ	Micro
2D	Two Dimensional
3D	Tridimensional
95PPU	95 Percent Prediction Uncertainty
ac	Acres
ADRF	Aerosol Direct Radiative Forcing
AHP	Analytic Hierarchy Process
AI	Aridity Index
ALPHA_BF	Base Flow Alpha Factor
ALPHA_BNK	Base Flow Alpha Factor For Bank Storage
AMD	Acid Mine Drainage
AMJ	April–June
ANN	Artificial Neural Network
AP	Andhra Pradesh
API	Application Programming Interface
APK	Android Package Kit
AQI	Air Quality Index
ASCAT	Advanced Scatterometer
ASMR	Advanced Microwave Scanning Radiometer-2
ASTER	Advanced Space Borne Thermal Emission and Reflection Radiometer
AT	Ambient Temperature
ATC	Air Traffic Control
AWBGT	Approximated Wet Bulb Globe Temperature
AWS	Annual Maximum Series
B	Blue
BBMP	Bruhat Bengaluru Mahanagara Palike
BMIC	Bengaluru Mysuru Infrastructure Corridor

BNU-ESM	Beijing Normal University Earth System Model
BOD	Biochemical Oxygen Demand
BSI	Bare Soil Index
BTA	Back Trajectory Analysis
BW	Bandwidth
C	Crop Management Factor
CA	Cellular Automata
Ca	Calcium
CaCO_3	Calcium Carbonate
CBD	Central Business District
CCME	Canadian Council Of Ministers of the Environment
CDD	Maximum Number of Consecutive Days When Daily Rainfall < 1 Mm
CGWB	Central Ground Water Board
CH_K2	Effective Hydraulic Conductivity in Main Channel Alluvium
CH_N2	Manning's N Value for the Main Channel
CI	Consistency Index
CIE	Commission for International Illumination
Cl	Chlorides
Cm	Centimeters
CMI	Crop Moisture Index
CMIP	Coupled Model Inter-Comparison Project
CMOS	Complementary Metal-Oxide-Semiconductor
CN	Curve Number
CNRM-CM5	Centre National for Meteorological Research Coupled Model Version 5
CO	Carbon Monoxide
Conc	Concentration
COVID	Corona Virus Disease
CP	Checkpoints
CPCB	Central Pollution Control Board
CR	Consistency Ratio
CRS	Coordinate Reference System
CRUTS	Climate Research Unit
CWC	Central Water Commission
CWD	Maximum Number of Consecutive Days When Rainfall > 1 Mm
CWT	Concentration Weighted Trajectory
D	Distance from Drainage Network
dB	Decibel
DEM	Digital Elevation Model
DIC	Digital Image Correlation
DN	Digital Number
DO	Dissolved Oxygen

DTM	Digital Terrain Model
EC	Electrical Conductivity
EDRIR	Earthquake Disaster Risk Index Report
EPCO	Plant Uptake Compensation Factor
EPSG	European Petroleum Survey Group
ERAinterim	European Reanalysis Interim
ERSCAT	European Remote Sensing Satellite Scatterometer
ESA	European Space Agency
ESCO	Soil Evaporation Compensation Factor
ESDAC	European Soil Data Center
ESDS	Earth Science Data System
ESRI	Environmental Systems Research Institute
ESSMI	Empirical Standardized Soil Moisture Index
ET	Evapotranspiration
ETM+	Enhanced Thematic Mapper Plus
EUFD	European Union Floods Directive
F	Flow Accumulation
FAO	Food and Agriculture Organization
FC	Field Capacity
FCC	False Colour Composite
FDM	Fused Deposition Modeling
Fig	Figure
FIM	Flood Inundation Mapping
FLOWA	Fuzzy Logic Ordered Weight Averaging
FR	Frequency Ratio
G	Green
G ratio	Green Ratio
GCC	Gnu Compiler Collection
GCM	General Circulation Models
GCP	Ground Control Point
GDVI	Green Difference Vegetation Index
GHG	Green House Gas
GHI	Global Horizontal Irradiance
GIMMS	Global Inventory Modeling and Mapping Studies
GIS	Geographic Information System
GNDVI	Green Normalized Difference Vegetation Index
GNSS	Global Navigation Satellite System
GPCC	Global Precipitation Climatology Centre
GPM	Ground Water Potential Mapping
GPS	Global Positioning System
GRD	Ground Range Detected
GRVI	Green Red Vegetation Index
GS	Ground Station
GSAVI	Green Soil Adjusted Vegetation Index
GSD	Ground Sample Distance

GUI	Graphical User Interface
GW	Ground Water
GW_DELAY	Groundwater Delay
GW_REVAP	Groundwater Revap Coefficient
GWDRVI	Green Wide Dynamic Range Vegetation Index
GWL	Ground Water Level
GWPI	Ground Water Potential Index
GWPZ	Ground Water Potential Zone
GWQMN	Threshold Depth of Water in the Shallow Aquifer Required For Return Flow (Mm)
H	Hue
H+	Hydrogen Ion
HCO_3	Bicarbonate
HD	High Definition
HEC	Hydrologic Engineering Centre
HEC-GeoRAS	Hydrologic Engineering Center's Geospatial River Analysis System
HEC-RAS	Hydrologic Engineering Center's River Analysis System
HI	Horizontal Irradiance
HMS	Hydrologic Modelling System
HR	Hazard Ranking
HRU	Hydrologic Response Unit
HWSD	Harmonized World Soil Database
HYSPLIT	Hybrid Single Particle Lagrangian Integrated Trajectory
IIASA	International Institute for Applied Systems Analysis
IMD	Indian Meteorological Department
IPCC	Intergovernmental Panel on Climate Change
IPSL-CM5A-MR	Institute Pierre-Simon Laplace Coupled Model Version 5A-Medium Resolution
IRDB	Irrigation Design and Research Board
ISC	Indian Standard Code
ISR	Impervious Surface Ratio
ISRIC	International Soil Reference and Information Centre
IT	Information Technology
IUCN	International Union for Conservation of Nature
JAS	July–August
JFM	January–March
K_c	Kappa Coefficient
K_{hist}	Kappa Histogram
kHz	Kilohertz
KIADB	Karnataka Industrial Areas Development Board
K_{loc}	Kappa Local
KM	Kilometers
KrRB	Krishna River Basin
KaRB	Karamana River Basin

KSB	Kijko-Sellevoll-Bayes
KSPCB	Kerala State Pollution Control Board
L	Length
Lat	Latitude
LAT Q	Lateral Flow
LC	Land Cover
Long	Longitude
LOS	Visible Line Of Sight
LR	Logistic Regression
LS	Slope Length and Slope Steepness Factor
LST	Land Surface Temperature
LULC	Land Use Land Cover
LWGNT	Surface Net Downward Longwave Flux
LWGNTCLN	Surface Net Downward Longwave Flux Assuming No Aerosol
LWTUP	Upwelling Longwave Flux at TOA
LWTUPCLN	Upwelling Longwave Flux at TOA Assuming No Aerosol
M	Meters
$M_{(max^obs)}$	Observed Maximum Magnitude
Max	Maximum
MC	Markov Chain
Mc	Magnitude Completeness
MCDA	Multiple-Criteria Decision Analysis
MCDM	Multi Criteria Decision Making
Mg	Magnesium
Mg/L	Milligram per Liter
MGSAVI	Modified Green Soil Adjusted Vegetation Index
MHz	Megahertz
Min	Minimum
MK	Mann & Kendall
ml	Microliters
MLC	Maximum Likelihood Classification
M_{max}	Probable Maximum Earthquake Magnitudes
MOLUSCE	Modules for Land Use Change Evaluation
MSAVI	Second Modified Soil Adjusted Vegetation Index
MT	Metric Tons
MtP	Meteorological Parameters
mW	Milliwatt
Mw	Moment Magnitude
Na	Sodium
NASA	National Aeronautics and Space Administration
NBBSLUP	National Bureau of Soil Survey and Land Use Planning
NCR	Net Change Ratio
NDBI	Normalized Difference Built-Up Index
NDMA	National Disaster Management Authority

NDT	Non Destructive Testing
NDVI	Normalized Difference Vegetation Index
NDWI	Normalized Difference Water Index
NEX-GDDP	NASA Earth Exchange Global Daily Downscaled Climate Projections
NGT	National Green Tribunal
NHAI	National Highway Authority of India
NIR	Near Infra-Red
NLOS	Non Visible or Beyond Line Of Sight
NO_2	Nitrogen Dioxide
NPCM	Normalized Pairwise Comparison Matrix
NPI	National Pollution Index
NRSC	National Remote Sensing Centre
NSE	Nash-Sutcliffe Simulation Efficiency
NSF	National Sanitation Foundation
OA	Overall Accuracy
OBD	Over Burden Dump
OLI	Operational Land Imager
OND	October–December
OSM	Open Street Map
P	Support Practice Factor
PA	Producers Accuracy
PBIAS	Percentage Bias
PCM	Pairwise Comparison Matrix
PDSI	Palmer Drought Severity Index
PET	Potential Evapotranspiration
PLA	Polylactic Acid
Plat	Platinum
PM	Particulate Matter
PMP	Probable Maximum Precipitation
Ppm	Parts Per Million
PRCPTOT	Total Amount of Rainfall in Wet Days
PSCF	Potential Source Contribution Factor
PSO	Particle Swarm Optimisation
$PtCl^{x}{}_{4}$	Chloroplatinate Ion
PTFs	Pedotransfer Functions
PTL	Power Transmission Lines
PV	Photovoltaic
PVIP	Periyar Valley Irrigation Project
QGIS	Quantum Geographic Information System
QGIS	Quantum GIS
qp	Peak Discharge Per Sq. Km Area of Catchment
Qp	Peak Discharge of the Unit Hydrograph for the Catchment Area
QSWAT	QGIS Interface for Soil and Water Assessment Tool

r	Correlation Coefficient
R	Rainfall Erosivity Factor
Rd	Red
R10MM	Number of Days When RR > 10 mm. Count the Number of Days Where Rr_{ij} > 10 mm
R^2	Coefficient of Determination
R20MM	Number of Days When RR > 20 mm. Count the Number of Days Where Rr_{ij} > 20 mm
R95PTOT	Total Rainfall When RR > 95p
RCP	Representative Concentration Pathway
RECHARGE_DP	Deep Aquifer Percolation Factor
REVAPMN	Threshold Depth of Water for Revap or Percolation to Occur
RF	Random Forest
RGB	Red Green Blue
RH	Relative Humidity
RI	Rainfall Intensity
RI	Redness Index
RMSE	Root Mean Square Error
ROI	Region of Interest
RRMSE	Relative Root Mean Square Error
RS	Remote Sensing
RSR	Root Mean Square Error to Standard Deviation Ratio
RTK	Real Time Kinematic
RUSLE	Revised Universal Soil Loss Equation
RX1DAY	Monthly Maximum 1-Day Precipitation
RX5DAY	Monthly Maximum Consecutive 5-Day Precipitation
S	Slope
S	Saturation
S/cm	Simen Per Centimeter
SO_4	Sulphate
SAR	Synthetic Aperture Radar
SAVI	Soil Adjusted Vegetation Index
SCS	Soil Conservation Service
SDG	Sustainable Development Goal
SDR	Sediment Delivery Ratio
SED YIELD	Sediment Yield
SEZ	Special Economic Zone
SfM	Structure from Motion
SLEUTH	Slope Land Use Excluded Urban Transport Hillshade
SLR	Stepwise Linear Regression
SM	Soil Moisture
SMA	Soil Moisture Accounting
SMADI	Soil Moisture Agricultural Drought Index
SMAP	Soil Moisture Active Passive

SMDI	Soil Moisture Deficit Index
SMI	Soil Moisture Index
SMOS	Soil Moisture and Ocean Salinity
SMU	Soil Mapping Unit
SNAP	Sentinel Application Platform
SNR	Signal to Noise Ratio
SO_2	Sulphur Dioxide
SOC	Soil Organic Carbon
SOL_AWC	Available Water Capacity of the Soil Layer
SOL_BD	Moist Bulk Density Mm Layer
SOL_BD	Moist Bulk Density (G/Cm3)
SOL_K	Saturated Hydraulic Conductivity
SPEI	Standardized Precipitation Evapotranspiration Index
SPI	Standardized Precipitation Index
SPSS	Statistical Package for the Social Sciences
SR1 SR2 SR3 & SR4	Scenario 1 2 3 & 4
SSA	Seismic Study Area
SSM	Surface Soil Moisture
SSR	Surface Soil Roughness
STD	Standard Deviation
STL	Social and Travel Lockdown
SUFI 2	Sequential Uncertainty Fitting 2
SUH	Synthetic Unit Hydrograph
SUR Q	Surface Runoff
SURLAG	Surface Runoff Lag Coefficient
SWAT	Soil and Water Assessment Tool
SWAT CUP	SWAT Calibration and Uncertainty Programs
SWDI	Soil Water Deficit Index
SWGNT	Surface Net Downward Shortwave Flux
SWGNTCLN	Surface Net Downward Shortwave Flux Assuming No Aerosol
SWIR	Short Wave Infra Red
SWMM	Storm Water Management Model
SWTNT	Toa Net Downward Shortwave Flux
SWTNTCLN	TOA Net Downward Shortwave Flux Assuming No Aerosol
SY_{obd}	Observed Sediment Yield
SY_{pred}	Simulated Sediment Yield
T	Temperature
TA	Total Alkalinity
TB	Time Base of SUH
TC	Total Coliform
TCU	True Colour Unit
TDS	Total Dissolved Solids
TH	Total Hardness

TIN	Triangular Irregular Network
TIRS	Thermal Infrared Sensor
TM	Thematic Mapper
T_m	Peak Time of SUH
TNN	Monthly Minimum Value of Daily Minimum Temperature
TNX	Monthly Maximum Value of Daily Minimum Temperature
TOA	Top of Atmosphere
t_p	Basin Lag
TSA	Theil-Sen Approach
TWI	Topographic Wetness Index
TXN	Monthly Minimum Value of Daily Maximum Temperature
TXX	Monthly Maximum Value of Daily Maximum Temperature
UA	Users Accuracy
UAS	Unmanned Aerial System
UAV	Unmanned Aerial Vehicle
UER	Urban Expansion Rate
UGM	Urban Growth Model
UHI	Urban Heat Island
UR	Urbanization Rate
USDA	United States Department of Agriculture
USGS	United States Geological Survey
UTM	Universal Transverse Mercator
V	Value
VCI	Vegetation Condition Index
VES	Vertical Electrical Sounding
VIC	Variable Infiltration Capacity
VSWI	Vegetation Supply Water Index
W50	Width of SUH at 50% of Peak Discharge Ordinate Q_p in Hours
W75	Width of SUH at 75% of Peak Discharge Ordinate Q_p in Hours
WAD	World Atlas of Desertification
WAWQI	Weighted Arithmetic W_{qi}
WBGT	Wet Bulb Globe Temperature
WDVI	Weighted Difference Vegetation Index
W_f	Weightage Factor
WG	Western Ghats
WGS	World Geodetic System
WHO	World Health Organization
WMO	World Meteorological Organization
WOA	Weighted Overlay Analysis
WOE	Weight of Evidence
WP	Wilting Point
WQ	Water Quality
WQI	Water Quality Index

WR50	Half Width of SUH at 50% of Peak Discharge Ordinate Q_p in Hours
WR75	Half Width of SUH at 75% of Peak Discharge Ordinate Q_p in Hours
WRIS	Water Resource Information System
WRS	Worldwide Reference System
WS	Wind Speed
X_m	Highest Value in the Series
XML	Extensible Markup Language
X_n	Mean of N Annual Maximum Values
$X_{n\text{-}m}$	Mean of the Series Excluding the Highest Value
XS	Cross Section
YLCD	Yearly Land Cover Dynamic
Z_MK	Standardized MK-Test Statistic

Analyzing the Potential Application of Low-Cost Digital Image Correlation in Direct Shear Test

G. Alhakim, C. Nuñez-Temes, J. Ortiz-Sanz, and M. Arza-García

Abstract Nowadays, Digital Image Correlation (DIC) has become a frequently used optical technique for testing lab materials due to its nature of non-contact method, allowing for full-field displacements and strains to be accurately measured. The purpose of this study is to examine the potential employment of DIC in performing the standard direct shear test by using a modified box to observe and measure the shear properties of a soil sample. Accessible tools (e.g., a consumer grade DSLR camera) and DIC open-source software were employed in this study in order to monitor every portion of the speckled pattern during deformation and tracking the relative displacements. While it is often difficult to separate the relative contributions of individual error sources in any optical technique, we propose the use of two different validation methods for assessing the accuracy of DIC: the noise-floor and the direct comparison with a second trusted source, like transducers. The results of the conducted study demonstrate that the DIC technique implemented on a properly prepared direct shear laboratory setup opens up new possibilities for the effective and accurate analysis of deformations in soil materials under direct shear loading conditions. The calculations of the noise floor of the setup and speckled pattern in terms of the mean and distribution of the displacements showed validated results with STD of 0.001 and 0.0012 mm for the horizontal and vertical planar components, respectively. The displacements measured by DIC showed good agreement with the results of the transducer with an average error of 0.1 mm.

Keywords Direct shear test · DIC analysis · Displacement measurement · NDT

G. Alhakim (✉) · C. Nuñez-Temes · J. Ortiz-Sanz · M. Arza-García
Civil and Geomatics Research Group, Department of Agroforestry Engineering, University of Santiago de Compostela, Santiago, Spain
e-mail: g.hakim@bau.edu.lb

G. Alhakim
Civil and Environmental Engineering Department, Faculty of Engineering, Beirut Arab University, Beirut, Lebanon

S. Mesapam et al. (eds.), *Developments and Applications of Geomatics*,
Lecture Notes in Civil Engineering 450, https://doi.org/10.1007/978-981-99-8568-5_1

1 Introduction

In recent years, Digital image correlation (DIC) has become one of the preferred optical techniques for the analysis of close-range displacements in several lab and in-field experiments, being able to capture the shape, motion, and deformation of object surfaces. Since the development of these DIC methods over the past decades, numerous applications have been found in many fields and areas of research, due to its ease of use. Hence, this technique has found a particularly interesting application niche in those fields related to civil engineering.

The basic principle of DIC is to track a speckle pattern in a sequence of images and to calculate the displacements of small portions of this pattern (often called subsets) between one reference image and the other (deformed) images. The variations of coordinates between the original and deformed subset centers provide the required displacement vector of the targeted point. Similarly, the full-field deformation of other interrogated points could be acquired. For this purpose, DIC usually requires an artificial speckle pattern to resolve unique issues [1]. Nonetheless, the pattern can also sometimes be naturally present over the surface of some materials (e.g., rocks, sand, etc.).

In general, there are now multiple software solutions, both commercial and open source, to perform the correlation between the images. Among others, we can find commercial systems like Aramis (GOM) or VIC-3D (Correlated Solutions), that are normally provided with their own specialized hardware and optical components. As the use of DIC becomes more and more widespread, diverse open-source projects are also emerging (e.g., nCorr [2], py2DIC [3], etc.) making this technique more accessible, even for small companies and laboratories with more limited resources. At a level of hardware, also industrial high-speed cameras have proven to be replaceable—at low cost—by consumer grade cameras in certain experimental setups where the frame rate is not a decisive issue [4]. The affordability of the approach and the simplicity of its implementation have democratized to some extent the use of DIC, bringing it closer to all practitioners.

DIC is currently considered a simple but reliable technique for analyzing surface displacements and deformations. However, the analysis of the results and their accurate assessment can be technically challenging. The main problem is a general lack of control over the final quality of the output, as the users are sometimes more fascinated by the detail of the full-field maps rather than aware of the metric quality of the results obtained. It is important to remark that DIC measurements, as many other optical techniques, are affected by different types of errors. Among others, the results can be affected by the stationary noise from camera sensors, errors related to the illumination or temperature, errors associated with speckle pattern, errors associated with the processing, external biases (e.g., vibrations), etc. [5, 6]. Therefore, the accuracy of every experiment requires careful verification.

1.1 DIC Applications in Lab Testing of Materials

The experimental technique of imaging correlation has gained widespread acceptance as a reliable tool for strain measurements. Thus, DIC has been appealing to researchers in various engineering applications as a non-destructive testing (NDT). Many studies have examined the use of the DIC technique to analyze the concrete behavior, such as the distribution of deformation and crack propagation of concrete [7], quantification of the fracture properties [8] and fatigue behavior [9] of reinforced concrete beams. Besides, the DIC optical method was employed to assess the behavior of road materials [10–12].

Furthermore, geotechnical engineering investigation relies greatly on DIC practices to determine subpixel displacements in the soil, through which strain is estimated [13]. Plé, Tourabi and Abuaisha [14] have investigated the strain deformation during a direct tensile test performed on a clayey soil by applying 3-D DIC technique. They stated that this method could be adapted for the strain determination of the cap cover in landfills composed of compacted clay. A modified consolidation mold has been developed to monitor the micro-structure behavior of unsaturated soil with axial drainage conditions. By using this half mold oedometer apparatus, Liu et al. [15] studied the deformation field of silty clay soil by applying digital image analysis. They declared the importance of this study in terms of understanding the correlation between micro and macro mechanical properties of the soil. Ko et al. [16] conducted an experiment of a small-scale retaining wall by reproducing the collapse behavior of the structure and measured the deformation and displacement of the whole surface of the wall by using 3D-DIC. Tong and Yoo [17] discussed the application of DIC technology on geotechnical small-scale models such as a retaining wall, trapdoor, shallow footing, and tensile test on geogrid. They revealed the effectiveness of DIC in monitoring the deformation and strain field during these laboratory tests.

1.2 Direct Shear Test

The direct shear test is one of the oldest experimental procedures implemented in geotechnical engineering practice, and it is the most widely used geotechnical shear device due to its simplicity. This experimental procedure is conducted in order to determine the shear strength and shear parameters of soil materials. The accurate determination of these parameters represents a key issue in the design of various structures, e.g., retaining walls, earth buildings, dams, and roads, as well as the design of foundations beneath any architectural and/or engineering structure. However, one of the demerits of the direct shear test could be the non-uniform distributions of stress and strain on the failure plane.

Accordingly, numerical methods have been employed to analyze the details of stress and strain occurrence within a shear box, that would be difficult to measure in the experimental study. Dirgelienė, Skuodis and Grigusevičius [18] carried out experimental and numerical investigations of the direct shear test performed under constant vertical stress and constant sample volume. The results of the finite element method showed that the vertical stress and displacements were non-uniformly distributed in the sample in both conditions. Zhang and Thornton [19] simulated the direct shear test by using the discrete element method and illustrated the heterogeneous distributions of stress and strain. In addition, they demonstrated that the shear strain as well as the dilation concentrated in the mid-height of the specimens (within the shear zone). To study the soil behavior at the microscopic level, Yan [20] presented a 3-D numerical model of granular soil with elongated particles in a direct shear test. It was observed that a localized dilative shear zone was revealed along the failure plane and the change in particle orientation was clearly noticed.

While these non-uniform distributions of stress and strain can be numerically modelled, the employment of a full-field measurement NDT method like DIC could be very useful for tracking experimentally the changes in displacements on the lateral surface of the sheared samples and for validating the models [21]. The main limitation for applying DIC in the direct shear test is that the mold used in the test apparatus is a metal closed box, which makes it impossible to monitor the real behavior of the soil. Kong, Cheng, and Hua [22] applied digital image technique to direct shear test by employing a modified shear box and determined the displacement and strain fields as well as the particle orientation. They concluded that the maximum shear strain occurred at the interface between the upper and lower shear boxes and the formation of shear band could be clearly detected. However, in their research they did not consider any method to validate the accuracy of DIC.

The review of the mentioned studies enlightens the benefits of DIC application in the direct shear test to monitor the strain field and displacement instead of assuming uniform distribution of stress and strain during shearing. Despite this, the usage of the DIC technique in the direct shear test of soil has not been fully investigated yet, especially when it comes to quantify the precision of this technology in this experimental test. Hence, the main objective of this research is to study the potential application of DIC in conducting the direct shear test on soil. Accordingly, a modified shear box and mold were designed for this purpose and accessible tools (e.g., a consumer grade digital single-lens reflex (DSLR) camera and open-source software) were employed. The relative contributions of individual error sources are hardly separated in any optical practice. Therefore, we suggest the use of two different validation methods to quantify the accuracy of DIC: the noise-floor, and the direct comparison with a conventional contact measurement device.

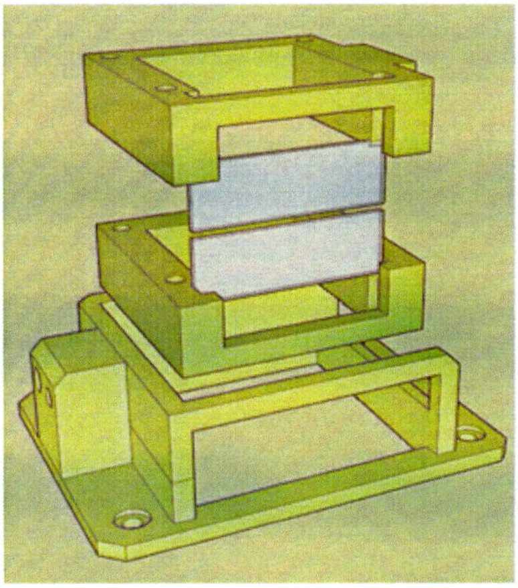

Fig. 1 Shear mold and box in 3-D model

2 Materials and Method

2.1 Shear Mold

In order to monitor the soil grains displacement and the deformation during shearing, the original shear box must be modified by cutting one of the sides of the upper and lower molds. For experimental purposes, and to prevent the damage of the original metal mold, a new cost-effective one was produced by applying Fused Deposition Modeling (FDM) printing. A 3-D printed mold was manufactured with a Prusa i3 (BQ, Spain) 3-D printer by using a bioplastic filament material, namely, Polylactic Acid (PLA). The original dimensions of the shear box (60 × 60 mm) were maintained. Moreover, the open sides of the upper and lower boxes were replaced with two plexiglass plates of 3 mm thickness, as illustrated in Fig. 1.

2.2 2D-DIC Fundamentals

To be able to analyze displacements with 2D-DIC in the image series, the first stage consists in selecting a subset or portion of the speckled pattern to be tracked. The subset size should be chosen in a way that allows to distinguish every subset in the region of interest from all other subsets. There is no one-size-fits-all rule in this sense, but it is commonly accepted between DIC practitioners to choose a subset size that allows facets containing at least 3 speckles. Larger subsets typically result in lower displacement noise, but often at the cost of increased spatial smoothing [23]. A process of image correlation is performed to detect homologous subsets between the reference image and the deformed ones.

As illustrated in Fig. 2, the coordinates of the central point of every initial subset can be mapped in the target subset by using the computed image displacement and first-order shape functions, representing translations, rotations, normal and shear strains (Eqs. 1 and 2).

$$\varsigma 1 = \mathrm{u} + (\partial \mathrm{u} \partial \mathrm{x})\mathrm{dx} + (\partial \mathrm{u} \partial \mathrm{v})\mathrm{dy} \tag{1}$$

$$\eta 1 = \mathrm{v} + (\partial \mathrm{v} \partial \mathrm{x})\mathrm{dx} + (\partial \mathrm{v} \partial \mathrm{y})\mathrm{dy} \tag{2}$$

where $\varsigma 1$ and $\eta 1$ are the displacements of the subset; u and v represent the translations; $\partial u/\partial x$ and $\partial v/\partial y$ represent the normal strains; $\partial u/\partial y$ and $\partial v/\partial x$ represent the shear strains; and dx and dy represent the distances from the subset center to an arbitrary point within the same subset in the x and y directions, respectively.

As DIC software works with pixel units, the displacement maps resulting from processing all the points of interest do not have an inherent scale. For this reason, DIC requires a calibration procedure to establish the image scale, relating pixel units with real units (what can be even done with a simple known distance). However, at the same time, the calibration procedure can be also used to correct for lens distortions if a calibration plate is used. In this study, we performed the 2D-DIC calibration by taking images of a special calibration plate (i.e., symmetric dot grid target) prior to the experiments. By tracking the displacements of the points of the grid in along the images of the plate, the parameters of the camera can be calibrated. If the intrinsic and extrinsic parameters of a 2D, single camera system are calibrated, then an out-of-plane tilt of the test piece can be determined and corrected. In any case, when only one camera is used for DIC (2D-DIC) is crucial to ensure that the translation remain strictly perpendicular to the optical axis; otherwise, false displacements due to out-of-plane motion will be produced.

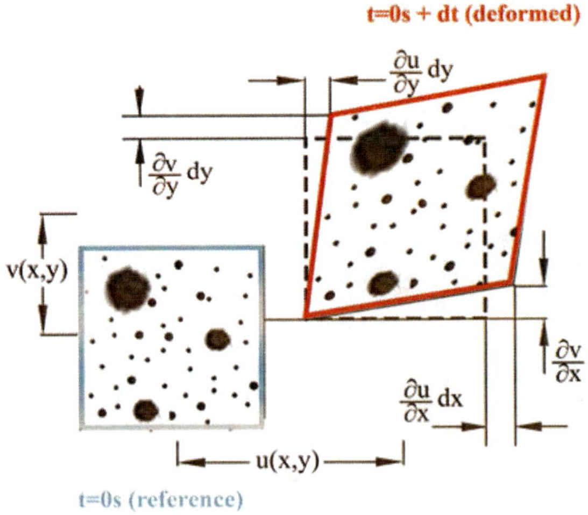

Fig. 2 Schematic illustration of the basic principles in 2-D DIC

2.3 *Experimental Setup*

The first step to process a DIC experiment is to prepare a pattern on the surface of the analyzed sample. Therefore, to get a reference for the box displacements, a half of the transparent plate was painted with white paint and speckled randomly with black spots, as is typical for DIC, in order to maximize the contrast of the speckles and to improve the reliability of the tracked displacements. To monitor the experiments, image series were taken by using a digital single-lens reflex camera (DSLR) Canon EOS 1200D with a resolution of 18.1-megapixel, equipped with Canon EF lens (image settings: focal length 21 mm, ISO-100, exposure time 1/25s, f/6). The camera and the modified shear mold were fixed on a wooden bench with its optical axis normal to the studied surface, and the proper lighting to the tested specimen was applied. The setup of the experimental work is illustrated in Fig. 3. The modified shear mold was subjected to a vertical load in correspondence with the conventional direct shear test.

The captured images of the examined sample are recorded before and during motion and/or deformation. In the center of the box, a transparent area was also kept in order to monitor the behavior of the soil sample when it is introduced into the box as shown in Fig. 3b. The area of interest was then selected on the investigated surface as highlighted in Fig. 4. During deformation, every portion of the pattern (i.e., subsets) is matched between the initial (reference) image and the deformed images, while the software is able to calculate the relative displacements. The studied image series were processed by using the open-source software DICe (SNL, Albuquerque, USA) that can be directly run on most operating systems.

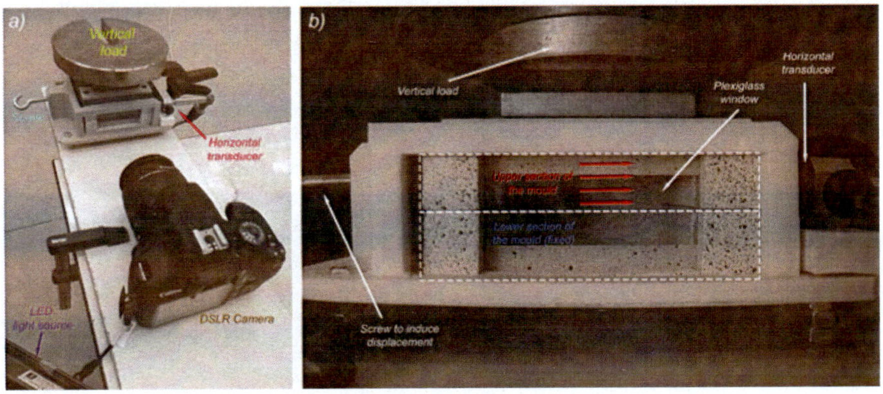

Fig. 3 Experimental setup **a** test bench for direct shear, and **b** modified shear box

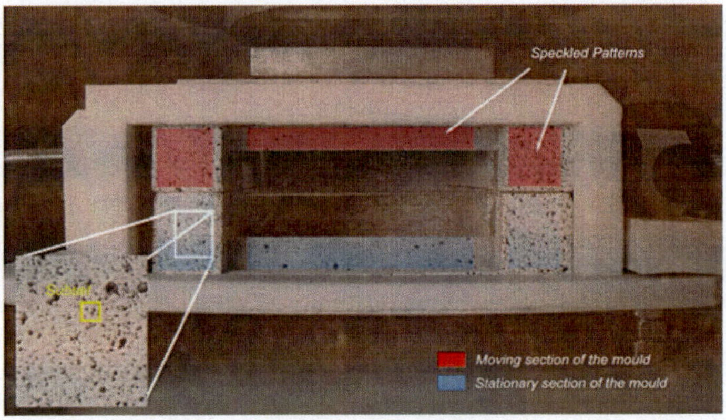

Fig. 4 Area of interest monitored using digital image correlation and representation of the subset size

2.4 *Validation of Results*

DIC results can vary widely among different setups, speckled patterns, matching parameters, etc. Therefore, it is important to use validation methods to check the obtained measurements. Two main methods were employed here in order to evaluate the DIC errors. On the one hand, the noise floor was calculated by taking and processing a series of 100 "stationary" (undeformed) images of the box to get an estimate of the threshold value of displacements under which measurements are indistinguishable from noise. On the other hand, the measured displacements between DIC and the transducers were compared. The induced controlled displacements to the shear boxes were achieved in this case by turning a screw that pushes the upper shear box, while the lower one is fixed.

3 Results and Discussion

3.1 *Noise Floor Analysis*

The main source of random error in digital image correlation is camera noise. This is the temporal fluctuations of the gray levels that each pixel of the camera sensor observing a fixed object will perceive. The quantification of uncertainty considering the variance errors (or noise) is critical for rational assessment of DIC results. In this case, to study the noise floor, a series of 100 pre-test images for the static box were captured. Without any movement in the setup, the predictable displacement computed by DIC analysis is zero, and any measured displacement is noise. Eight

Fig. 5 Selected check points on the upper and lower shear boxes

points were chosen on the monitored surface of the direct shear box within the desired field of view as displayed in Fig. 5.

The results of noise in both X and Y directions are presented in Fig. 6. As it can be noticed, the average peak-to-peak ranges for all the selected points, or in other words the average difference between the lowest and highest detected displacement, are 0.0035 and 0.0052 mm for X and Y displacements, respectively. The standard deviation is generally computed to quantify the variance errors, and typically it should be similar for both X and Y directions. In this case, the standard deviation attained the values of 0.001 and 0.0012 in the horizontal (X) and vertical (Y) directions, respectively. It is worth mentioning that considering the maximum or average standard deviation between both directions, indicates that the values tend to be close to the mean of the set with a small amount of dispersion, thus, a low noise level.

Moreover, the standard deviation was calculated for the check points selected on the plexiglass and the plastic mold distinctly. Particularly, points 0, 1, 2, and 3 correspond to the shear mold while points 4, 5, 6, and 7 refer to the plexiglass as shown in Fig. 5. In the first place, the standard deviation of the noise X-displacement reached 0.0011 and 0.0008 mm of the points selected on the mold and the glass, respectively. In addition, for the noise Y-displacement, the standard deviation of the chosen points attained 0.0013 for the mold, and 0.0011 for the glass. It can be noted that the standard deviation of the noise of the points selected on the mold surface is slightly greater than that of the glass in both directions. This could probably be due to the different planar positions of both materials, more precisely, the normal distance between the camera lens and the mold surface is relatively less than that of the glass plane.

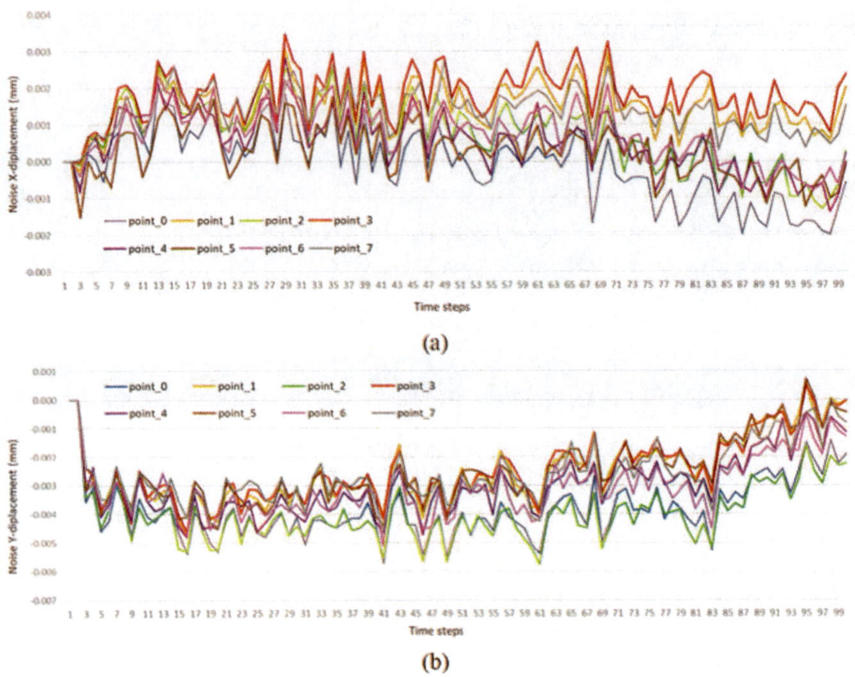

Fig. 6 Noise baseline in 8 checkpoints for **a** X-displacement, and **b** Y-displacement

3.2 Accuracy Assessment

Figure 7 shows the displacements of the upper shear box measured by the extensometer from one side and processed by DIC method from the other side. To analyze the potential bias that could be introduced by the plexiglass, DIC results were also obtained for the glass and the plastic mold separately. To track these displacements, four points were selected on the speckled surface of each material, and it can be noticed that the results are in good agreement.

Furthermore, Fig. 8 illustrates the relation between the displacements determined by the transducer and image processing technique. The correlation coefficients displayed on the graph depict a high fit between the results from DIC and LVDT for both the glass and the shear mold. The scatter plots visualizing bivariate data conform to a linear, strong relationship between both variables. The average estimated error between the results is about 0.1 mm. Thus, it can be concluded that the application of DIC in direct shear test is a reliable technique to study the mechanical behavior of soils particularly the shear strength and parameters.

For a better illustration of the results, Table 1 presented the links for videos of the previously discussed results, namely, the noise floor analysis and the DIC measured displacements, in the horizontal and vertical directions, as well as the

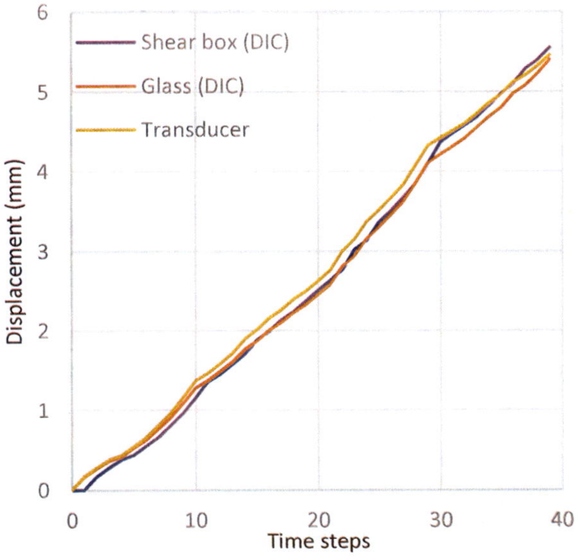

Fig. 7 Displacements versus time steps

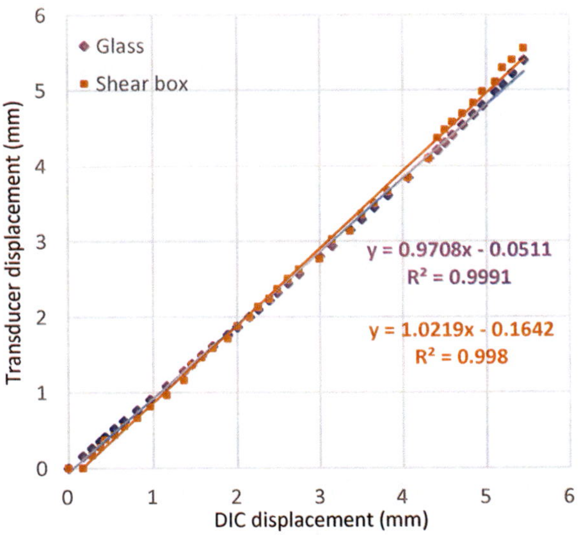

Fig. 8 Comparison of displacements measured by transducer and DIC

component measurements. For instance, Fig. 9 showed a screenshot of the horizontal displacement captured at a certain time interval and processed by DIC technique.

Table 1 Video links showing DIC results for noise floor and displacements

	Description	Video links
Noise floor **(stationary images)**	Absolute noise	https://doi.org/10.6084/m9.figshare.20154491.v1
	Horizontal noise	https://doi.org/10.6084/m9.figshare.20154494.v1
	Vertical noise	https://doi.org/10.6084/m9.figshare.20154488.v1
Displacement measured by DIC **(moving shear box)**	XY displacement component	https://doi.org/10.6084/m9.figshare.20154479.v1
	Horizontal displacement	https://doi.org/10.6084/m9.figshare.20154482.v1
	Vertical displacement	https://doi.org/10.6084/m9.figshare.20154485.v1

Fig. 9 Screen capture for X-displacement measured by DIC

4 Conclusions

The results of the present research reveal the significance of applying the DIC technique on a properly prepared direct shear laboratory setup with the use of a modified shear box. The DIC application creates new possibilities for the effective and accurate analysis of deformations in soil materials when subjected to shear loading. And consequently, it compensates for the limitation of the standard direct shear test, particularly by measuring the non-uniform strain field and displacement within the soil sample rather than assuming a uniform distribution on the failure plane. The calculations of the noise floor of the setup and speckled pattern in terms of the mean and distribution of the displacements showed validated results with STD of 0.001 and 0.0012 mm in horizontal and vertical directions, respectively. Moreover, the displacements of the shear box, that was subjected to a lateral force introduced by turning a screw, measured by DIC are in conformity with the results of transducer deformations with an average error of 0.1 mm. The displacements were also measured separately for the plexiglass and the plastic box, and the results showed good agreement.

Acknowledgements This work was supported by the EU Commission under the program Erasmus+ KA107 and Xunta de Galicia (Spain) under the grant "Financial aid for the consolidation and structure of competitive units of investigation in the universities of the University Galician System (2020-22)" Ref. ED341B 2020/25.

References

1. Dong YL, Pan B (2017) A review of speckle pattern fabrication and assessment for digital image correlation. Exp Mech 57:1161–1181. https://doi.org/10.1007/s11340-017-0283-1
2. Blaber J, Adair B, Antoniou A (2015) Ncorr: open-source 2D digital image correlation matlab software. Exp Mech 55:1105–1122. https://doi.org/10.1007/s11340-015-0009-1
3. Belloni V, Ravanelli R, Nascetti A, Di Rita M, Mattei D, Crespi M (2019) Py2dic: a new free and open source software for displacement and strain measurements in the field of experimental mechanics. Sensors (Switzerland) 19:1–19. https://doi.org/10.3390/s19183832
4. Arza-García M, Nuñez-Temes C, Lorenzana JA, Ortiz-Sanz JP, Castro A, Portela-Barral M, Gil-Docampo M, Bastos G (2022) Evaluation of a low—cost approach to 2—D digital image correlation vs . a commercial stereo—DIC system in Brazilian testing of soil specimens. Arch Civ Mech Eng 22:1–13. https://doi.org/10.1007/s43452-021-00325-0
5. Reu PL (2013) Uncertainty quantification for 3D digital image correlation. Conf Proc Soc Exp Mech Ser 3:311–317. https://doi.org/10.1007/978-1-4614-4235-6_43
6. Dematteis N, Giordan D (2021) Comparison of digital image correlation methods and the impact of noise in geoscience applications. Remote Sens 13:1–25. https://doi.org/10.3390/rs13020327
7. Huang Y, He X, Wang Q, Xiao J (2019) Deformation field and crack analyses of concrete using digital image correlation method. Front Struct. Civ Eng 13:1183–1199. https://doi.org/10.1007/s11709-019-0545-3
8. Fayyad TM, Lees JM (2014) Application of digital image correlation to reinforced concrete fracture, procedia. Mater Sci 3:1585–1590. https://doi.org/10.1016/j.mspro.2014.06.256
9. Mahal M, Blanksvärd T, Täljsten B, Sas G (2015) Using digital image correlation to evaluate fatigue behavior of strengthened reinforced concrete beams. Eng Struct 105:277–288. https://doi.org/10.1016/j.engstruct.2015.10.017
10. Romeo E (2013) Two-dimensional digital image correlation for asphalt mixture characterisation: interest and limitations. Road Mater Pavement Des 14:747–763. https://doi.org/10.1080/14680629.2013.815128
11. Núñez-Temes C, Bastos G, Arza-García M, Castro A, Lorenzana Fernández JA, Ortiz-Sanz J, Portela M, Gil-Docampo M, Prego FJ (2022) Assessment of pavement deflection under vehicle loads using a 3D-DIC system in the field. Sci Rep 12:1–15. https://doi.org/10.1038/s41598-022-13176-3
12. Górszczyk J, Malicki K, Zych T (2019) Application of digital image correlation (DIC) method for road material testing, Materials (Basel) 12. https://doi.org/10.3390/ma12152349
13. Eichhorn GN, Bowman A, Haigh SK, Stanier S (2020) Low-cost digital image correlation and strain measurement for geotechnical applications. Strain 56. https://doi.org/10.1111/str.12348

14. Plé O, Tourabi A, Abuaisha MS (2013) 3-Dimensional digital image correlation for strains determination in clayey soil. Appl Mech Mater 353–354:463–466. https://doi.org/10.4028/www.scientific.net/AMM.353-356.463
15. Liu W, Cao L, Li X, Li Y, Wang P (2011) Experimental study of silty clay's deformation fields based on the principles of oedometer tests. Adv Mater Res 261–263:1539–1543. https://doi.org/10.4028/www.scientific.net/AMR.261-263.1539
16. Ko Y, Seo S, Jin T, Chung M (2021) Feasibility evaluation of the 3D-DIC non contact measurement system using small-scaled model test of earth retaining wall. Int J Geo-Eng 12. https://doi.org/10.1186/s40703-021-00141-8
17. Tong BC (2022) Application of digital image correlations (DIC) technique on geotechnical reduced-scale model tests. J Korean Geosynth Soc 21:33–48. https://doi.org/10.12814/jkgss.2022.21.1.033
18. Dirgelienė N, Skuodis Š, Grigusevičius A (2017) Experimental and numerical analysis of direct shear test. Procedia Eng 172:218–225. https://doi.org/10.1016/j.proeng.2017.02.052
19. Zhang L, Thornton C (2007) A numerical examination of the direct shear test. Geotechnique 57:343–354. https://doi.org/10.1680/geot.2007.57.4.343
20. Yan WM (2009) Fabric evolution in a numerical direct shear test. Comput Geotech 36:597–603. https://doi.org/10.1016/j.compgeo.2008.09.007
21. Shen J, Wang X, Liu W, Zhang P, Zhu C, Wang X (2020) Experimental study on mesoscopic shear behavior of calcareous sand material with digital imaging approach. Adv Civ Eng 2020. https://doi.org/10.1155/2020/8881264
22. Kong L, Chen F, Hua L (2014) Meso-direct-shear test of sands based on the digital image method. In: American society of civil engineers (ASCE), pp 265–274. https://doi.org/10.1061/9780784413388.027
23. IDICS, a good practices guide for digital image correlation. Int Digit Image Correl Soc 94. 10.32720/idics/gpg.ed1.%0Ahttp://idics.org/guide/

Applications of GIS in Estimating the Probable Maximum Earthquake Magnitude for Amaravati Region, Andhra Pradesh, India

M. Madhusudhan Reddy, R. Siddhardha, G. Kalyan Kumar, and R. Suresh

Abstract An earthquake magnitude and their sizes have a significant impact on damage at a particular site. Hazard analysis provides the insights for vulnerability statistics of the area that is considered under study. Several important parameters including probable maximum earthquake magnitudes (M_{max}) are required to quantify the seismic risk. In this study to estimate the M_{max} an area of 500 km radial distance has been considered as a seismic study area keeping a center latitude of 16° 52′N and longitude of 80° 51′E. An earthquake catalogue prepared for the period of 221 years, the earthquake data collected from 1801 to 2022, and the historical earthquake data has been collected from several data sources in order to investigate the seismicity of the area. The declustering techniques are used to remove the dependent events from earthquake data. All different earthquake magnitudes are converted into a moment magnitude (M_w) using empirical relations. To estimate the seismic hazard parameters the catalogue completeness analysis was carried out; for this entire catalogue grouped into six different magnitude ranges with a constant bin width of 0.5 $M_{w.}$ The whole extended study region has been separated into four source zones and designated with zone 1, 2, 3, and 4 based on estimated hazard parameters. The M_{max} was estimated considering the data from each zone using Gupta and Kijko-Sellevoll-Bayes (KSB) methods. It has been observed that the obtained M_{max} is varying from 6.1 to 6.9 M_w according to Gupta's method whereas as per the KSB method the estimated M_{max} varies from 6.0 to 7.7 M_w. The subzone-2 is contributing the highest value of M_{max} compared to other zones. The estimated M_{max} can be used further to quantify the seismic hazard risk of the Amaravati region.

Keywords Earthquake catalogue · Maximum magnitude (M_{max}) · The incremental increase method · Kijko-Sellevoll-Bayes method

---

M. M. Reddy (✉) · R. Suresh
Civil Engineering Department, Institute of Aeronautical Engineering, Dundigal, Hyderabad, India
e-mail: madhu.mrc@gmail.com

R. Siddhardha · G. K. Kumar
Civil Engineering Department, National Institute of Technology Warangal, Hanamkonda, Telangana, India

S. Mesapam et al. (eds.), *Developments and Applications of Geomatics*,
Lecture Notes in Civil Engineering 450, https://doi.org/10.1007/978-981-99-8568-5_2

1 Introduction

The risks that come along with earthquakes almost always result in considerable damage to the buildings, and in some cases, it may even be the cause of fatalities that occur as a direct result of earthquakes in different parts of the globe. In recent decades, a number of significant earthquakes have struck India, causing enormous damage throughout the nation [1], even in the country's stable continental region. According to the seismic microzonation map of India (2011), the whole of the nation has been separated into four distinct seismic zones, which have been designated as Zones II, III, IV, and V accordingly [2]. The seismic zonation map of India is given in Fig. 1, in which zones V and IV are classed as extremely high-risk and high-risk zones, whereas zones III and II are classified as moderate-risk to low-risk zones. On the other hand, the 2019 Earthquake Disaster Risk Index Report (EDRIR) states that a total of 59% of Indian lands are designated as being sensitive to earthquakes, and the intensity of the earthquakes would fall somewhere in the range of moderate to catastrophic. An experimental investigation was carried out by the National Disaster Management Authority (NDMA), India, and the International Institute of Information Technology, Hyderabad to study the risk index of 50 cities across the India and one complete district [3]. The cities that have been selected for evaluation risk are primarily based on the three major factors which include population density, housing threat factor followed by the smart cities as a factor. Initially, the risk associated with the selected cities has been categorised into low, medium, and high based on the seismic hazard map of India and other past research studies. The past few major earthquakes are actually considered as a strong evident to understand the influence of the local geological, geomorphological and other regional factors on damage during the earthquake. Some of major earthquakes that occurs across India are the Manipur (6.7 M_w) in 2016, Nepal (7.8 M_w) in 2015, Sikkim (6.9 M_w) in 2011, Kashmir (7.6 M_w) in 2005, Bhuj (7.7 M_w) in 2001, Chamoli (6.8 M_w) in 1999, and Jabalpur (5.8 M_w) in 1997. After experiencing several major earthquakes the various governing bodies which are associated with the Indian government have made a decision to investigate subsoil stratification and identifying the earthquake prone areas across the country is very much needed and considered as most important and hence concluded that the outcomes of the planned investigations are effectively used as a guiding tool for land users for safe construction practises [4]. It has been observed by many researchers that the microzonation study will provide a best solution to estimate and control the seismic hazard during the earthquakes. Many researchers across the country are working towards making the microzonation procedure simple and realistic. The primary step in microzonation studies is the characterization of the local site conditions and estimation of seismic hazard analysis. In the current study, after making earthquake catalogue free from dependency events, the seismic hazard parameters such as a-value, b-value, and magnitude completeness (M_c), β and M_{max} have been estimated for the selected study region of area 217.3 km^2, where M_{max} is nothing but it is the maximum magnitude of an earthquake that might be take place in the future. Thus, estimation of the expected M_{max} is very important for parameters in seismic hazard analysis and hence these estimated M_{max} values are significantly used to obtain the safe design criteria by the structural engineers. The estimated M_{max} values are very commonly used by agencies like disaster management, builders, and investors to make sure the proposed structure to be safe throughout their life time [5, 6].

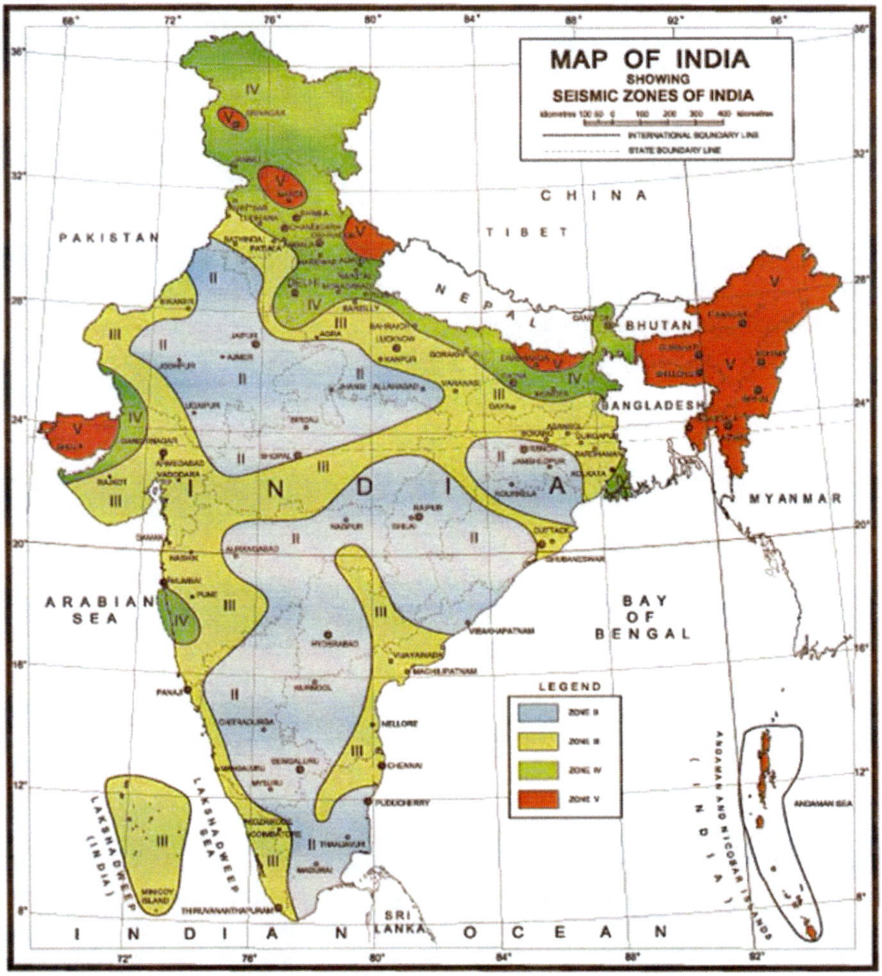

Fig. 1 The seismic zonation map of India (*Source* IS 1893: part (1)—2016)

2 Details of the Study Area

In this current study, the expected M_{max} for Amaravati region of Andhra Pradesh is evaluated by collecting earthquake data from the control region of 500 kms from Velagapudi as furnished in Fig. 2 and this extended boundary is considered as the seismic study area (SSA). The SSA is covers Telangana, Andhra Pradesh, Odisha, Chhattisgarh, Maharashtra, and Karnataka states of India. Among the six states four states covered partially and two states (Telangana, Andhra Pradesh) covered completely. All the necessary details of previous earthquakes have been collected from multiple data source centres which include the United States Geological Survey, the Indian Meteorological Department, the Amateur Seismic Centre, and the Geological Survey of India, since the earthquake data is the prime source to characterise the seismicity of the study region. The details of an historical earthquakes are compiled from the published articles of Chandra (1977), Rao and Rao (1984), and Guha and Basu (1993) [7–9]. The histogram of the earthquake data shows that the latest events including minor events are completed more accurately than the historical events. The observed maximum earthquake magnitude is considered as the reference magnitude to estimate the expected M_{max} as per the Gupta's method.

3 Characterization of Seismicity

Seismicity of the study region characterised by compiling the earthquake data from multiple sources and for the mentioned period a total of 634 events was collected from the SSA of 500 km area. As a result of the instability of an earthquake measurement scale, particularly in the instrumental period, the data on earthquakes are published using a variety of scales which includes body wave magnitude, surface wave magnitude, local magnitude, and moment magnitude (M_w). In order to keep uniform in the earthquake magnitudes, all of these earthquakes with different magnitudes are converted into moment magnitudes. The earthquake magnitudes converted using the widely popular relationships across the world were presented by Scordilis (2006) [10] and Heaton et al. [11]. The magnitudes of the earthquakes that are usually associated with the pre-instrumental era are often notified using an intensity scale and magnitude assigned purely based on the damage level of that particular earthquake; to convert such type of magnitudes into Mw the relationship proposed by Gutenberg-Richter (1956) has been used [12]. The declustering procedure was used in order to make the main event from its chain of dependent events. After the declustering process, the catalogue was prepared with 386 main events that were less than 3.0 M_w. Most of the situations measuring the smaller size earthquake magnitudes neglected particularly before 1976 AD, because there were no accurate instruments that are specially used to measure smaller size magnitude earthquakes, hence it is concluded that the completeness period of the earthquakes magnitudes related to the instrumental era is well established compared to pre-instrumental data. Figure 3 presents

Fig. 2 Selected seismic study area

the histogram of the seismicity decay wise, which was developed to understand the pattern of seismicity over the selected study region. Further, the data accordingly decay wise has been used to identify the completeness period of the earthquake data for all different magnitude ranges; for this the methods of CUVI and Steeps (1972) have been used [13, 14].

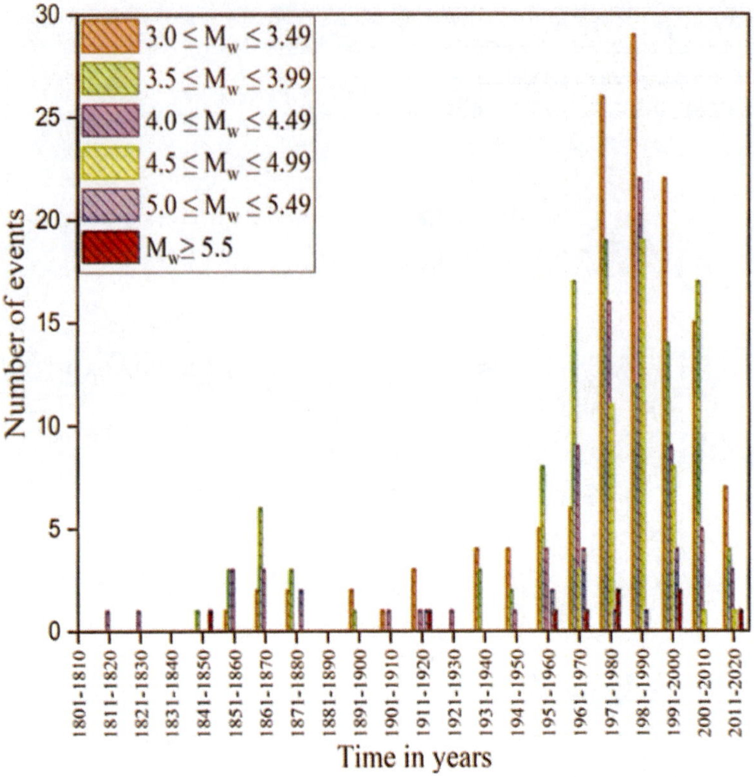

Fig. 3 Histogram showing the seismicity pattern within the seismic study region decay wise

4 Estimation of Maximum Magnitude (M_{max})

Evaluation of M_{max} is one of the essential criteria in seismic hazard analysis since there exist factors that are used to predict the degree of ground shaking that may occur during earthquakes. Also, it is utilised as an essential input parameter while computing the seismic hazard analysis. It has been observed from the seismicity pattern that the earthquake data is not evenly distributed over the region and in order to understand the pattern of seismicity intensity wise, four subzones were created using GIS software and by importing the earlier created shapefiles related to seismic sources a seismotectonic map of the Amaravati region has been generated. The M_{max} is computed using two different scenarios. In the first scenario, all the events in the catalogue have been taken into account. Whereas, in the second scenario, the seismic study area of 500 km (Fig. 4) has been divided into four subzones (zone 1, zone II, zone III, and zone IV) based on the group of seismicity; catalogue completeness analysis, and seismic hazard parameters are estimated for the considered earthquake's scenarios separately. Figure 4 presents the distribution of an earthquake's magnitude over the region, divisions of subzones of the region under investigation and this map also illustrates the orientation of various seismic sources such as faults, lineaments, and shear zones. Next, for the purpose of conducting additional research, the catalogue has been roughly segmented into six magnitude group ranges, each bin with a constant width of magnitude 0.5 M_W beginning with 3.0 M_W. This is due to the fact that earthquakes with magnitudes that are lower than 3.0 M_W will, in most of the cases, not cause any damage to the structure. The histogram of all of the incidents broken down by zone can be seen in Fig. 5.

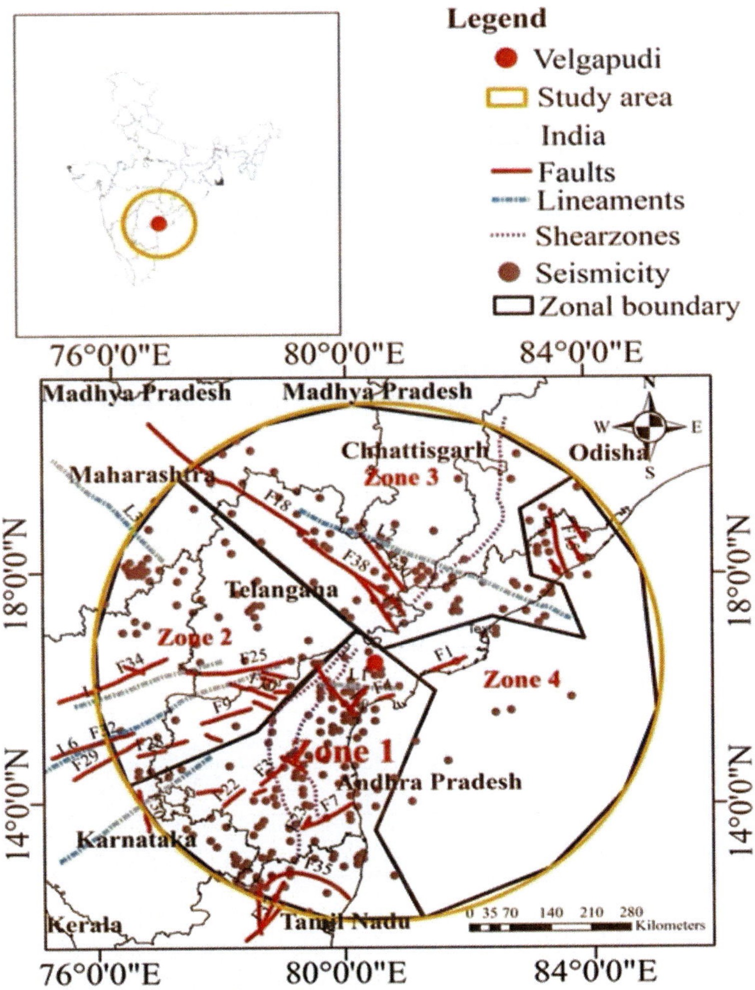

Fig. 4 Seismotectonic map of the Amaravati region including considered subzones

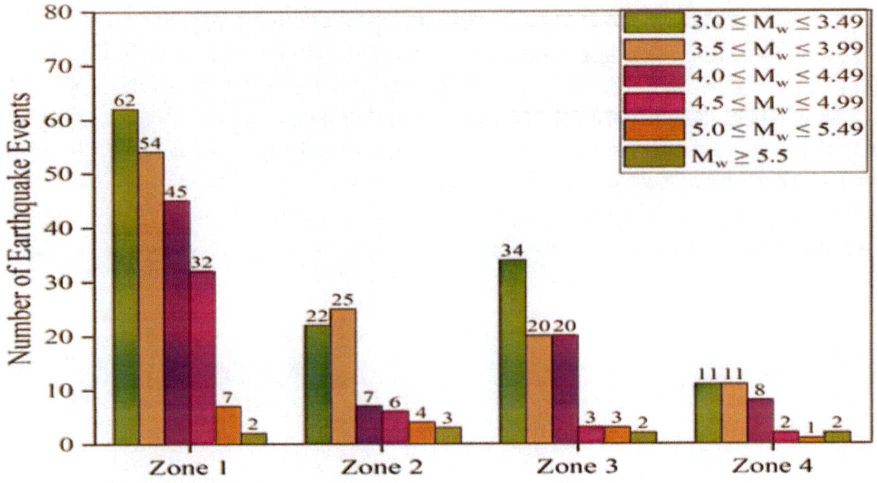

Fig. 5 Histogram of earthquake magnitudes related to four subzones

4.1 Gupta (2002) (Magnitude Incremental Method)

According to the method suggested by Gupta's (2002) an increment of 0.5 M_w is to be added to the observed maximum magnitude M_{max}^{obs} of that particular magnitude range, after adding the 0.5 M_w thar will be the expected maximum M_{max} of the same magnitude range of the earthquake group[15]. Likewise, for all different magnitude ranges of the earthquakes groups an increment of 0.5 M_w is to be added for estimating the M_{max} of each zone. This method is widely popular and preferred by many researchers across the world (Wheeler 2009) due to the wide range of users. The expected M_{max} values for both scenarios according to Gupta's method are summarized in Table 1.

Table 1 Maximum magnitude by Gupta and KSB procedure

Categories	Gupta method		M_{max} by KSB procedure
	Observed M_{max}	M_{max}	
500 km area	6.4	6.9	6.52 ± 0.28
Zone 1	5.6	6.1	5.67 ± 0.26
Zone 2	6.4	6.9	7.70 ± 1.32
Zone 3	5.7	6.2	6.02 ± 0.40
Zone 4	5.6	6.1	0.90

4.2 Kijko-Sellevoll-Bayes (KSB) Method

Further, the M_{max} is estimated as per the KSB method proposed by Kijko and Sellevoll [16]. According to the Kijko, the probable maximum earthquake of that region is called as M_{max} of that respective zone. Due to the inconsistency in magnitude scales, uncertainties are involved in both the pre-instrumental and instrumental data. The Kijko procedure will consider the complete and incomplete earthquake catalogue during the estimation of M_{max} using the following equation.

$$M_{max} = M_{max}^{obs} + \frac{\delta^{1/q+2}\exp[n \cdot r^q/1 - r^q]}{\beta}\left[\Gamma\left(-1/q,\, \delta \cdot r^q\right) - \Gamma(-1/q,\, \delta)\right] \quad (1)$$

where, M_{max}^{obs} = observed maximum magnitude, $\beta = 2.303b$, $\delta = nC\beta$, n = number of earthquake events, r = p/(p + mmax- mmin), $C\beta$ = normalizing coefficient of β-value, $q = (\frac{\bar{\beta}}{(\sigma_\beta)^2})$, $\sigma\beta$ = standard deviation of β- value describes its uncertainty, $\bar{\beta}$ = mean value of β, $p = (\frac{\bar{\beta}}{(\sigma_\beta)^2})$ and $\Gamma_{(.,.)}$ = Complementary incomplete gamma function. The other seismic hazard parameters like b- value and λ—value was also estimated for both scenarios using MATLAB software HA3 code. The b-value varies from 0.35 to 0.67 and the range of λ—value 0.14 ± 0.03 to 1.66 ± 0.41. The M_{max} values according to the KSB procedure as given in Table 1.

5 Conclusions

In this study, an estimation of the possible maximum (M_{max}) earthquake magnitude for the Amaravati region was carried out using the incremental method that was proposed by Gupta (2002) and the Kijko-Sellevoll-Bayes (KSB) method for two different scenarios. These methods were used to calculate the magnitude of an earthquake using the Kijko-Sellevoll-Bayes (KSB) method. These two approaches were contrasted and compared in order to ascertain which one yielded the most precise findings. According to the findings, it has been determined that the incremental technique and the K-S-B method both state that the greatest earthquake took place in the subzone 2 (6.9 Mw according to the incremental method and 7.70 ± 1.32 Mw according to the K-S-B method). This conclusion was reached as a result of the findings. There have been a total of seventy earthquakes registered in Subzone 2, three of which have had magnitudes greater than 5.5. In addition to this, and particular information about a variety of seismic sources, including faults, lineaments, and shear zones, is shown in Fig. 4. It has been found that the Gundla–Kamma fault, the Addanki–Nujividu fault, and the Karempudi–Nakirekallu lineament are the most prominent active faults in this area. This conclusion was reached by determining which faults had the shortest distance between the event and the seismic source. In addition, subzone 2 is home to both the Wajrakarur fault and the Kumadavati–Narihalla fault, and all of these faults can be found in subzone 2. In the future, both probabilistic and deterministic evaluations of the seismic hazard will make use of the value that was determined for Mmax. These analyses will be carried out in the future.

References

1. Reddy M, Konda RR, Kumar GK, Asadi SS (2020) Site characterization and evaluation of seismic sources for amaravati region. Int J Geotechn Earthq Eng (IJGEE) 11(1):71–86
2. Indian Standard, IS 1893 (Part I), 2016. Criteria for earthquake resistance design of structures, Part-I, Bureau of Indian Standard, New Delhi
3. Earthquake Disaster Risk Index Report 50 Towns and 1 Distrcit in Seismic Zone III, IV and V, 2019. Ministry of Home Affairs Government of India, NDMA Bhawan A-1, New Delhi
4. Anbazhagan P, Sitharam TG (2008) Seismic microzonation of Bangalore, India. J Earth. Syst Sci 117(S2):833–852
5. Anbazhagan P, Ketan B, Satyajit P (2015) Seismic hazard maps and spectrum for Patna considering region-specific seismotectonic parameters. Nat Hazards 78(2):1163–1195
6. Raghukanth STG (2011) Seismicity parameters for important urban agglomerations in India. Bull Earthq Eng 9(5):1361–1386
7. Chandra U (1977) Earthquakes of peninsular India-a seismotectonic study. Bull Seismol Soc Am 67(5):1387–1413
8. Ramalingeswara RB, Sitapathi RP (1984) Historical seismicity of Peninsular India. Bull Seismol Soc Am 74(6):2519–2533
9. Guha SK, Basu PC (1993) Catalogue of earthquakes (=> M 3.0) in peninsular India (No. AERB-TD-CSE--1). Atomic Energy Regulatory Board
10. Scordilis EM (2006) Empirical global relations converting Ms and mb to moment magnitude. J Seismol 10(2):225–236
11. Heaton TH, Tajima F, Mori AW (1986) Estimating ground motions using recorded accelerograms. Surv Geophys 8(1):25–83
12. Gutenberg B, Richter CF (1956) Earthquake magnitude, intensity, energy, and acceleration (second paper). Bull Seismol Soc Amer 46(2):105–145
13. Mulargia F, Tinti S (1985) Seismic sample areas defined from incomplete catalogues: an application to the Italian territory. Phys Earth Planet Inter 40(4):273–300
14. Stepp JC (1972) Analysis of completeness of the earthquake sample in the Puget Sound area and its effect on statistical estimates of earthquake hazard. In: Proceedings of the 1st international conference on microzonazion, Seattle, 2, 897–910
15. Gupta ID (2002) The state-of-the-art in seismic hazard analysis. ISET J Earthq Technol 39(4):311–346
16. Kijko A, Sellevoll MA (1992) Estimation of earthquake hazard parameters from incomplete data files, Part II, Incorporation of magnitude heterogeneity. Bull Seismol Soc Am 82(1):120–134

Assessing the Effect of Land Use Land Cover Change on the Water Quality Index of a River Basin Using GIS and Remote Sensing Techniques

W. S. Adhima, J. S. Gouri, Pooja N. Raj, P. S. Riya, and Lini R. Chandran

Abstract Water pollution is a major issue faced in both developed and developing countries. Land use land cover changes in urbanization, industrialization, and agricultural processes tend to have negative impacts on water quality at all scales. The water quality index serves as a commonly employed instrument for addressing data organization issues and assessing the effectiveness of management approaches aimed at enhancing water quality. The objectives of this study are estimation of water quality index (WQI) from water quality parameters for the study site using various methods and in determining the influence of land use land cover (LULC) change in water quality for the site. Using WQI, the classification of water quality can be done. GIS and remote sensing technology is used for estimating the LULC change for two time periods. The water quality parameters for calculating the WQI are collected from the Kerala State Pollution Control Board. Estimating WQI and identifying the relationship between LULC and water quality is important for effective and sustainable surface water quality management especially in reducing the pollutant concentration in water body.

Keywords Water quality · Land use land cover · GIS · Remote sensing

1 Introduction

Earth is covered with 75% of water, of which 96.5% is ocean, 2.5% is freshwater, and about 1% is salty water present in the ground. Freshwater sources are nearly 70% locked up in ice and the rest lies is in ground. Only 1.3% is surficial freshwater found mostly in lakes. Water Quality Index (WQI) comprises the data from different water quality parameters into a mathematical equation that evaluates and classifies the quality of water body. WQI was estimated using different methods with the aim to give a single value to the water quality of a source along with reducing higher number

W. S. Adhima · J. S. Gouri · P. N. Raj · P. S. Riya · L. R. Chandran (✉)
Govt. Engineering College, Barton Hill, Trivandrum, India
e-mail: lini.sam@gecbh.ac.in

S. Mesapam et al. (eds.), *Developments and Applications of Geomatics*,
Lecture Notes in Civil Engineering 450, https://doi.org/10.1007/978-981-99-8568-5_3

of parameters [1]. The quality of water resources has been seriously deteriorating over years as a result of the Land Use Land Cover (LULC) change. Urban settlement areas and industries have a serious impact on water quality parameters. There was significant negative spatial relationship between water quality parameters and land use types [2].

This study estimates the WQI of a river basin in South Kerala using selected water quality parameters of the basin. This river basin serves as the major source of drinking water for Thiruvananthapuram, one of the metropolitanized cities located in the south western seacoast of the state of Kerala. The study enables us to: (1) assess the quality of surface water by establishing a WQI using different methods of its estimation (2) analyse the impacts of contaminants on water sources (3) establish the relationship between LULC change and water quality which leads to alteration of hydrological system. Finally, these results can provide a valid reference for the land use optimization and control of water pollution.

1.1 Objectives

The objectives of the study are: (1) To evaluate the surface water quality of intake points of Karamana river using water quality parameters. (2) Developing WQI from the selected parameters using suitable methods. (3) Utilize GIS and Remote Sensing (RS) techniques to determine variations in LULC. (4) To assess the relationship between water quality and land use land cover changes in the Karamana River Basin (KRB).

1.2 Study Area

The study area chosen was the Karamana river basin with the sample collecting stations located at Peppara, Aruvikkara, Mangattukadavu and Thiruvallam. It hosts the water supply to the Thiruvananthapuram city with its reservoirs at Peppara and Aruvikkara. Karamana River is a west flowing river through the city of Thiruvananthapuram in Kerala, India. The river originates near the southern end of the Western Ghats at Agastyar Koodam. The location map of the study area is given in Fig. 1 which was prepared using the watershed map of Karamana river prepared using GIS.

1.3 Water Quality Index

Water Quality Index is a widely used tool to solve the problems of data management and to assess success and failures in management strategies for improving water quality. WQI was first developed by Horton in 1965 as an index rating system for

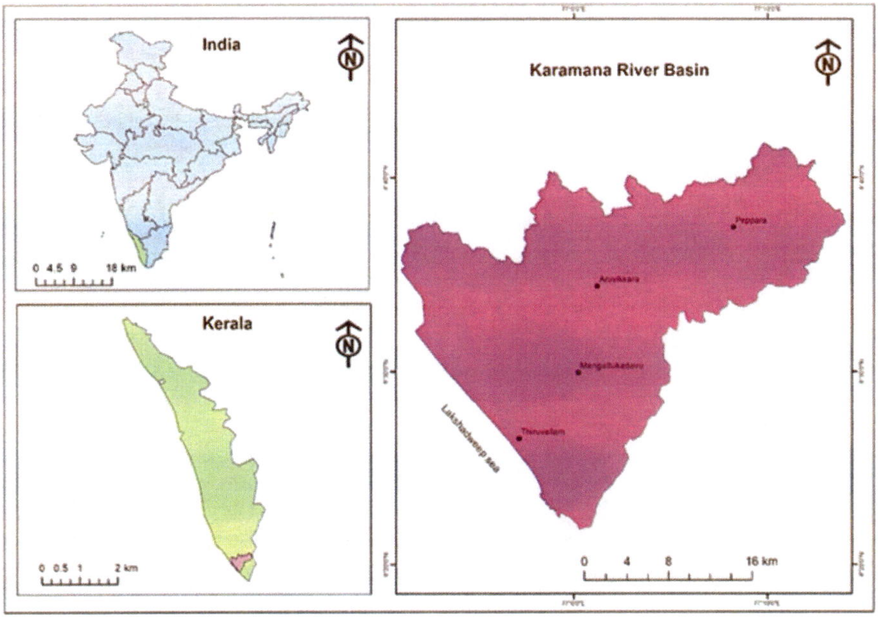

Fig. 1 Location Map of the study area

representing the water quality. WQI is a dimensionless number that incorporates various water quality parameters into a single number by normalizing the values and thus enabling easy interpretation of evaluating data. Estimating WQI includes three main steps: (1) selection of parameters (2) defining quality function for each parameter (3) aggregation through mathematical equation [3]. There are various methods to estimate WQIs according to the varying types and number of water quality parameters chosen. Some of the commonly used WQI methods are Weighted Arithmetic WQI (WAWQI) method, Nemerow's Pollution Index (NPI) method for water quality assessment, National Sanitation Foundation WQI (NSFWQI) and Canadian Council of Ministers of the Environment WQI (CCMEWQI).

The WQI of the present study area was estimated using WAWQI and NPI methods.

Weighted Arithmetic Water Quality Index Method. In WAWQI method, the water quality status is ascertained using water quality rating and grading proposed by Brown et al. [4], and Chatterji and Raziuddin [12]. According to the Weighted Arithmetic Water Quality Index method, WQI is calculated as:

$$WQI = \sum_{i=1}^{n} \frac{Q_i W_i}{W_i} \quad (1)$$

where Q_i = quality rating of ith water quality parameter, W_i = unit weight of ith water quality parameter. Q_i describes value of the water quality parameter in polluted water to standard permissible value, which is obtained as:

$$Q_i = 100\left(\frac{v_i - v_{io}}{s_i - v_{io}}\right) \quad (2)$$

where v_i = estimated value of the ith parameter, v_{i0} = ideal value of the ith parameter, s_i = standard permissible value of the ith parameter. In majority of cases, $v_i \neq 0$ except for pH and Dissolved Oxygen (DO). For pH, $v_{i0} = 7$; for DO, $v_{i0} = 14.6$ mg/l. The unit weight (W_i) which is inversely proportional to values of the recommended standards is obtained as:

$$W_i = \frac{k}{s_i} \quad (3)$$

where,

$$k = \frac{1}{\sum_{i=1}^{n} \frac{1}{s_i}} \quad (4)$$

The rating of the water quality using WAWQI method is shown in Table 1

Nemerow's Pollution Index (NPI) method. The NPI is one of the simplest WQI evaluation method compared to other methods. Dawood [5] successfully accessed the water quality for drinking and irrigation in five monitoring stations of Basrah marshes in the south of Iraq using the NPI. It is mathematically expressed as:

$$NPI = \frac{Ci}{Li} \quad (5)$$

where C_i is the revealed concentration of ith parameter and L_i is the allowable limit of allowable parameter. The unit of C_i and L_i must be the same. The NPI value represents the general pollution provided by a single parameter. NPI has no units. When the NPI value exceeds 1, it indicates contamination of water. The parameters vary according to the study region.

Table 1 Water Quality Rating and their Grading according to WQI values [4]

WQI	Water quality rating	Grading
0–25	Excellent	A
26–50	Good	B
51–75	Poor	C
76–100	Very poor	D
> 100	Non-potable	E

1.4 LULC Mapping Using GIS and Remote Sensing

LULC refers to the categorization or classification of human activities and natural factors on the landscape within a specific time frame based on established statistical and scientific methods of analysis of source materials. It largely depends upon the ecological conditions, altitudes, geological structure and slope. In addition to the above factors, increase in population density, technological and other socio-economic factors also affect the LULC pattern. LULC studies tend to explain (1) what land cover types are changing, (2) where change is occurring, (3) the types of alteration occurring, (4) the rates or amounts of land change, and (5) the driving forces and proximate causes of change [6].

LULC mapping is enormously important for analysis of urban development and consequent land degradation.

Satellite-based Remote Sensing, by virtue of its ability to provide comprehensive information regarding land use and land cover at a particular location and time, has significant impacts in the study of LULC. GIS help to integrate the acquired spatial and attribute data and to analyze the changes. There exist different image classification algorithms based on their rationality and complexity. Among those, Supervised classification is the most widely used method. In this method, the user can choose sample pixels in an image that are representative of specific classes and then direct the image processing software that is instructed to use these training sites as references for the classification of all other pixels in the image. Training data sets or input classes are chosen based on the user's knowledge.

The ERDAS Imagine software is used for image processing such as creating the False Colour Composite (FCC) and ArcGIS software is used for the supervised classification and to prepare land use land cover map and to determine the area under various features. The image classification can be done using multiple remote sensing features such as spatial, spectral, multi-sensor and multi-temporal. The numbers of LULC classes are preferred based on the requirement of a specific project for a particular application. The combination of water quality assessment method and spatial analysis tool in ArcGIS software greatly enhances the visualization of the research results.

Remote Sensing and GIS provide an effective tool for LULC changes investigation. However, LULC changes based exclusively on Remote Sensing and GIS may not be reliable for particular environmental application at a local level.

2 Methodology

2.1 General

The study area chosen is Karamana river basin with the sample collecting stations located at Peppara, Aruvikkara, Mangattukadavu and Thiruvallam. Karamana River is a west flowing river through the city of Thiruvananthapuram in Kerala, India. The river originates near the southern tip of the Western Ghats at Agastyar Koodam. It hosts the water supply to Thiruvananthapuram city with its reservoirs at Peppara and Aruvikkara.

2.2 Overview of Procedure

The water quality analysis data for the stations were collected from the Kerala State Pollution Control Board (KSPCB) for the years 2012, 2018, 2019 and 2020. The water data regarding various parameters of surface water compiled by the River Rejuvenation Committee (RRC) of KSPCB was obtained to characterize the water quality. From the water quality data, the WQI was calculated using the WAWQI and NPI methods for the years 2012, 2018, 2019 and 2020. The parameters selected for the calculation of NPI are pH, EC, DO, TC and BOD. All parameters except TC is selected for WAWQI calculation. In the WAWQI method, after computing the WQI for the stations, the water quality status is ascertained using water quality rating and grading proposed by Brown et al. [4] and Chatterji and Raziuddin [12]. Meanwhile, in the NPI method the water quality parameters which have exceeded its standard limits are recognizable.

The boundary of Karamana river basin was collected from Centre for Environment and Development Archive. The LULC changes in KRB was estimated from an input of Landsat 7 and Landsat 8 remote sensing images (30m resolution) for the years 2001 and 2020 respectively through the USGS Earth Explorer. False Color Composite (FCC) images are created by combining band 4 (NIR), band 3 (Red) and band 2 (Green) in Landsat 7, and band 5 (NIR), band 4 (Red) and band 3 (Green) in Landsat 8. Further, supervised image classification was done by ERDAS IMAGINE IV software and was reclassified in ArcGIS 10.2. The accuracy of the classified results was evaluated with the corresponding Google Earth images. The analysis of the results obtained using the above methodology helps in establishing a qualitative relationship between LULC and water quality.

3 Results and Discussion

3.1 Estimation of WQI

The WQI was calculated using WAWQI and NPI method. The parameters selected for the calculation of WAWQI were: pH, EC, DO, and Biochemical Oxygen Demand (BOD). While for the calculation of NPI the selected WQ parameters were pH, EC, DO, BOD and TC. The selected years for WQI estimation are 2012, 2018, 2019 and 2020. The WAWQI calculations are given in the Tables 2, 3, 4 and 5.

It was observed that the WQI of all the stations were consistently increasing from 2012 to 2019 with WQ rating of Peppara and Aruvikkara being 'Good' with a B grading throughout the years. This can be attributed to its location relatively at the upstream side of the basin. The water quality of Mangattukadavu, located just before entering the city, slightly deteriorated over the years. Thiruvallam, being the atmost downstream station and located within the corporation limits has a WQI value exceeding 100 for all the chosen years and thus falls under 'non-potable' category

Table 2 WAWQI Calculations for 2012

Stations	Parameter	V_i	S_i	W_i	Q_i	WQI
Peppara	pH	7.09	6.5–8.5	0.23	6	**35.64**
	EC (μS/cm)	15	250	0.01	6	
	DO (mg/l)	7.37	5	0.38	75.31	
	BOD (mg/l)	0.7	5	0.38	14	
				1.00		
Aruvikkara	pH	6.6	6.5–8.5	0.23	26.67	**43.38**
	EC (μS/cm)	28	250	0.01	11.2	
	DO (mg/l)	6.8	5	0.38	81.25	
	BOD (mg/l)	0.8	5	0.38	16	
				1.00		
Mangattukadavu	pH	7.42	6.5–8.5	0.23	28	**41.22**
	EC (μS/cm)	51.92	250	0.01	20.77	
	DO (mg/l)	6.84	5	0.38	80.83	
	BOD (mg/l)	0.49	5	0.38	9.8	
				1.00		
Thiruvallam	pH	7.38	6.5–8.5	0.23	25.33	**101.59**
	EC (μ S/cm)	1008.92	250	0.01	403.57	
	DO (mg/l)	4.25	5	0.38	107.81	
	BOD (mg/l)	6.71	5	0.38	134.2	
				1.00		

Table 3 WAWQI Calculations for 2018

Stations	Parameter	V_i	S_i	W_i	Q_i	WQI
Peppara	pH	6.9	6.5–8.5	0.23	6.67	**42.31**
	EC (μS/cm)	28	250	0.01	11.20	
	DO (mg/l)	6.9	5	0.38	80.21	
	BOD (mg/l)	1.3	5	0.38	26.00	
				1.00		
Aruvikkara	pH	6.7	6.5–8.5	0.23	20.00	**48.21**
	EC (μS/cm)	102	250	0.01	40.80	
	DO (mg/l)	7	5	0.38	79.17	
	BOD (mg/l)	1.70	5	0.38	34.00	
				1.00		
Mangattukadavu	pH	6.6	6.5–8.5	0.23	26.67	**50.13**
	EC (μS/cm)	95.33	250	0.01	38.13	
	DO (mg/l)	6.7	5	0.38	82.29	
	BOD (mg/l)	1.60	5	0.38	32.00	
				1.00		
Thiruvallam	pH	6.5	6.5–8.5	0.23	33.33	**123.58**
	EC (μ S/cm)	225	250	0.01	90.00	
	DO (mg/l)	2.80	5	0.38	122.92	
	BOD (mg/l)	8.90	5	0.38	178.00	
				1.00		

with an E grading. However, in 2020, there was a slight decline in the WQI values of all the stations indicating an improvement of water quality.

The NPI calculations are shown in Tables 6 and 7. The results from NPI calculations specify water quality in terms of parameter concentrations. The NPI values of pH, EC and BOD were less than unity, indicating their concentrations confine within the standard limits. It was found that for all the chosen years, the NPI values of DO and TC exceeded unity inferring that they transcended their standard limits. At Thiruvallam, the NPI value of TC was severely high which implies further need for immediate treatment. In 2020, the NPI values decreased considerably for all the parameters at all the stations.

3.2 LULC Supervised Classification

The LULC analysis for the years 2001 and 2020 was done with the help of ArcGIS and the classes for this analysis are forest, vegetation, built-up, water body and open area. The supervised classification provides the results of land use variations of these

Table 4 WAWQI Calculations for 2019

Stations	Parameter	V_i	S_i	W_i	Q_i	WQI
Peppara	pH	6.68	6.5–8.5	0.23	21.67	**58.51**
	EC (μS/cm)	70	250	0.01	28.00	
	DO (mg/l)	6.35	5	0.38	85.94	
	BOD (mg/l)	0.6	5	0.38	12.00	
				1.00		
Aruvikkara	pH	6.9	6.5–8.5	0.23	6.67	**82.32**
	EC (μS/cm)	67	250	0.01	26.80	
	DO (mg/l)	6.43	5	0.38	85.16	
	BOD (mg/l)	0.93	5	0.38	18.50	
				1.00		
Mangattukadavu	pH	6.9	6.5–8.5	0.23	6.67	**65.34**
	EC (μS/cm)	94.75	250	0.01	37.90	
	DO (mg/l)	4.6	5	0.38	104.17	
	BOD (mg/l)	1.85	5	0.38	37.00	
				1.00		
Thiruvallam	pH	7.03	6.5–8.5	0.23	1.67	**124.86**
	EC (μ S/cm)	337.5	250	0.01	135.00	
	DO (mg/l)	2.35	5	0.38	127.60	
	BOD (mg/l)	7.5	5	0.38	150.00	
				1.00		

years. The LULC classification of KRB for the year 2001 is conveyed in Fig. 2 and its corresponding pictorial representation is shown in Fig. 3.

Figure 4 represents the LULC classification of KRB for the year 2020 and its corresponding pictorial representation is conveyed in Fig. 5

The percentage of change in area according to various LULC classes is shown in Table 8. It was found that the forest area depleted by one percent, i.e., from 17 percent during 2001 to 16 percent during 2020 leading to negative 8.7 percent change in area. The same trend was observed for vegetation whose percentage change in area was negative 29.33. Such unusual change was seen in the case of water body as its area increased by 8.27 percent. This is related to the COVID pandemic effect that resulted in lockdown. Researches conducted by Chakraborty et al. [7] and Aswathy et al. [8] proved the positive influence of lockdown in the case of water quality. Hence, the water demand drastically decreased due to shut down of industrial and commercial establishments as well as the disposal to river bodies came to rest for a couple of months leading to slight increase in quality and quantity. In case of open areas, the percentage change in area showed an increase of 22.65% with 5.66 km^2 in 2001 to 6.94 km^2 in 2020. It is obvious that the reduction in forests as well as vegetation will lead to increase in the built-up area. The LULC results backed up this

Table 5 WAWQI Calculations for 2020

Stations	Parameter	V_i	S_i	W_i	Q_i	WQI
Peppara	pH	6.68	6.5–8.5	0.23	21.67	**42.65**
	EC (μS/cm)	70	250	0.01	28.00	
	DO (mg/l)	6.35	5	0.38	85.94	
	BOD (mg/l)	0.6	5	0.38	12.00	
				1.00		
Aruvikkara	pH	6.9	6.5–8.5	0.23	6.67	**41.45**
	EC (μS/cm)	67	250	0.01	26.80	
	DO (mg/l)	6.43	5	0.38	85.16	
	BOD (mg/l)	0.93	5	0.38	18.50	
				1.00		
Mangattukadavu	pH	6.9	6.5–8.5	0.23	6.67	**55.92**
	EC (μS/cm)	94.75	250	0.01	37.90	
	DO (mg/l)	4.6	5	0.38	104.17	
	BOD (mg/l)	1.85	5	0.38	37.00	
				1.00		
Thiruvallam	pH	7.03	6.5–8.5	0.23	1.67	**107.84**
	EC (μ S/cm)	337.5	250	0.01	135.00	
	DO (mg/l)	2.35	5	0.38	127.60	
	BOD (mg/l)	7.5	5	0.38	150.00	
				1.00		

assumption because the percentage change in area for built-up was about 186.96. The built-up area was found to have upsurged from 81.14 km^2 to 232.82 km^2.This results in Urban Heat Islands in which cities replace natural land cover with dense pavements, buildings and other heat absorbing surfaces. This makes the water in rivers, lakes, ponds and streams around the cities to be warmer than usual. Warm water holds less DO than cool water. This may be the cause of depletion of DO at Thiruvallam. Huang et al. [9] had analyzed the influence of LULC changes on the water quality parameters of the Chaohu lake basin and it was found that the built-up area and WQI were negatively related. Similarly, Permatasari et al. [10] analysed the LULC changes in Ciliwung watershed and found that the forest dominated and urban dominated areas had affected the water quality in a positive and negative manner respectively.

The population growth of Thiruvananthapuram with respect to Kerala and India is indicated in Fig. 6. From this figure, it is inferred that the population of Thiruvananthapuram is increasing and according to an article published by The Hindu, certain studies showed that Kerala's population will see an increase from 3.34 crores to 3.69 crores during 2011 to 2036.The population and the proportionate need for proper shelter leads to deforestation and clearing of vegetation for built up areas. These

Table 6 NPI Calculations for 2012 and 2020

		2012			2018		
Stations	**Parameters**	C_i	L_i	**NPI**	C_i	L_i	**NPI**
Peppara	pH	7.09	8.5	0.83	6.9	8.5	0.81
	EC (μS/cm)	15	250	0.06	28	250	0.11
	DO (mg/l)	7.37	5	**1.47**	6.9	5	**1.38**
	BOD (mg/l)	0.7	5	0.14	1.3	5	0.26
	TC (cfu/100ml)	293	50	**5.86**	100	50	**2.00**
Aruvikkara	pH	6.6	8.5	0.78	6.7	8.5	0.79
	EC (μS/cm)	28	250	0.11	102	250	0.41
	DO (mg/l)	6.8	5	**1.36**	7	5	**1.40**
	BOD (mg/l)	0.8	5	0.16	1.7	5	0.34
	TC (cfu/100ml)	500	50	**10**	110	50	**2.20**
Mangattu-kadavu	pH	7.42	8.5	0.87	6.6	8.5	0.78
	EC (μS/cm)	51.92	250	0.21	95.33	250	0.38
	DO (mg/l)	6.84	5	**1.37**	6.7	5	**1.34**
	BOD (mg/l)	0.49	5	0.1	1.6	5	0.32
	TC (cfu/100ml)	77	50	**1.54**	2500	50	**50.0**
Thiruvallam	pH	7.38	8.5	0.87	6.5	8.5	0.76
	EC (μS/cm)	1008.92	250	**4.04**	225	250	0.90
	DO (mg/l)	4.25	5	0.85	2.8	5	0.56
	BOD (mg/l)	6.71	5	**1.34**	8.9	5	**1.78**
	TC (cfu/100ml)	11,075	50	**221.50**	52,000	50	**1040**

results go hand in hand with the population studies. In a recent study by Prasood et al. [11] it was found that the built-up area expanded progressively from 2001.

Table 7 NPI Calculations for 2019 and 2020

		2019			2020		
Stations	Parameters	C_i	L_i	NPI	C_i	L_i	NPI
Peppara	pH	6.66	8.5	0.78	6.675	8.5	0.79
	EC (μS/cm)	64	250	5	70	250	0.28
	DO (mg/l)	6.06	5	**1.21**	6.35	5	**1.27**
	BOD (mg/l)	2.49	5	0.50	0.6	5	0.12
	TC (cfu/100ml)	283.33	50	**5.67**	300	50	**6.00**
Aruvikkara	pH	6.57	8.5	0.77	6.9	8.5	0.81
	EC (μS/cm)	31.92	250	0.13	67	250	0.27
	DO (mg/l)	0.72	5	**0.14**	6.425	5	**1.29**
	BOD (mg/l)	2.65	5	0.14	0.925	5	0.19
	TC (cfu/100ml)	325.83	50	**6.52**	450	50	**9.00**
Mangattu-kadavu	pH	6.48	8.5	0.76	6.9	8.5	0.81
	EC (μS/cm)	71.35	250	0.29	94.75	250	0.38
	DO (mg/l)	5.8	5	**1.16**	4.6	5	**0.92**
	BOD (mg/l)	2.89	5	0.58	1.85	5	0.37
	TC (cfu/100ml)	1233.33	50	**24.67**	1525	50	**30.5**
Thiruvallam	pH	6.45	8.5	0.76	7	8.5	0.82
	EC (μS/cm)	321.67	250	**1.29**	337.5	250	**1.35**
	DO (mg/l)	1.14	5	0.23	2.35	5	0.47
	BOD (mg/l)	8.06	5	**1.61**	7.50	5	**1.50**
	TC (cfu/100ml)	5608.33	50	**112.2**	4050	50	**81.0**

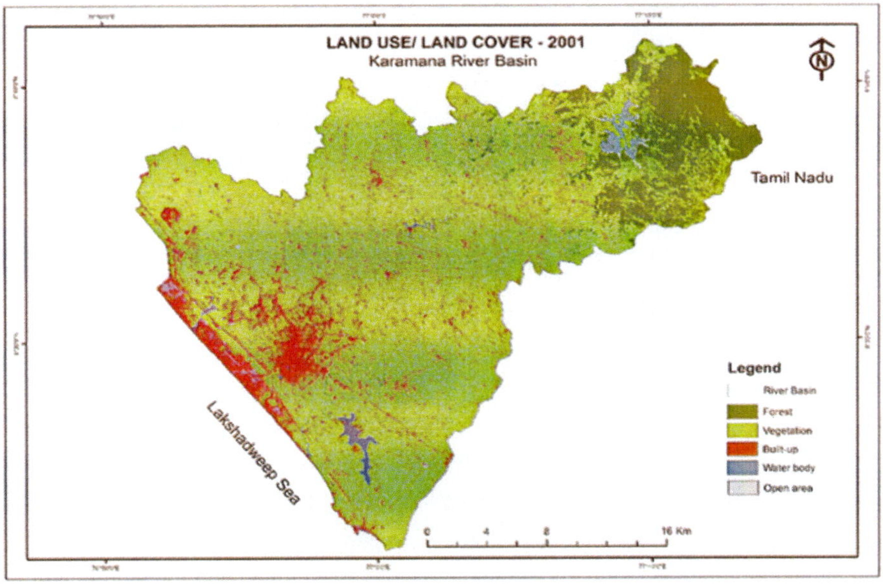

Fig. 2 LULC Supervised classification of KRB (2001)

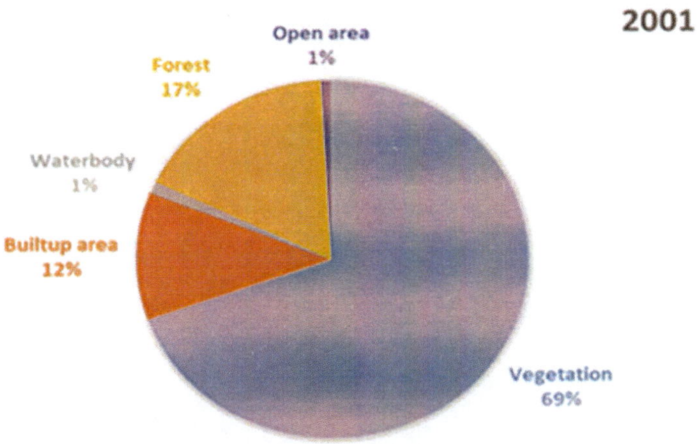

Fig. 3 Pictorial representation of LULC for the year 2001

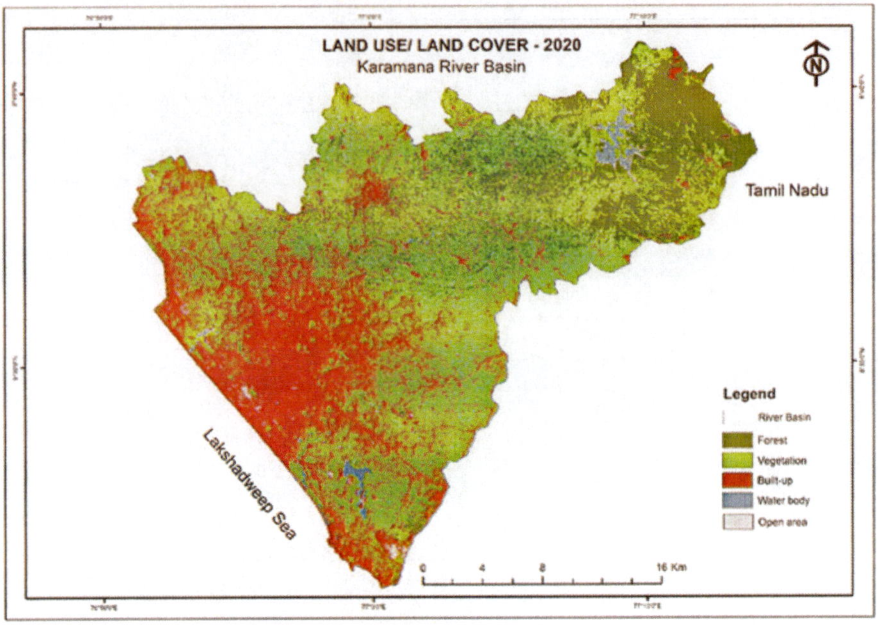

Fig. 4 LULC Supervised classification of KRB (2020)

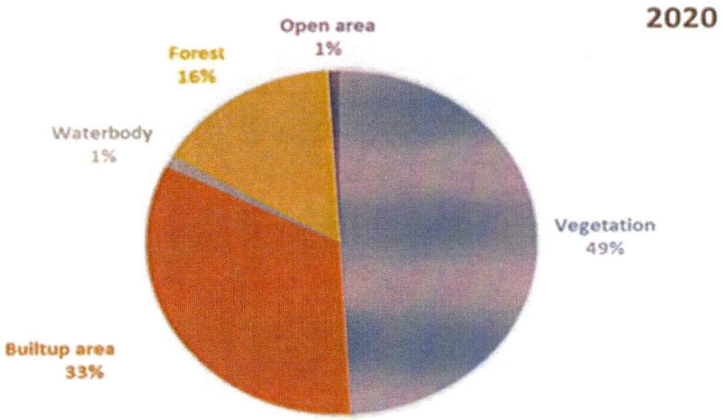

Fig. 5 Pictorial representation of LULC for the year 2020

Table 8 LULC Changes of KRB during 2001 and 2020

Year	2001		2020		% Change in area
LULC classes	Area (km^2)	Area (%)	Area (km^2)	Area (%)	
Vegetation	488.23	69.54	345.03	49.14	−29.33
Built-up area	81.14	11.56	232.84	33.16	186.96
Water body	7.46	1.06	8.07	1.15	8.27
Forest	119.62	17.04	109.21	15.55	−8.70
Open area	5.66	0.81	6.94	0.99	22.65
Total	702.10	100	702.10	100	

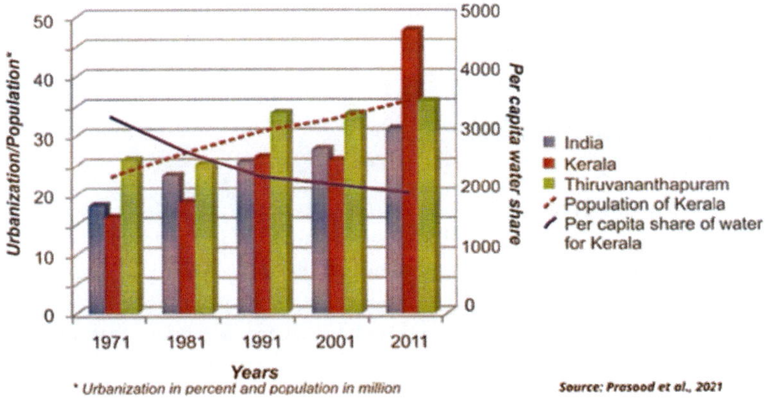

Fig. 6 Population growth of Thiruvananthapuram with respect to Kerala and India

4 Conclusion

The main objective of the present research was to establish a correlation between the land use land cover change and water quality. A qualitative negative relationship was inferred between them. Water quality Index clearly reveals the effect of urbanization in the river. Samples collected from Thiruvallam, which lies in the corporation limits shows maximum level of pollution in terms of BOD and a low DO content. Peppara reservoir which lies far off from the city and covered with its evergreen forests is not in danger of immediate pollution.

Despite the various precautionary and action plans undertaken, the continuously declining water quality was clearly due to environmental stresses that resulted from rapid and dense urbanization and consequent increase in the built-up area over the observed span of years. There was a huge surge in the built-up area which expanded to nearly three times to that of 2001 within a span of two decades. According to various reports, the Thiruvananthapuram city, will be witnessing rigorous infrastructural development in the upcoming years which will result in further expansion of the urban population. This will certainly catalyse the environmental deterioration. Also, localized pollution occurs from industrial units and slums located near the banks of the river.

An observable enrichment of water quality was observed during 2020 owing to the reduction of industrial effluent discharge and shift in domestic and commercial water demand–supply patterns during the COVID-19 lockdown. However, restoration of all the activities post lockdown would result in the continuing trend of declining water quality. Thus, proper controlling measures are crucial to preserve the natural areas as well as to improve the quality of water.

Policy makers and stakeholders must ensure checks for water quality at regular intervals of time. Action plans for reduction of net effluent discharge, regulation of channel precipitation and urban runoff must be enforced. The capacity of Sewage Treatment Plants must be increased proportionally with increase in population.

4.1 Scope of Future Work

At present, there are diverse methods for the estimation of WQI. The absence of a universal WQI affects the uniformity of research works in the field of water quality. Thus, a well-established WQI which considers the vital WQ parameters and whose calculation is free from complex mathematical computations is to be formulated. It would be of great use if it can also be applied anywhere irrespective of the study area. Hence, more researches should be conducted in order to invent a suitable method for properly finding out the exact relationship.

GIS proves to maximize the efficiency of decision making and planning in organizations and industries of all genres. This study has emphasized the application of Remote Sensing and GIS to develop LULC changes over the time for the river water quality based on the water quality parameters. LULC mapping has considerable importance in scientific, scholarly research, planning and management of water quality. GIS proves to be an effective means for data distribution and handling and helps to escalate the efficiency of planning and decision making. It has the ability to consolidate information from many sources and helps in the analysis of information involving geographical reference data at minimum cost.

References

1. Tyagi S, Sharma B, Singh P, Dobhal R (2013) Water quality assessment in terms of water quality index. Amer J Water Resour 1(03):34–38
2. Chen D, Elhadj A, Xu H, Xu X, Qiao Z (2020) A study on the relationship between land use change and water quality of the Mitidja watershed in Algeria based on GIS and RS. Sustainability 12(09):01–20
3. Uddin MG, Nash S, Olbert AI (2021) A review of water quality index models and their use for as-sessing surface water quality. Ecol Indicators 122(107218):1–21

4. Brown RM, McCleiland NI, Deininger RA, O'Connor MF (1972) A water quality index crossing the psychological barrier, indicators of environmental quality. In: 6th annual conference, advances in water pollution research, pp 787–794
5. Dawood AS (2017) Using of Nemerow's Pollution Index (NPI) for water quality assessment of some Bas-rah Marshes, South of Iraq. J Babylon Univ/Eng Sci 25(05):1708–1720
6. Loveland TR, Acevedo W (2006) Land cover change in the Eastern United States, US Geological survey. https://landcovertrends.usgs.gov/east/regionalSummary.html. Accessed 28 June 2022
7. Chakraborty B, Bera B, Adhikary PP, Bhattacharjee S, Roy S, Saha S, Ghosh A, Sengupta D, Kumar P, Positive effects of COVID-19 lockdown on river water quality: evidence from River Damodar, India.https://doi.org/10.1038/s41598-021-99689-9
8. Aswathy TS, Francis S, Achu AL, Gopinath G (2021) Assessment of water quality in a tropical ram-sar wetland of southern India in the wake of COVID-19. Remote Sensing Appl Soc Environ 23(6)
9. Huang J, Zhan J, Yan H, Wu F, Deng X (2013) Evaluation of the impacts of land use on water quality. Scientific World J, 1–7
10. Permatasari PA, Setiawan Y, Khairiah RN, Effendi H (2017) The effect of land use change on water quality: a case study in Ciliwung watershed. In: IOP conference series: earth and environmental science, pp 1–7. IOP Publishing, Bogor, Indonesia
11. Prasood SP, Mukesh MV, Rani VR, Sajinkumar KS, Thrivikramji KP, Urbanization and its effects on water resources: scenario of a tropical river basin in South India. Remote Sen Appl Soc Environ.https://doi.org/10.1016/j.rsase.2021.100556
12. Chatterjee C, Raziuddin M (2002) Determination of water quality index of a degraded river in Asanol In-dustrial area, Raniganj, Burdwan, West Bengal. Nat Environ Pollut Technol 01(02):181–189

Assessment of Fluctuations in Pre-monsoon and Post-monsoon Ground Water Levels in Kurukshetra, Haryana

Vikas Singh and A. K. Prabhakar

Abstract Analyzing the trajectory of hydrological indicators is crucial due to significant reduction in water supplies, as well as the rise in demand. The aim of the current study is to evaluate the status of ground water levels in Kurukshetra district at all the seven Blocks viz. Thanesar, Shahbad, Pipli, Babain, Ladwa, Pehowa, Ismailabad for past one decade i.e., from 2010 to 2020. The current study examines ground water variation and trend analysis during the Pre-Monsoon and Post-Monsoon seasons. Also, Inversed Distance Weight (IDW) interpolation approach was used for creating ground water contour maps by using the Arc-GIS 10.8. Results depict that in the Pre-Monsoon study Ismailabad block reflected maximum fluctuation and Pehowa Block reflected minimum fluctuation. In the Post-monsoon study Pehowa Block reflected maximum fluctuation and Ladwa block reflected minimum fluctuation. For Pre-Monsoon, the rate of Groundwater depletion for one decade at seven blocks of Kurukshetra viz. Thanesar 38.88%, Shahbad 27.34%, Pipli 31.86%, Pehowa 6.44%, Ladwa 25.67%, Ismailabad 29.55%, Babain 29.52%. For the Post-Monsoon the rate of Groundwater depletion for one decade for seven blocks of Kurukshetra viz. Thanesar 46.97%, Shahbad 30.32%, Pipli 28.47%, Pehowa 51.58%, Ladwa 25.57%, Ismailabad 36.14%, Babain 27.97%. Average rate of Groundwater Depletion during Pre-Monsoon and during Post-Monsoon seasons for one decade in Kurukshetra district is 27.03%.and 35.28% respectively. As a result, regional entities must design a framework for sensible use of ground resources, with effort made to increase groundwater recharge in order to ensure enough groundwater in the future.

Keywords Groundwater · Pre-Monsoon · Post-Monsoon · Fluctuations · GIS

V. Singh (✉) · A. K. Prabhakar
Civil Engineering Department, NIT Kurukshetra, Kurukshetra, Haryana, India
e-mail: Vikassinghsailor65@gmail.com

A. K. Prabhakar
e-mail: akprabhakar@nitkkr.ac.in

S. Mesapam et al. (eds.), *Developments and Applications of Geomatics*,
Lecture Notes in Civil Engineering 450, https://doi.org/10.1007/978-981-99-8568-5_4

1 Introduction

1.1 General

Water is one of the most basic requirements of all living beings. It is essential for life, whether it is in the form of groundwater, surface water, or precipitation. It is impossible to imagine life on Earth without water [1]. Groundwater among the available sources plays a critical role due to demands for drinking water, irrigation, livestock water, and industrial activities. Currently, 90% of rural water supply, 50% urban water supply, and 70% agricultural water supply in India depend on groundwater trapped in porous and permeable layers of the soil, referred to as aquifers [2].

Aside from human demands, groundwater is required to maintain the biological water cycle since it serves as a source of recharging for lakes, river, and wetlands. Secular variation refers to changes in ground water level over a long period of time [6]. Long-term fluctuations in level are caused by alternating seasons of wet and dry years, in which rainfall is above and below the mean, respectively. The regulating factor is recharge, which is determined by rainfall intensity and dispersion, as well as the amount of surface runoff. A downward trend in ground water level may persist for many years in overdeveloped basins where draught exceeds recharge [3]. Seasonal variations caused by factors such as rainfall and irrigation, as well as pumping discharge, all of which follow well-defined seasonal cycles [4]. The beginning and end of the irrigation seasons are marked by the highest and lowest levels, which occur around April and September, respectively.

Groundwater is chosen over surface water because it is more cost-effective, more convenient, and less polluted than surface water [5]. India is approaching a point when meeting the rising water demands of a geometrically growing and unevenly distributed population would be difficult. Rapid urbanization, caused by economic and social developments, has altered the consuming network and whole water requirement, affecting particularly places with hard and impervious terrain, where groundwater access is problematic [7]. The main problem is unrestricted access to groundwater, as well as waste during agricultural operations. Because of the higher consumption rate than the usual sluggish rate of groundwater recharge, the groundwater level is rapidly diminishing. Given the approaching shortage of available groundwater sources in the near future, it has become critical for water scientists and planners to quantify available water in order to make wise use of them [9].

1.2 Significance of Present Study

- This study will be helpful in predicting trends in the groundwater depth with respect to the development of the area.
- To suggest some preventive measures for the future decades to use the groundwater effectively will be the major scope of the research.

1.3 Objective of the Study

- To analytically study the recent trends of the Groundwater Fluctuation levels in Kurukshetra district of Haryana that is majorly affected due to development.
- To assess long-term impacts of the groundwater table depth and the fluctuations in Pre-Monsoon and Post-Monsoon conditions.

2 Study Area

The Kurukshetra district is situated in the north section of Haryana and is surrounded by the Haryana districts of Ambala, Karnal, Yamunanagar, and Kaithal, as well as the Punjab districts of Patiala, with area geographically covering around 1530 km^2. Between 29°52′ to 30°12′ N latitude and 76°26′ to 77°04′E longitude is the location of the District. The district's major rivers are the Markanda and Saraswati. The Saraswati River was once considered to flow across Haryana, but it has since vanished. Between the Yamuna and the Sutlej, the Ghaggar rises in the lower Shivalik highlands of the outer Himalayas and reaches Haryana near Pinjore district Panchkula. It passes via Ambala and Hisar on its way to Bikaner in Rajasthan, where it runs for 290 kms before disappearing into the Rajasthan deserts. During the rainy season, it also affects a large portion of the Kurukshetra District. Kurukshetra district lacks a perennial river. After emerging from the Nahan hills, the river Markhanda flows through the district's northwestern portion. Yamuna and Saraswati, two holy rivers, flow along the eastern and northern borders, respectively (Fig. 1).

3 Methodology

3.1 General

The research work was carried out in Arc GIS 10.8 version at *Siemens* Centre of Excellence (SCOE) in Water Resources Engineering Lab of the Civil EngineeringDepartment, National Institute of Technology, Kurukshetra, Haryana, India. GIS software illustrates geographical information through layer building maps, such as temperature data or trade flows, and enables the handling and analysis of geographic information. Like the majority of GIS programs, ArcGIS produces maps with categories arranged as layers. Each layer is geographically recorded, allowing the computer to precisely align the layers when they are layered on top of one another to create a complex data map [8]. Almost always, the base layer is a geographic map that has been assembled from many sources depending on the type of presentation needed (satellite, road map, etc.). The first three layers are known as feature or vector

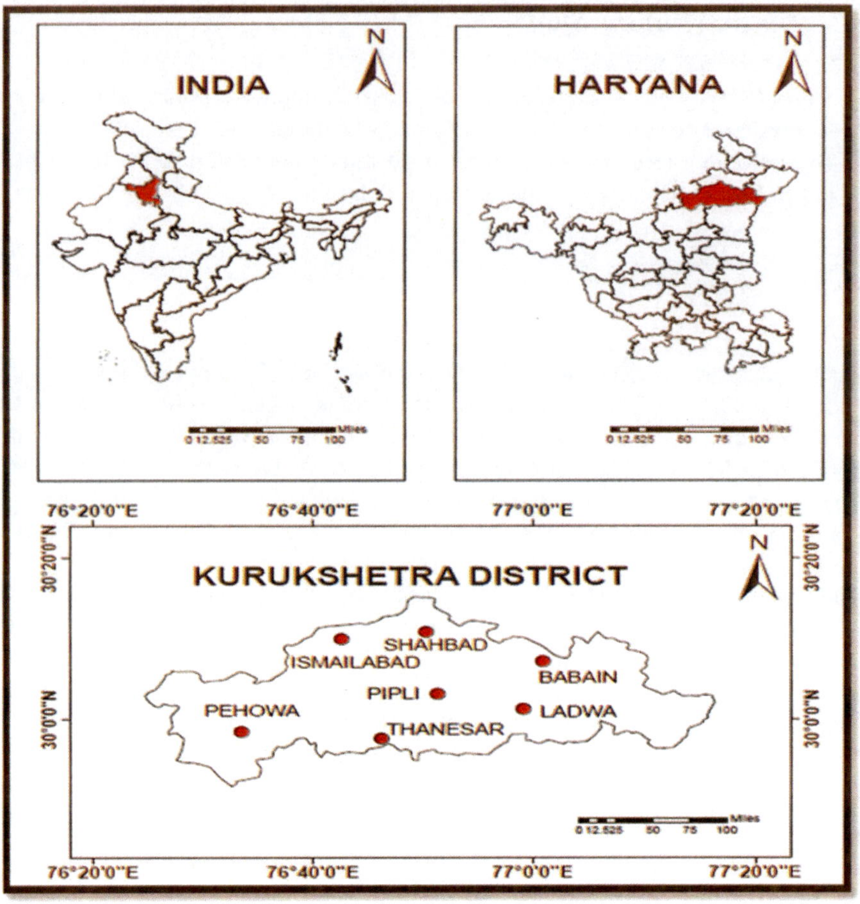

Fig. 1 Location of the study area

layers, and each one contains distinct functions that are distinguished by the platform. These are points, lines, and polygon and raster images. Data can be mapped and analyses done using at least one of these spatial layers, and can be connected with factors like demographic changes or data tables (Fig. 2).

3.2 IDW Contours in ArcGIS

IDW i.e., Inverse Distance Weighted Interpolation implicitly assumes things that are close to one another that have more in common than things that are far apart. IDW predicts a value for any unknown point using the measured data in the vicinity of the prediction location. The expected value is more affected by the measured values

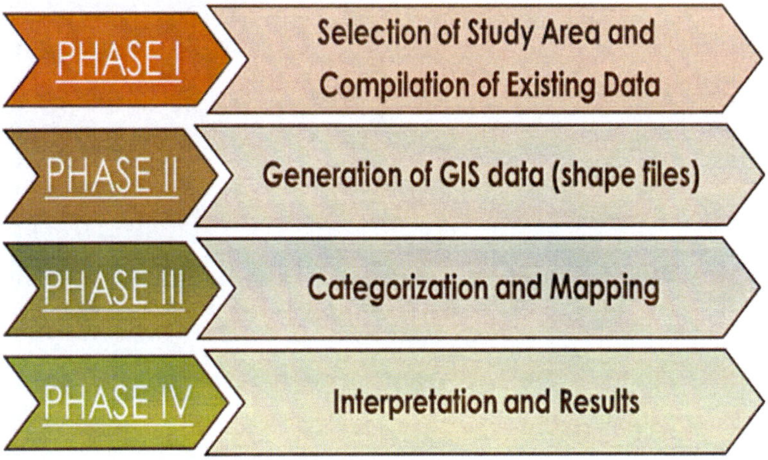

Fig. 2 Steps involved in research work

that are in close proximity to the prediction location than by those that are farther away. According to IDW, each measured point has a local impact that gets smaller as you move away. It offers higher weights to points that are closer to the prediction location, and the weights decrease as the distance increases, therefore known as Inverse Distance Weighted [10]. The weights are determined by taking the inverse of the distances (between the data point and the projected point) raised to the power p. As a result, as the distance increases, the weights fall quickly. Depending on the value of p, the weights decrease at various rates. If $p = 0$, the forecast will be the mean of all the data values in the search region because each weight I is equal. As p increases, the weights for distant points decrease quickly. If the p value is really high, only the nearby surrounding points will have an impact on the projection.

Geostatistical analysts employ power levels greater than or equal to one. When $p = 2$, the inverse distance squared weighted interpolation technique is applied. Although there is no theoretical rationale for selecting $p = 2$ above other values, the impact of changing p should be investigated by looking at the cross-validation statistics and the output preview. Because objects that are close to one another are more similar than those that are farther apart, the measured values will have little relevance to the value of the forecast location as the locations go farther apart. You can eliminate the more distant sites that will have little impact on the prediction to speed up calculations.

3.3 Box Plot

In simple terms, the box plot can be defined in terms of descriptive statistics ideas. That is, a box or whiskers plot is a graphical representation of numerical data

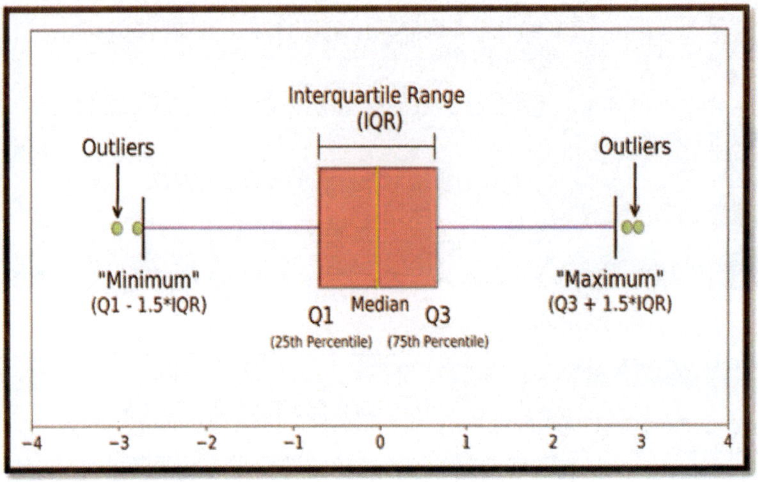

Fig. 3 Details of box plot

grouped through their quartiles. The words box-and-whisker plot and box-and-whisker diagram refer to the variability outside the lower and higher quartiles, which is indicated by lines extending from the boxes or whiskers. Individual points can be used to indicate outliers. Graphs can be used to determine how much the data values change or spread out. The box plot comes in handy when we need more information than only the measures of central tendency. It also takes up less space. It's a form of pictorial depiction as well. Because the Centre, spread, and overall range are instantly visible, the distributions can be easily compared using these box plots (Fig. 3).

4 Result and Discussion

4.1 Groundwater Table Level Average in the Pre- Monsoon Season in 2010

Groundwater levels during the Pre -Monsoon period in the year 2010 varied from 26.08 m to 37.11 m below the ground level. Water Table was highest in theThanesar Block and lowest in the Pehowa Block of the district (Fig. 4).

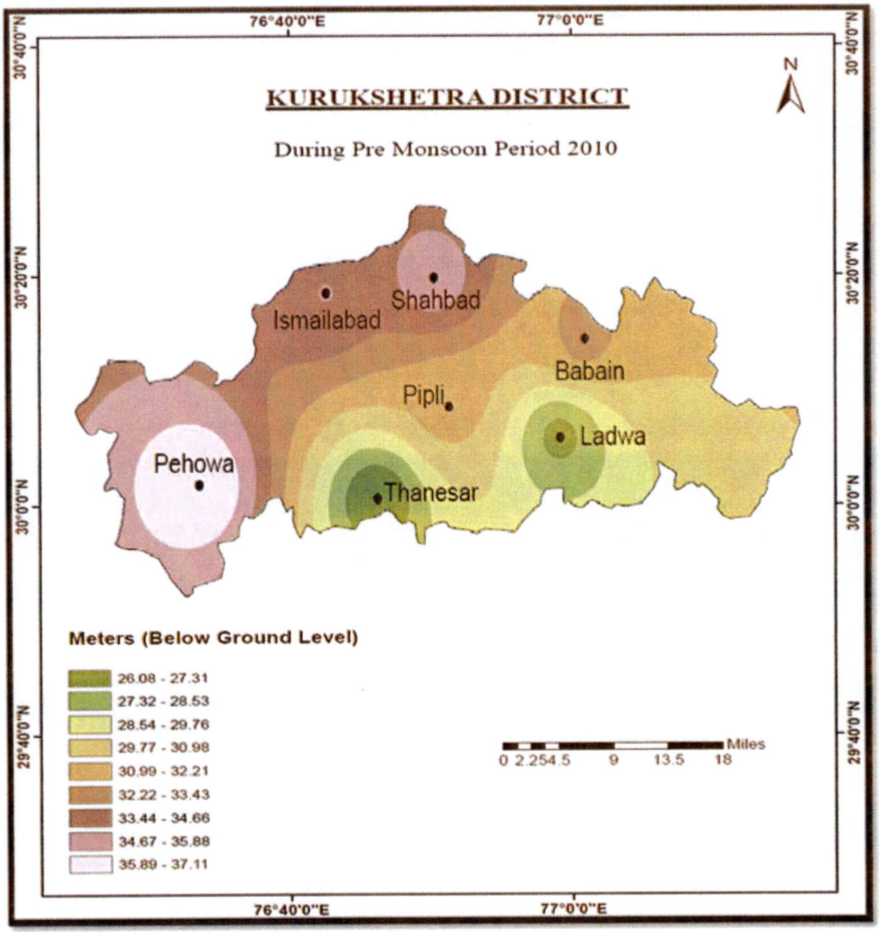

Fig. 4 Pre-Monsoon 2010 average ground water level

4.2 *Groundwater Table Level Average in the Pre- Monsoon Season in 2020*

Groundwater levels during the Pre - Monsoon period in the year 2020 varied from 33.82 m to 45.22 m below the ground level. Water Table was lowest in Shahbad Block and highest in the Ladwa Block of the district (Fig. 5).

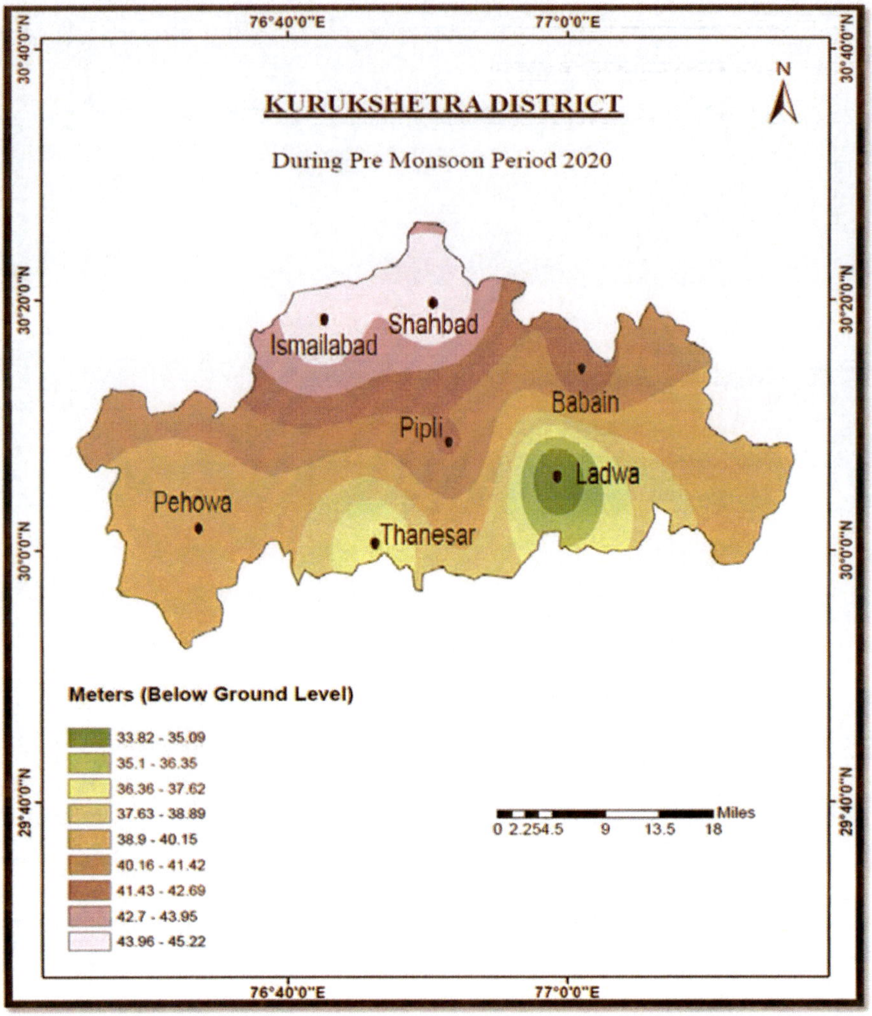

Fig. 5 Pre-Monsoon 2020 average ground water level

4.3 Groundwater Table Level Average During Post Monsoon Period in 2010

Ground water levels during the Post-Monsoon period in the year 2010 varied from 25.65 m to 34.92 m below the ground level. Water Table was highest in the Thanesar and lowest in the Shahbad block of the district (Fig. 6).

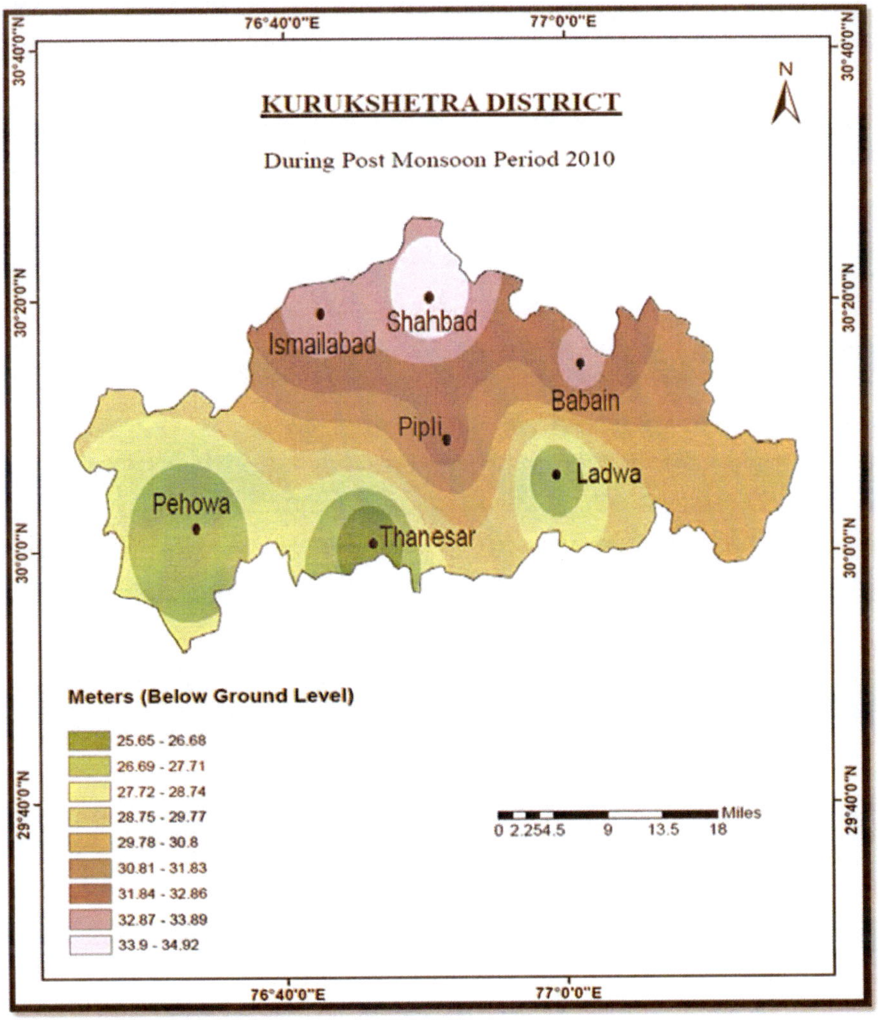

Fig. 6 Post-Monsoon 2010 average ground water level

4.4 *Groundwater Table Level Average During Post Monsoon Period in 2020*

Ground water levels during the Post-Monsoon period in the year 2020 varied from 33.68 m to 45.51 m below the ground level. Water Table was highest in the Ladwa Block and lowest in the Shahbad block of the district (Fig. 7).

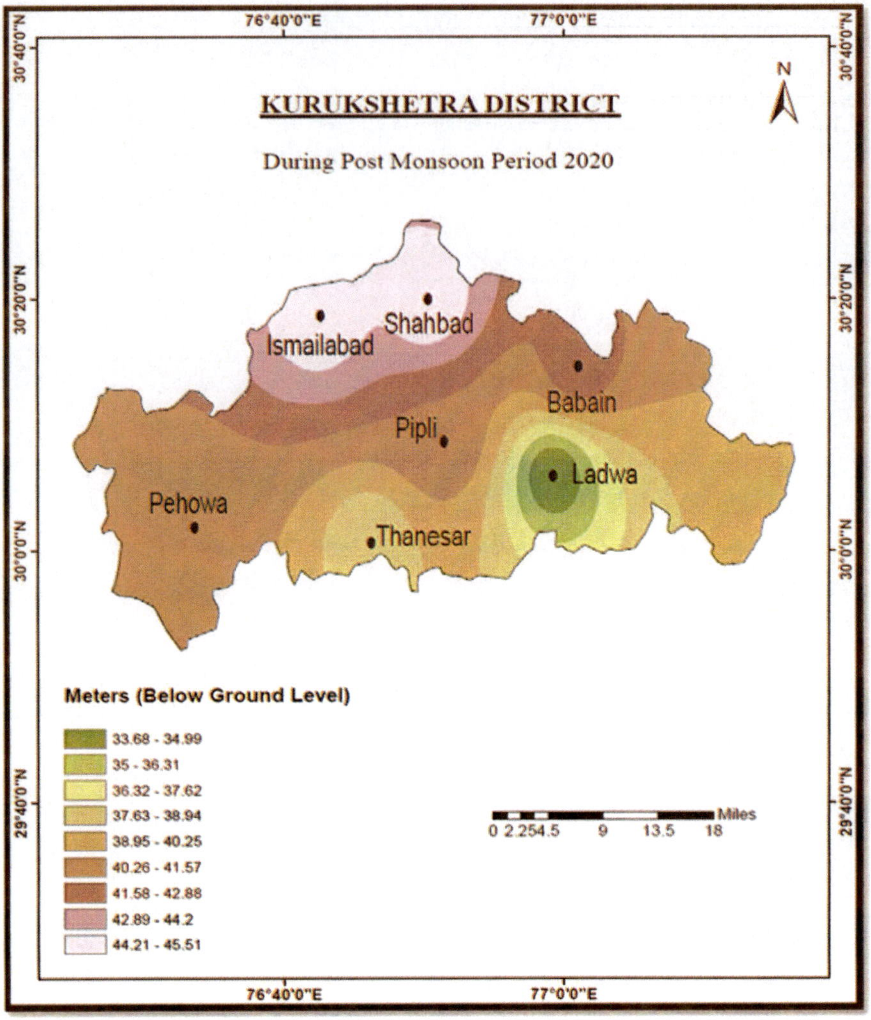

Fig. 7 Post-Monsoon 2020 average ground water level

5 Conclusion

For a better understanding, long-term Seasonal average groundwater levels were interpreted (Pre-Monsoon and Post-Monsoon), from the above done works, the following key conclusions were drawn:

- In the study it was found that during the Pre-Monsoon Season over the decade 2020 the fluctuations were maximum at the Thol station in Ismailabad Block i.e., 10.25 m and the fluctuations were minimum at the Galedwa station in Pehowa

Block i.e., 2.39 m. During the Post-Monsoon Period over the decade from 2010 to 2020 the fluctuations were maximum at the Galedwa station in Pehowa Block i.e., 13.87 m and the fluctuations were minimum at the Lohara station in Ladwa Block i.e., 6.86 m.

- For the Pre-Monsoon period, the rate of Groundwater depletion for one decade at Thanesar block 38.88%, Shahbad block 27.34%, Pipli block 31.86%, Pehowa block 6.44%, Ladwa block 25.67%, Ismailabad block 29.55%, Babain block 29.52%. For the Post-Monsoon period, the rate of Groundwater depletion for one decade at Thanesar block is 46.97%, Shahbad block 30.32%, Pipli Block 28.47%, Pehowa block 51.58%, Ladwa block 25.57%, Ismailabad block 36.14%, and Babain block 27.97%.
- Average rate of Groundwater Depletion during the Pre-Monsoon Period and during the Post-Monsoon Period for one decade in Kurukshetra district is 27.03%.and 35.28% respectively.
- During Pre-Monsoon ground the southern and western areas of the district had higher water levels than the remaining of the district. In the Post-Monsoon Period groundwater table level was high in Thanesar and Ladwa blocks i.e., the southern and eastern parts of Kurukshetra.

The analysis of water level patterns is crucial for determining the future scenario of accessible Groundwater resources of the state. The success of various groundwater recharge strategies could also be determined via trend analysis. As a result, regional entities must design a framework for sensible use of present ground resources, with every effort made to increase Groundwater recharge in order to ensure enough supplies in the future.

References

1. Aneja R (2017) Ground water level in Haryana: a challenge for sustainability. Int J Res Analyt Rev 4(3):43–48
2. Devi K, Singh M (2021) Evaluating ground water fluctuations in district Hisar, Haryana: a Temporal Analysis (1974–2018). PalArch's J Archaeol Egypt/Egyptol 18(4):6985–6994
3. Duhan AK (2017) Groundwater pumping irrigation in Haryana: issues and challenges. Int J Res Geogr 3(2):18–21
4. Goyal SK, Chaudhary BS (2010) GIS based study of Spatial-Temporal changes in groundwater depth and quality in Kaithal district of Haryana, India. J Indian Geophys Union 14:75–87
5. Goyal SK, Chaudhary BS, Singh O, Sethi GK, Thakur PK (2010) Variability analysis of groundwater levels—AGIS-based case study. J Indian Soc Remote Sens 38(2):355–364
6. Indhulekha K, Chandra Mondal K, Jhariya DC (2019) Groundwater prospect mapping using remote sensing, GIS and resistivity survey techniques in ChhokraNala Raipur district, Chhattisgarh, India. J Water Supply Res Technol AQUA 68(7):595–606
7. Joshi SK, Gupta S, Sinha R, Densmore AL, Rai SP, Shekhar S, van Dijk WM (2021) Strongly heterogeneous patterns of groundwater depletion in Northwestern India. J Hydrol 598:126492
8. Krishan G, Sharma LM, Yadav BK, Ghosh NC (2016) Analysis of water level fluctuations and TDS variations in the groundwater at Mewat (Nuh) district, Haryana (India). Current World Environ 11(2):388

9. Kumar PJ (2022) GIS-based mapping of water-level fluctuations (WLF) and its impact on groundwater in an Agrarian District in Tamil Nadu, India. Environ Dev Sustain 24(1):994–1009
10. Kumar S, Rani M, Verma S (2020) Assessment of pre-monsoon and monsoon groundwater level and fluctuation at regional level in south of Haryana; Reference to Rewari District. Int J Appl Sci Eng 8(2):123–129

Assessment of Land Use—Land Cover Changes in District Dehradun (1991–2021)

Madhusudan Thapliyal and A. K. Prabhakar

Abstract The region of district Dehradun in Uttarakhand, India has shown a rapid growth in population and urbanization in the last few decades. An attempt was made to assess the land use-land cover (LULC) changes in the entire district in the last 30 years from 1991 to 2021. Satellite imagery from the Landsat—5, 8 with a spatial resolution of 30 m was used as the remotely sensed data and the open-source software QGIS was used to process the data and produce the LULC maps and other important statistics using the maximum likelihood algorithm. The study has shown that during 1991–2021, dense forest, active cropland and sparse vegetation areas have decreased by 17.52%, 62.11% and 7.50% respectively whereas built-up and bare soil areas have increased by 112.23% and 44.32% respectively. No definite conclusions regarding change in areas could be reached for water and dry river bed but in general, the total stream flow area was observed to be decreasing. The average overall accuracy of the classifications was found to be 85.01% with a kappa coefficient of 0.794.

Keywords GIS · Landuse · Temporal and spatial LULC change analysis · Classification & accuracy assessment

1 Introduction

Remote sensing is a very useful technique for detecting land cover changes in rapidly developing regions so as to provide useful data for landuse planning and policy making. Landuse changes are a serious concern for developing cities due to rapid increase in population which increases the requirement of land viz. housing, road, agricultural and water etc. and causes issues such as land degradation and problems in the water cycle [4, 7, 14]. Several others researchers [5, 6, 10, 15] found that changes

M. Thapliyal (✉)
Climate Hydrology Division, National Institute of Hydrology, Roorkee, India
e-mail: madhu2412380@gmail.com

A. K. Prabhakar
Department of Civil Engineering, National Institute of Technology, Kurukshetra, India

S. Mesapam et al. (eds.), *Developments and Applications of Geomatics*,
Lecture Notes in Civil Engineering 450, https://doi.org/10.1007/978-981-99-8568-5_5

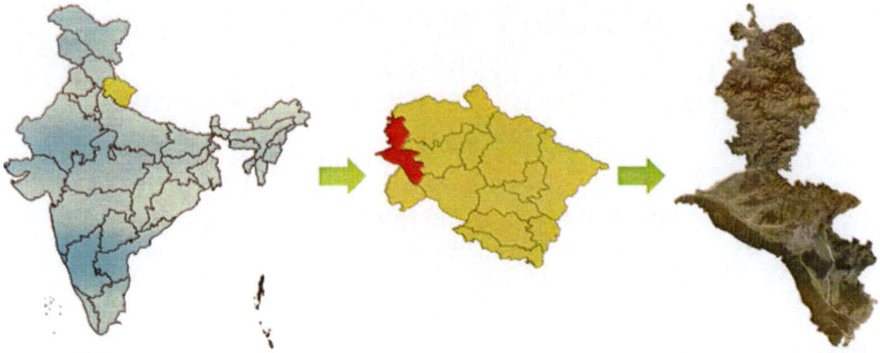

Fig. 1 The study area (District—Dehradun, State—Uttarakhand, Country—India) (Right to Left)

in landuse such as decrease in forest area, increase in urbanization and agricultural area are adversely impacting the water balances. In the current study, proper methods have been adopted and land use-land cover maps have been generated for the region of district Dehradun in Uttarakhand for different time frames with satisfactory to good levels of accuracy and hence, spatio-temporal LULC change analysis has been done to analyze the changes in the landuse pattern.

1.1 The Study Area

The entire district of Dehradun in Uttarakhand, India was chosen as the study area of the research. Dehradun lies in the extreme western part of its state Uttarakhand. Its geographical extent lies between 29° 58′ N to 31° 02′ N and 77° 34′ E to 78° 19′ E. The district has river Yamuna flowing at its west and river Ganga at its east. A large part of the district is under forest cover and its topography is a mixture of hills and valleys. The region is also known for its good climatic conditions. The study region can be seen in the following figures (Figs. 1 and 2).

1.2 Need for the Study

The region of the Dehradun district has gone through a rapid urbanization in the last few decades due to the gradual migration of a large number of people from the hilly districts of the state (for various reasons like education, healthcare facilities and ease of living) as well as from other states. Over the years, there has been a lot of infrastructure development in terms of highway expansions, construction of buildings, residential colonies, commercial centres etc. which has led to the crucial loss of some agricultural lands and forest cover of the region. As per the literature

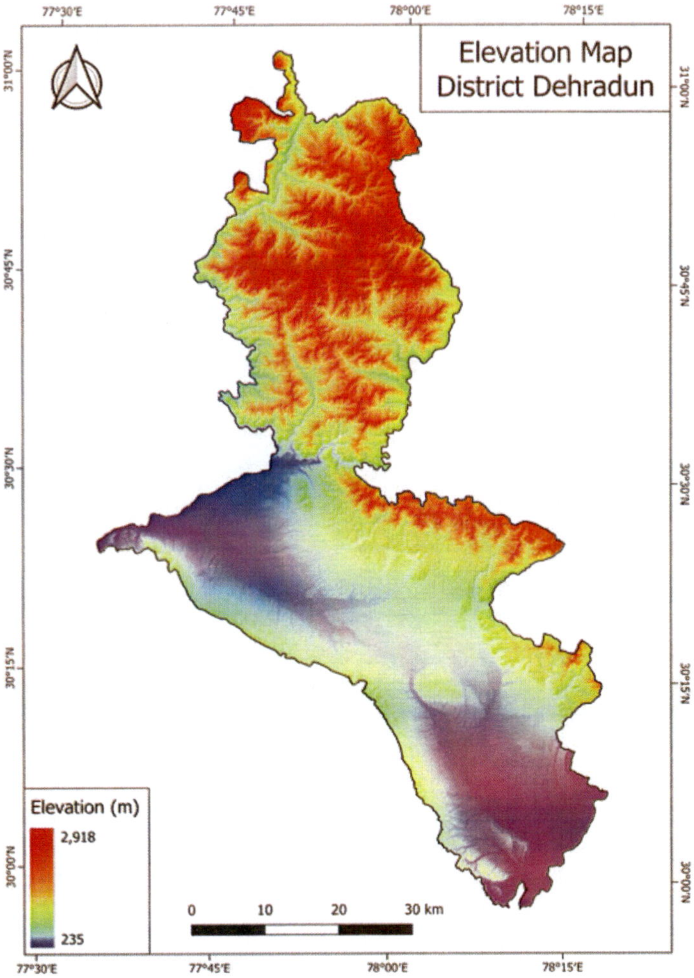

Fig. 2 Elevation Map—District Dehradun

review done, the land use-land cover study for the entire district had not been done considering a long period of time. So, the study was found necessary and was carried out for the entire Dehradun district for a time period of 30 years from 1991 to 2021 in order to analyze the long-term changes in areas of various land cover features. The study might be useful for any further research work or atleast for the availability of generated LULC maps and statistics of the adopted study area.

2 Data and Methodology

2.1 Data Used

The satellite imagery data was obtained from United States Geological Survey's website earthexplorer.usgs.gov. The considered study area was a part of the region falling in the Worldwide Reference System (WRS-2)'s Path-146 and Row-39. All the images were dated between 15th Dec (of previous year) and 20th Jan (of the study year) and had cloud covers between 2 and 9%. The images were in the CRS—EPSG: 32644 (WGS-84 / UTM Zone-44N). For years 1991, 2001 and 2011, the image data was from the satellite 'Landsat-5' with the sensor 'Thematic Mapper (TM)'. For year 2021, the data was from the satellite 'Landsat-8' with sensors 'Operational Land Imager (OLI) and the Thermal Infrared Sensor (TIRS)'. The satellite images used, had a spatial resolution of 30 m.

2.2 Methodology

LULC maps have been generated for the region of district Dehradun for different time frames viz. 1991, 2001, 2011 and 2021. Atmospheric correction was not performed on the satellite images as the cloud cover for all images was less than 10% and was negligible over the study area. Although, the values of data sets were converted to spectral reflectance (in %) in order to obtain the spectral signature plots of various LC classes.

The entire process of mapping the land use-land cover of the study area was performed in the open source software 'QGIS Desktop-3.20.1'. The study region was extracted from the satellite imagery using 'Clip Raster by Mask Layer' function of QGIS. Then by using the 'Semi-Automatic Classification Plugin', FCCs were generated using different band combinations to identify various LC classes. For classifying, training samples (initially 15 to 20 in numbers) were provided for all LC classes. The classification was performed using 'Land Cover Signature' classification and by applying the 'Maximum Likelihood' classification algorithm for the unclassified pixels. More number of training samples were subsequently added to increase the mapping accuracy upto a good extent in terms of observation. In order to remove the pixels appearing in the form of noise due to misclassification, classification sieve provided in QGIS was used. The classified area and accuracy statistics were generated from QGIS itself and some raster-based area analysis was also done using QGIS's 'Raster Calculator' and later the required statistical analysis and other interpretations were carried out using Microsoft Excel. The complete process has been shown in Fig. 3.

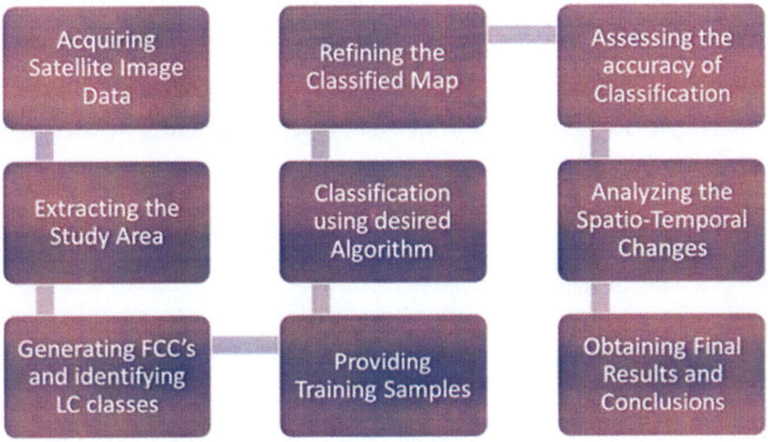

Fig. 3 Process flowchart of the study

2.3 *Accuracy Assessment*

For accuracy assessment, samples in the form of individual pixels were created (total 120–150 in numbers) from the classified map and then the accuracy was checked by comparing the sample pixel areas with google satellite images and the FCCs of the respective years. Stratified random sampling approach was used to take the samples from the LULC maps for accuracy assessment and for deciding the total number of samples, the following equation was used from Cochran [3].

$$n = \frac{(\sum W_i S_i)^2}{\left[S(\hat{O})\right]^2 + (1/N)\sum W_i S_i^2} \approx \left(\frac{\sum W_i S_i}{S(\hat{O})}\right)^2$$

where, 'n' is the total number of sample units, 'N' is the number of units in the ROI (total number of pixels), 'W_i' is the mapped area proportion of class 'i', 'S_i' is the standard deviation in mapping class 'i', $S_i = \sqrt{Ui(1 - Ui)}$, 'U_i' is the conjectured value of user's accuracy for class 'i' and '$S(\hat{O})$' is the standard error in the estimated overall accuracy.

Considering the standard error of overall accuracy as 0.03 (or 3%), the total number of samples was calculated. For distributing it among the various LC classes, three different sample size allocations were made [8] for all the land cover classes based on equal sample size for all the classes, mapped area proportion of the classes and standard deviation in mapping of classes. Then, the final number of samples for each class was calculated by simply taking average of the three allocations.

3 Results and Discussions

The following results (shown in Table 1) have been obtained after the classification of satellite imageries of the considered study years viz. 1991, 2001, 2011 and 2021 for the district Dehradun:

As the mapping of 'Dry River Bed' and 'Water' was not very accurate and also as they shared a low proportion of area in the district, leaving them, all other classes (forest, cropland, built-up, bare soil and sparse vegetation) were considered and referred to as 'major' land cover classes.

In order to remove the snow cover from the maps, the 1991 map was compared with the 2001 map using the raster calculator and it was concluded that as per the increase/decrease of LC class proportions, 25%, 43% and 32% of the snow cover area should be replaced with forest, bare soil and sparse vegetation respectively. After doing the same, the finally estimated percentage areas for the major LC classes along with their rate of change have been given and shown below in Table 2 and Fig. 4.

The results (as per Table 2) have shown that during 1991–2021, dense forest, active cropland and sparse vegetation areas have decreased by 17.52%, 62.11% and 7.50% respectively whereas built-up and bare soil areas have increased by 112.23% and 44.32% respectively. Area wise (as per Table 3 ahead), there has been a loss of

Table 1 Year wise percentage areas of different LC classes

S. No	LULC class type	1991	2001	2011	2021
1	Forest	40.036	39.401	37.012	33.937
2	Cropland	2.185	1.682	1.148	0.828
3	Built-up	2.421	3.109	4.367	5.138
4	Water	1.144	1.207	0.687	1.937
5	Dry river bed	3.324	2.322	1.744	2.092
6	Bare soil	17.410	22.095	25.693	27.891
7	Sparse/dry vegetation	28.997	30.147	29.349	28.137
8	Snow cover	4.482	0.038	0.000	0.040

Table 2 Finally estimated percentage areas of major LC classes

LULC class type	1991	2001	2011	2021	Rate of change (%)			Overall change (%) (1991–2021)
					1991–2001	2001–2011	2011–2021	
Forest	41.157	39.411	37.012	33.947	−4.24	−6.09	−8.28	−17.52
Cropland	2.185	1.682	1.148	0.828	−23.04	−31.70	−27.90	−62.11
Built-up	2.421	3.109	4.367	5.138	+28.38	+40.50	+17.65	+112.23
Bare soil	19.337	22.111	25.693	27.908	+14.35	+16.20	+8.62	+44.32
Sparse vegetation	30.431	30.159	29.349	28.150	−0.89	−2.69	−4.09	−7.50

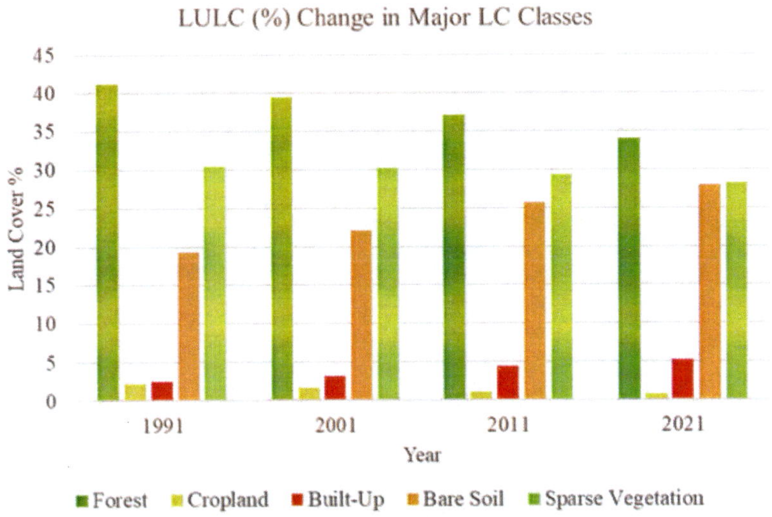

Fig. 4 Change in percentage areas of major LC classes with the years

222.63 km^2, 41.9 km^2, and 70.45 km^2 of forest, cropland and sparse vegetation areas respectively and the gain of 83.9 km^2 and 264.68 km^2 of built-up and bare soil areas respectively, considering the whole district of Dehradun.

The following LULC maps have been generated for the considered time frames viz. 1991, 2001, 2011 and 2021 for the Dehradun district (Fig. 5) and the decadal changes in LULC features are given in Table 3.

It can be inferred from the land cover classification results that the agricultural lands are decreasing at a rapid rate whereas the natural vegetation areas are also decreasing, but gradually. The built-up regions are increasing very fast due to the increase in migration and settlement of people into the district which is leading to the requirement of more construction. Bare soil regions also, have been found to increase. This is due to the fact that agricultural lands are getting converted into residential plots which remain vacant for a long time and also due to clearance of forests for road construction and other development works. From the results, the river area also seems to be decreasing probably due to the encroachment around the stream courses. Similar results have been reported in other studies [2, 9, 12, 13].

Using the raster calculator provided in QGIS, some areas were identified where urban/built-up areas were not present in 1991, but appeared in 2021. They can be seen in the map (Fig. 6).

Table 3 Year wise (decadal) absolutes areas of LC classes

LULC class type	1991	2001	2011	2021	Change in area (km^2)			Overall change (km^2)
					1991–2001	2001–2011	2011–2021	1991–2021
Forest	1270.91	1217.00	1142.92	1048.28	−53.92	−74.08	−94.64	−222.63
Cropland	67.47	51.93	35.46	25.57	−15.54	−16.46	−9.89	−41.9
Built-up	74.77	95.99	134.87	158.67	+21.22	+38.87	+23.80	+83.9
Bare soil	597.13	682.80	793.41	861.81	+85.66	+110.62	+68.39	+264.68
Sparse vegetation	939.72	931.31	906.29	869.27	−8.40	−25.03	−37.02	−70.45

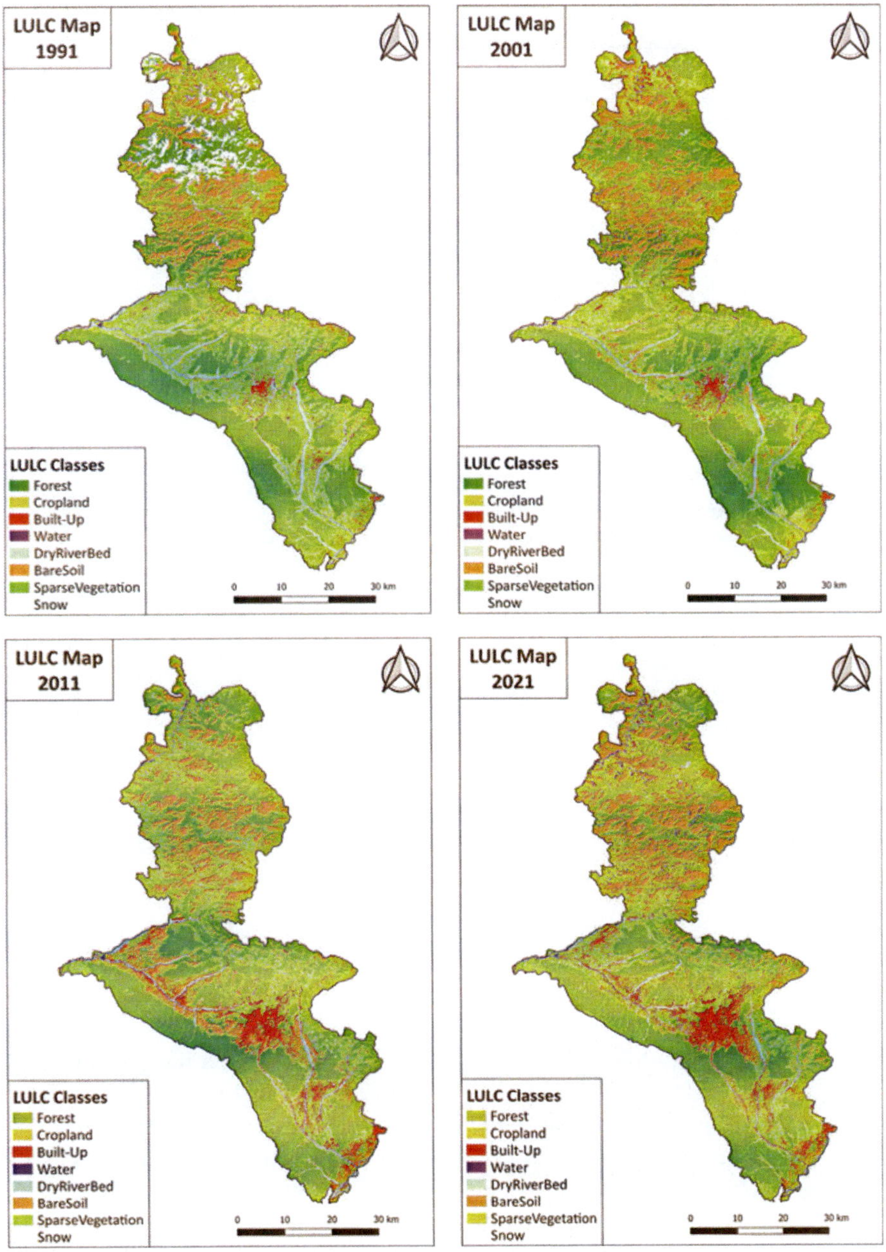

Fig. 5 Generated LULC Maps of District Dehradun

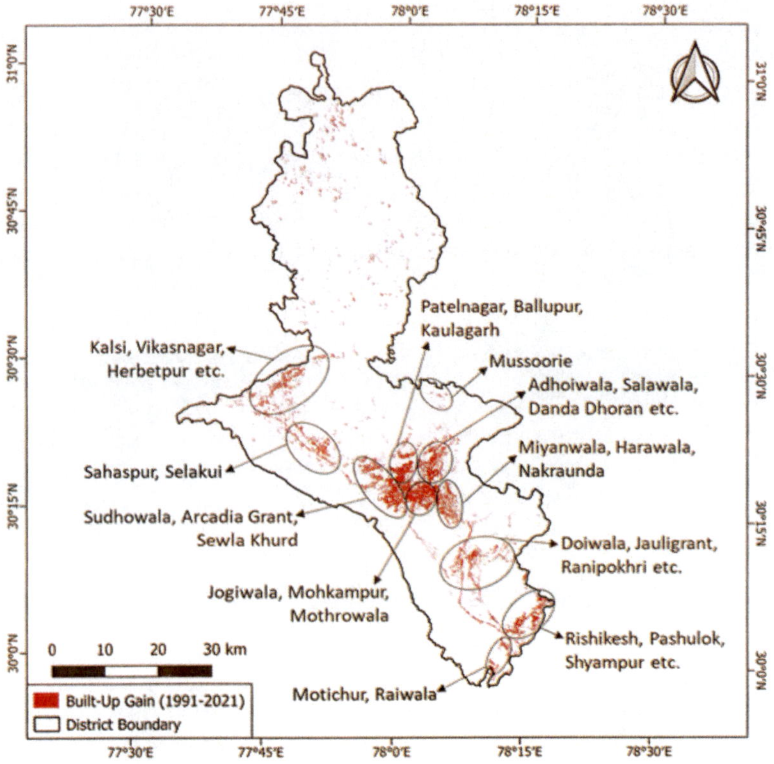

Fig. 6 Hotspots of urbanization

3.1 Problems Faced During Classification

Following issues were faced during the classification of land use-land cover for the region of district Dehradun:

- A large amount of area of the district in the northern part was mountainous having high elevation and slopes and therefore the satellite imageries of these areas contained large shadow regions which were difficult to map as they distorted (darkened) the spectral signature of the land cover features. Similarly, high regions facing the sensor brightened up and also contributed to the difficulty in classification. Also, topographic corrections through algorithms were unsuccessful in solving the problem due to high slopes.
- Mapping of croplands and built-up areas in mountainous regions was very difficult due to topographic variations, shadows etc. as stated above.
- In 1991, a significant amount of area was under snow cover which is temporary, so some analysis was required to know what was the actual land cover beneath it.
- The river sand or 'Dry River Bed' had a spectral signature very similar to 'Built-Up' areas and hence they both often got cross-classified. Large buildings with

Table 4 Classification accuracies

LULC class no	LULC class type	Average accuracy		Average Kappa Hat
		PA	UA	
1	Forest	99.26	90.88	0.862
2	Cropland	53.71	82.20	0.818
3	Built-up	71.83	65.84	0.647
4	Water	64.84	37.52	0.368
5	Dry river bed	70.70	71.63	0.711
6	Bare soil	82.48	92.09	0.896
7	Sparse vegetation	79.91	78.41	0.699
8	Snow	100.00	94.87	0.947
	Overall	**85.01**		**0.794**

smooth finish or shiny material showed higher reflectance and were classified as dry river bed and some of the river sand was classified as built-up.

- Also, the dark shadow areas often got misclassified as water. Therefore, the mapping of water (bodies) was not much accurate. Mostly, it was over-estimated.
- As the spatial resolution of the satellite sensors was 30 m, so small features such as minor roads, small unclustered buildings, small fields etc. couldn't be mapped accurately.

3.2 Classification Accuracy

Inspite of the problems listed above, good levels of overall accuracy were reached as the problem was mostly with the classes having lesser percentage of area, which is reflected in the accuracy assessment results (Table 4).

The accuracy of classification was overall 85.01% with a kappa coefficient of 0.794 which can be considered good taking into account the spatial resolution of the satellite data. It can be seen in Table 4, that for forest, and bare soil, the accuracy was good; for cropland, built-up areas, dry river bed and sparse vegetation, the accuracy was fair; for water, the accuracy was satisfactory to poor whereas, for snow cover, the accuracy was excellent.

3.3 Relation of LULC with Population

As Dehradun district has experienced a population explosion for past decades, it was also tried to find the relation of land-use with the growing population. Table 5 shows the population stats from official census data.

Table 5 Population data—District Dehradun (as per official Census data)

Year	Population	Growth Rate
1961	4,29,014	–
1971	5,77,306	34.57
1981	7,61,668	31.93
1991	10,25,679	34.66
2001	12,82,143	25.00
2011	16,96,694	32.33
2021 (expected)	22,34,512	31.70

Table 6 Relation of LULC changes with population

Parameter/Year	1991	2001	2011	2021	Correlation with population
Population	10,25,679	12,82,143	16,96,694	22,34,512	NA
Forest (km^2)	1270.91	1217.00	1142.92	1048.28	−0.9994
Sparse vegetation (km^2)	939.72	931.31	906.29	869.27	−0.9939
Cropland (km^2)	67.47	51.93	35.46	25.57	−0.9697
Built-up (km^2)	74.77	95.99	134.87	158.67	+0.9856
Bare soil (km^2)	597.13	682.80	793.41	861.81	+0.9785

It can be inferred from Table 6 that forest, sparse vegetation and croplands have a very high negative correlation with population whereas built-up area and bare soil have a very high positive correlation with the population of the district. It means that, with the growth of population, the natural vegetation cover and the agricultural lands are decreasing rapidly whereas the bare soil regions and built-up areas are increasing so as to fulfil the needs of space and infrastructure for the large number of people in the district.

3.4 Impact of LULC Changes

There have been various studies over the impact of landuse changes. Considering the effect of urbanization, i.e. the increase in built-up, open areas and the decrease in agricultural, vegetation areas (as can be seen in this study), there can be two, very prominent impacts related to hydrology. One can be the increase in runoff due to increase in built-up/impermeable areas, and the other can be the decrease in groundwater recharge as more water will flow away as runoff. Similar changes have been observed in Dehradun. The study by Anh et al. [1] shows that due to the increase in built-up area by 88.65% in the Asan watershed (major changes in sub-basins 20 and 16), and the decrease in cropland and forest areas by 6.61% and 0.25%, respectively, there was an average increase of 70% and 65% in the surface runoff from the above sub-basins during the 13 years' time period due to the rapid growth of urban industries in the area. Also, the study by Shukla et al. [11] shows that the historic well hydrograph data for Dehradun city for the period 1997–2008, indicates that groundwater levels in the most of the wells have shown a declining trend. Such findings are indicative of the impact of land cover changes on the hydrology of a region.

4 Conclusion

Remote Sensing and GIS are definitely very helpful ways of tracking the changes in land use-land cover for developing regions that need study over a large time frame. Areas with plain topography can be mapped with a good accuracy even with open source softwares and freely available satellite data. The study was quite successful in assessing the LULC changes in the study region of district Dehradun. Statistically, the study has shown that during 1991–2021, dense forest, active cropland and sparse vegetation areas have decreased by 17.52%, 62.11% and 7.50% respectively whereas built-up and bare soil areas have increased by 112.23% and 44.32% respectively.

The results also show that a lot of natural vegetation cover of the district has been lost and the rate of decline of it is growing every decade. Another crucial observation was regarding the rapid loss of agricultural land due to building of residential colonies. The district is very well known for the production of various crops including the famous basmati rice, but it might soon lose that fame. Proper landuse planning is required to limit the urbanization so as to preserve the natural heritage of the district. Also, the people with farming lands should be encouraged and supported so that they can help to conserve the agricultural areas of the district. Landuse planning is also important to keep a check on the water resource availability as the increase of built-up/impermeable areas also seems to have an impact on the hydrological characteristics such as runoff and groundwater storage of the district.

References

1. Anh NN, Chouksey A, Aggarwal SP (2016) Assessment of land use/land cover change impact on the hydrology of Asan River Watershed of Dehradun District, Uttarakhand. Int J Current Eng Technol 6(4):1125–1131
2. Bhat PA, ul Shafiq M, Mir AA, Ahmed P (2017) Urban sprawl and its impact on landuse/land cover dynamics of Dehradun City, India. Int J Sustain Built Environ 6(2):513–521
3. Cochran WG (1977) Sampling techniques, 3rd edn. Wiley, New York
4. Farida & Noordwijk MV (2004) Analysis of river discharge due to land-use change and application of GenRiver model in Way Besai watershed, Sumberjaya. World Agroforestry Centre-ICRAF

5. Li C, Sun G, Caldwell PV, Cohen E, Fang Y, ... Meentemeyer RK (2020) Impacts of urbanization on watershed water balances across the conterminous United States. Water Resour Res 56(7):e2019WR026574
6. Mawardi I (2010) River Basin Watershed damage and decrease the carrying capacity of water resources in Java as well as efforts to handle. Hidrosfir Indonesia J
7. Nugroho P, Marsono D, Sudira P, Suryatmojo H (2013) Impact of land-use changes on water balance. Procedia Environ Sci 17:256–262
8. Olofsson P, Foody GM, Herold M, Stehman SV, Woodcock CE, Wulder MA (2014) Good practices for estimating area and assessing accuracy of land change. Remote Sens Environ 148:42–57
9. Patidar S, Sankhla V (2015) Change detection of Land–use and Land-cover of Dehradun City: a spatio-temporal analysis. Int J Adv Remote Sens GIS 4(1):1170–1180
10. Prabhakar A, Tiwari H (2015) Land use and land cover effect on groundwater storage. Model Earth Syst Environ 1:1–10
11. Shukla AK, Ojha CSP, Garg RD, Pal L, Satyavati SP (2018) Groundwater appraisal of Dehradun City, Uttarakhand, India using remote sensing and geographical information system technology. Conference: STIWM at IIT Roorkee, India
12. Thapa R, Bahuguna V (2021) Monitoring land encroachment and land use & land cover (LULC) change in the Pachhua Dun, Dehradun District using landsat images 1989 and 2020. JGISE: J Geospatial Inf Sci Eng 4(1):71–80
13. Tiwari K, Khanduri K (2011) Land use/land cover change detection in Doon valley (Dehradun tehsil), Uttarakhand: using GIS& Remote sensing technique. Int J Geomat Geosci 2(1):34–41
14. Turkelboom F, Poesen J, Trébuil G (2008) The multiple land degradation effects caused by land-use intensification in tropical steeplands: a catchment study from northern Thailand. CATENA 75(1):102–116
15. Wang Y (2020) Urban land and sustainable resource use: unpacking the countervailing effects of urbanization on water use in China, 1990–2014. Land Use Policy 90:104307

Comparison of Streamflow Simulations for Different DEMs

Nagireddy Venkata Jayasimha Reddy and R. Arunkumar

Abstract Hydrological modeling of the rainfall-runoff process is crucial not only for understanding the characteristics of a basin but also for sustainable planning and management of water resources. Most of the hydrologic models require digital elevation model (DEM) as the prime input, from which the basin and its stream network are delineated by using GIS tools. DEMs from various satellite imageries like Cartosat, SRTM, ASTER, IKONOS, and others have been widely used for such purposes. However, the primary challenge is the selection of an appropriate source of DEM, since the accuracy produced from satellite imageries varies. This study aims to compare the streamflow simulations in the Chaliyar basin using a HECHMS model developed from two different DEMs. The Cartosat-DEM and ASTER-DEM are considered for comparison. Results of the study show that Cartosat-DEM gives a clear watershed boundary and area of 2912 km^2 which is very close to the India-WRIS data, whereas the ASTER-DEM gives slightly better results in simulating the streamflow based on performance measures of the model.

Keywords HEC-HMS · ASTER-DEM · Cartosat-DEM · Chaliyar basin

1 Introduction

Hydrological models are simplified replication of the real world system [8]. They not only simulate various hydrological processes but also a better choice in predicting catastrophic events like floods and droughts [10]. In the previous two decades, hydrological models have evolved in a number of ways, from conceptual to physically based distributed models. Within a hydrological model, complexity varies from simple empirical models to complicated process-based numerical models [9]. Physically based hydrological models are developed based on complex hydrological processes, as well as regionally varying information regarding watershed topography

N. V. J. Reddy (✉) · R. Arunkumar
Department of Civil Engineering, National Institute of Technology Calicut, Kozhikode 673601, Kerala, India
e-mail: nvjreddy1729@gmail.com

S. Mesapam et al. (eds.), *Developments and Applications of Geomatics*,
Lecture Notes in Civil Engineering 450, https://doi.org/10.1007/978-981-99-8568-5_6

such as soil properties and land use patterns. Several tools like SWAT, HEC-HMS, VIC, SWMM, MIKE-SHE, etc. are available for developing hydrological models. However, the selection of the appropriate tool is governed by the basin characterstics and the objectives of the study [4]. Accurate simulation of streamflow in a basin is always challenging because of the complexity associated with the hydrological processes [6].

Prediction of streamflow is one of the most important tasks in hydrological modelling [5]. When it comes to hydrological assessments and streamflow forecasting, it's crucial to use the proper model. However, the complexity lies in selecting an appropriate model and this aspect has received little attention. The hydrological model selection mainly depends on the data availability, spatial and temporal scales, model complexity, cost, convenience of use, and level of knowledge.

Verma et al. [10] used the HEC-HMS model for simulating the streamflow in the upper Sabarmati basin, Gujarat. The model performance was very good with coefficient of determination and Nash–Sutcliffe Efficiency as 0.88 and 0.84, respectively. [1] estimated the peak flows and abstraction losses by integrating GIS with HEC-HMS and calibrated with the actual flow. It was reported the simulated flows were close to actual flows. From this, HEC-HMS is one of the most suitable rainfall run-off models to simulate streamflow for Indian conditions. DEM is the main input for the rainfall runoff model which can be taken from many sources. The primary objective of the study is to compare the simulation of streamflow using the HEC-HMS model developed from different DEMs. Thus, this study concentrates on comparison of Cartosat-DEM and ASTER-DEM that can be used as an input to the HMS model to generate streamflow.

1.1 Hydrologic Engineering Center-Hydrologic Modeling System (HEC-HMS) Model

The HEC-HMS is a physically-based semi-distributed model developed by the US Army Corps of Engineers and is widely used to simulate a variety of hydrological processes [2]. Mostly the rainfall-runoff processes are modeled using the HMS. It is helpful for addressing a wide variety of challenges in a broad range of geographic situations. The hydrology of floods, small urban or natural watershed drainage, and the water supply from huge river basins are all included in this. Studies on water availability, urban drainage, flow forecasting, impacts of urbanisation, design of reservoir spillways, mitigation of flood damages, floodplain control, and system operation can all be studied using the HEC-HMS model, either on their own or in conjunction with other tools.

HEC-HMS model has mainly three components, a basin model, meteorological model and control specifications. Basin model transforms the atmospheric conditions into streamflow in the watershed. To achieve this the watershed is separated into smaller unit is using hydrologic principles. These smaller units are linked together to form a dendritic network in order to model the stream network [2]. HMS has various elements like junctions, sub-basins, sources, sinks, reaches, etc. through which the basin is represented in a spatial context. The meteorological model specifies the boundary conditions that have an impact on the watershed during a simulation. If the basin model includes sub-basins, then meteorological model must be specified for each sub-basin. Control specifications are one of the most important aspects of an HMS model as they control the duration of simulations and their beginning and termination. Runs of simulation are the primary method used to calculate results. One set of meteorological conditions, and one set of control requirements constitute each run. Using the results in HEC-HMS, graphs, summary tables, and time-series tables can be made using the basin map and the watershed explorer.

1.2 Digital Elevation Model (DEM)

The three-dimensional space is represented by digital elevation models (DEMs) in a discrete form. DEM is useful for characterizing topography and determining basin boundaries as well as studying the terrain inside the watershed. DEM from satellite imageries such as ASTER, SRTM, Cartosat etc. have evolved over the years with various spatial and temporal resolutions and used for a wide range of applications.

The DEM derived from Cartosat is widely used in India for many applications. ISRO launched Cartosat-1 in 2005 which renders stereo data for entire India. Using digital photogrammetric techniques, DEM (1 arc Sec) are produced for the entire country. Photogrammetric blocks were created to produce seamless DEMs, and DEMs were edited throughout the scenes with few break-lines. Utilizing ground control points, through dense feature matching, triangulated irregular network modelling, and automated strip mosaicking, seamless DEM data creation is carried out automatically. To identify distortions, the quality verification technique uses panning and draped visualization. Height accuracy for stereo pairs with overlapping sections is further validated.

The other DEM that is widely used globally is the ASTER DEM, a 3A1 product that is obtained from the ERSDAC, Japan-based ASTER science team. This georeferenced satellite imagery includes visible, infrared, thermal, bands and a DEM. The nadir and backward-looking infrared bands are used to create the DEM. The spatial resolution of ASTER DEM is 30 m. DEMs serve a variety of functions and are necessary prerequisite for numerous applications. They are especially helpful in areas without comprehensive topographic maps. Geomorphometry, which focuses on surface processes like landslides and may be directly represented from a DEM, is one area of study where DEMs have been proven to be helpful [3].

2 Materials and Methods

2.1 Data

The data required for the study includes observed daily precipitation which is collected from the IMD for 20 years from 1995 to 2014. DEM of 30 m resolution is obtained from NASA Earth data (ASTER) and Bhuvan (Cartosat). The observed daily streamflow for a period 1995 to 2014 is obtained from India-WRIS.

2.2 Methodology

HEC-HMS is a physically-based semi-distributed model which is widely used to simulate the streamflow. The essential components are basin model, meteorological model and control specifications which can simulate the streamflow. The basin is delineated with the help of ArcGIS using both ASTER-DEM and Cartosat-DEM. The inputs for the HEC-HMS model are delineated watershed, daily observed precipitation, sub basin parameters and observed streamflow. In HMS many methods are available to calculate canopy loss, surface loss, infiltration loss, baseflow, transformation and routing.

One component that may be applied to the sub-basin element and used to reflect the presence of vegetation in the landscape is the canopy. The amount of precipitation that reaches the ground surface is decreased by vegetation as it intercepts precipitation. Between storm events, intercepted water evaporates. Transpiration is the term for the process by which vegetation draws water from the soil. Evaporation and transpiration are frequently combined which is known as evapotranspiration. Although choosing a canopy approach is optional, it should be used for applications involving continuous simulation. HMS offers three different methods: simple canopy, gridded simple canopy, and dynamic canopy. Canopy loss calculations in this study are done using a simple canopy. One of the elements that can be a part of the sub-basin element is the surface. It represents the ground surface where water may accumulate in surface depression storage. Gridded simple surface and simple surface are the methods that are available in HMS. In this model, the simple surface method is employed.

Although a sub-basin element conceptually represents infiltration, surface runoff, and subsurface processes, the actual infiltration calculations are performed by a loss method that is integrated into the sub-basin. There are total of twelve distinct loss methods offered, including deficit and constant, layered Green and Amptdeficit, exponential Green and Ampt, gridded Green and Ampt, gridded SCS curve number, initial and constant, etc. In this work, the infiltration loss is estimated using the deficit and constant approach. A baseflow algorithm integrated within the subbasin executes the real subsurface calculations. There are six different baseflow techniques available, including recession baseflow, linear reservoir baseflow, and constant monthly baseflow. The recession method is used to determine baseflow in this study. Within the sub-basin, a transform approach is used to calculate the real surface runoff. Several unit hydrograph approaches, a kinematic wave implementation, a linear quasi-distributed method, and a two-dimensional (2D) diffusion wave method are some of the popular techniques. In this model, the excess precipitation is converted to runoff using a Clark unit hydrograph.

Although a reach element conceptually represents a segment of stream or river, a routing method is placed within the reach actually performs the calculations. There are nine different routing options available. To route the reaches to sink, the Muskingum method is applied. The calculated streamflow is calibrated with the observed streamflow once the model parameters and control specification are provided.

2.3 *Study Area*

The study area, as shown in Fig. 1, is the Chaliyar River Basin in Kerala located in India. With a length of 169 km, the Chaliyar is the fourth-longest river in Kerala and flows through the districts of Malappuram and Kozhikode. It starts 2066 m above sea level in the Elambalari Hills in Gudalur taluk of Nilgiris district in Tamil Nadu and drains into the Lakshadweep Sea at Beypore. Its coordinates are 75° 35′ and 76° 45′ East longitude and 11° 05′ to 11° 40′ North latitude. The area of the basin as a whole is 2933 km^2, of which 2545 km^2 are in Kerala and the rest 388 km^2 are in Tamil Nadu.

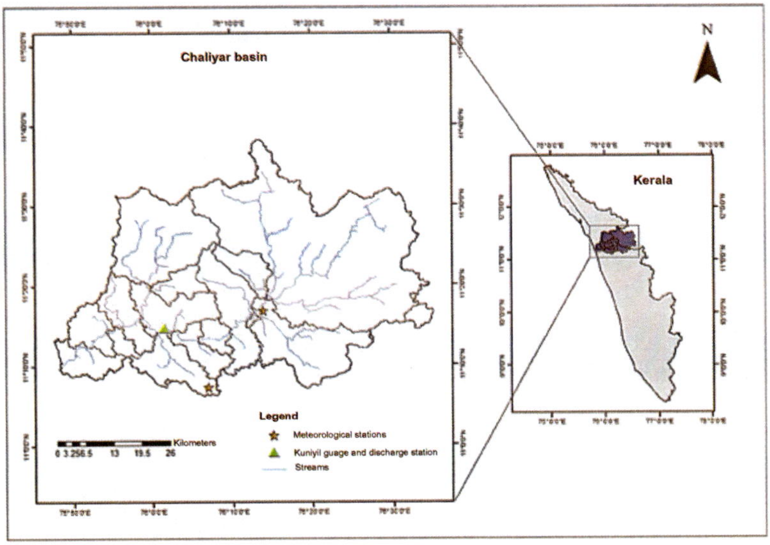

Fig. 1 Location of Chaliyar basin

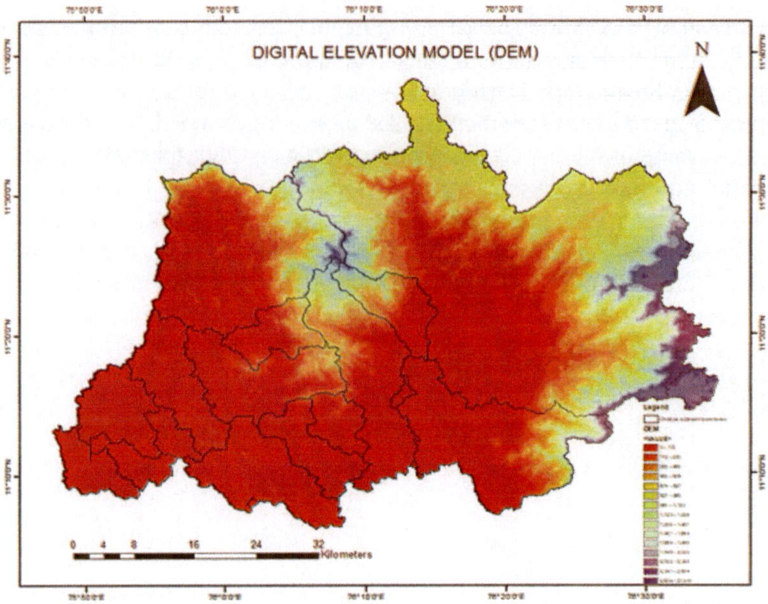

Fig. 2 Digital Elevation Model of Chaliyar basin

In the basin, the average annual rainfall is about 3012 mm. The southwest (June–September) and northeast (October–November) monsoons are the two main rainy seasons. Major thunderstorm activity is prevalent during the pre-monsoon months (March to May), whereas throughout the winter months (December-February). The four main soil types in the river basin, which account for 60.73%, 24.56%, 9.85%, and 4.86% of the total area, are clay, gravelly loam and loam, gravelly clay. The overall area is made up of urban areas, water bodies and rocky areas with agricultural land making up approximately 74.26 percent of the total area and forests making up 14.21 percent. There is one CWC hydrological observation station at Kuniyil and two IMD meteorological stations in the watershed located at Nilambur and Manjeri as shown in Fig. 1. Figure 2 shows the ASTER DEM with the elevation ranges from 0 to 2604 m.

3 Results and Discussions

3.1 Results

HEC-HMS model calibration for streamflow is performed using observed daily streamflow data for eight years (1995–2002) and validated for 3 years (2011–2013). The RSR, NSE, R^2, PBIAS values of model simulations run during calibration are evaluated; the estimated and observed streamflows are graphically assessed for agreement. The parameter values are modified until the observed streamflow and computed streamflow are relatively close to each other. Both manual and automatic calibration are performed. For automatic calibration, simplex and univariate methods are used. Both the Cartosat-DEM and ASTER-DEM final calibrated values for the model parameters are included in Table 1.

The performance parameters adopted in the study are RMSE-observations standard deviation ratio (RSR), Nash–Sutcliffe efficiency (NSE), coefficient of determination (R^2) and percentage bias (PBIAS). Moriasi et al. [7] criteria are used for model rating. The model performance during the calibration and validation are listed in Table 2.

Cartosat-DEM and ASTER-DEM are compared with the observed streamflow at the Kuniyil for 1995–2002 to check how simulated and observed streamflow are closely related as shown in Fig. 3 and Fig. 4 respectively.

From Fig. 3, it is clear that the day i.e., 12-07-1997 of simulated peak flow is matched with the observed peak flow and the peak flows are 1859.2 m^3/s, 2193 $m^3/$

Table 1 Calibrated model parameters for both Cartosat-DEM and ASTER-DEM

Model	Parameter	Calibrated value (Cartosat-DEM)	Calibrated value (ASTER-DEM)
Canopy	Initial storage Maximum storage (mm) Crop coefficient	14% 10 0.90	16% 7 0.80
Loss	Initial deficit (mm) Maximum deficit (mm) Constant rate (mm/hr)	6.77 21.55 0.48	4.73 14 0.46
Transform	Time of concentration (hr) Storage coefficient (hr)	16.33 56	16.33 56
Baseflow	Initial discharge (m3/s) Recession constant Ratio to peak	19 0.90 0.42	17 0.90 0.42
Routing	K (hr) X	0.32 0.20	0.32 0.10

Table 2 Comparison of performance indices of simulated and observed streamflow during calibration and validation for Cartosat-DEM and ASTER-DEM

Performance measures	Cartosat-DEM		ASTER-DEM	
	Calibration period	Validation period	Calibration period	Validation period
RSR	0.55	0.68	0.54	0.66
NSE	0.70	0.53	0.71	0.56
R^2	0.58	0.59	0.58	0.59
PBIAS	−3.81	2.50	−5.92	2.76

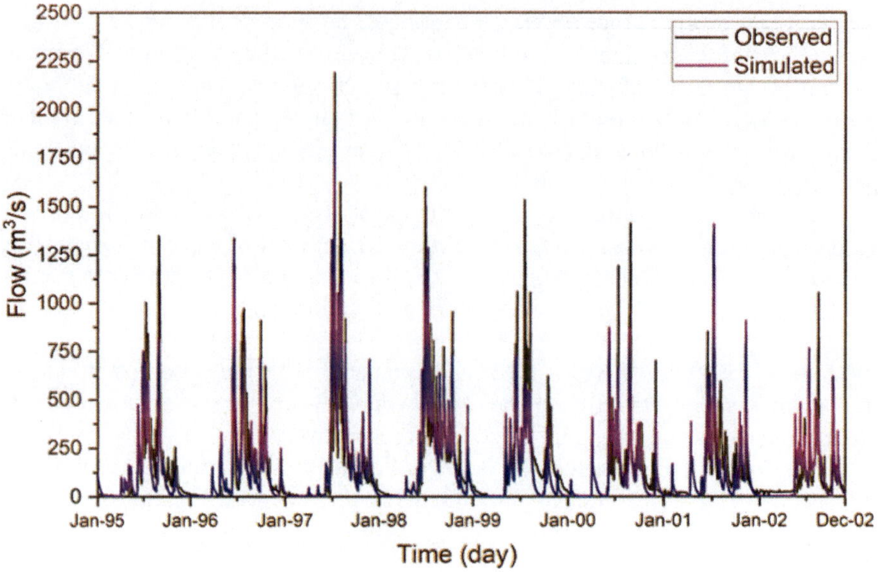

Fig. 3 Comparison of simulated and observed daily streamflow for Cartosat-DEM

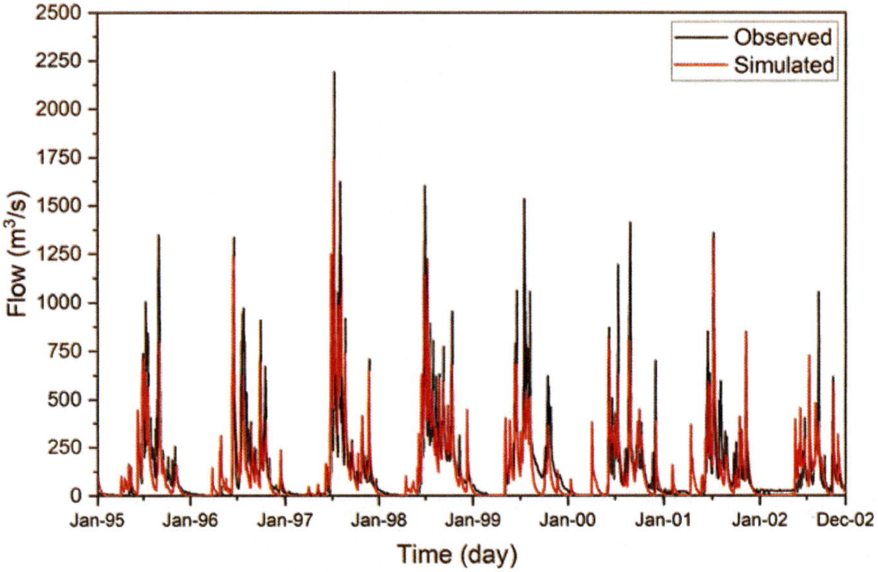

Fig. 4 Comparison of simulated and observed daily streamflow for ASTER-DEM

s respectively. The model predicts 15.22% lesser than the observed value when the peak flow is compared for the Cartosat-DEM.

From Fig. 4, it is clear that the day i.e., 12-07-1997 of simulated peak flow is matched with the observed peak flow and the peak flows are 1739.9 m^3/s, 2193 m^3/s respectively. The model predicts 20.66% lesser than the observed value when the peak flow is compared for the ASTER-DEM.

3.2 Conclusion

In this study, the streamflow simulations are compared for two DEMs. The streamflow is simulated through HEC-HMS model. During the calibration of HEC-HMS model, it is observed that time of concentration, initial deficit, maximum storage and constant rate are highly sensitive parameters as small change in the parameters affects the streamflow. From the results, it is noticed that the HEC-HMS model has simulated the streamflow close the observed values. The performance of the model is very good during calibration and good during validation for both DEMs. It is observed that the Cartosat-DEM and ASTER-DEM give the all most similar results. On comparing the performance measures of both DEMs R^2, NSE and RSR do not show significant change whereas the PBIAS has slight change. Cartosat-DEM gives the clear watershed boundary of 2912 km^2 which is very close the India-WRIS, Chaliyar river basin boundary. Over all, comparing the performance measures of model, ASTER-DEM gives slightly better results over Cartosat-DEM.

References

1. Alfy M (2016) Assessing the impact of arid area urbanization on flash floods using GIS, remote sensing, and HEC-HMS rainfall–runoff modeling. Hydrol Res Nordic Assoc Hydrol 47(6):1142–1160
2. Feldman AD (2022) Hydrologic modeling system HEC-HMS user's manual
3. Gajalakshmi K, Anantharama V (2015) Comparative study of cartosat-DEM and SRTM-DEM on elevation data and terrain elements. Cloud Publ Int J Adv Remote Sens GIS, 1361–1366
4. Halwatura D, Najim MM (2013) M, Application of the HEC-HMS model for runoff simulation in a tropical catchment. Environ Model Softw 46:155–162
5. Hamdan ANA, Almuktar S, Scholz M (2021) Rainfall-runoff modeling using the HEC-HMS model for the Al-Adhaim River Catchment, Northern Iraq. Hydrology, MDPI AG 8(2):58
6. Kumari N, Srivastava A, Sahoo B, Raghuwanshi NS, Bretreger D (2021) Identification of suitable hydrological models for streamflow assessment in the Kangsabati River Basin, India, by using different model selection scores. Nat Resour Res Springer 30(6):4187–4205
7. Moriasi DN, Arnold JG, van Liew MW, RL Bingner, Harmel RD, Veith TL (2007) Model evaluation guidelines for systematic quantification of accuracy in watershed simulations. Trans ASABE Amer Soc Agric Biological Eng (ASABE) 50(3):885–900
8. Motovilov YG, Gottschalk L, Engeland K, Rodhe A (1999) Validation of a distributed hydrological model against spatial observations. Agric Forest Meteorol 98:257–277

9. Sanjay Shekar NC, Vinay DC (2021) Performance of HEC-HMS and SWAT to simulate streamflow in the sub-humid tropical Hemavathi catchment. J Water Climate Change IWA Publ 12(7):3005–3017
10. Verma R, Sharif M, Husain A (2022) Application of HEC-HMS for hydrological modeling of upper Sabarmati River Basin, Gujarat, India. Model Earth Syst Environ, 1–9

Comprehensive Analysis of Impact of COVID-19 Lockdown on Air Quality in Andhra Pradesh, India

Donthi Rama Bhupal Reddy and Ramannagari Bhavani

Abstract The Novel Corona Virus Disease (COVID-19) initially observed in Wuhan city of China in 2019 induced a severe risk globally. In a drastic response to the COVID-19 pandemic, the entire country was under social and travel lockdown from 25/03/2020 to 14/04/2020 (21 days), which was extended until May 3, 2020. This lockdown had a significant influence on both the local and global economies, and it will take some time for things to return to normal. However, one major advantage of this lockdown has emerged an increase in the air quality of cities around the world. Therefore, regarding the current situation, in our study, we have analyzed the impact of lockdown on air quality in the state of Andhra Pradesh (Amravati, Rajamahendravaram, Tirupati and Vizag) and also source identification to above mentioned areas during lockdown period. The study results emphasized a statistically significant decline in pollutant concentration. Average Air Quality Index of Vizag declined more by ~34% (96.67–63.52) when compared with Amaravati by ~26% (66–48.78), Tirupati by ~31% (83.8–52.75), Rajamahendravaram by ~4% (60.95–56.59). Future this study identified the potential sources during lockdown towards study areas with the application of HYSPLIT Back Trajectory Analysis (BTA) found that majorly from Bay of Bengal, within the state and partially from surrounding regions Odisha, Tamilnadu, Chhattisgarh, and Telangana.

Keywords COVID-19 · Air quality · Back trajectory analysis

1 Introduction

COVID-19 originated from Wuhan city of China in 2019 induced a severe risk globally and declared as a major disaster due to widespread global infection by the World Health Organization (WHO) [1]. According to WHO, as of May 13, 2020, ~4,369,933 COVID-19 cases have been confirmed globally including 15% deaths (https://covid19.who.int/). In case of India, first case was identified on January 30th,

D. Rama Bhupal Reddy (✉) · R. Bhavani
JNTUACEA, Anantapur, Andhra Pradesh, India
e-mail: ramabhupalreddy1@gmail.com

S. Mesapam et al. (eds.), *Developments and Applications of Geomatics*,
Lecture Notes in Civil Engineering 450, https://doi.org/10.1007/978-981-99-8568-5_7

2020 and it's been increased gradually. As of 2020 May 13, India crossed 75,048 COVID cases including 2440 death [2]. In a drastic response from the COVID-19 pandemic, Prime minister of India on 22 March 2020 declared a Janata curfew. After that, the entire country was under social and travel lockdown (STL) from 25/03/2020 to 14/04/2020 (21 days), which was extended until May 3, 2020, with conditional relaxation and followed by the III phase of lockdown till 17/05/2020. All businesses were forced to close as a result of social and travel restrictions, which comprises of hotels and restaurants, shops, entertainment facilities (like Malls, theatres, etc.,), transportation, etc., excluding small number of essential services like medicines, groceries, etc.

Despite the lockdown, power plants continued to operate, albeit at a reduced capacity as a result of the drop in the demand for electricity from industrial and commercial units [3]. The local and global economies have already been significantly impacted by this catastrophe, and it will take some time for things to return to normal. Argumentatively, each nation's oil usage has been significantly reduced as a result of the suspension of local transportation and routine social activities [4]. For e.g., Due to the lockdown in March 2020, resulted in a drop in fuel consumption in India by ~60–70%, which was witnessed the biggest blow in the recent 20 years [5]. Many other nations, like India, have implemented various restrictions and observed lockdowns with differing degrees of severity and implementation. The global decline in anthropogenic activity has strengthened a number of scholars to look at the environmental effects of the lockdown. Although the primary function of these restrictions was to flatten the COVID-19 infection curve, they also produced a novel experiment to examine how human activities affect air pollution. It is noticed that, there is an improvement in the air quality worldwide because of Lockdown and Travel restrictions [6, 7]. Many Countries like China, Brazil, Italy, etc., across the world reported declines in air pollution levels. In Brazil, during lockdown, there is a decline in the concentration of NO, CO, and NO_2 by ~77%, ~65%, and ~54% respectively when compared with average concentration values of the same months from 2015 to 2019 reported in [8]. Muhammad et al. [4] reported there was a reduction of 20–30% in the concentration of NO_2 in Italy, China, and Spain. Further, due to lockdown there is a significant decline in traffic related pollutants like Particulate matter (PM2.5, PM10), CO, NO_X, and Benzene in the city of Milan [9]. A similar scenario has been noticed in the Indian context. Sharma et al. [10] reported during the lockdown reduction in concentration levels of PM2.5, PM10, CO, NO_2 by 43, 31, 10, and 18% respectively by comparing with 2017–2019 pollutant concentrations to the same period. Further, [5] also reported that pollution levels in Delhi during lockdown reduced by ~41%, ~52%, ~51%, and ~28% of PM2.5, PM10, NO_2, and CO respectively, same trend has been followed in case of Kolkata Mumbai, Bangalore, and Chennai. Similar decline in the air pollution levels have been reported by [11] in India (Pune, Ahmadabad), [12] considered lockdown period (25th March to 03rd May 2020) and reported a significant decline in the concentrations of NO_2, PM2.5, and PM10 by comparing the results with 2017–2019 concentration levels over India from 134 stations.

According to [12] report, every 9 out of 10 people respire polluted air that leads to an increase in premature mortality, and around 4.2 million deaths occur globally which is expected to rise to 6–9 million by 2060. And very few studies have been done on South-East Indian region and most of them are not focused on the Source identification during lockdown. Therefore, considering the current situation, in our study we analyzed consequence of Social and Travel Lockdown (STL) on air quality in the state of Andhra Pradesh (Amravati, Rajamahendravaram, Tirupati, and Vizag) and also source identification to above mentioned areas during lockdown period.

2 Methodology

Figure 1 represents the overview of methodological approach adopted to conduct the study. The present study assessed the impact of STL during the Novel COVID-19 break out on the air quality on major locations of Andhra Pradesh (AP) state, India as shown in Fig. 2.

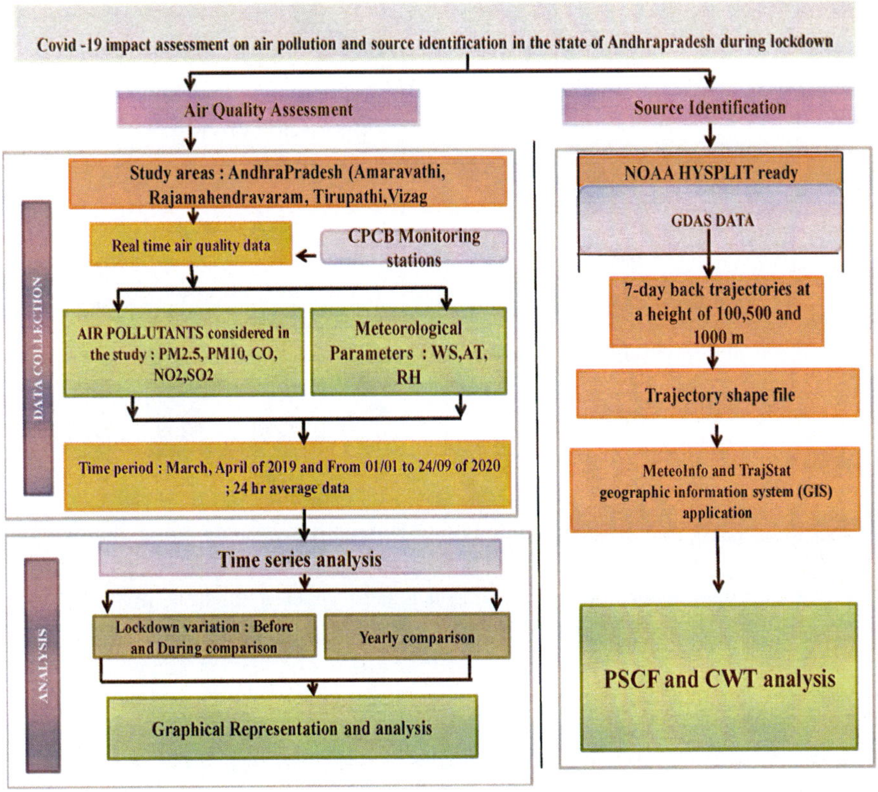

Fig. 1 Methodological approach used in the present study

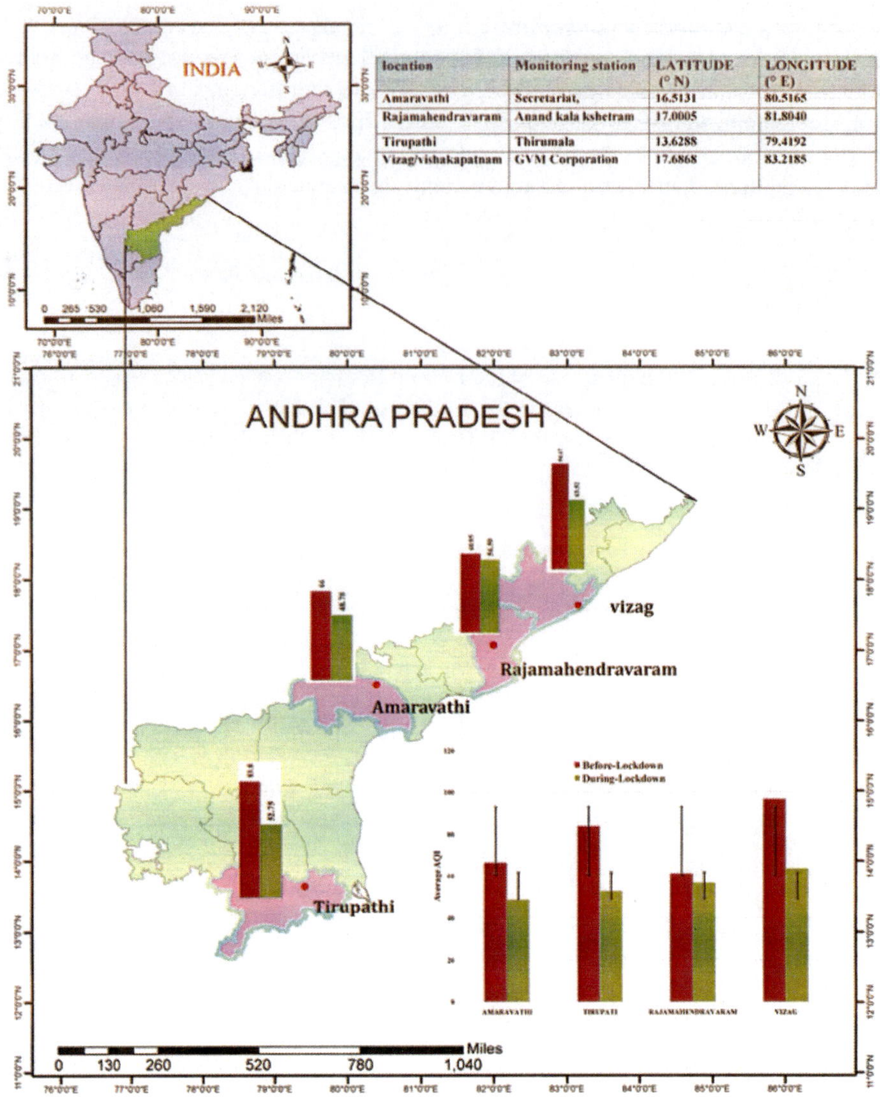

location	Monitoring station	LATITUDE (° N)	LONGITUDE (° E)
Amaravathi	Secretariat,	16.5131	80.5165
Rajamahendravaram	Anand kala kshetram	17.0005	81.8040
Tirupathi	Thirumala	13.6288	79.4192
Vizag/vishakapatnam	GVM Corporation	17.6868	83.2185

Fig. 2 Map of study regions (Amaravati, Tirupati, Rajamahendravaram, Vizag) in Andhra Pradesh State, India along with variation in average AQI for the period before lockdown and after lockdown

2.1 *Study Area and Data Source*

The current study assesses the variation in air quality due to STL on various locations namely Amaravati, Rajamahendravaram, Tirupati, Vizag of AP state, India. From 01/01/2020 to 24/03/2020 was considered as before lockdown, from 25/03/2020 to 31/05/2020 as during lockdown, and from 01/06/2020 to 24/09/2020 after-lockdown in this study. Required air quality data of study areas for analysis was obtained from "Central Pollution Control Board (CPCB) online portal (https://app.cpcbccr.com/ccr/#/caaqm-dashboardall/caaqm-landing)". The 24-h average concentration data of 5 major air pollutants i.e., particulate matter with a diameter $\leq$ 2.5 μm (PM2.5), particulate matter with a diameter $\leq$ 10 μm (PM10), nitrogen dioxide (NO_2), sulfur dioxide (SO_2), carbon monoxide (CO) and also 3 Meteorological Parameters (MtP) Relative humidity (RH), Ambient Temperature (AT), and Wind speed (WS) were obtained from monitoring stations of each location. The above mentioned parameters continuous 24 h average data were collected for the entire study duration i.e., 01/03/2018 to 30/04/2018, 01/03/2019 to 30/04/2019, 01/01/2020 to 24/09/2020.

2.2 *Data Analysis*

Trend Analysis: The following two time periods were used to examine how STL affected the air quality of study areas: Scenario 1: Lockdown period comparison was made from 01/01/2020 to 24/03/2020 (before lockdown period), 25/03/2020 to 31/03/2020 (During Lockdown period) 01/06/2020 to 24/09/2020 (after lockdown).

Scenario 2: Yearly comparison: Air quality data obtained from CPCB for the time period of March–April 2020 was compared with data from March–April 2019. Along with these, the effect of MtP on air quality with the help of time series data was analyzed. For the entire analysis in the study, 24 h average data was considered.

2.3 *HYSPLIT Backward Trajectory Analysis (BTA)*

The HYSPLIT ("Hybrid Single Particle Lagrangian Integrated Trajectory") Model developed by the NOAA (The "National Oceanic and Atmospheric Administration") Air Resources Laboratory used for calculating "air parcel trajectories, dispersion, complex transport, chemical transformation, deposition simulations by using puff or particle approaches" [13]. The model is built with a modular library structure that consists of programs for trajectories and air concentration applications. Required meteorological data fields are obtained from the forecast or meteorological models (like WRF, MM5, etc.) for calculations [14]. BTA is one of the common applications of the model, from the knowledge of paths of air mass transport from origin or source region to receptor locations that might be qualitatively measured. Back trajectories provide a mode of tracing the route of an air parcel (i.e., the origin of air masses) and develop a relationship between the source and receptor [15]. Stein et al. [16] explained model evolution, development of HYSPLIT model and also highlighted the recent applications of HYSPLIT Model.

In this study, we adopted this model to identify the sources which are impacting the study areas during lockdown period. The data obtained from HYSPLIT Model is been visualized and analyzed with the help of Meteoinfo an ArcGIS tool [17]. Trajstat is a plugin tool used to conduct "Concentration-Weighted Trajectory" (CWT) and "Potential Source Contribution Factor" (PSCF) analysis in identifying potential sources [18]. CWT value indicates the strength of the source [19]. In CWT method, the trajectories arriving at the receptor location were weighted based on the mean concentration detected there during the trajectory's arrival. PSCF defines "the probability that a receptor area impacted from identified regions". Dimitriou et al. [20] used PSCF model for assessing the contribution of regional sources to urban air pollution in Athens.

3 Result and Discussion

Obtained data from CPCB was analyzed in terms of daily and yearly time periods. It was observed that the concentration of pollutants varied daily and yearly due to the implementation of lockdown in the study areas (see Fig. 3). Time period of lockdown in entire India started on 24th March 2020 to 31st May 2020, and air quality data obtained from CPCB was analyzed for different time phases i.e., before-, during- and after-lockdown. Figure 3 shows the concentration of pollutants in the study region before-, during- and after-lockdown. The concentrations of all the air pollutants significantly decreased over time, according to temporal analysis of changes in their levels from the period before to during lockdown, it showed a decline in the concentration from the period before to during to after lockdown as shown in Fig. 3.

3.1 Effects of Lockdown on Air Quality

When compared in the average concentration of pollutant PM2.5 before and during lockdown decrease in the concentration observed in Amaravati from 35.73 to 20.13 $\mu g\ m^{-3}$ (43.66%), Tirupati from 30.66 to 18.55 $\mu g\ m^{-3}$ (39.5%), Rajamahendravaram from 39.18 to 16.45 $\mu g\ m^{-3}$ (58.01%), Vizag from 46.2 to 18.38 $\mu g\ m^{-3}$ (60.22%). Same trend was followed for PM10 from 59.28 to 47.36 $\mu g\ m^{-3}$ (20.11%), 54.67 to 40.86 $\mu g\ m^{-3}$ (40.73%), 73.05 to 43.3 $\mu g\ m^{-3}$ (55.74%), 100.26 to 62.27 $\mu g\ m^{-3}$ (37.89%) in Amaravati, Tirupati, Rajamahendravaram, and Vizag respectively. Average Concentration of NO_2 also decreased from 12.21 to 5.72 $\mu g\ m^{-3}$ (53.15%) in Amaravati, 29.37 to 7.42 $\mu g\ m^{-3}$ (74.73%) in Tirupati, 17.98 to 7.95 $\mu g\ m^{-3}$ (55.78%) in Rajamahendravaram, and 34.36 to 25.33 $\mu g\ m^{-3}$ (26.28%) in Vizag. For SO_2 decrease in the concentration from 19.11 to 14.04 $\mu g\ m^{-3}$ (26.53%), 5.83 to 4.78 $\mu g\ m^{-3}$ (18.01%), 8.8 to 7.05 $\mu g\ m^{-3}$ (19.89%), 8.45 to 8.03 $\mu g\ m^{-3}$ (4.97%) in Amaravati, Tirupati, Rajamahendravaram, and Vizag respectively. It was observed that pollutant CO average concentration before and during lockdown was decreased in Amaravati from 0.58 to 0.26 $mg\ m^{-3}$ (55.17%), Tirupati

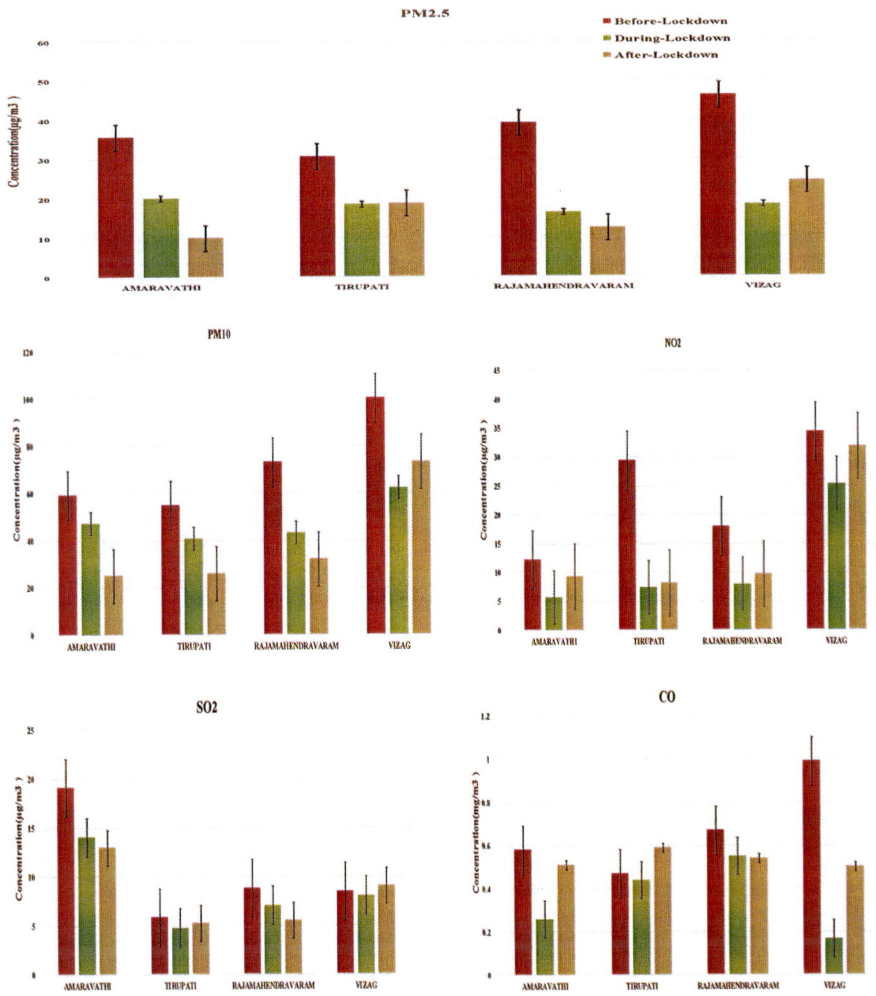

Fig. 3 Impact of lockdown on air quality (before-, during-, after-lockdown)

from 0.47 to 0.44 mg m^{-3} (6.38%), Rajamahendravaram from 0.67 to 0.55 mg m^{-3} (17.91%), Vizag from 0.99 to 0.17 mg m^{-3} (82.82%) which is the highest % of reduction in CO concentration Percentage decline in the average concentration of CO, SO_2, NO_2, PM2.5, and PM10 in all study regions during lockdown and before lockdown is shown in Fig. 4. Highest percentage of reduction in the average concentration is observed in Vizag for CO pollutant i.e., 82.83% (see Fig. 4).

Variation in the concentrations between During-Lockdown to After-Lockdown. Even after lockdown to a certain extent, it was observed that few pollutant concentrations had reduced when compared with during lockdown concentration Fig. 3. Average PM2.5 concentration after lockdown period in Amaravati and

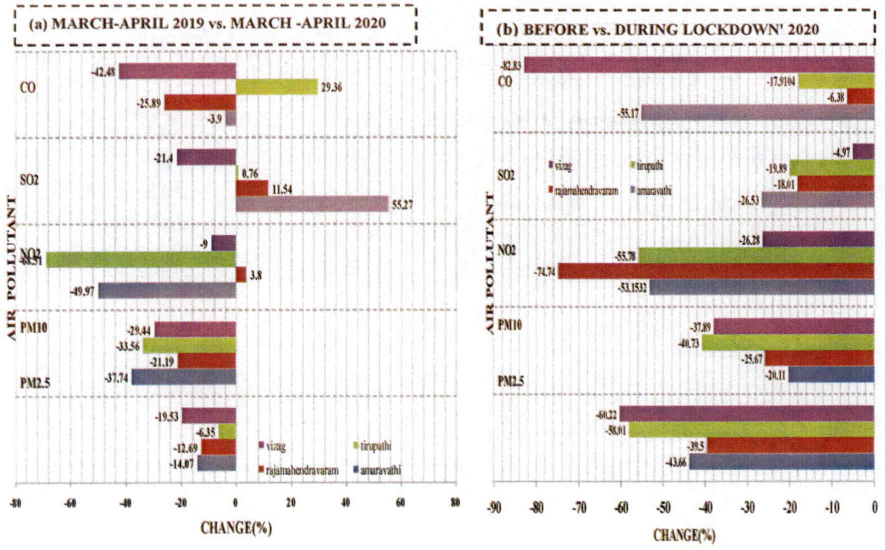

Fig. 4 The % change in the concentration of pollutants **a** between March and April months of 2019 and 2020, **b** between before lockdown and after lockdown

Rajamahendravaram decreased from 20.12 to 9.99 μg m^{-3} and 16.45 to 12.51 μg m^{-3} respectively. For PM10 expect Vizag remaining Amaravati, Tirupati, and Rajamahendravaram decreased from 47.36 to 25.06 μg m^{-3}, 40.86 to 25.99 μg m^{-3}, and 43.3 to 31.99 μg m^{-3} respectively Fig. 3. For SO_2 average concentration decreased in Amaravati and Rajamahendravaram (see Fig. 3). All the study region average pollutant concentrations showed no violation of "National Ambient Air Quality Standards (NAAQS: PM10 = 100 μg m^{-3}, PM2.5 = 60 μg m^{-3}, NO_2 = 80 μg m^{-3}, SO_2 = 80 μg m^{-3} based on 24 h average CO = 2 mg m^{-3} (based on 8 h average))" during lockdown.

Variation in Pollutant Levels in March and April months of 2019 and 2020. Figure 5 shows the concentrations of PM2.5, PM10, NO_2, CO, and SO_2 in the study regions for March and April months of 2019 and 2020. A significant reduction in PM2.5 concentration from 23.46 to 20.16 μg m^{-3}, PM10 from 64.09 to 39.91 μg m^{-3}, NO_2 from 11.88 to 5.94 μg m^{-3}, CO from 0.53 to 0.51 mg m^{-3}, amounting to 14.06%, 37.73%, 49.97% and 3.89% reductions, respectively, whereas a rise in SO_2 concentration ~55.26% (12.42–19.22 μg m^{-3}) was observed in Amaravati during the months of March and April in 2019 and 2020. In case of Rajamahendravaram, the concentration of pollutants decreased from 20.40 to 17.81 μg m^{-3} (12.68%) for PM2.5, 52.87 to 41.67 μg m^{-3} (21.18%) for PM10, 0.74 to 0.54 mg m^{-3} (25.88%) for CO, whereas there was a rise in the concentration of NO_2 from 10 to 10.38 μg m^{-3} (3.79%) and for SO_2 from 6.53 to 7.28 μg m^{-3} (11.53%). A similar trend in decline of pollutant concentrations from 23.39 to 21.91 μg m^{-3} (6.34%) for PM2.5, 63.31 to 42.07 μg m^{-3} (33.55%) for PM10, 40.31 to 12.7 μg m^{-3} (68.51%) for NO_2, but in case of SO_2 and CO there is a rise from 5.85 to 5.89 mg m^{-3} (0.76%) and from 0.37 to 0.47 $\mu g\ m^{-3}$ respectively for the location Tirupati. In case of Vizag, there is a significant decline in PM2.5 concentration from 28.92 to 23.27 $\mu g\ m^{-3}$, PM10 from 93.44 to 65.93 $\mu g\ m^{-3}$, NO_2 from 32.36 to 29.44 $\mu g\ m^{-3}$, SO_2 from 10.59 to 8.32 $\mu g\ m^{-3}$, CO from 0.74 to 0.43 mg m^{-3}, amounting to 19.53%, 29.43%, 9%, 21.4%, and 42.47% reductions, respectively was observed during month of march and April in 2019 and 2020.

AQI Values in Study Areas during the Lockdown Period. AQI values were determined for March and April months of 2019 and 2020 years (see Fig. 6). As per [21] AQI is "good for the range (0–50), Satisfactory (51–100), moderately polluted (101–200), Poor (201–300) Very poor (301–400) and severe (401–500)". In 2019 Amaravati's AQI in the good range was 27.87% whereas it was 73.77% in 2020 and 8.2% in the moderately pollutant category in 2019 whereas zero % in 2020, which signifies a decline in the air pollutant concentrations (Fig. 6). Similarly, criteria of decline in pollutant concentration were occurred in Rajamahendravaram and Tirupati also. In Vizag there was a huge decline in the pollutant concentration in the year 2020 compared to 2019 i.e., moderately polluted category decreased from 27.6 to 6.52%. Same scenario was observed in New Delhi, Kolkata, and Mumbai, decline in AQI from 225.52, 93.8, and 65.56 in 2019 to 125.4, 65.79, and 69.79 respectively during lockdown period [22].

Mean AQI variations. Average AQI variations of study regions for the time period before and during Lockdown are presented in Fig. 2. It is clearly observed from Fig. 1 Average AQI of Vizag declined more ~34% (96.67–63.52) when compared with Amaravati ~26% (66–48.78), Tirupati ~31% (83.8–52.75), Rajamahendravaram ~4% (60.95–56.59) for Before-Lockdown to After-Lockdown period.

3.2 *Meteorology and Air Quality*

The concentration of a pollutant during different phases of the air pollution cycle, such as pollutants' release at the source, transport, diffusion, and their reception, is significantly influenced by meteorological factors [23]. Specifically, conditions like relative humidity (RH), wind speed (WS), and air temperature (AT) play crucial roles.

In situations where RH is low and both WS and AT are high, the conditions are favorable for greater dispersion of air pollutants compared to calm weather conditions [10]. Time series analysis of all criteria air pollutants along with MtP like considered in the study for Vizag region is shown in Fig. 7. Before lockdown average values of meteorological parameters at Vizag location are AT = 31.44 °C, RH = 73.73% and WS = 1.92 m/s, During lockdown AT = 32.89 °C, RH = 71.89% and WS = 2.31 m/s. When we compare both before and during lockdown values, in case of during lockdown AT and WS are higher and RH is lower than before lockdown which are favorable for more dispersion of air pollutants. Jayamurugan et al. [24] reported the statistical correlation between RH and Particulate matter is negatively

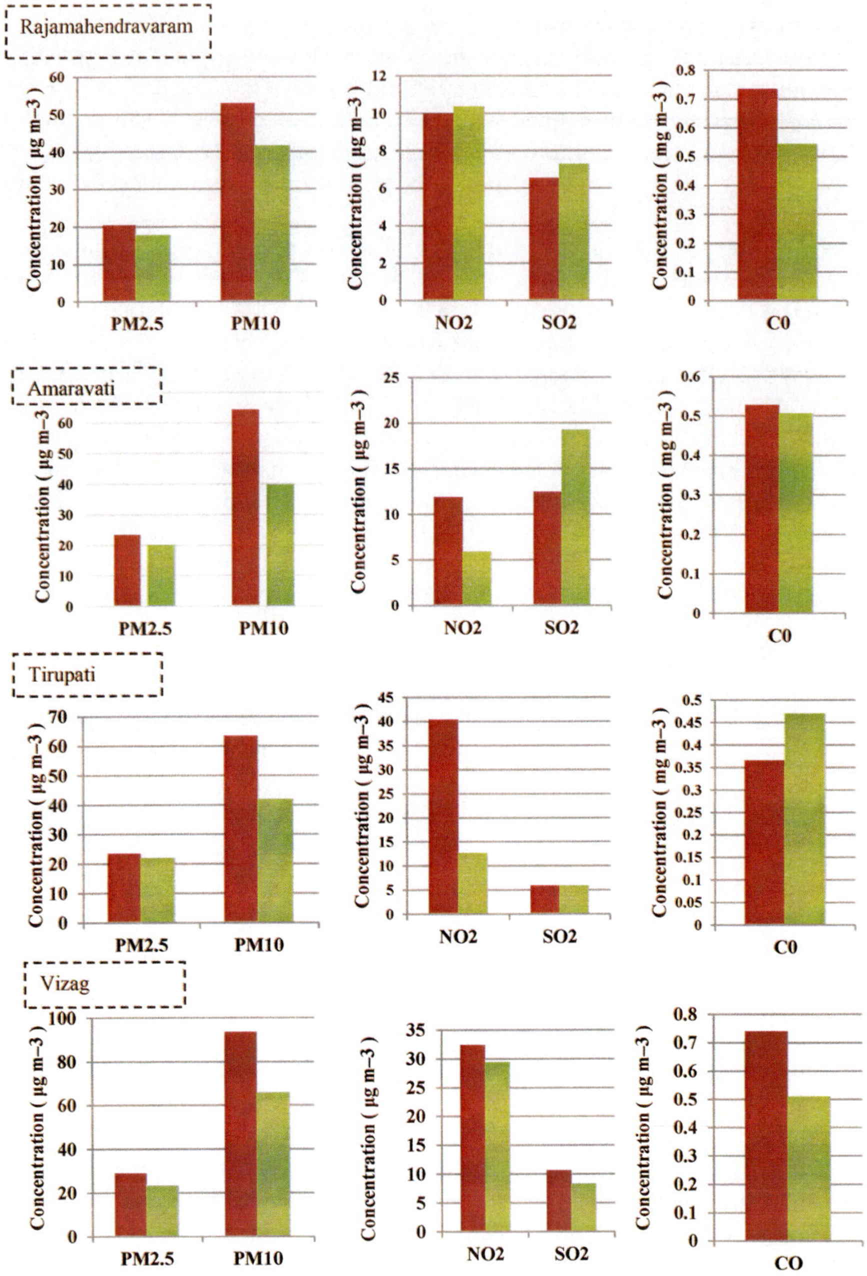

Fig. 5 Variation in pollutants levels in 2019 and 2020 for March and April months for Rajamahendravaram, Amaravati, Tirupati, and Vizag

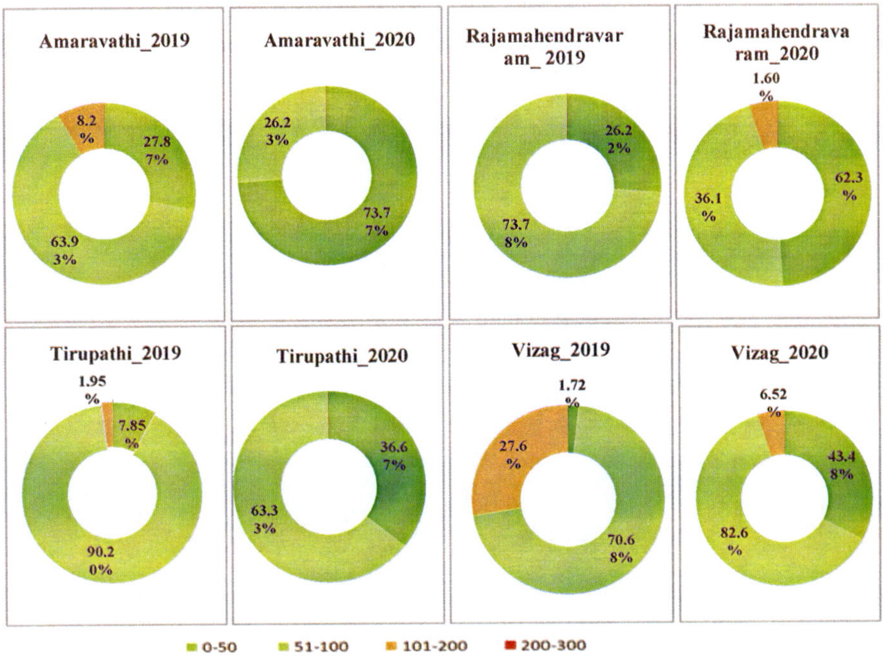

Fig. 6 AQI comparison for the period of March–April months of 2019 and 2020

related. Sharma et al. [10] reported the average WS in North India from predominant directions (South and southwest) as 1.5 m/s and WS observed during March and April months of 2020 as ~1 m/s for southern India and ~0.7 m/s for Eastern India. Srivastava [25] investigated the relation between Meteorology, Air pollution, and COVID-19, stated that parameters like RH and AT are negatively correlated whereas solar radiation (SR) and WS are positively correlated to the coronavirus cases. And also stated there is a significant decline in the concentration of SO_2, CO, PM10, PM2.5, BC, and NO_X, due to the effect of lockdown. Future [26], observed that initially there was a decline in the O_3 levels due to almost stable meteorology, but in later days it increased as an increase in AT and SR which enhances the photochemical activity between nitrogen oxides and volatile organic compounds (VOC) results in ozone formation. Thus, apart from STL, (MtP) also played a significant role in reducing air pollution. Furthermore, there was a notable increase in the rainfall concentration in the state of Andhra Pradesh, from 532.9 mm (in 2019) to 694.0 mm (in 2020), specifically during the southwest monsoon period (June to September) (http://apsdps.ap.gov.in/WeatherPages/history.html).

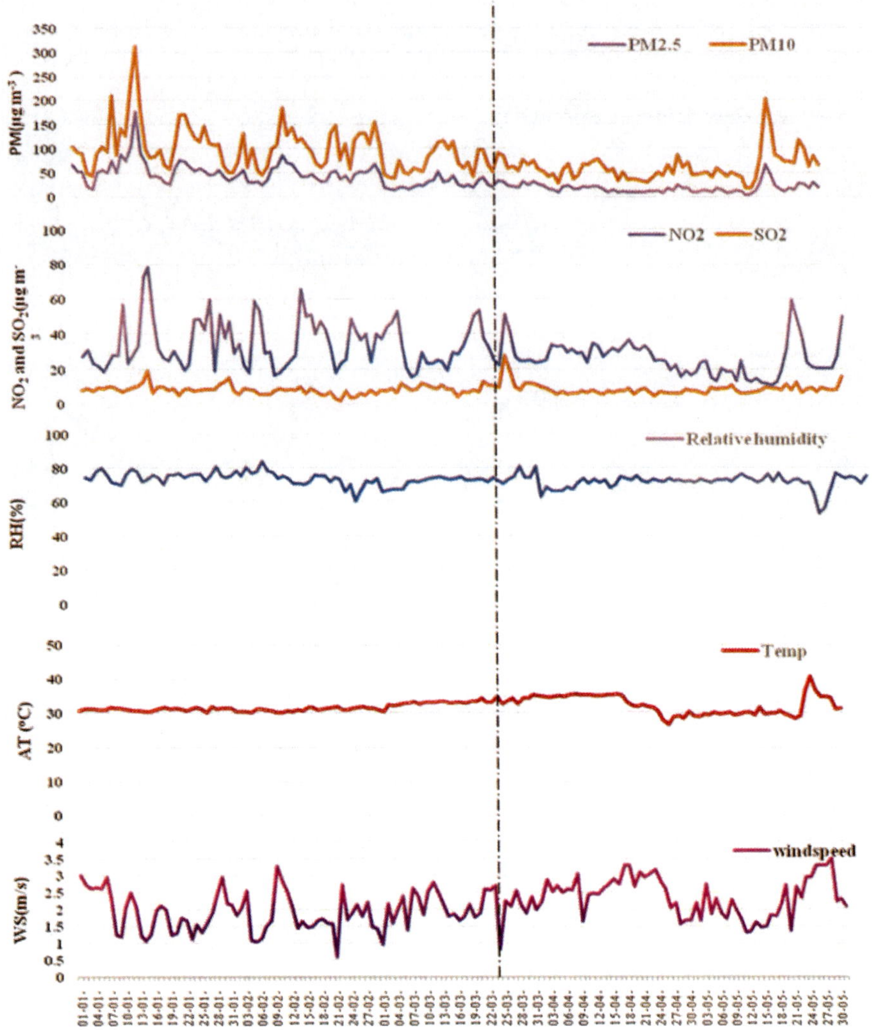

Fig. 7 Trend analysis representing the impact of lockdown period and meteorology on air quality in Vizag

3.3 *Source Identification*

Even though there was travel and social lockdown imposed in the AP certain quantity of pollutant concentration been observed. To know from where the source of air mass reaching the study locations can be identified with the help of BTA. Waked et al. [27] used BTA for identifying the geographical origin of PM10 particles impacting Northern France. Similarly, [28] used BTA to know the behavior of aerosols in Spain. Limited studies done on source identification in India [29]. Mahapatra et al. [30]

observed that PM2.5 at receptor location are significantly impacted by sources near to study reason and long range transport of pollutants. Ravindra et al. [31] stated vehicle emissions and Crop residue burnings are predominant sources in Indo Gangetic Plain. However, all these were before COVID-19 and there were no studies done on source identification during lockdown period. The PSCF approach primarily focuses on source identification using BTA to calculate and describe potential source locations. Whereas CWT method is a concentration weighting algorithm which clearly distinguishes the source strength by assigning the concentration values at the receptor region to their corresponding trajectories.

Figure 8a, c, e, g shows the Map of PSCF analysis for the study regions Amaravati, Rajamahendravaram, Tirupati, Vizag which shows the potential sources of air pollution in lockdown period. Over Amaravati, air masses are majorly from Bay of Bengal region and within the surrounding regions. Over Rajamahendravaram, major air mass reaches from Bay of Bengal and partially from neighbor states Telangana and Odisha. In lockdown period over Tirupati, air parcels are coming from south-east regions of India i.e. Tamilnadu, Bay of Bengal majorly, and a few potential sources from Karnataka and Sri Lanka. Major potential sources over Vizag location are from Bay of Bengal and the surroundings of the location, Partially from the states of Odisha and Chhattisgarh. From PSCF analysis potential source towards the study areas are majorly from Bay of Bengal and surrounding Regions. The results of the CWT analysis are shown in Fig. 8b, d, f, h creating a grid with the same resolution as the PSCF analysis. Singh and Chauhan [32] used HYSPLIT model for source identification during lockdown for the Locations New Delhi, Kolkata, Chennai, Hyderabad, and Mumbai.

4 Conclusion

In response to the Novel COVID-19 outbreak, the Indian Government implemented a lockdown from March 24th, 2020, to May 31st, 2020. This measure effectively curbed the spread of the virus and led to a notable decrease in air pollution levels, however, the imposed lockdown in reaction to the COVID-19 outbreaks has affected the lives of hundreds of millions of Indians. The present study evaluated the variation in the concentration of pollutants in four major regions of Andhra Pradesh state, India. Scenario 1: On Comparison of Air quality between before lockdown to after lockdown for the study regions a significant decline in the pollution levels, Average AQI of Vizag declined more ~34% (96.67–63.52) when compared with Amaravati ~26% (66–48.78), Tirupati ~31% (83.8–52.75), Rajamahendravaram ~4% (60.95–56.59). Scenario 2: Comparison between the months of March–April 2019 and March–April 2020 showed that decline in the concentration levels of PM2.5, PM10, NO_2 and CO pollutants considered in the study in all regions except CO in Tirupati, NO_2 in Rajamahendravaram. Whereas, SO_2 declined only in Vizag and in the rest of regions it is increased. Along with COVID lockdown, meteorological parameters also impacted the air quality.

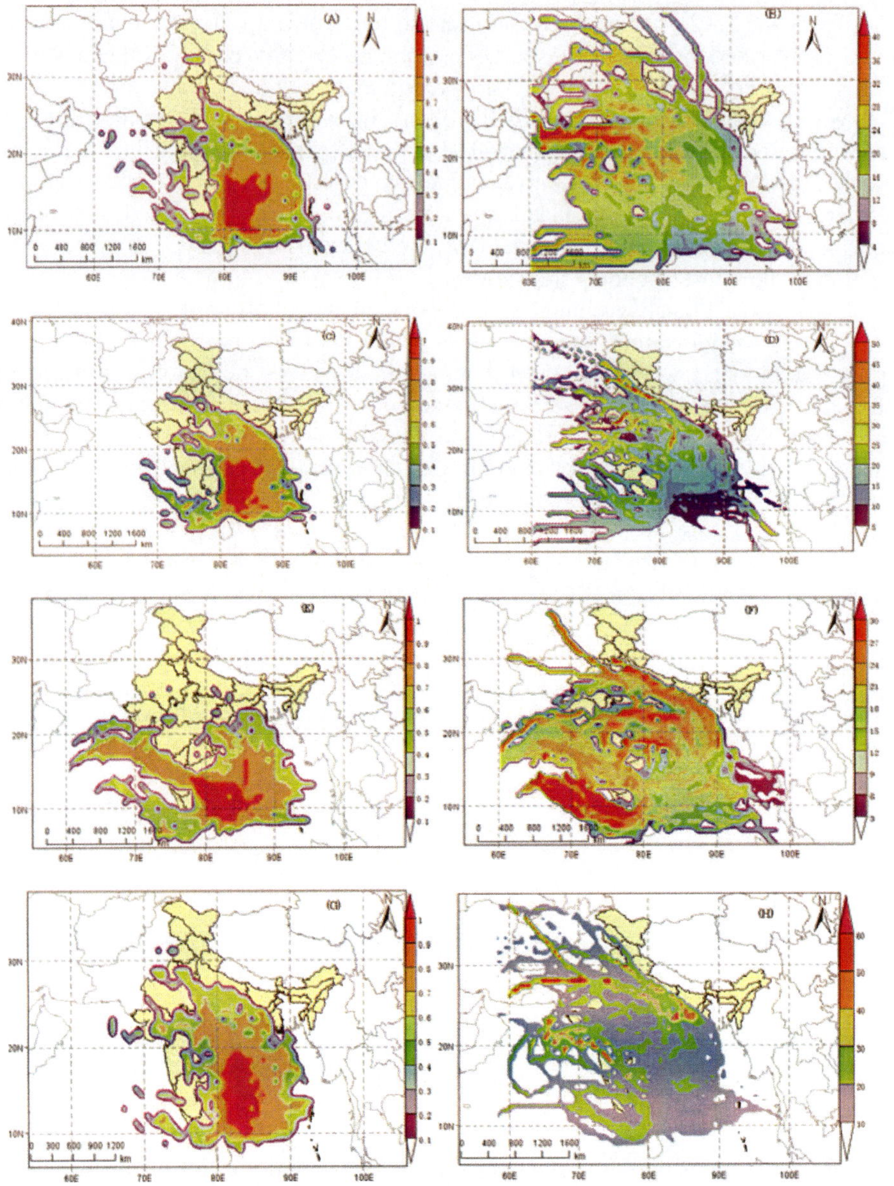

Fig. 8 HYSPLIT back trajectory PSCF analysis (**a**, **c**, **e**, **g**) and CWT analysis (**b**, **d**, **f**, **h**). Amaravati (**a**, **b**), Rajamahendravaram (**c**, **d**), Tirupati (**e**, **f**), and Vizag (**g**, **h**)

In identifying the potential sources towards study areas with the application of HYSPLIT back trajectory analysis (PSCF and CWT) found that majorly within the state (local sources) and partially from surrounding regions Odisha, Tamilnadu, Chhattisgarh, Telangana. Long range transport from South-East coastal regions and Sri Lanka region. The findings would prompt the Andhra State as well as Indian government to consider ways to strictly reduce biomass burnings, and industrial and vehicular pollution in order to improve air quality and support better public health in India.

References

1. WHO (2020) Coronavirus disease (COVID-19) pandemic. https://www.who.int/emergencies/diseases/novel-coronavirus-2019
2. MoHFW (2020) COVID-19 INDIA. https://www.mohfw.gov.in/
3. CRISIL (2020) COVID-19 corollaries. CRISIL Limited, Mumbai, India
4. Muhammad S, Long X, Salman M (2020) COVID-19 pandemic and environmental pollution: a blessing in disguise? Sci Total Environ 138820
5. Jain S, Sharma T (2020) Social and travel lockdown impact considering coronavirus disease (COVID-19) on air quality in megacities of India: present benefits, future challenges and way forward. Aerosol Air Qual Res 20(6):1222–1236
6. Saadat S, Rawtani D, Hussain CM (2020) Environmental perspective of COVID-19. Sci Total Environ 728:138870. https://doi.org/10.1016/j.scitotenv.2020.138870
7. Bao R, Zhang A (2020) Does lockdown reduce air pollution? Evidence from 44 cities in northern China. Sci Total Environ 731:139052. https://doi.org/10.1016/j.scitotenv.2020.139052
8. Nakada LYK, Urban RC (2020) COVID-19 pandemic: impacts on the air quality during the partial lockdown in São Paulo state, Brazil. Sci Total Environ 730:139087. https://doi.org/10.1016/j.scitotenv.2020.139087
9. Collivignarelli MC, Abbà A, Bertanza G, Pedrazzani R, Ricciardi P, Miino MC (2020) Lockdown for CoViD-2019 in Milan: what are the effects on air quality? Sci Total Environ 732
10. Sharma S, Zhang M, Gao J, Zhang H, Kota SH (2020) Effect of restricted emissions during COVID-19 on air quality in India. Sci Total Environ 728:138878
11. Yadav R, Korhale N, Anand V, Rathod A, Bano S, Shinde R, Latha R, Sahu SK, Murthy BS, Beig G (2020) COVID-19 lockdown and air quality of SAFAR-India metro cities. Urban Clim 34:100729
12. WHO (2018) World health statistics 2018: monitoring health for the SDGs. https://www.who.int/gho/publications/world_health_statistics/2018/en/
13. Draxler RR, Hess GD (1998) An overview of the HYSPLIT_4 modelling system for trajectories. Aust Meteorol Mag 47(4):295–308
14. Draxler R, Stunder B, Rolph G, Stein A, Taylor A (2018) HYSPLIT4 user's guide 254
15. Verma S, Venkataraman C, Boucher O, Ramachandran S (2007) Source evaluation of aerosols measured during the Indian Ocean Experiment using combined chemical transport and back trajectory modeling. J Geophys Res 112:D11210. https://doi.org/10.1029/2006JD007698
16. Stein AF, Draxler RR, Rolph GD, Stunder BJB, Cohen MD, Ngan F (2015) NOAA's HYSPLIT atmospheric transport and dispersion modeling system. Bull Am Meteorol Soc 96:2059–2077. https://doi.org/10.1175/BAMS-D-14-00110.1
17. Wang YQ (2014) MeteoInfo: GIS software for meteorological data visualization and analysis: meteorological GIS software. Meteorol Appl 21(2):360–368. https://doi.org/10.1002/met.1345

18. Wang YQ, Zhang XY, Draxler RR (2009) TrajStat: GIS-based software that uses various trajectory statistical analysis methods to identify potential sources from long-term air pollution measurement data. Environ Model Softw 24(8):938–939. https://doi.org/10.1016/j.envsoft.2009.01.004
19. Cheng I, Zhang L, Blanchard P, Dalziel J, Tordon R (2013) Concentration weighted trajectory approach to identifying potential sources of speciated atmospheric mercury at an urban coastal site in Nova Scotia, Canada. Atmos Chem Phys 13(12):6031–6048. https://doi.org/10.5194/acp-13-6031-2013
20. Dimitriou K, Grivas G, Liakakou E, Gerasopoulos E, Mihalopoulos N (2021) Assessing the contribution of regional sources to urban air pollution by applying 3D-PSCF modeling. Atmos Res 248:105187
21. National Air Quality Index - Report of the Expert Committee (2014) Control of Urban Pollution Series (CUPS/82/2014–15)—Central Pollution Control Board, Ministry of Environment, Forest and Climate Change, Government of India. https://app.cpcbccr.com/ccr_docs/FINAL-REPORT_AQI_.pdf
22. Tripathi A (2021) Air pollution in four Indian cities during the Covid-19 pandemic. Int J Environ Stud 78(4):696–717
23. Turner DB (1973) Effects of meteorological parameters on transport and diffusion. EPA. Air Pollution Training Institute Control Programs Development Division Office of Air and Water Programs
24. Jayamurugan R, Kumaravel B, Palanivelraja S, Chockalingam MP (2013) Influence of temperature, relative humidity and seasonal variability on ambient air quality in a coastal urban area. Int J Atmos Sci 2013:264046. https://doi.org/10.1155/2013/264046
25. Srivastava A (2021) COVID-19 and air pollution and meteorology-an intricate relationship: a review. Chemosphere 263:128297
26. Garg A, Kumar A, Gupta NC (2021) Comprehensive study on impact assessment of lockdown on overall ambient air quality amid COVID-19 in Delhi and its NCR, India. J Hazard Mater Lett 2:100010
27. Waked A, Bourin A, Michoud V, Perdrix E, Alleman LY, Sauvage S, Delaunay T, Vermeesch S, Petit J-E, Riffault V (2018) Investigation of the geographical origins of PM10 based on long, medium and short-range air mass back-trajectories impacting Northern France during the period 2009–2013. Atmos Environ 193:143–152. https://doi.org/10.1016/j.atmosenv.2018.08.015
28. Piñero-García F, Ferro-García MA, Chham E, Cobos-Díaz M, González-Rodelas P (2015) A cluster analysis of back trajectories to study the behaviour of radioactive aerosols in the south-east of Spain. J Environ Radioact 147:142–152. https://doi.org/10.1016/j.jenvrad.2015.05.029
29. Banerjee T, Murari V, Kumar M, Raju MP (2015) Source apportionment of airborne particulates through receptor modeling: Indian scenario. Atmos Res 164:167–187. https://doi.org/10.1016/j.atmosres.2015.04.017
30. Mahapatra PS, Sinha PR, Boopathy R, Das T, Mohanty S, Sahu SC, Gurjar BR (2018) Seasonal progression of atmospheric particulate matter over an urban coastal region in peninsular India: role of local meteorology and long-range transport. Atmos Res 199:145–158. https://doi.org/10.1016/j.atmosres.2017.09.001
31. Ravindra K, Singh T, Mor S, Singh V, Mandal TK, Bhatti MS, Gahlawat SK, Dhankhar R, Mor S, Beig G (2019) Real-time monitoring of air pollutants in seven cities of North India during crop residue burning and their relationship with meteorology and transboundary movement of air. Sci Total Environ 690:717–729. https://doi.org/10.1016/j.envpol.2017.08.016
32. Singh RP, Chauhan A (2020) Impact of lockdown on air quality in India during COVID-19 pandemic. Air Qual Atmos Health 13(8):921–928

Development of Mobile Application for Assessing Urban Heat Island (UHI) Using Geospatial Techniques a Case Study of Chennai City

S. Jayalakshmi

Abstract Urban Heat Island (UHI) is the phenomenon where urbanization results in an increase in surface temperature among different locations within the city. UHI hotspots not only lead to poor air quality and make people's health at higher risk, but they also tend to magnify the heat stress and level of thermal discomfort experienced by the people. This study aims to find the UHI spots using thermal remote sensing based on satellites, for the estimation of surface temperature, over a continuous spatial and temporal scale and to develop a mobile application indicating the spatial pattern of UHI and heat stress. Wet Bulb Globe Temperature (WBGT) data collected at various locations across Chennai city was evaluated to obtain the indices reflecting risk levels of heat stress in each area. This was subsequently analyzed in a GIS environment, along with the disaggregated Land Surface Temperature (LST) data, to arrive at valuable information that was used to delineate the hotspots of high heat stress and UHI intensity in the city. Finally, this data was exported to a mobile platform (Android) and an application indicating the spatial pattern of UHI and heat stress was developed, which shows the heat risk zones, mitigation measures, etc. This study confirmed the existence of UHI effect in Chennai city during summer. Temperature difference was found to be even as high as 6–7 °C in many parts of the city. The intensity of UHI was established to be strongly dependent on urban factors such as the density of built-up areas, vegetation cover and presence of water bodies. It was shown that such adverse heat conditions deteriorated the urban environment causing health problems. The results of this study indicate that the highest thermal stress is found in the South-Western and Northern part of the city, which is predominantly crowded, constructed (built-up), industrial and commercial areas.

Keywords Urban heat island · Thermal data · Wet bulb globe temperature · Land surface temperature · Mobile application

S. Jayalakshmi (✉)
Institute of Remote Sensing, College of Engineering Guindy, Anna University, Chennai, India
e-mail: j_lak2001@yahoo.co.in

S. Mesapam et al. (eds.), *Developments and Applications of Geomatics*,
Lecture Notes in Civil Engineering 450, https://doi.org/10.1007/978-981-99-8568-5_8

1 Introduction

As urban regions develop, modifications occur in their landscape. Man-made features like roads, buildings, and other infrastructure replace natural land cover such as vacant land and vegetation. Areas that were once permeable and moist become impermeable and dry [1]. These changes result in urban areas becoming warmer than their rural surroundings. This leads to the formation of an "island" of higher temperatures in the urban landscape, termed as a surface 'Urban Heat Island' (UHI).

Surface temperature indicates the heat energy given off by features such as vacant land, buildings, and other surfaces. Thermal instruments mounted on satellites measure this surface temperature of features on the ground, and provide better geographic coverage than ground-based instruments which are normally used for recording air temperatures [2]. There are some drawbacks in satellite data also. A combined use of satellite data for surface temperatures and data from monitoring stations for air temperatures offers the most complete picture of a city's heat island [3].

Due to industrialization, urban landscapes are often characterized by the substitution of natural surfaces like vegetation by man-made surfaces such as high albedo parking spaces, concrete areas, asphalt roadways etc. Such an urbanization process has been found to affect the thermal environment in cities [4]. The magnitude of UHI intensity is generally found to be greater at night than during the day [5]. The main cause of this phenomenon is the higher rate of nocturnal cooling of the vacant areas around cities when compared with that of densely built-up localities [6]. Satellite-based thermal remote sensing data can provide temperature information of different land use categories at regional levels. This can be utilized to identify, map and study the spatial distribution of the surface temperature of UHI hot spots. Voogt and Oke [7] stated that atmospheric UHI is observed from ground-based air temperature measurements taken from weather stations, whereas surface UHI is observed from thermal remote sensors that record the upwelling thermal radiance emitted by the surface area that lies within the instantaneous field of view of the sensor.

Schwarz et al. [8] found that the eleven indicators that were analyzed individually reveal diurnal and seasonal patterns, but show rather low correlations over time and for single points in time, the different indicators show only weak correlations, although they are supposed to quantify the same phenomenon. UHIs not only have fundamental impacts on urban climate, air quality and economics, but also on the public health [9]. Steeneveld et al. [10] applied an approximated 'wet bulb globe temperature' (AWBGT) index, that is based on the measurement of basic environmental variables, to quantify the measure of human comfort. While the scientific community has embraced the use of satellite imagery as a tool for phenological studies, BeeBox, a web-based application developed by Arundel et al. [11] attempts to make this same information available to a more general audience (beekeepers).

Finer spatial resolution thermal data comes with the drawback of poorer temporal resolution was used in most of the studies [12–15]. This study focuses on mapping the spatial distribution of UHI and heat stress levels in the metropolitan city of

Chennai, and analyzing the correspondence between them. It incorporates an automated approach to downscale satellite-based surface temperature data, for delineation of heat island hotspots. This provides the basis for subsequent development of a useful mobile application featuring heat stress in various heat zones, and supplying vital precautionary information to alleviate the adverse impacts of heat stress on human health.

1.1 Study Area

Chennai is one of the four metropolitan cities in India. It is the capital of Tamil Nadu state which is also a very famous south Indian eastern coastal city located in North-Eastern part of Tamil Nadu. Its geographical extent is 12° 51′ 3″ N to 13° 18′ 2″ N latitude and 80° 00′ 00″ E to 80° 14′ 15″ E longitude (Fig. 1). The city is divided as North, Central, South and West. North Chennai and Central Chennai are industrial and commercial centers of the city respectively. The known residential places of South and West Chennai are now turned into commercial hubs at rapid manner. The city is growing dramatically with diversified economic base along with industrial and commercial development. The city is one of the known automobile assembly and production centers in the world and also it is becoming an IT center and back office. Due to the dramatic increase in economic growth of the city, the built-up area was increased exponentially which resulted in the exhaustive change in land use/land cover. It is observed that the average population density was also arbitrarily increased in the city for the past few years which is the cause for the increased concrete structures. The climatic and weather condition of Chennai is determined by its geographical location. The city's weather and climate is reliable and has mild variation in the seasonal temperature due to its close proximity to the sea. Though Chennai comes across three seasonal variations such as summer, monsoon and winter, mostly it experiences hot and humid weather conditions only. The city experiences the maximum temperature variations from 38 to 42 °C during April to June i.e. summer season.

2 Data Sets Used

2.1 Satellite Datasets

The methodology consists of remotely sensed satellite thermal data analysis using data from the OLI/TIRS and MODIS sensors on-board NASA's Landsat-8 and Terra satellites respectively. Surface reflectance (MOD09GA) data corrected for atmospheric errors and LST data products (MOD11A1) of MODIS for Chennai (H25V07

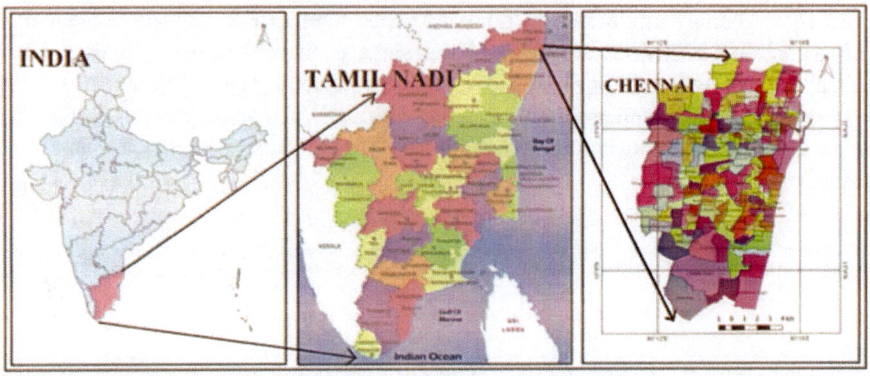

Fig. 1 Location of Chennai

MODIS tile). The images are in HDF-EOS format, in sinusoidal projection. Resampled LST data at sampling intervals of 960 and 240 m for surface reflectance products using geometrically registered UTM projection Zone 44N, WGS 1984 Datum of MODIS Reprojection Tool; thus, the pixel dimensions are ensured to be in multiples of 30 m. The output images are to be stored in GeoTIFF format. Concurrent Landsat-8 images for Chennai (Path 142, Row 51), with the same geographic projection as MODIS, are resampled to 60 m, by changing 'Cell Size' in 'Reproject' option in ERDAS Imagine software (for NL-disTrad algorithm). The datasets are acquired for the date 12 May 2017, for which both sensor data are available.

2.2 *Other Datasets*

Administrative ward boundary of Chennai City is used to identify urban and rural areas. It is procured in shapefile format, in UTM Zone 44N (WGS84) co-ordinate system. Field measurements to estimate WBGT were carried out using WBGT monitoring devices across selected locations in the city during summer season (May 2017). It recorded certain essential parameters such as temperature, relative humidity, etc. A perception social survey was also conducted, to ascertain self-related symptoms of heat-related illnesses. It was carried out among individuals residing in different parts of the urban area of Chennai.

3 Methodology

The overall methodology of the study is depicted in Fig. 2 which encompasses two components of work. The first component is concerned with mapping of UHI in the study area, using MODIS LST that has been disaggregated to Landsat-8 scale. The second component involves mapping the heat stress index (WBGT) and heat stress-related illnesses, in the study area. Finally, the development of a mobile application using the results of the overlay analysis, to issue alerts and suggest mitigation measures.

3.1 UHI Mapping

The fine-resolution temperature map was prepared using LST dataset obtained by manual processing of Landsat-8 thermal data product. The urban and rural area temperature difference was selected as the indicator to assess UHI intensity. To identify urban/rural areas in the temperature data layer, the raster was overlaid with

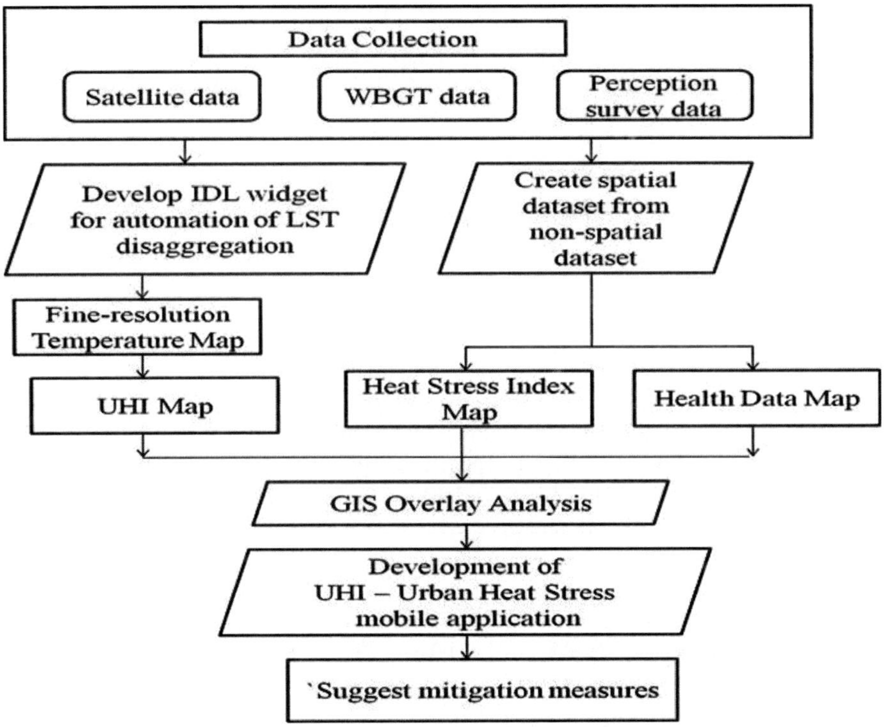

Fig. 2 Overall methodology

an administrative boundary map of Chennai. UHI intensity indicator was computed using Eq. (1):

$$\text{UHI}_{\text{intensity}} = \text{LST}_{\text{urban}} - \text{LST}_{\text{rural}} \quad (1)$$

where:

$\text{LST}_{\text{urban}}$ LST of a urban pixel.
$\text{LST}_{\text{rural}}$ LST of a rural pixel (taken as 25 °C).
$\text{UHI}_{\text{intensity}}$ UHI intensity.

Temperature was classified in four different classes such as <30 °C, 30–31 °C, 31–32 °C and >32 °C for hotspot identification. The mean LST within each ward was determined using Zonal Statistics tool in QGIS.

3.2 Heat Stress Index (WBGT) Mapping

For human beings, the experience of thermal discomfort is related to the UHI effect [10]. Wet Bulb Globe Temperature (WBGT) was used to assess the heat stress. The WBGT combines the effect of the four main thermal components affecting heat stress: air temperature, humidity, air velocity and radiation, as measured by the dry bulb, wet bulb and globe temperatures [16]. The index can be estimated from the Wet Bulb Globe Temperature using Eq. (2) [17]:

$$\text{WBGT} = 0.7T_w + 0.2T_g + 0.1T_a \quad (2)$$

where:

T_w Natural wet bulb temperature.
T_g Black globe temperature.
T_a Air temperature (dry bulb temperature).

A WBGT monitoring device was used to assess WBGT accurately in several locations across Chennai. Model QuesTemp°34 heat stress monitor with dry bulb temperature accuracy of ±0.5 °C for the temperature range of 0–120 °C and ±5% accuracy for relative humidity (RH) in the range between 20 and 95% RH manufactured by Quest Technologies, USA was used.

3.3 Heat Stress-Related Illnesses Mapping

Assessments of heat vulnerability are done based on self-rated health questions in the health sciences and other fields because, over decades of research, they have been found to be strong correlates of clinical conditions [18]. A perception–social survey was conducted across various localities in Chennai, to study the interactions between humans and environment. Opinions of the respondents about the changes in quality of life in their neighborhoods of the natural environment and in their location are collected by asking questions emphasizing climate and land use. Questions on experiencing heat-related symptoms and self-reported heat illnesses were also included.

3.4 Development of Mobile Application

With the different maps that were obtained in the study, a UHI—heat stress mobile application was developed, using which a user can see if they might be in a heat stress prone area. It can serve as a visual indicator of the heat stress index and 'risk level' specific to their current geographical location. The application was developed on the Android platform using Android Studio (Version 3.0.1) software. Google Maps Android API, was used to include Google Maps and add customized mapping information (LST, WBGT, Health Impact, NDVI, NDBI, NDWI, Landuse Proportion) in the application. This was done by embedding maps into an activity as a fragment with a simple XML snippet. The APK file can be installed and run on any Android device.

4 Results and Discussion

4.1 Spatial Distribution of UHI Hotspots

UHIs were identified (Fig. 3) from the thermal map, by the difference between the temperature of each pixel with that of a rural pixel (taken as 25 °C). This was accomplished using Raster Calculator tool in QGIS. Temperature difference was found to be even as high as 6–7 °C in many parts of the city.

The UHI hotspots were demarcated (Fig. 4a) by categorizing the temperature classes and highlighting the higher value classes. The places which were known to have congested built-up concrete areas/crowded urban areas and also, poor vegetation cover, such as Old Washermanpet (George Town), Choolaimedu, Virugambakkam, were observed to be within the UHI hotspots.

Temperatures were cooler than these areas, in central and southern regions of the city, where urbanization process was recognized to be more recent, and hence were of less building density. Both of these zones were known to be of less built-up concrete area. The occurrence of green cover was also known to be higher (such as Guindy), or proximity to water bodies was known to be closer (such as Velachery, Neelankarai). Dense greenery showed its positive impact to the thermal environment,

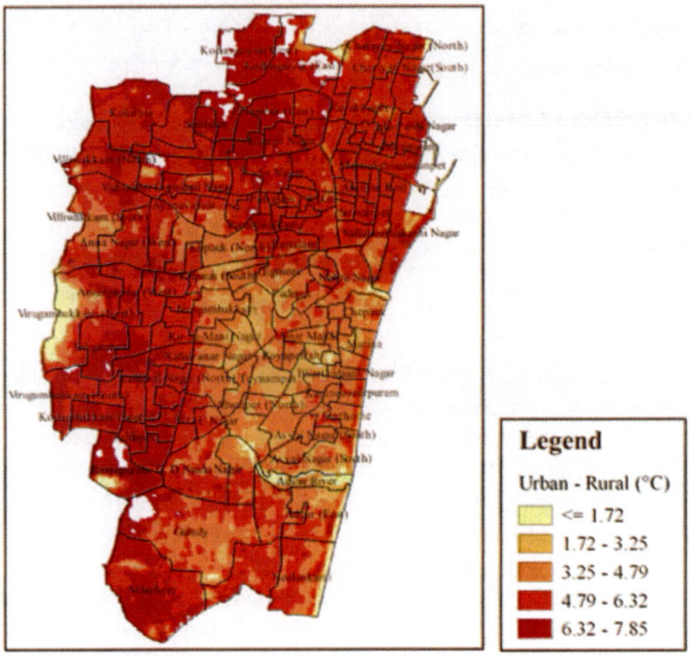

Fig. 3 Urban–rural temperature difference indicator: UHI

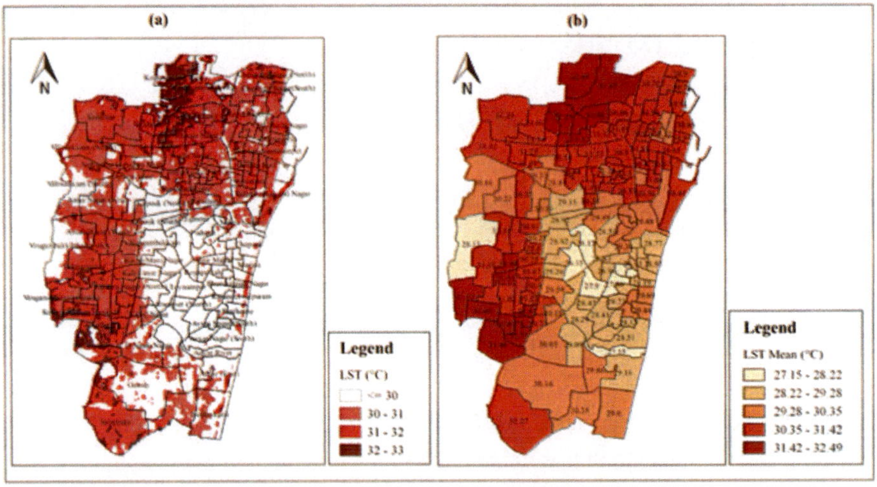

Fig. 4 Temperature maps: **a** UHI hotspots **b** ward-wise LST mean

as these places were outside the UHI hotspots. The mean LST within each ward was computed using Zonal Statistics tool in QGIS (Fig. 4b).

4.2 *Spatial Distribution of WBGT*

World's widely used heat stress index of WBGT index was used in this study. It combines the influence of four environmental parameters into a single number. Measurements were taken at different categories of localities to cover all weather conditions, such as near park/dense trees (Guindy, AC Tech, Viyasarpadi), close to water bodies (Kottur Garden, Velachery, Ramapuram), crowdly constructed area (George town, Cholaimedu, Virugambakkam), near hilly region (St. Thomas mount), near sea shore (Triplicane, Besant nagar, Neelankarai).

Simultaneous measurements of natural wet temperature, air temperature, air velocity, relative humidity and globe temperature were taken for five days at half-hourly intervals, at selected locations in the city, using the instrument 'Questemp 34'. The point measurements at a site were generalized for the corresponding entire ward in GIS. In this way, point information was converted to regional information. Figure 5 graphically represents the results of summer heat stress on a map, prepared using QGIS.

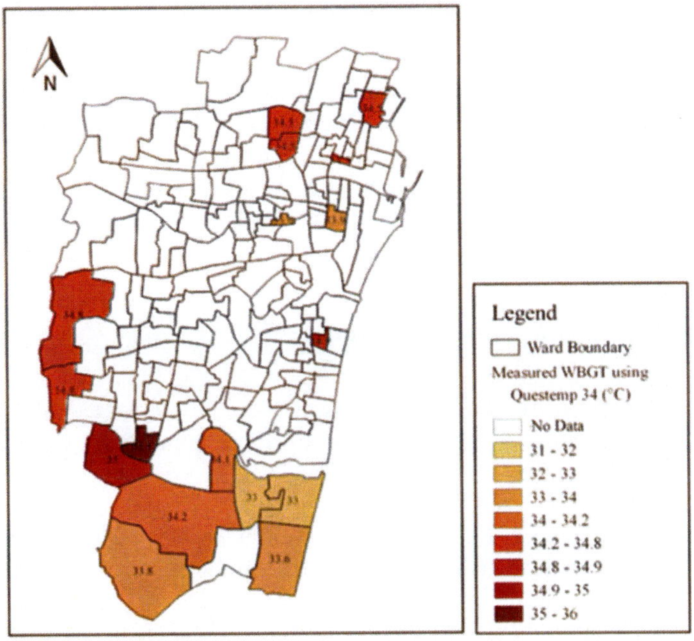

Fig. 5 WBGT map

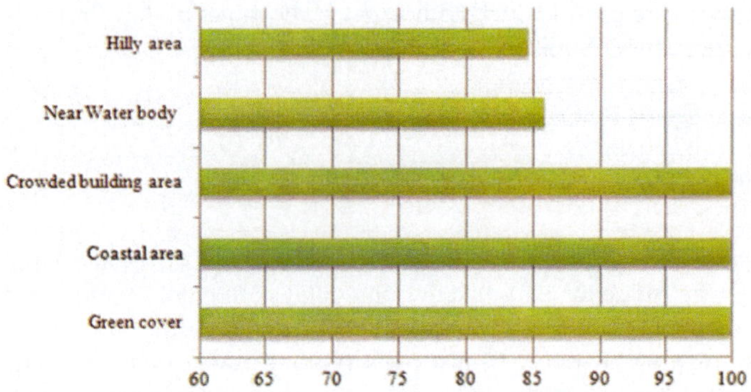

Fig. 6 Results of health impact study

According to ISO-7243, the temperature ranges of 33 °C, 33–35 °C and more than 35 °C were classified as safe, caution and stress respectively. In summer, most parts of Chennai measured higher value of WBGT than the reference value of 33 °C. Only very few safe conditions were found throughout the city, in the wards where WBGT was studied. This situation was characterized by a warning status of heat stress.

4.3 Examination of Health Impact

Based on the questionnaire perception survey conducted across the various localities, it was determined that a majority of the working class people in the city felt that heat did induce some health problems and caused an adverse effect on their overall well-being. The results of the health impact of heat stress have been represented as a graphical bar chart in Fig. 6.

4.4 Interpretation from GIS Analysis

A land use/land cover (LULC) map of Chennai (Fig. 7) was prepared using Landsat-8 OLI bands 5,4,3. The projection and extent of the resulting raster were changed to be the same as that of the other rasters (such as LST).

Satellite based temperature measurement is found to be a potential way to monitor the temperature in a highly human influenced urban environment. The thermal UHI hotspot map which indicated that northern and central western areas were the hottest of the city, was overlaid with the WBGT map (Fig. 8).

It is observed from the LULC map (Fig. 7) that the regions which were known to have dense built-up areas and subsequently, poor vegetation cover, fell within the

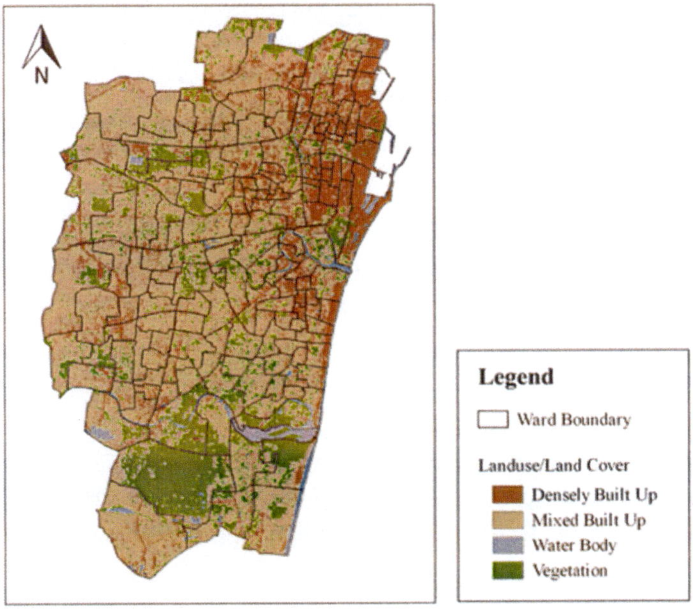

Fig. 7 LULC map—Chennai

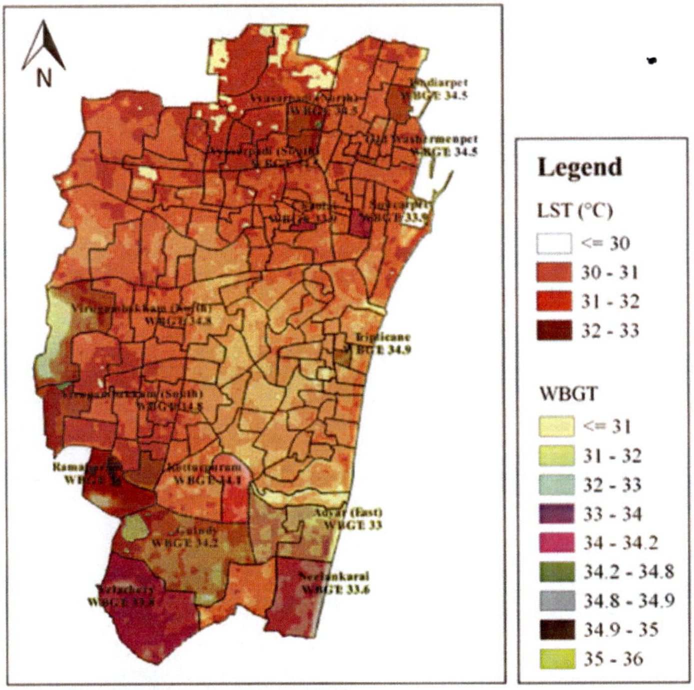

Fig. 8 Overlay of UHI hotspot map and WBGT map

UHI hotspots and had higher WBGT values. The reverse case: lower temperatures, outside the UHI hotspots, and lower WBGT values were found in recently urbanized wards, where vegetation cover was higher and also most of them were closer to inland water bodies or the coast.

The results of this study, indicate that the highest thermal stress is found in the South-Western and Northern parts of the city, which are predominantly crowded, constructed (built-up), industrial and commercial areas, such as George Town (Old Washermanpet), Virugambakkam, etc. The map also depicts that in the southern part of the city and to some extent, in its central parts corresponding to green regions (park areas, hilly areas, etc.) or blue regions (areas closer to water bodies, coastal areas, etc.), respectively, heat stress condition is relatively lesser, as the WBGT values are lower.

The significant relationship between UHI hotspots, WBGT, health impacts and land use/land cover, especially in summer shows that some specific localities of the city suffer the most heat stress. Addressing heat stress and UHI within continuous space has thus been explored to be an effective approach for better planning and management of the adverse effects of heat that can be adopted by the authorities.

4.5 *Information from Mobile Application*

The rich maps provided by Google, when augmented with custom markers and custom geographic data (such as LST, WBGT, etc.), allow the users to explore the area of interest in much depth and obtain valuable, easy-to-read information. Thus, in order to present the results of this UHI-WBGT study to an end-user without GIS knowledge, the Google Maps Android API was used to develop a simple mobile application. If the GPS is switched on, the user's current location is detected by the app, and a marker is placed on the Google Maps screen. Google Maps is designed to zoom into the study area (Chennai), with a ward boundary map overlaid on it. The map is color coded to indicate the 3 levels of heat stress within the city—high, moderate and low, based on the results of GIS analyses, on LST, WBGT and Health Impact. The display of the app icon and home screen is shown in Fig. 9.

5 Conclusion

In this study, satellite-derived temperature data was found to be suitable for studying the spatial pattern of UHI intensity distribution. The implications of UHIs on human heat stress were also addressed from the perspective of continuous space. From this study, it is proved that Chennai is affected by UHI during summer. South-Western and Northern part of the city which have dense buildings shows UHI. The developed mobile app will give information about UHI affected areas and mitigation measures to the users. Surface temperature measured using satellite-based instruments, can be

Fig. 9 Launching the app: **a** app icon **b** home screen

utilized to estimate air temperature indirectly, this estimates are less reliable than direct ground-based measurements. WBGT can't provide an accurate assessment of heat stress; it will give only a general guideline. More accurate information about heat stress can be obtained by measuring the specific elements of the thermal environment.

References

1. Alfraihat R, Mulugeta G, Gala TS (2016) Ecological evaluation of urban heat island in Chicago City, USA. J Atmos Pollut 4(1):23–29
2. Meng F, Liu M (2013) Remote-sensing image-based analysis of the patterns of urban heat islands in rapidly urbanizing Jinan, China. Int J Remote Sens 34(24):8838–8853
3. Dousset B, Gourmelon F, Laaidi K, Zeghnoun A, Giraudet E, Bretin P, Mauri E, Vandentorren S (2010) Satellite monitoring of summer heat waves in the Paris metropolitan area. Int J Climatol 31(2):313–323
4. Roth M, Oke TR, Emery WJ (1989) Satellite-derived urban heat islands from three coastal cities and the utilization of such data in urban climatology. Int J Remote Sens 10(11):1699–1720
5. Oke TR (1982) The energetic basis of the urban heat island. Q J R Meteorol Soc 108:1–24
6. Nichol JE, Hang TP (2012) Temporal characteristics of thermal satellite images for urban heat stress and heat island mapping. ISPRS J Photogramm Remote Sens 74:153–162
7. Voogt JA, Oke TR (2003) Thermal remote sensing of urban climates. Remote Sens Environ 86:370–384
8. Schwarz N, Lautenbach S, Seppelt R (2011) Exploring indicators for quantifying surface urban heat islands of European cities with MODIS land surface temperatures. Remote Sens Environ 115:3175–3186
9. Beniston M (2004) The 2003 heat wave in Europe: a shape of things to come? An analysis based on Swiss climatological data and model simulations. Geophys Res Lett 31:2022–2026
10. Steeneveld GJ, Koopmans S, Heusinkveld BG, Hove LWA, Holtslag M (2011) Quantifying urban heat island effects and human comfort for cities of variable size and urban morphology in the Netherlands. J Geophys Res 116(D20129)
11. Arundel J, Winter S, Gui G, Keatley M (2016) A web-based application for beekeepers to visualise patterns of growth in floral resources using MODIS data. Environ Model Softw 83:116–125

12. Chapman S, Thatcher M, Salazar A, Watson JEM, McAlpine CA (2019) The impact of climate change and urban growth on urban climate and heat stress in a subtropical city. Int J Climatol 39:3013–3030
13. Keikhosravi Q (2019) The effect of heat waves on the intensification of the heat island of Iran's metropolises (Tehran, Mashhad, Tabriz, Ahvaz). Urban Clim 28:100453
14. Masek JG, Wulder MA, Markham B, McCorkel J, Crawford CJ, Storey J, Jenstrom DT (2020) Landsat 9: empowering open science and applications through continuity. Remote Sens Environ 248:111968
15. Jabbar HK, Hamoodi MN, Al-Hameedawi AN (2023) Urban heat islands: a review of contributing factors, effects and data. IOP Conf Ser: Earth Environ Sci 1129:012038
16. Song Y, Wu C (2017) Examining human heat stress with remote sensing technology. GIScience & Remote Sens 54
17. Chinnadurai J, Venugopal V, Kumaravel P, Paramesh R (2016) Influence of occupational heat stress on labour productivity—a case study from Chennai India. Int J Prod Perform Manag 65(2):245–255
18. Belanger D, Gosselin P, Valois P, Abdous B (2015) Neighbourhood and dwelling characteristics associated with the self-reported adverse health effects of heat in most deprived urban areas: a cross-sectional study in 9 cities. Health Place 32:8–18

Drones as an Alternate Communication System During Calamities

D. S. Vohra, Pradeep Kumar Garg, and Sanjay Kumar Ghosh

Abstract In this age of rapidly advancing technology, numerous concepts have captured the attention of researchers, and drone technology is no exception. Researchers are significantly captivated by the myriad uses of drones, encompassing civil applications such as the scrutiny of power infrastructure, surveillance of wildlife, transportation of medical supplies to remote locales, identification of forest fire outbreaks, and assessment of landslides. Additionally, they are intrigued by the military potentials, including real-time monitoring, surveillance operations, patrolling activities, and demining efforts. Despite the wide range of applications already explored, some countries still have untapped drone potential. One such area is the utilization of drones as communication relays during natural disasters when conventional communication lines are disrupted. This approach could prove highly beneficial in rescuing affected individuals, as the aerial node created by drones would enable people to communicate with rescue teams using mobile phones or ordinary landline telephones, even when devastating natural calamities like tsunamis, earthquakes, or floods have destroyed traditional communication towers.

Keywords Communication relay · Path loss model · Signal-to-Noise Ratio (SNR)

D. S. Vohra (✉) · P. K. Garg · S. K. Ghosh
Indian Institute of Technology (IIT) Roorkee, Roorkee, India
e-mail: ds_vohra@ce.iitr.ac.in

P. K. Garg
e-mail: p.garg@ce.iitr.ac.in

S. K. Ghosh
e-mail: sanjay.ghosh@ce.iitr.ac.in

S. Mesapam et al. (eds.), *Developments and Applications of Geomatics*,
Lecture Notes in Civil Engineering 450, https://doi.org/10.1007/978-981-99-8568-5_9

1 Introduction

While most people commonly associate drones with aerial photography and videography, only a few have considered their potential as communication relays. Particularly during natural disasters like tsunamis, floods, earthquakes, and landslides, traditional communication lines often get destroyed, leaving the affected areas without reliable communication. In our research, we propose using drones as relays to bridge the communication gap between ground devices and terrestrial base stations far away from the affected region. The idea is to strategically position the drone in the air, allowing it to act as an intermediary link, reconnecting the ground communication devices to the terrestrial network until regular communication facilities are fully restored. This approach could offer a crucial lifeline to maintain communication during times of crisis and help coordinate rescue and relief efforts efficiently [7, 9, 19].

2 Preceding Work

Over time, many researchers have delved into diverse strategies for using drones as communication relays. These strategies frequently center around two primary facets: the mitigation of transmit power to economize energy and the enhancement of quality of service (QoS) to ensure proficient communication [18, 25]. Several scholars have grounded their investigations in path loss models to scrutinize the viability of employing drone communication relays. Furthermore, specific scholars have directed their focus toward augmenting drones' energy efficiency, aiming to refine their efficacy as communication relays. They have undertaken inquiries into the deployment of drones to enhance the dependability of their services. To systematically categorize the wide array of deployment techniques and roles that drones assume as communication relays, Fig. 1 presents a classification scheme grounded in diverse criteria.

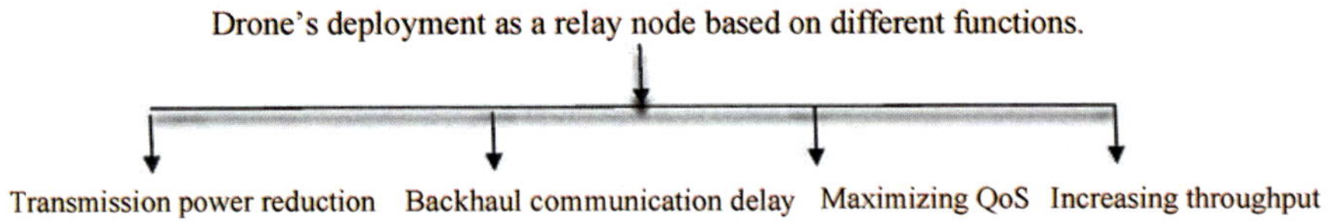

Fig. 1 Deployment classification of the drone as a relay [22]

3 Mathematical Model

3.1 Logical Settings for the System

To optimize the system settings for effective communication, we begin by considering a total of M wireless devices that are positioned far from the central communication base station, and due to their limited power capacity, they cannot establish direct communication with the remote base station [5]. To address this issue, we aim to strategically position a drone at coordinates (X_u, Y_u, Z_u) to serve as a relay for the M wireless devices, enabling them to establish communication with the remote base station [4]. Figure 2 illustrates this concept pictorially. To ensure successful communication, we propose using Frequency Division Multiple Access (FDMA) as the multiple access technique for drones. FDMA allows the drone to transmit information on a specific sub-channel, ensuring that the Signal-to-Noise Ratio (SNR) for communication is greater than or equal to a predefined minimum threshold value (SNR_{th}). This approach minimizes interference and facilitates reliable communication for each user with one dedicated sub-channel per drone [23]. Furthermore, we assume the maximum communication power the aerial drone can transmit is denoted as p_{max}. These system settings and techniques will enable efficient communication between the M wireless devices and the remote base station, leveraging the drone as a reliable relay.

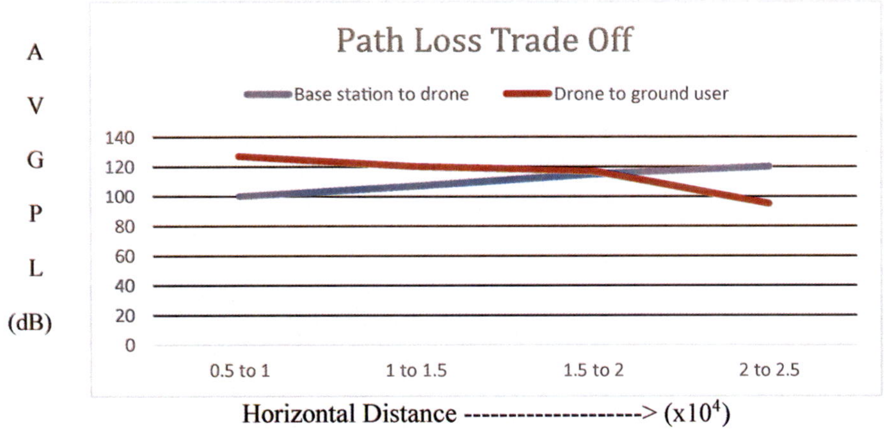

Fig. 2 Path loss model equilibrium [22]

3.2 Investigation of Path Deficit Models

For a comprehensive assessment of path loss models, we initiated the computation of these communication models connecting the primary remote communication station with the drone and subsequently between the drone, serving as a relay, and the terrestrial users. Moreover, determining the path loss model is also feasible between the drone and the ground-based user [20, 27]. Equation 1 embodies the path loss model that links the base station and an airborne relay node. This equation serves as a mathematical representation that encapsulates path loss, taking into cognizance variables like distance, frequency, and other pertinent parameters between the base station and the airborne relay drone. The pertinence of this model is underscored by its pivotal role in the comprehension and enhancement of communication performance within the drone relay system [22].

$$L_{\mathrm{BS-UAV}}(d_{\mathrm{2D}}, \theta) = 10\alpha\log(d_{\mathrm{2D}}) + A(\theta - \theta_o)\exp(-(\theta - \theta_o)/B) + \eta_o + N(0, a\theta + \sigma_o) \quad (1)$$

where

d_{2D}	$\sqrt{\left[(X_u - X_{\mathrm{BS}})^2 + (Y_u - Y_{\mathrm{BS}})^2\right]}$ represents the flat-plane separation between a base remote communication station and a drone,
θ	Receding Angle,
α	Path Attenuation Example,
A	Amplifier for Substantial Path Attenuation,
θ_o	Angular Deviation,
B	Angle Multiplier,
η_o	Compensation for Significant Path Attenuation,
$N\ (0, a\theta + \sigma_o)$	is a Random variable (Gaussian) in which 'a' = shadowing of drone slope, σ_o = offset for shadowing of drone.

Equation 2 delineates the path loss model connecting drones and ground users (denoted as $i \in M$), where the variable i represents individual ground users, with the total count not exceeding M [22].

$$\mathrm{LUAV}_{-i}(d_{\mathrm{3D}}, \phi) = P(\mathrm{LOS}) \cdot L_{\mathrm{LOS}} + P(\mathrm{NLOS}) \cdot L_{\mathrm{NLOS}} \quad (2)$$

where

d_{3D}	$\sqrt{\left[(Xu - Xi)^2 + \left(Yu - Yi^2 + Zu^2\right]\right.}$ denotes the spatial separation in three dimensions between a drone and a ground user,
P (LOS)	Ensuring the likelihood of a clear line of sight (LOS) at a specific elevation angle ϕ,
P (NLOS)	Ensuring the non-likelihood of a clear line of sight (LOS) at angle ϕ,
L_{LOS}	Mean path attenuation for line of sight (LOS) trajectories,
L_{NLOS}	Average path attenuation for non-line of sight (NLOS) scenarios.

3.3 Path Loss Model Equilibrium

Attaining an optimal equilibrium between the distances spanning from the base remote communication station to the drone and from the drone to the ground user stands as a pivotal endeavor, bearing profound significance in harnessing path loss and enacting the most influential power allocation mechanism [10, 17]. This equilibrium is paramount in securing a seamless and dependable communication environment within a drone relay system. As the distance between the drone and the ground user dwindles, the path loss originating from the base station to the drone accentuates [11]. Conversely, when the gap between the drone and the ground user broadens, the path loss between the base station and the drone abates. To maximize advantages for ground users, identifying a strategic vantage point for the drone becomes imperative, where the collective impacts of path losses from the base station to the drone and from the drone to the ground users are minimized. Once attained, this equilibrium empowers the drone to function proficiently as a relay, adeptly catering to an expanded spectrum of ground users, extending the scope of communication, and ushering in enhanced coverage. The meticulous selection of an optimum drone placement catalyzes the overall communication performance, elevates connectivity, and bolsters reliability for users in locales marked by limited communication infrastructure. Consequently, a methodical contemplation of path loss and judicious power allocation emerges as a linchpin in the triumphant deployment of drone communication relays across diverse applications (Fig. 3).

4 Devising of Mathematical Problem

Initially, the focus was on triple-play communication services, encompassing data, voice, and video transmission between a geographically distant communication base station and a terrestrial user. This intricate communication process is made possible through the intermediary role of a drone functioning as a relay. The quantification of the communication link's capability between the distant communication base station and the drone is expressed by Eq. 3. Equation 3 quantifies the data rate or throughput of the communication link between the base station and the drone. The capacity of this link plays a critical role in determining the overall performance of the communication system. It depends on various factors, including the available bandwidth, signal-to-noise ratio, modulation scheme, coding efficiency, and other

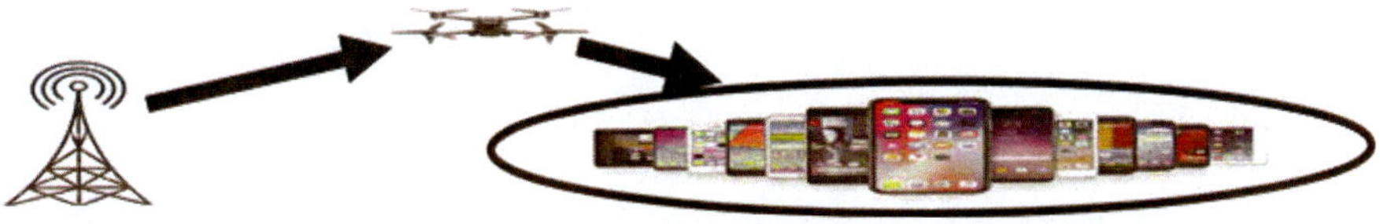

Fig. 3 Framework configuration [22]

channel characteristics. By optimizing the link's capacity, we can enhance data transmission efficiency between the base station and the drone, thus ensuring seamless communication for triple-play services to the ground users. Equation 3 and the parameters involved are necessary to understand the communication link's capacity and its implications for the triple-play services [22].

$$C_{\mathrm{BS-UAV}} = B_{\mathrm{BS}} \cdot \log_2(1 + \mathrm{SNR}_{\mathrm{BS-UAV}}) \tag{3}$$

where

B_{BS} Bandwidth emitted by the ground base station,
$\mathrm{SNR}_{\mathrm{BS-UAV}}$ Drone's SNR.

The data capacity is [22]

$$C_{\mathrm{UAV}-i} = B_{\mathrm{UAV}-i} \cdot \log_2(1 + \mathrm{SNR}_{\mathrm{UAV}-i}) \tag{4}$$

where

$B_{\mathrm{UAV}-i}$ Drone's originating bandwidth,
$\mathrm{SNR}_{\mathrm{UAV}-i}$ Ground user's SNR.

The ascribed bandwidth ceiling for a ground user is estimated to stand at BUAV/ M, with BUAV signifying the drone's bandwidth and M denoting the upper limit of wireless devices [16, 21]. Subsequently, the requisite data rate Ri for each wireless device is expounded in Eq. 5 [22].

$$p_i = \left(2^{Ri\ M/\mathrm{BUAV}-1}\right) N \cdot L_{\mathrm{UAV}-i} \tag{5}$$

The succeeding lines elucidate the meanings of the assorted variables in Eq. 5.

$L_{\mathrm{UAV}-i}$ Average attenuation experienced between the drone and the ground user i.
N Power via noise.

The objective is to strategically position drones to guarantee optimal communication for the maximum number of affected ground users [2, 3, 13]. This optimization challenge is thus outlined through Eqs. 6 and 7 [22].

$$\text{Maximize} \sum (i = 1 \text{ to } M)\ B_{\mathrm{UAV}-i} \cdot \log_2(1 + \mathrm{SNR}_{\mathrm{UAV}-i})$$
$$(x_u, y_u, z_u, P) \tag{6}$$

$$\text{subject to} \sum (i = 1 \text{ to } M) B_{\mathrm{UAV}-i} \cdot \log_2(1 + \mathrm{SNR}_{\mathrm{UAV}-i}) \le \mathrm{B}_{\mathrm{BS}} \cdot \log_2(1 + \mathrm{SNR}_{\mathrm{BS-UAV}})$$
$$\mathrm{SNR}_{\mathrm{UAV}-i} \ge \mathrm{SNR}_{\mathrm{th}}, \forall i \in M,$$
$$\sum (i = 1 \text{ to } M) p_i \le p^{\max},$$
$$pi \ge 0, \forall i \in M$$
$$x_{\min} \le X_u \le x_{\max}, y_{\min} \le Y_u \le y_{\max}, Z_{\min} \le Z_u \le Z_{\max} \tag{7}$$

4.1 Constraints for Mathematical Problem

The constraints described in the scenario play a crucial role in determining the optimal placement and operation of the drone as a communication relay. These constraints are elaborated in brief in succeeding lines.

1. **Capacity Constraint**. The communication link's capacity established between the remote communication base station and the drone ought to adhere to a principle wherein it remains either lower than or equal to the capacity inherent in the communication link uniting the drone with the ground user.
2. **Signal-to-Noise Ratio (SNR) Constraint**. The SNR of the ground user's communication should always be greater than or equal to the threshold value SNR_{th}. This constraint ensures that the received signal quality is sufficient for reliable communication.
3. **Power Constraint**. The total power consumed by the drone should not exceed the maximum emanating level p_{max}. This constraint limits the drone's power usage to ensure energy efficiency and avoid overwhelming the system.
4. **Non-Negative Power Constraint**. The power consumed by the drone should always be greater than or equal to zero, indicating that the drone cannot consume negative power.

The constraints above cited earlier collectively serve as guiding principles that steer the decision-making process for the placement of the drone and the allocation of power. This strategic decision-making aims to maximize communication efficiency while staying within the bounds of physical and operational limitations. This scenario is visually depicted in Fig. 4, showcasing the intricate interplay involving the remote communication base station, the drone, and the ground users. The

Fig. 4 Scenario description of remotely placed base station, drone, and ground users [22]

illustration underscores the drone's role as a relay, orchestrating the transmission of signals between the base station and the ground users, all while diligently considering the established constraints. The overarching goal of optimizing both the placement and operation of the drone, in alignment with these constraints, holds paramount importance in ensuring an effective and reliable communication framework. This optimization is pivotal in facilitating the successful delivery of triple-play services to ground users [26].

5 Methodology

The Particle Swarm Optimization (PSO) algorithm is widely employed for tackling optimization challenges, encompassing domains like drone placement. Its essence lies in iteratively adjusting particle velocities and positions to determine optimal configurations or positions. Within the context of this research, the drone was statically positioned at a fixed altitude, and the analysis revolved around the throughput of users impacted by disasters. Various power allocation algorithms were explored, including equal power distribution, water-filling, and modified water-filling algorithms. The outcomes disclosed that the water-filling algorithm consistently outperformed its counterparts, even the modified variant, across diverse distances between ground users and drones. Notably, this superiority persisted beyond 3000 m, exemplifying the robustness and effectiveness of the water-filling algorithm in heightening user throughput. These assertions are corroborated by Fig. 5, which consistently underscores the supremacy of the water-filling algorithm, regardless of the distance between ground users and drones. This research effectively

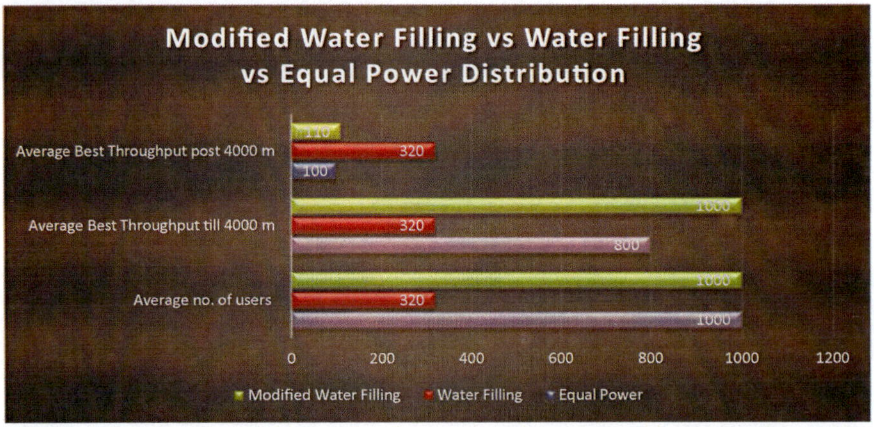

Fig. 5 A comparative analysis of the modified water-filling algorithm, the traditional water-filling algorithm, and the equal power distribution algorithm reveal distinct approaches in power allocation strategies [22]

fosters efficient and optimal communication in disaster-affected zones by synergizing the PSO algorithm for drone placement and the water-filling algorithm for power allocation. These zones necessitate dependable communication for practical rescue and relief endeavors. Taken as a whole, the findings, coupled with judicious employment of optimization and power allocation methodologies, contribute to informed decision-making and the identification of prime strategies for deploying drones as relay media in disaster-prone or impacted regions.

6 Technical Parameters for Simulation

The simulation parameters established for drawing conclusions are detailed in Table 1.

The algorithmic parameters established for drawing conclusions are detailed in Table 2.

Table 1 Simulation parameters [22]

Measurements of area *R*	(X_1, X_2, Y_1, Y_2)	(0,1000,0,1000)
The highest quantity of impacted users resulting from a disaster	*M*	100–1000
The maximum transmission capability of the drone	*P* (drone)	30 mW
Maximum transmission capacity for the affected users	*P* (user)	46 mW
Bandwidth of drone	BW (drone)	50 MHz
Bandwidth of user	BW (user)	75–100 MHz
Position of the distant communication base station	Loc (remote_base_stn)	(7000, 500)

Table 2 Algorithmic parameters [22]

Frequency of the carrier signal	Fc	2 GHz
PSO size	PSO (population)	50
Quantity of iterations	Iteration (PSO)	50
Arbitrary parameters	(a, b)	(9.2, 0.8)
Magnitude of noise power	Noise (power)	−120 mW
Threshold signal-to-noise	SN_{th}	35 dB

7 Water-Filling Algorithm

Utilizing a water-filling algorithm has demonstrated its efficacy in assessing optimal power distribution among drones and ground users. The arrangement of users within the impacted region "*R*" is depicted in Fig. 5. In this scenario, the aggregate transmission bandwidth amounts to "*X*" megahertz (MHz), with each user being allocated a fraction of "*X*/1000" bandwidth to ensure satisfactory transmission and reception capabilities. When considering a total bandwidth of 50 MHz and a user being assigned 50 Kbps, we assume this allocation allows for effective ground-user communication [14]. For comparison with alternative power allocation methodologies, the equal power allocation algorithms were also examined alongside the water-filling algorithm [1]. Findings indicated that the water-filling algorithm consistently surpasses the ergodic capacity exhibited by comparable communication channels [8]. The core concept encompassed identifying a threshold value dictating the point at which the drone should initiate transmission. The strategy involves channeling additional power to favorable communication channels between the drone and the ground user while still according to due consideration to unfavorable channels [12]. In this approach, attaining a balance between the judicious allocation of optimal power for favorable channels and the acknowledgment of unfavorable ones is paramount. In analyzing the water-filling algorithm, efforts were made to allocate power below the threshold value. However, transmitting negative power lacked practical significance and was thus excluded from consideration. The water-filling algorithm, in contrast to other power allocation techniques, secures heightened coverage for users, rendering it a preferred choice [15]. All in all, the adaptive nature of the water-filling algorithm empowers it to optimize power distribution, thereby fostering efficient communication and comprehensive coverage for ground users operating within the drone relay system (Fig. 6).

However, as per the users consuming power till 4000 m, the Modified Water-Filling Algorithm performed very well. But here in this work, the concentration is on a scenario where there is a high probability that the remote communication base towers will be far off, more than 4000 m. The comparison between the modified water-filling algorithms and equal power distribution is summarized post-simulation as per the pictorial diagram in Fig. 7.

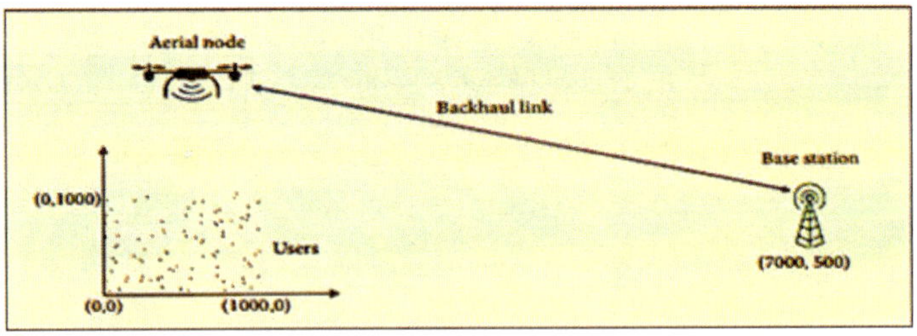

Fig. 6 Arrangement of users within the impacted subsection [22]

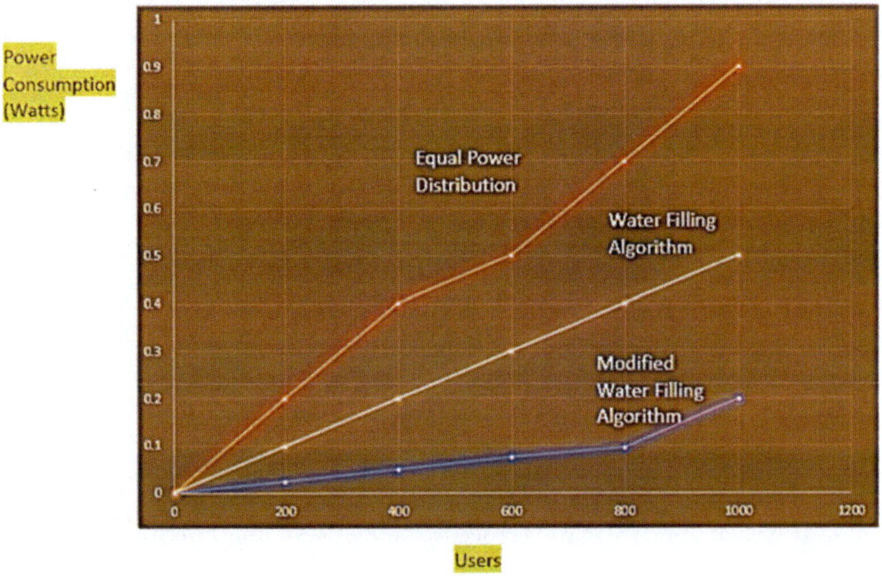

Fig. 7 Power consumption versus users in a simulated environment till 4000 m. *Source* Prepared by authors

8 Results and Discussion

The study's outcomes unmistakably reveal the water-filling algorithm's superior performance when distances exceed the 4000 m threshold. However, within the domain of 4000 m and lower distances, the advanced and equal power distribution algorithms exhibited better results than the water-filling algorithm. While these findings may yield theoretical insights, it is paramount to account for pragmatic considerations. As mentioned, disaster-affected regions typically span a breadth of 5000–10,000 m in disaster-prone areas. In such contexts, the relevance of utilizing the water-filling and equal power distribution algorithms becomes more pronounced. Given the expansive nature of disaster-hit territories, the practical implications derived from the research point toward the water-filling algorithm retaining greater relevance for situating drones as relay conduits within these areas. This research approach avoids drawing incorrect theoretical inferences by aligning with the practical dimensions of disaster-stricken zones. The spotlight is placed on the necessity of making astute decisions grounded in real-world circumstances to ensure the effective implementation of drone relay systems, serving communication and disaster response objectives. Acknowledging the practical realities surrounding disaster-hit areas is pivotal, as it underscores the significance of tailoring the selection of the apt power allocation algorithm according to the specific application and geographical context. This tactful approach guarantees the triumphant deployment of drone relay systems, adeptly addressing communication requisites within disaster-vulnerable regions.

9 Conclusion

Leveraging drones as airborne relay mediums offers a promising solution for establishing connectivity in areas lacking pre-existing communication infrastructure or facing disruptions like disasters [7]. However, before embracing this concept, an intricate examination of path loss models becomes imperative, as these models constitute the foundational underpinning of any communication framework. The cellular-to-drone path loss model was meticulously investigated to establish a backhaul link connecting the terrestrial base station with the drone. Moreover, to establish the downlink connection between the drone and ground users, an exhaustive exploration into the path loss model between the drone and the ground users was undertaken [6]. After the comprehensive analysis of both path loss models, a judicious equilibrium was achieved to discern the optimal positioning of drones. This intricate optimization dilemma, incorporating various constraints, was methodically formulated as a convex mathematical challenge. The primary goal encompassed enhancing the throughput for ground users by strategically locating drones as relays. In this context, the water-filling power allocation algorithm proved apt for orchestrating effective drone placement, amplifying the system's overall performance. Building upon this approach, a symmetrical exploration of the reverse scenario involving the uplink communication link—where ground users engage with drones—can also be scrutinized meticulously [24]. By factoring in the communication link between ground users and drones, a holistic evaluation of the complete drone relay system can be executed, effectively guaranteeing seamless bidirectional communication connecting ground users and the base station, all orchestrated through strategically positioned drones. This research underscores the pivotal role played by path loss models, optimization methodologies, and power allocation algorithms in crafting efficient drone relay systems. These systems bear the potential to extend communication coverage, amplify connectivity, and yield substantial benefits across diverse scenarios.

References

1. Al-Hourani A, Gomez K (2017) Modeling cellular-to-UAV path-loss for suburban environments. IEEE Wirel Commun Lett 7(1):82–85
2. Al-Hourani A, Kandeepan S, Jamalipour A (2014) Modeling air-to-ground path loss for low altitude platforms in urban environments. In: IEEE global communications conference, Austin, TX, USA, pp 2898–2904
3. Al-Hourani A, Kandeepan S, Lardner S (2014) Optimal lap altitude for maximum coverage. IEEE Wirel Commun Lett 3(6):569–572
4. Alnagar SI, Salhab AM, Zummo SA (2020) Unmanned aerial vehicle relay system: performance evaluation and 3D location optimization. IEEE Access 8:67635–67645
5. Chen Y, Feng W, Zheng G (2017) Optimum placement of UAV as relays. IEEE Commun Lett 22(2):248–251
6. Demir U, Toker C, Ekici O (2020) Energy-efficient deployment of UAV in v2x network considering latency and backhaul issues. In: IEEE international Black Sea conference on communications and networking (Black Sea Com), Odesa, Ukraine, pp 1–6

7. Garg PK (2021) Introduction to unmanned aerial vehicles. New Age International Pvt Ltd, Delhi
8. He Y, Zhai D, Wang D, Tang X, Zhang R (2020) A relay selection protocol for UAV-assisted VANETs. Appl Sci 10(23):876
9. Hildmann H, Kovacs E (2019) Using unmanned aerial vehicles (UAVs) as mobile sensing platforms (MSPs) for disaster response, civil security, and public safety. Drones 3(3):59
10. Kalantari E, Yanikomeroglu H, Yongacoglu A (2016) On the number and 3D placement of drone base stations in wireless cellular networks. In: 2016 IEEE 84th vehicular technology conference (VTC-Fall), Montreal, QC, Canada, pp 1–6
11. Khuwaja AA, Chen Y, Zhao N, Alouini MS, Dobbins P (2018) A survey of channel modeling for UAV communications. IEEE Commun Surv Tutor 20(4):2804–2821
12. Kumar S, Suman S, De S (2018) Backhaul and delay-aware placement of UAV-enabled base station. In: IEEE INFOCOM 2018-IEEE conference on computer communications workshops (INFOCOM WKSHPS), Honolulu, HI, USA, pp 634–639
13. Mozaffari M, Saad W, Bennis M, Debbah M (2015) Drone small cells in the clouds: design, deployment and performance analysis. In: 2015 IEEE global communications conference (GLOBECOM), San Diego, CA, USA, pp 1–6
14. Palomar DP, Fonollosa JR (2005) Practical algorithms for a family of water filling solutions. IEEE Trans Signal Process 53(2):686–695
15. Perera TDP, Jayakody NK, Garg S, Kumar N, Cheng L (2020) Wireless-powered UAV assisted communication system in Nakagami-m fading channels. In: IEEE 17th annual consumer communications & networking conference (CCNC), Las Vegas, NV, USA, pp 1–6
16. Sawalmeh AH, Othman NS, Shakhatreh H, Khreishah A (2017) Providing wireless coverage in massively crowded events using UAVs. In: 2017 IEEE 13th Malaysia international conference on communications (MICC), Johor Bahru, Malaysia, pp 158–163
17. Sawalmeh AH, Othman NS, Shakhatreh H, Khreishah A (2019) Wireless coverage for mobile users in dynamic environments using UAV. IEEE Access 7:126376–126390
18. Shakhatreh H, Sawalmeh AH, Al-Fuqaha A (2019) Unmanned aerial vehicles (UAVs): a survey on civil applications and key research challenges. IEEE Access 7:48572–48634
19. Shakoor S, Kaleem Z, Baig MI, Chughtai O, Duong TQ, Nguyen LD (2019) Role of UAVs in public safety communications: energy efficiency perspective. IEEE Access 7:140665–140679
20. Ullah H, Nair NG, Moore A, Nugent C, Muschamp P, Cuevas M (2019) 5G communication: an overview of vehicle-to-everything, drones, and healthcare use-cases. IEEE Access 7:37251–37268
21. Ullah Z, Al-Turjman F, Mostarda L (2020) Cognition in UAV aided 5g and beyond communications: a survey. IEEE Trans Cogn Commun Netw 6(3):872–891
22. Vohra DS, Garg PK, Ghosh SK (2023) Communication via quad/hexacopters during disasters. Int J Intell Unmanned Syst
23. Yan C, Fu L, Zhang J, Wang J (2019) A comprehensive survey on UAV communication channel modeling. IEEE Access 7:107769–107792
24. Yin S, Qu Z, Li L (2018) Uplink resource allocation in cellular networks with energy-constrained UAV relay. In: 2018 IEEE 87th vehicular technology conference (VTC Spring), Porto, Portugal, pp 1–5
25. Zeng Y, Wu Q, Zhang R (2019) Accessing from the sky: a tutorial on UAV communications for 5G and beyond. Proc IEEE 107(12):2327–2375
26. Zhan P, Yu K, Swindlehurst AL (2011) Wireless relay communications with unmanned aerial vehicles: performance and optimization. IEEE Trans Aerosp Electron Syst 47(3):2068–2085
27. Zhang L, Zhao H, Hou S (2019) A survey on 5g millimeter-wave communications for UAV-assisted wireless networks. IEEE Access 7:117460–117504

Drought Analysis of an Area Using Google Earth Engine

Jyothsna Devi Adapa and Keesara Venkatareddy

Abstract Drought is a long period of low rainfall which affects the growth of plants and living organisms. Drought occurs when an area or region suffers below-average rainfall or a water deficit for a prolonged time period. It is one of the most complex environmental disasters across the world. Drought also creates the same impact like floods or cyclones. Drought is categorized as meteorological, hydrological, agricultural and socio-economic. The area which experiences the precipitation deficit subjects to meteorological drought. Reduction in stream flow, ground water, and reservoir and lake levels indicates Hydrological drought. Soil water depletion causes Agricultural drought. The effect caused by the meteorological, hydrological and agricultural drought on people and economic activities is termed as socio-economic drought. This study is focusing on meteorological drought. Standard Precipitation Index (SPI) and Drought Severity Index (DSI) are used to calculate meteorological drought.

Keywords Drought · SPI · Precipitation · DSI

1 Introduction

1.1 Drought

When an area or region experiences abnormally low rainfall or a shortage of water for a prolonged period of time, then the area is said to be under drought. The amount of rainfall of an area varies from year to year. The lack of rainfall reduces the ground water or soil moisture. Due to reduction of ground water or soil moisture, there is crop damage. It takes weeks or months to identify drought. We can't say exactly how long the drought lasts. It may be for weeks, months or years.

J. D. Adapa (✉) · K. Venkatareddy
Department of Civil Engineering, National Institute of Technology, Warangal, India
e-mail: ja931806@student.nitw.ac.in

S. Mesapam et al. (eds.), *Developments and Applications of Geomatics*,
Lecture Notes in Civil Engineering 450, https://doi.org/10.1007/978-981-99-8568-5_10

1.2 Drought in India

India has been witnessing drought since the eighteenth century. About 28% of the geographical area of India is subject to drought. From 1870 to 2018, India witnessed 26 major droughts out of which the drought that occurred in 2016–2018 was severe [1]. The drought in India is due to the delay of monsoon or failure of monsoon. During drought, India witnessed several deaths due to hunger and even suicides too. In order to avoid these incidents in future, drought monitoring is necessary.

1.3 About Google Earth Engine

Google Earth Engine (GEE) is a cloud-based geospatial analysis platform that enables users to visualize and analyze satellite images of our planet. GEE uses JavaScript or Python programming language. Deforestation detection, Identification of drought, classifying land cover and Flood monitoring are some of the applications of GEE. GEE helps us to analyze the future conditions based on the past available data. GEE contains parameters which are useful for the assessment of drought like precipitation, temperature, evapotranspiration, soil moisture and vegetation. With the help of these parameters, we can calculate the standard precipitation index (SPI) and Drought Severity Index (DSI) of a particular area.

2 Literature Review

Pai et al. [2] conducted a study to determine district-wide drought climatology in India over south-west monsoon season (June–September) by calculating the Percent of Normal Precipitation (PNP) and SPI for the period 1901–2003. In this study, rainfall data of all 458 districts in India is used. This study concluded that SPI is a better drought index as compared to PNP for calculating drought as PNP depends on the aridity of the region.

Jahangir Alam et al. [3] conducted a study to monitor meteorological and agricultural drought in Bangladesh, Barind region using SPI and Markov method. They collected the 12 stations' rainfall data from 1971 to 2008. Using gamma function, the precipitation data was normalized to compute SPI. They concluded that meteorological drought existed in the area in 1982, 1994 and 2010. Among these, the drought which occurred on 2010 was devasting. In a span of 5 years (2006–2010), meteorological drought occurred in the Barind region four times.

Shah et al. [4] conducted a study for determining drought in Surat, Gujarat by calculating the SPI index. Forty years of precipitation data was collected for this study i.e., from 1971 to 2010 and used gamma distribution method to find SPI values. They assessed that droughts of moderate to exceptional severity occurred in 1974, 1985,

1987, 1989, 1991, and 1995. 1976, 1983, 1988, 2003, 2004, 2005, and 2007 were all judged to be moderately to highly wet. Surat district was determined to be in normal condition in the remaining years.

Sazib et al. [5] conducted an experimental study to find out drought characteristics and comparison between meteorological and agricultural drought indicators using GEE. This study was conducted in Ethiopia and South Africa. The satellite-based soil moisture datasets were used to calculate surface and Root Zone Soil Moisture (RZSM) and their anomalies. Standard Precipitation Index (SPI) i.e., SPI3, SPI6, SPI9 was calculated. They concluded that in comparison to meteorological drought indicators, RZSM anomalies have longer drought duration but lower drought intensity.

Zhao et al. [6] conducted an experimental study to examine the drought status and the trend change of Yellow River Basin from 2003 to 2019 in GEE. In this study, Moderate Resolution Imaging Spectroradiometer (MODIS) is used as a data source. MODIS Land Surface Temperature (LST) and Normalized Difference Vegetation Index (NDVI) are used to calculate the Temperature Condition Index (TCI), Vegetation Condition Index (VCI), Vegetation Health Index (VHI), and Temperature-Vegetation Drought Index (TVDI). In high vegetation coverage areas, NDVI is more sensitive to drought chances so, for such regions drought monitoring through VCI is more preferable. In deserts, drought monitoring through TCI is better. They concluded that the TCI and TVDI show almost the same drought characteristics and VCI and VHI exhibit same drought characteristics.

Khan and Gilani [7] conducted a study to determine drought of North America, South America, Africa, Europe, Asia, and Australia from 2001 to 2019 using GEE. In this study, they used MODIS satellite terra sensor evapotranspiration (ET), potential evapotranspiration (PET) 8-Day Global 500 m and MODIS Terra NDVI 8-Day Global 500 m data. They concluded that Australia, Africa, and Asia are undergoing frequent and extreme drought events. Europe is the least affected when compared to other continents.

Khan and Gilani [8] conducted a study for drought monitoring using GEE in North America, South America, Europe, Asia, Africa, and Australia from 2001 to 2019. In this study, NDVI, LST, Soil moisture and Precipitation and Land cover mask data were used. With these data Vegetation Condition Index (VCI), Temperature Condition Index (TCI), Soil Moisture Condition Index (SMCI), and Precipitation Condition Index (PCI) were computed. VCI is used for agricultural drought, PCI is used for meteorological drought, and SMCI is used for agricultural and hydrological drought. They concluded that every continent is indicating signs of severe to extreme drought.

Table 1 Condition of an area with SPI values (NDMC)

SPI values	Condition of the area
≤−2	Extremely dry
−1.5 to −1.99	Severely dry
−1.0 to −1.49	Moderately dry
−0.99 to +0.99	Near normal
1.0–1.49	Moderately wet
1.5–1.99	Very wet
>2.0	Extremely wet

3 Procedure

3.1 Procedure for Calculation of SPI

Firstly, the monthly average precipitation data for 2000–2021 is collected from GEE using CHIRPS dataset. Using Excel, Mean and Standard deviation of the precipitation data is computed. Using Eq. (1) the value of SPI is calculated. The condition of the area in every month of a year is determined with SPI values from Table 1

$$\text{SPI} = \frac{X_i - X_{\text{avg}}}{X_{\text{sd}}} \quad (1)$$

where X_i = Precipitation data for the particular month, X_{avg} = Mean/Average Rainfall, X_{Sd} = Standard Deviation of Rainfall.

3.2 Procedure for Calculation of DSI

Firstly, the data of ET and PET is collected from 2000 to 2021 from GEE using MODIS Global Terrestrial Evapotranspiration 8-Day Global 1 km. Using Excel, ratio of ET and PET is calculated and is mentioned as transpiration ratio (tran). $Z1$ is calculated which is the normalized value of tran. The equation for $Z1$ is as follows:

$$Z1 = (\text{tran}_i - \text{tran}_{\text{mean}})/\text{tran}_{\text{sd}} \quad (2)$$

where tran_i = transpiration ratio for a particular month, $\text{tran}_{\text{mean}}$ = average transpiration ratio and tran_{sd} = standard deviation of transpiration ratio. NDVI data is obtained from MODIS Terra Vegetation Indices 16-Day Global 500 m in GEE. The normalized value of NDVI is $Z2$, the calculation of which is done in Excel. The equation of $Z2$ is as follows:

$$Z2 = (\text{NDVI}_i - \text{NDVI}_{\text{mean}})/\text{NDVI}_{\text{sd}} \quad (3)$$

where $NDVI_i$ = NDVI for a particular month, $NDVI_{mean}$ = average NDVI and $NDVI_{sd}$ = Standard deviation of NDVI. The value Z is the summation of $Z1$ and $Z2$

$$Z = Z1 + Z2 \tag{4}$$

DSI is the normalized value of Z. The equation for the calculation of DSI is as follows:

$$\mathrm{DSI} = \frac{Z_i - Z_{\mathrm{mean}}}{Z_{\mathrm{sd}}} \tag{5}$$

where $Z_i = Z$ for a particular month, Z_{mean} = average value of Z and Z_{sd} = standard deviation of Z.

4 Database Collection from GEE

Jangaon, Karimnagar, Khammam, Mahbubnagar, and Nizamabad shape files were independently collected and uploaded to GEE. Climate Hazardous Group Infrared Precipitation with Station data (CHIRPS) was used to collect precipitation for the above mentioned districts in the form of pentad. Figures 1 and 2 show the research area's boundaries and satellite imagery. For the above districts, total rainfall and mean rainfall were calculated and plotted from 2000 to 2020. The monthly average rainfall for Jangaon district is shown in Fig. 3. The collection of data from NDVI, ET and PET for Jangaon district are shown in Figs. 4, 5, and 6 respectively.

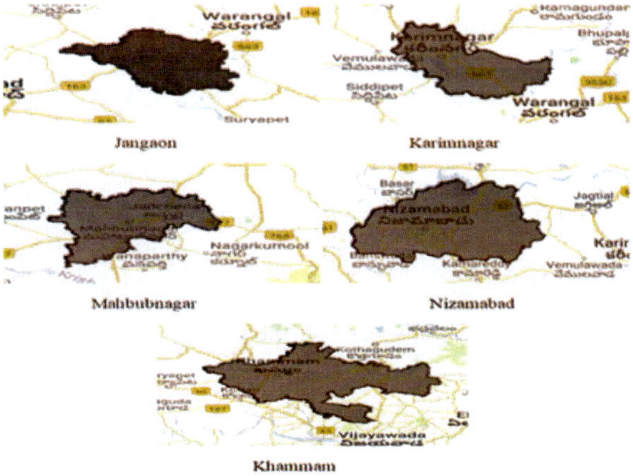

Fig. 1 Study area boundaries

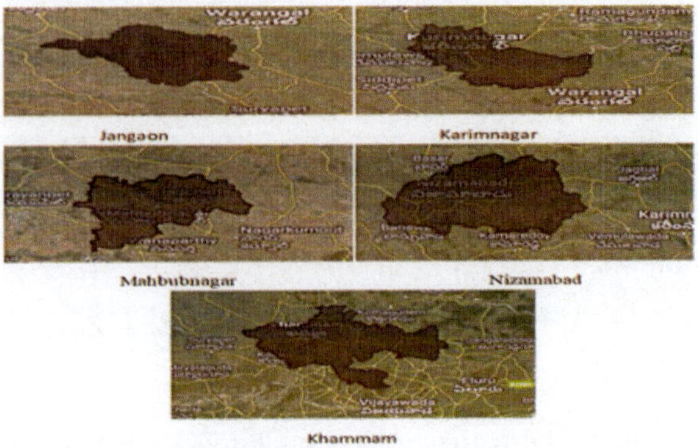

Fig. 2 Satellite image of study area

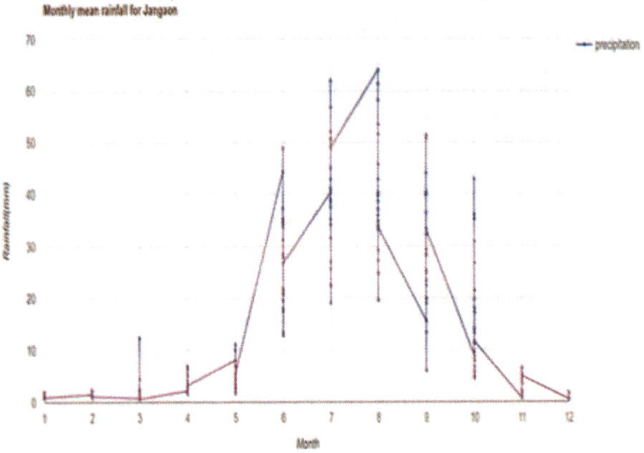

Fig. 3 Mean monthly rainfall for Jangaon district

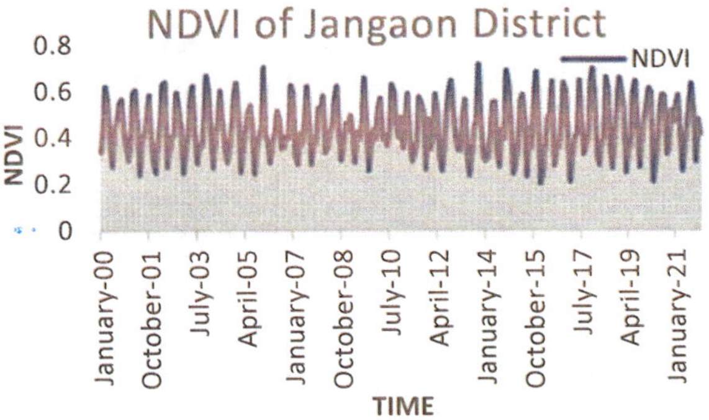

Fig. 4 NDVI data for Jangaon district

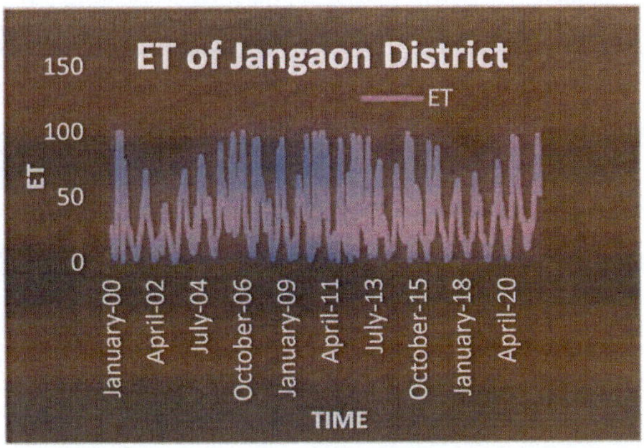

Fig. 5 ET data for Jangaon district

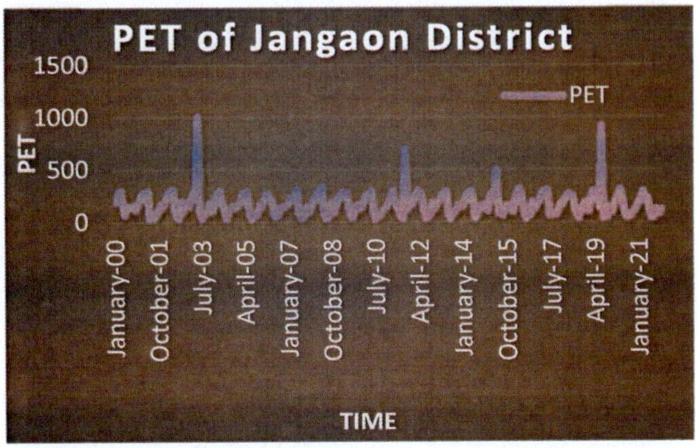

Fig. 6 PET data for Jangaon district

5 Results and Discussions

Based on SPI values, the condition of the area is classified into seven categories. Near normal condition indicates that the particular area is neither subjected to drought nor received excess rainfall and for this condition, SPI values lies between −0.99 and +0.99. SPI values greater than 0.99 indicate wet condition and the value greater than 1.99 indicates extremely wet condition of the area. SPI values less than −0.99 indicates dry condition of the area and a value less than −1.99 indicates extremely dry condition of the area.

Based on DSI values, the condition of the area is divided into eleven categories. For near normal conditions, DSI values lie in the range of 0.49 to −0.49. DSI values greater than 0.49 indicate the wet condition and a value greater than 3.99 indicates extremely wet condition of the area. DSI values less than −0.49 indicates dry condition of the area and the value less than −3.99 indicates extremely dry condition of the area.

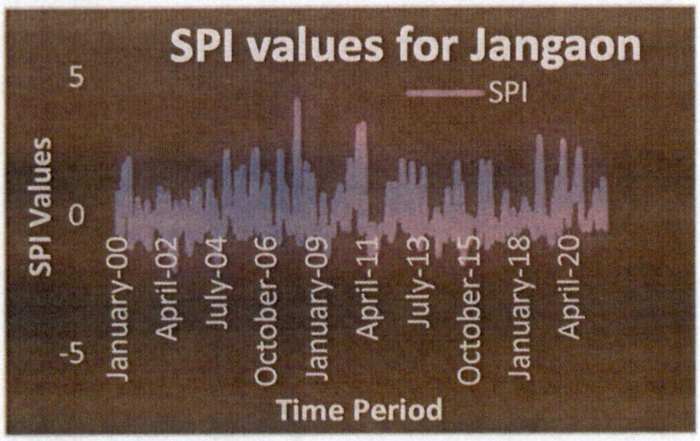

Fig. 7 SPI values for Jangaon district

5.1 *Jangaon District*

The condition of the area based on SPI values for Jangaon district from 2000 to 2021 for all months is tabulated in Table 1. From the results, it is concluded that the Jangaon district was in moderately dry condition in April 2002, 2016, May 2003, 2008, 2015, June 2011, 2014, July 2001, 2002, 2009, 2019, August 2013, 2015, 2016, September 2000, 2011 and in October 2015. The district has experienced severe drought in July 2015, August 2004, and September 2002. The SPI values for Jangaon district from 2000 to 2021 are shown in Fig. 7.

The condition of an area based on DSI values for Jangaon district from 2000 to 2021 is shown in Table 2. From the results, it is observed that the Jangaon district has Mild drought in January 2001, 2003, 2012, in February 2016, 2017, in March 2016, 2018, 2019, in April 2009, 2020, 2021, in May 2011, 2012, 2013, in June 2001, 2002, 2012–2014, in July 2002, 2015, 2016, in August 2005, in September 2018, 2019, in October 2008, 2018, in November 2000, 2009, 2011. The district has experienced moderate drought in September 2002 and October 2009. The DSI values for Jangaon district are shown in Fig. 8 (Tables 3 and 4).

5.2 *Karimnagar District*

The Karimnagar district was in moderately dry condition in May 2003, 2007, 2012, June 2011, 2014, July 2009, August 2005, 2013, 2015, 2016, September 2000, 2003, October 2006 and 2015 whereas it experienced severe drought in July 2002 and September, August 2004, July 2015. The SPI values for Karimnagar district are shown in Fig. 9.

The Karimnagar district is in mild drought condition in January 2001, 2004, 2006, 2012, in February 2005, 2010, 2016, in March 2005, 2007, 2010, 2016, in April 2003, 2005, 2010, 2012, in May 2001, 2005, in June 2001, 2002, 2006, in July 2011, in August 2001, 202, 2004, 2016, in September 2000, 2002, 2008, 2009, in October

Table 2 Condition of an area with PDSI values (NDMC)

PDSI values	Condition of the area
≤−4.0	Extreme drought
−3.0 to −3.99	Severe drought
−2.0 to −2.99	Moderate drought
−1.0 to −1.99	Mild drought
−0.5 to −0.99	Incipient dry spell
0.49 to −0.49	Near normal
0.5–0.99	Incipient wet spell
1.0–1.99	Slightly wet
2.0–2.99	Moderately wet
3.0–3.99	Very wet
≥4.0	Extremely wet

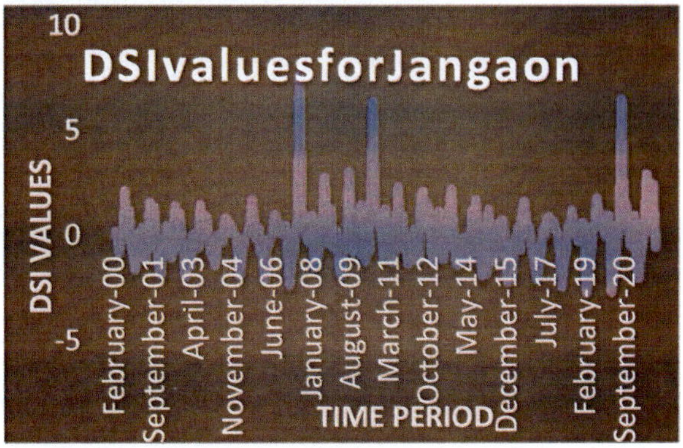

Fig. 8 DSI values for Jangaon district

2008, 2018, in November 2002, 2004, 2007, 2008, 2015, 2018, in December 2002, 2004, 2015 and it has moderate drought in July 2000. The DSI values for Karimnagar district are shown in Fig. 10.

5.3 *Mahbubnagar District*

Mahbubnagar district is in moderately dry condition in April 2000, May 2001, 2003 in April, May and September, October 2006, 2008 in April and May, July 2009, June 2011, 2012 in May and September, June 2014, July 2015, April 2016, April 2017, 2018 in July and September and in Severe drought in September 2002, August 2004 and September 2011. The SPI values for Mahbubnagar district are shown in Fig. 11.

Mahbubnagar district had mild drought in January 2016, February 2001, 2005, March 2000, 2016, April 2001, 2003, 2005, 2019, May 2003, 2005, June 2002,

Table 3 Condition of Jangaon district using SPI values

Month	Condition of the area						
	Extremely wet condition	Very wet condition	Moderately wet condition	Near normal area condition	Moderately dry condition	Severely dry condition	Extremely dry condition
January	2005, 2020		2019	2000–2021 (except 2005, 2019, 2020)			
February	2011	2005, 2010		2000–2021 (except 2005, 2010 and 2011)			
March	2008			2000–2021 (except 2008)			
April	2006	2013, 2015	2004, 2012	2000–2021 (except 2002, 2004, 2006, 2012, 2013, 2015, 2016)	2002, 2016		
May	2006	2016	2000	2000–2021 (except 2000, 2003, 2006, 2008, 2015, 2016)	2003, 2008, 2015		
June	2007	2000	2015	2000–2021 (expect 2000, 2007, 2011, 2014, 2015)	2011, 2014		

(continued)

Table 3 (continued)

Month	Condition of the area						
	Extremely wet condition	Very wet condition	Moderately wet condition	Near normal area condition	Moderately dry condition	Severely dry condition	Extremely dry condition
July		2005, 2010		2000–2021 (except 2001, 2002, 2005, 2009, 2010, 2015, 2019)	2001, 2002, 2009, 2019	2015	
August	2000	2008, 2010	2007, 2020	2000–2021 (except 2000, 2004, 2007, 2008, 2010, 2013, 2015, 2016, 2020)	2013, 2015, 2016	2004	
September		2016, 2020	2005	2000–2021 (except 2000, 2002, 2005, 2011, 2016, 2020)	2000, 2011	2002	
October	2020	2005, 2013	2019	2000–2021 (except 2005, 2013, 2015, 2019, 2020)	2015		

(continued)

Table 3 (continued)

Month	Condition of the area						
	Extremely wet condition	Very wet condition	Moderately wet condition	Near normal area condition	Moderately dry condition	Severely dry condition	Extremely dry condition
November	2010	2012	2006, 2008, 2021	2000–2021 (except 2006, 2008, 2010, 2012, 2021)			
December	2010, 2018			2000–2021 (except 2010, 2018)			

2003, 2011, July 2000, 2009, August 2001, 2002, 2006, 2016, September 2001, 2002, 2008, 2011, October 2004, 2006, 2016, November 2004, 2015, 2018 and in December 2000, 2004, 2015 and it has experienced moderate drought in May 2001. The DSI values for Mahbubnagar district are shown in Fig. 12.

5.4 *Nizamabad District*

The Nizamabad district is in Moderately dry condition in September 2000, July 2001, September 2002, 2003 in May and September, 2004 in June and August, October 2006, May 2007, June 2008, July 2009, June 2011, May 2012, August 2013, 2017 in April and July, September 2018 and the district has experienced severe dry condition in July 2002, June 2014, July 2015 and August 2016. The SPI values for Nizamabad district are shown in Fig. 13.

The Nizamabad district has experienced mild drought in January 2003, 2016, February 2003, 2004, 2016, March 2016, April 2003, 2010, 2016, 2019, May 2001, 2019, June 2001, 2002, 2008, July 2002, 2011, August 2002, 2004, 2010, 2013, September 2000, 2003, 2015, October 2000, 2014, 2015, 2018, November 2000 and in December 2000, 2002, 2015. The district was in moderate drought conditions in July 2000 and August 2016. The DSI values for Nizamabad district are shown in Fig. 14.

Table 4 Condition of an area based on DSI values for Jangaon district

Month	Condition of the area									
	Extremely wet	Very wet	Moderately wet	Slightly wet	Incipient wet spell	Near normal	Incipient dry spell	Mild drought	Moderate drought	Severe drought	Extreme drought
January		2020		2005, 2019	2007, 2021	2002, 2004, 2006, 2008, 2009, 2010, 2011, 2014, 2017, 2018, 2020	2000, 2015, 2016	2001, 2003, 2012			
February			2010, 2011	2008, 2020, 2021		2001, 2002, 2005, 2006, 2009, 2012, 2013, 2014, 2019	2003, 2004, 2007, 2015, 2018	2016, 2017			
March			2008	2011	2000, 2001, 2014, 2021	2002, 2006, 2010, 2012, 2013, 2020	2003, 2004, 2005, 2007, 2009, 2015, 2017	2016, 2018, 2019			
April			2013, 2015	2012	2002, 2006	2000, 2001, 2003, 2004, 2005, 2007, 2011, 2014, 2016	2008, 2010, 2017, 2018, 2019	2009, 2020, 2021			
May			2006, 2016	2002, 2004	2003, 2014, 2015	2000, 2001, 2005, 2008, 2009, 2010, 2017	2007, 2018, 2019, 2020, 2021	2011, 2012, 2013			
June				2000, 2007, 2015	2016, 2017, 2018	2004, 2005, 2006, 2008, 2019, 2020, 2021	2003, 2009, 2010, 2011	2001, 2002, 2012, 2013, 2014			

(continued)

Table 4 (continued)

Month	Condition of the area										
	Extremely wet	Very wet	Moderately wet	Slightly wet	Incipient wet spell	Near normal	Incipient dry spell	Mild drought	Moderate drought	Severe drought	Extreme drought
July			2010	2012	2008, 2018, 2019, 2020, 2021	2000, 2001, 2004, 2005, 2006, 2007, 2009, 2017	2003, 2011, 2013, 2014	2002, 2015, 2016			
August				2007, 2020	2008, 2010, 2021	2000, 2001, 2002, 2009, 2011, 2012, 2014, 2019	2003, 2004, 2006, 2013, 2015, 2016, 2017, 2018	2005			
September				2001, 2005, 2010, 2013, 2016, 2020	2009, 2021	2000, 2003, 2004, 2008, 2012, 2014, 2017	2006, 2007, 2011, 2015	2018, 2019	2002		
October				2013, 2020	2000, 2001, 2012, 2019	2002, 2003, 2004, 2005, 2010, 2011, 2014, 2015, 2016, 2017	2006, 2007, 2021	2008, 2018	2009		

(continued)

Table 4 (continued)

Month	Condition of the area										
	Extremely wet	Very wet	Moderately wet	Slightly wet	Incipient wet spell	Near normal	Incipient dry spell	Mild drought	Moderate drought	Severe drought	Extreme drought
November			2021	2003, 2012	2008, 2015, 2017	2004, 2006, 2010, 2013, 2014, 2016, 2019, 2020	2001, 2002, 2005, 2007, 2018	2000, 2009, 2011			
December				2005, 2018, 2019	2003, 2004, 2006, 2017	2007, 2009, 2016, 2020, 2021	2002, 2008, 2010, 2011, 2012, 2014, 2015				

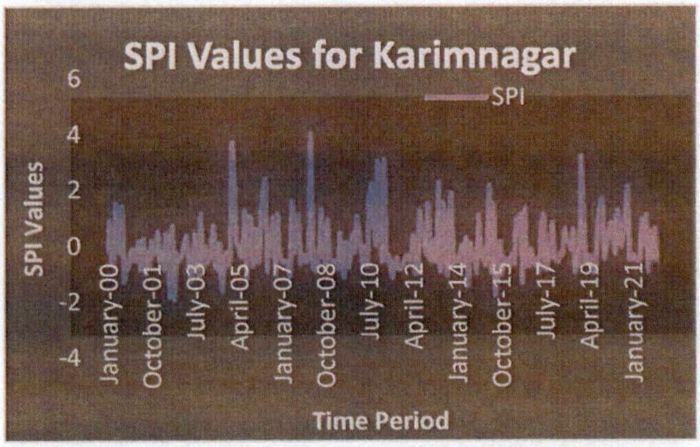

Fig. 9 SPI values for Karimnagar district

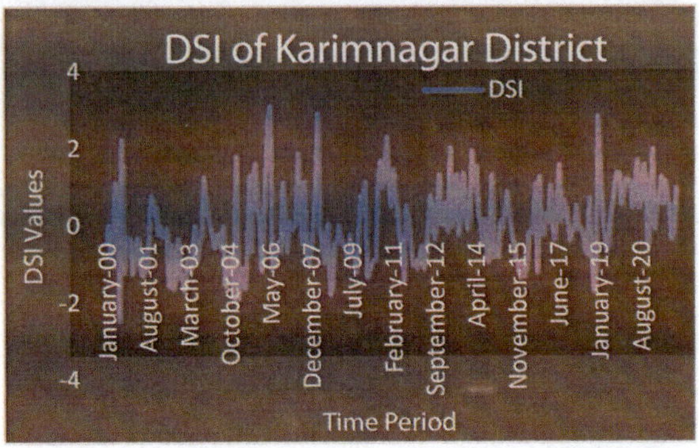

Fig. 10 DSI values of Karimnagar district

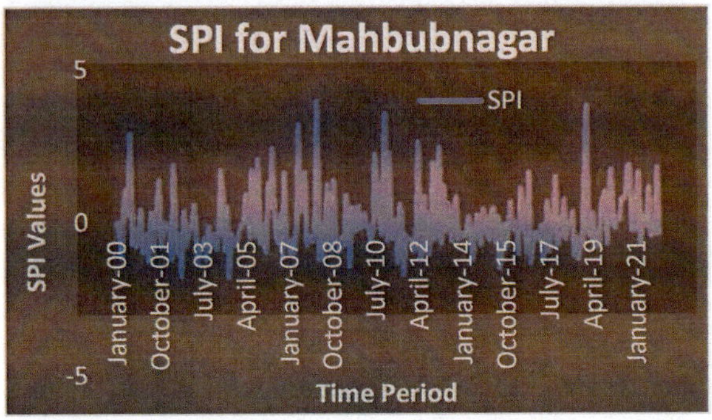

Fig. 11 SPI values for Mahbubnagar district

5.5 *Khammam District*

Khammam district is in moderately dry condition in September 2000, 2001 in May and July, April 2002, May 2003, May 2007, 2008 in April, May, September, 2009 in July and October, 2011 in June and September, August 2013, April 2014, 2015 in

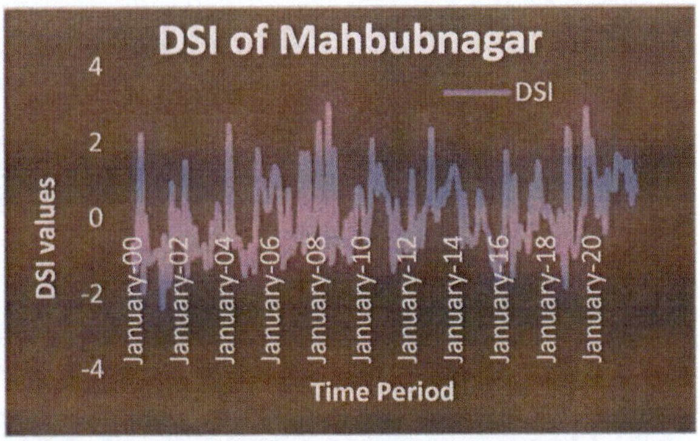

Fig. 12 DSI values of Mahbubnagar district

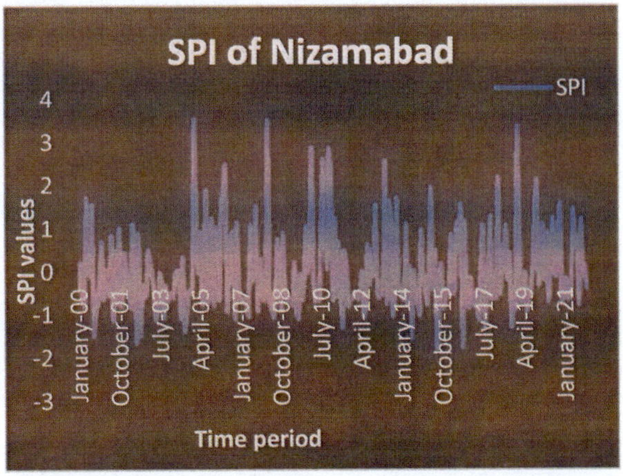

Fig. 13 SPI values of Nizamabad district

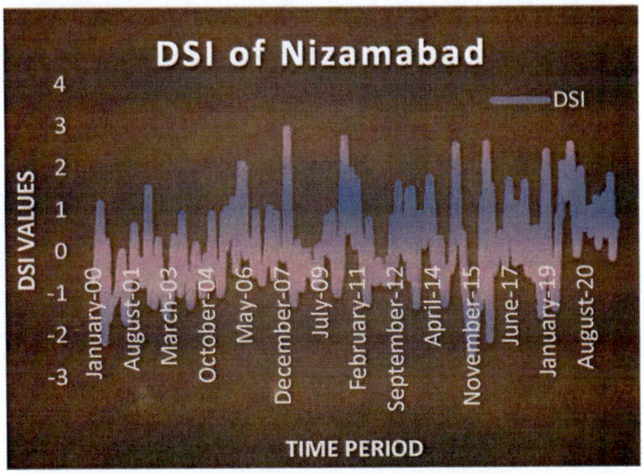

Fig. 14 DSI values of Nizamabad district

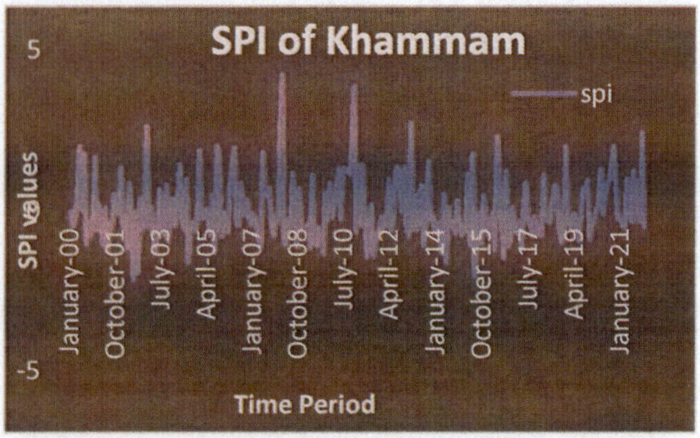

Fig. 15 SPI values of Khammam district

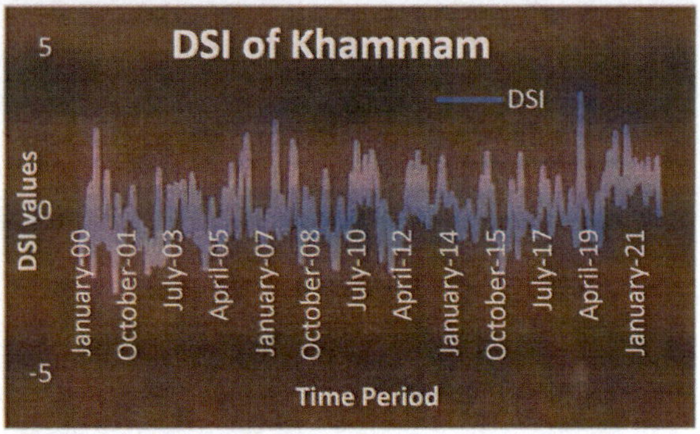

Fig. 16 DSI values of Khammam district

May, August and October, 2016 in April, August and October, April 2017 and Severe dry condition in September 2002, August 2004, June 2014 and in extreme drought in July 2002 and 2015. SPI values for Khammam district is shown in Fig. 15.

Khammam district is in mild drought conditions in July 2000, 2001 in February and August, 2002 in April, August, September, November and December, 2003 in February, March, April, May, 2004 in August and October, 2005 in May and June, Jan 2007, June 2008, 2009 in September and October, 2011 in July and November, 2012 in February and March, 2016 in January, March, April and August, Oct 2018, September 2019 and has experienced moderate drought in May 2001. DSI values for Khammam district are shown in Fig. 16.

6 Summary and Conclusions

GEE has user friendly environment where accessing the data is easy and fast. The aim of this research is to determine the area's meteorological drought. From 2000 to 2021, the SPI and DSI figures for Jangaon, Karimnagar, Mahbubnagar, Nizamabad, and Khammam were calculated. Based on SPI and DSI values the condition of the district is determined from Tables 1 and 2. The condition of the area based on SPI and DSI values is almost the same, but the SPI values which are close to the range are changing in DSI. Based on observations from our study, the district which is getting more repeated droughts needs immediate water resource management. The available water resources can be augmented by reservation of the existing water bodies and transferring water from the nearest reservoirs which are from Godavari and Krishna basins.

References

1. https://india.mongabay.com/2021/05/southern-indias-2016-2018-drought-was-the-worst-in-150-years/
2. Pai DS, Sridhar L, Guhathakurta P, Hatwar HR (2011) District-wide drought climatology of the southwest monsoon season over India based on standardized precipitation index (SPI). Nat Hazards. https://doi.org/10.1007/s11069-011-9867-8
3. Jahangir Alam ATM, Sayedur Rahman M, Saadat AHM (2013) Monitoring meteorological and agricultural drought dynamics in Barind region Bangladesh using standard precipitation index and Markov chain model. Int J Geomat Geosci 3(3). ISSN 0976-4380
4. Shah R, Bharadiya N, Manekar V (2015) Drought index computation using standardized precipitation index (SPI) method for Surat District, Gujarat. Aquat Procedia 4:1243–1249. https://doi.org/10.1016/j.aqpro.2015.02.162
5. Sazib N, Mladenova I, Bolten J (2018) Leveraging the Google Earth Engine for drought assessment using global soil moisture data. Remote Sens 10:1265. https://doi.org/10.3390/rs10081265
6. Zhao X, Xia H, Pan L, Song H, Niu W, Wang R, Li R, Bian X, Guo Y, Qin Y (2021) Drought monitoring over Yellow River Basin from 2003–2019 using reconstructed MODIS land surface temperature in Google Earth Engine. Remote Sens 13:3748. https://doi.org/10.3390/rs13183748
7. Khan R, Gilani H (2021) Global drought monitoring with drought severity index (DSI) using Google Earth Engine. Theor Appl Climatol 146:411–427. https://doi.org/10.1007/s00704-021-03715-9
8. Khan R, Gilani H (2021) Global drought monitoring with big geospatial datasets using Google Earth Engine. Environ Sci Pollut Res 28:17244–17264. https://doi.org/10.1007/s11356-020-12023-0
9. https://drought.unl.edu/ranchplan/DroughtBasics/WeatherandDrought/MeasuringDrought.aspx

Effects of Urbanization on Land Use Land Cover of Warangal Region Using RS and GIS

Ch. Sree Laxmi Pavani, Keesara Venkatareddy, and S. Joshmitha

Abstract Rapid expansion of urbanization has increased the population and economic growth of towns and cities in various parts of the country. This increase in urbanization has affected the natural resources such as vegetation and water bodies. In the present study, the effects of urbanization on LULC (Land Use Land Cover) changes of Warangal urban and rural areas are studied for the years 2014, 2017 and 2020 for a span of 6 years using the RS and GIS techniques. LANDSAT imagery of 3 years is collected and enhanced and LULC data is extracted for four classes i.e., urbanization (built-up land), agriculture cum forest (vegetation), bear soil (barren land) and water bodies. RS and GIS tools are used to compare and analyze the effect of urbanization. LULC feature extraction is made for urban and rural areas of Warangal region using supervised classification to study the growth of urban areas and its effect on LULC changes. The results indicate a drastic increase in urbanization and barren land, severe decrease in vegetation and a very slight increase in waterbodies. Increase in urbanization and decrease in agriculture cum forest indicates vegetation reduction and built-up area increment which directly affects the climate and environment. Warangal region is being affected due to the rising population and the improper usage of land. The quantitative results obtained explain the effect of urbanization on LULC changes of Warangal rural and urban areas, which helps in the best management and planning of land use for Warangal region.

Keywords Classification · Land use land cover · RS and GIS · Urbanization · Warangal

Ch. Sree Laxmi Pavani (✉) · K. Venkatareddy
Civil Engineering Department, National Institute of Technology, Warangal, Telangana, India
e-mail: slpavani@student.nitw.ac.in

S. Joshmitha
Civil Engineering Department, Kakatiya Institute of Technology and Sciences, Warangal, Telangana, India

S. Mesapam et al. (eds.), *Developments and Applications of Geomatics*,
Lecture Notes in Civil Engineering 450, https://doi.org/10.1007/978-981-99-8568-5_11

1 Introduction

1.1 Effects of Urbanization on LULC

Land Use Land Cover (LULC) changes are affected in the cities and towns due to Rapid Urbanization [1]. This occurs due to the relocation of people from rural areas to urban areas. Monitoring Land use Land cover changes regularly will help the urban planners and decision makers to overcome the urban sprawl problem and will help in sustainable development [2]. Since 1950, water usage has increased more than tripled globally and one out of every six persons does not have regular access to safe drinking water [3]. For any type of sustainable development program LULC changes on the earth's surface are enormously important. Due to various human activities on earth's surface undesirable changes in LULC have taken place which led to land degradation [4]. Expansion of urbanization is a historic and speedy conversion of human life on a universal scale, whereby rural tradition is being quickly replaced by urban civilization. Many people from rural areas relocate to the city in search of financial success and social flexibility. Though, the picture of urbanization growth is not as great as it appears to be. As a result of rapid industrial development, modernistic cities have evolved in an unplanned and disorganized manner. Cities in emerging nations become overcrowded and overpopulated for a variety of reasons, including migration and long-term increase in population [5]. Understanding the characteristics of landscape is made imaginable by remote sensing techniques such as digital change detection using multitemporal Landsat-imagery. Remote sensing and GIS methods are essential technologies for the temporal quantification and investigation of spatial trends or else it is impossible when compared with traditional mapping methods. With the aid of RS and GIS technologies, change detection enables users with faster, affordable and more accurate results [6]. Visual interpretation provides a broad overview of the land cover changes and its classes across the specified time period. The effectiveness of change detection from images may vary depending on the type of change involved and the efficiency of the image pre-processing as well as the classification operations [7]. Using remote sensing, characterizing natural and urban ecosystems is important for change detection techniques. Accurate radiometric and geometric rectification is crucial for the analysis of changes in land use and land cover using multitemporal remote sensing data [8]. The worrying aspects are that land degradation is getting worse due to deforestation, farming on slopes and land fragmentation. GIS software served as a platform for data analysis and product management [9]. In the present study effect of urbanization on LULC of Warangal region of Telangana, India using RS and GIS techniques is carried out. Warangal rural and urban areas were considered to analyze the urbanization effects on land use land cover. The urban growth and its impacts are detected using RS and GIS tools and techniques. The study reveals that the major land covered in the study area is vegetation which is slowly being occupied by built-up and barren land. Effects of urbanization bring a lot of change in vegetation and water bodies which can lead to the occurrence of floods and fluctuations in ground water levels.

1.2 Study Area

Study area map of Warangal region of Telangana is shown in Fig. 1. Warangal urban and rural areas are situated in Telangana, India with its geographical extent as 18° 16′ 45.278″ to 17° 37′ 4.0296″ N and 79° 13′ 41.412″ to 80° 2′ 28.478″ E. Warangal district was reformed in the year 2016 by carving out Warangal rural district from erstwhile Warangal district. Warangal Urban area was renamed in the year 2021 as Hanamkonda. Warangal Urban covers an area of 1294.4 km^2 with 14 mandals and Warangal Rural covers an area of 2139 km^2 with 13 mandals. Warangal region feels a dry and hot, seasonally wet, tropical climate. The average annual temperature of Warangal region is 27.2 °C. The monsoon season is from June to September and the rainfall recorded on an average is about 1020 mm.

2 Materials and Methods

To obtain the Vector files of Warangal region, India and Telangana boundary shape files are collected to extract Warangal urban and rural areas using GIS techniques. The Landsat images are collected from the USGS Earth Explorer from band1 to band7 and are combined using the composite band tool in ArcGIS. To enhance the image, supervised classification is performed using Remote Sensing techniques. Supervised classification is performed by selecting training areas for each land cover class to be created. Signature file is generated with the collected training sample data. Now, using GIS signature file is taken as input to classify the study area. In Classify tool, maximum likelihood supervised classification method is applied to create LULC maps. The LULC maps are classified as urbanization, waterbodies, agriculture cum

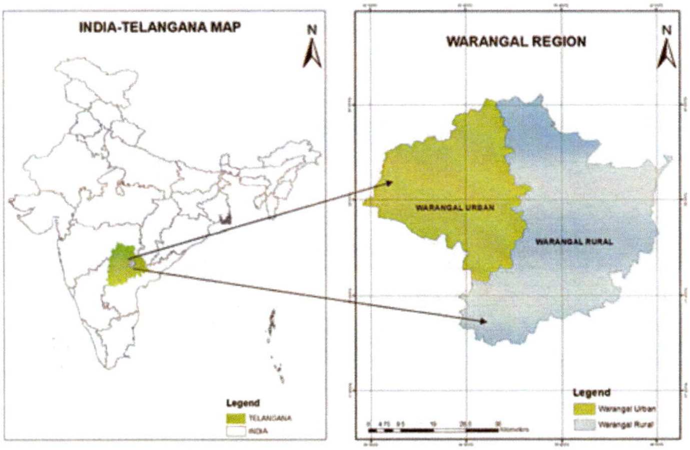

Fig. 1 Study area map showing Warangal urban and rural regions located in Telangana, India

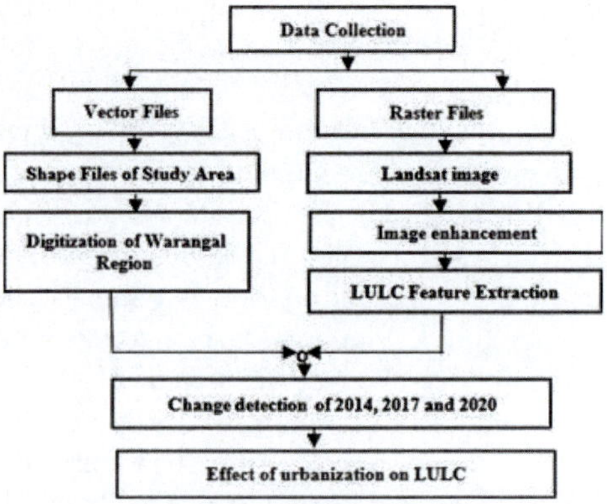

Fig. 2 Methodology of LULC change analysis

forest and bear soil for Warangal urban and rural areas. The changes are detected for 3 recent years 2014, 2017 and 2020 to compare the effect of urbanization on LULC. Methodology followed for the study area is shown in Fig. 2.

Urbanization of Warangal urban and rural areas for the years 2014, 2017 and 2020 is shown in Figs. 3 and 4. In 2014, 2017 and 2020 urbanization occupied an area of 21.61%, 28.73% and 32.65% of the total area. This shows how rapidly the urbanization has increased by 11.04% in Warangal urban area from 2014 to 2020 as shown in Fig. 3. In rural areas of Warangal region urbanization has covered an area of 10.66%, 14.31% and 18.76% in the years 2014, 2017 and 2020. These values indicate that urbanization is growing by 8.1% for rural areas of Warangal region also.

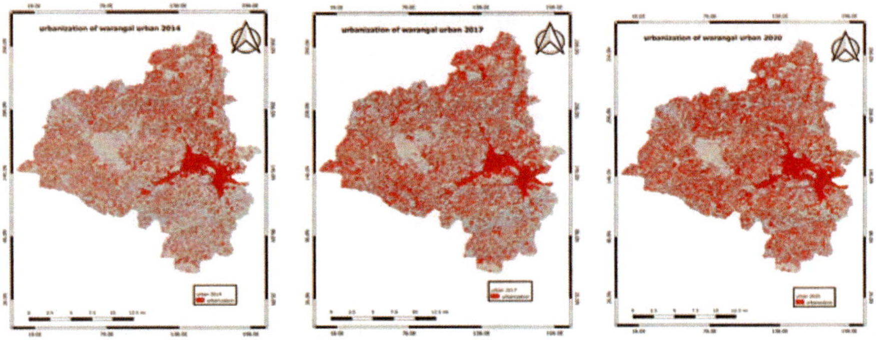

Fig. 3 Urban area in 2014, 2017 and 2020

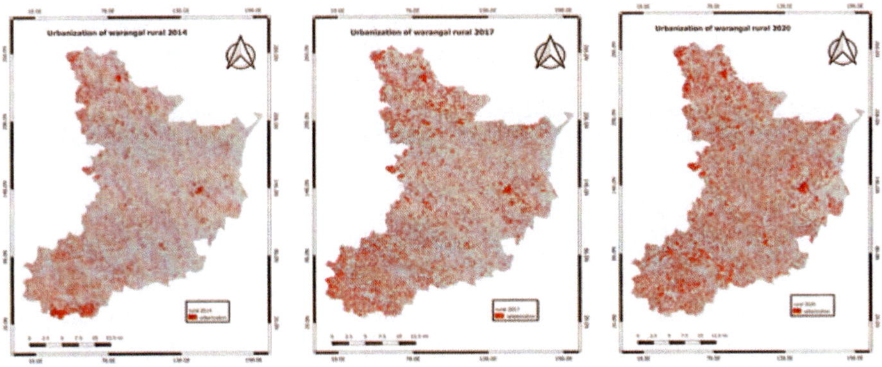

Fig. 4 Rural area in 2014, 2017 and 2020

3 Results

LULC maps of Warangal Urban and Rural areas for the years 2014, 2017 and 2020 are shown in Fig. 5. The results indicate that Urbanization has increased drastically, waterbodies increased in very less quantity and agriculture and forest areas decreased severely from 2014 to 2020.

3.1 LULC Details of Warangal Urban Region

The total area of Warangal urban is 1294.4 km^2, out of which in the year 2014, urbanization covered 279.80 km^2, agriculture cum forest is 957.5 km^2, water bodies covered 44.27 km^2 and bear soil is covered by 12.87 km^2. In the year 2017, urbanization increased from 279.80 to 371.97 km^2 which indicates an increase of 7.12% from 2014 to 2017. Agriculture and forest have decreased from 957.5 to 814.45 km^2 with a percentage decrease of 11.05%. Water bodies have a very minimal increase of 0.46% from 44.27 to 50.2 km^2 and bear soil has increased from 12.87 to 57.82 km^2 with a percentage increase of 3.44% from 2014 to 2017. In the year 2020, urbanization has increased to 422.74 km^2 with a percentage increase of 3.92%. Agriculture cum forest has decreased to 726.41 km^2 with a percentage decrease of 6.8%. Water bodies have a slight increase of 0.79% from 50.2 to 60.4 km^2. Bear soil has increased from 57.82 to 84.87 km^2 with a percentage increase of 2.09%. LULC details of Warangal Urban area are classified along with its percentages are arranged in Table 1. Percentage increase and decrease of each LULC feature of Warangal Urban area are shown in Table 6.

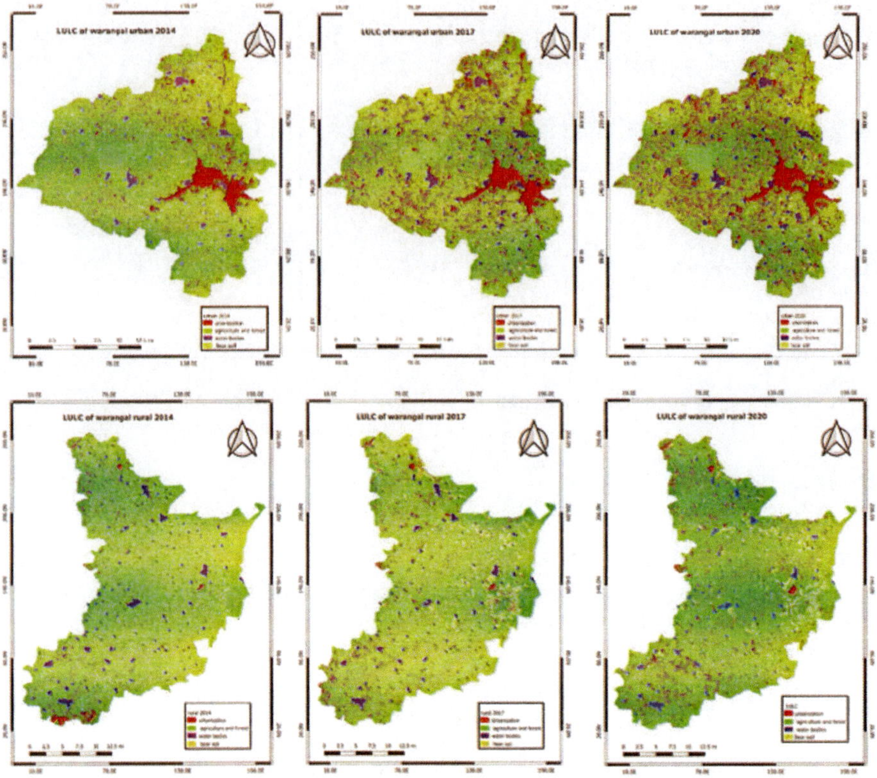

Fig. 5 LULC map of Warangal urban and rural areas for the years 2014, 2017 and 2020

Table 1 Warangal urban LULC classified area and percentage of the 4 categories i.e., urbanization, agriculture cum forest and bear soil for 2014, 2017 and 2020

Year	Urbanization (km^2)	Agriculture cum forest (km^2)	Water bodies (km^2)	Bear soil (km^2)	Total area (km^2)
2014	279.80 (21.61%)	957.5 (73.96%)	44.27 (3.41%)	12.87 (1.02%)	1294.4
2017	371.97 (28.73%)	814.45 (62.91%)	50.2 (3.87%)	57.82 (4.46%)	1294.4
2020	422.74 (32.65%)	726.41 (56.11%)	60.4 (4.66%)	84.87 (6.55%)	1294.4

3.2 *LULC Details of Warangal Rural Region*

The total area of Warangal Rural is 2139 km^2, from which in the year 2014, urbanization covered an area of 228.14 km^2, agriculture cum forest 1823.79 km^2, water bodies covered 69.34 km^2 and bear soil covered 16.79 km^2. In the year 2017, urbanization increased from 228.14 to 306.22 km^2 which indicates a percentage increase of 3.65% from 2014 to 2017. Agriculture cum forest decreased from 1823.79 to 1594.47 km^2 with a percentage decrease of 10.74%. Water bodies have a negligible decrease of 0.17% from 69.34 to 65.68 km^2 and bear soil has increased from 16.79 to 172.13 km^2 with a percentage increase of 7.26% from 2014 to 2017. In the year 2020, from 2017 to 2020 urbanization has enlarged to 401.39 km^2 with a percentage increase of 4.45%. Agriculture cum forest has diminished to 1497.39 from 1594.47 km^2 with a percentage decrease of 4.56%. Water bodies have a nominal increase of 1.03% i.e., from 65.68 to 87.91 km^2. Bear soil has decreased from 172.13 to 152.10 km^2 with a percentage decrease of 0.93%. LULC details of Warangal Rural classified area along with its percentages are tabulated in Table 2. Percentage increase and decrease of each LULC feature of Warangal Rural area is shown in Table 7.

Bar graph of Warangal urban and rural regions is shown in Fig. 6 for the years 2014, 2017 and 2020. From the graph, it is observed clearly that agriculture cum forest is decreasing and in urban areas, bear soil is increasing. On the other side, there is a very slight increase in water bodies, which may not be sufficient for the growing needs of the population.

Table 2 Warangal rural classified area and percentage of the 4 categories i.e., urbanization, vegetation, bear soil for 2014, 2017 and 2020

Year	Urbanization (km^2)	Agriculture cum forest (km^2)	Water bodies (km^2)	Bear soil (km^2)	Total area (km^2)
2014	228.14 (10.66%)	1823.79 (85.29%)	69.34 (3.24%)	16.79 (0.78%)	2139
2017	306.22 (14.31%)	1594.47 (74.55%)	65.68 (3.07%)	172.13 (8.04%)	2139
2020	401.39 (18.76%)	1497.39 (69.99%)	87.91 (4.10%)	152.10 (7.11%)	2139

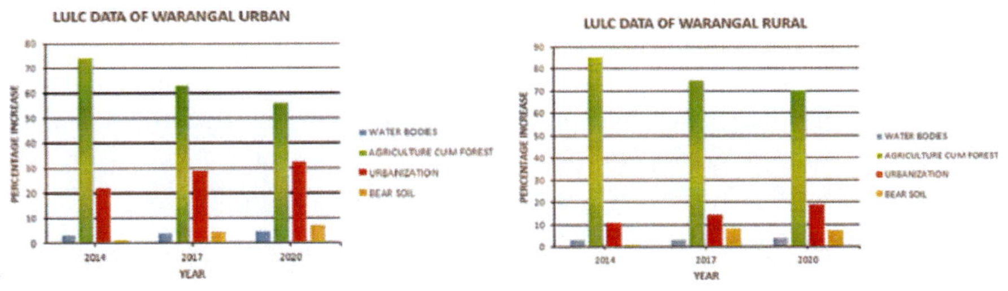

Fig. 6 Bar graph of Warangal urban and rural regions

Table 3 Population growth of Warangal region

Rainfall data of Warangal region	
Year	Annual rainfall (mm)
2014	740
2017	892
2020	1853

Source https://www.census2011.co.in and Telangana socio-economic outlook 2015–2016

3.3 Population Growth of Warangal

Table 3 indicates the population growth rate of Warangal region. As per the 2001 census the year 2001 population of Warangal region was 3,246,004 which increased to 3,512,576 in the year 2011 and as per Intensive Household Survey 2014, the population of Warangal region increased to 3,647,000. Sweeping increase in population has led to an increase in urbanization of Warangal region. This population growth not only affected urbanization but also decreased agriculture cum forest and increased bear soil of Warangal region.

3.4 Annual Rainfall of Warangal

Annual rainfall data of Warangal region is displayed in Table 4. It is observed that from 2014 to 2020 the rainfall has increased drastically from 740 to 1853 mm. Because of which there is a slight increase in water bodies. Due to the increase in population, humidity and moisture content may increase, because of which evaporation may increase and thereby as part of the hydrological cycle rainfall may also increase. Tables 4 and 5 show annual and monthly rainfall data recorded for the years 2014, 2017 and 2020. Average Annual rainfall data of Warangal region is collected from the central ground water board (CGWB) website to analyze the recorded rainfall. Yearly annual rainfall data shows an increasing trend from 2014 to 2020 (Table 4). Annual rainfall recorded in the years 2014, 2017 and 2020 are 740 mm, 892 mm and 1853 mm respectively. From 2014 to 2017 there was a rainfall increase of 152 mm. Whereas from 2017 to 2020 there is a sudden increase of extremely heavy rainfall of 961 mm.

Especially in the month of August 2020 (Table 5) rainfall has risen from 321.7 mm (July) to 810.2 mm (August). During this period, low lying areas of Warangal region were totally flooded due to very heavy rainfall. Because of which there was hefty loss of people and property. Therefore, proper measures are to be taken to control heavy rainfall and flooded areas. Therefore, heavy rainfall flooded water shall be guided in a proper sequence for storage of water. Table 5 shows the monthly rainfall information of Warangal region for the years 2014, 2017 and 2020 collected from CGWB. Hence, for a slight increase in water bodies, rainfall increase is one of the

Table 4 Annual rainfall data of Warangal region for the years 2014, 2017 and 2020

Population growth of Warangal region		
Year	Population	Growth rate (%)
2001	3,246,004	
2011	3,512,576	8.21
2014	3,647,000	3.82

Source http://cgwb.gov.in

Table 5 Monthly rainfall data of Warangal region for the years 2014, 2017 and 2020

Year	Jan	Feb	Mar	Apr	May	Jun	Jul	Aug	Sep	Oct	Nov	Dec	Annual
2014	0.5	0	42.1	5	105.6	53.9	172.1	201.7	118.3	29.5	11	0	740
2017	0	0	0	19.5	1	226	197.4	171	156	121	0	0	892
2020	0.8	14.3	12.1	16.6	28.4	206.4	321.7	**810.2**	240.4	191.5	10.5	0	1853

Source http://cgwb.gov.in

Table 6 Percentage change in Warangal urban region

Year	Urbanization	Agriculture and forest	Waterbodies	Bear soil
2014–2017	Increased by 7.12%	Decreased by 11.05%	Increased by 0.46%	Increased by 3.44%
2017–2020	Increased by 3.92%	Decreased by 6.8%	Increased by 0.79%	Increased by 2.09%

Table 7 Percentage change in Warangal rural region

Year	Urbanization	Agriculture and forest	Waterbodies	Bear soil
2014–2017	Increased by 3.65%	Decreased by 10.74%	Decreased by 0.17%	Increased by 7.26%
2017–2020	Increased by 4.45%	Decreased by 4.56%	Increased by 1.03%	Decreased by 0.93%

causes. Not only this, Telangana Government has implemented tank rejuvenation as part of mission Kakatiya, to improve the groundwater resources. This is also one of the reasons for the increase in water bodies.

From the results of Tables 1 and 2, it is observed that, urban development has rapidly increased in the span of 6 years. Urbanization has led to a decrease in vegetation. Waterbodies increased in very less quantity which is less than 1% and is not sufficient for the growing needs of the Warangal population. Necessary measures are being taken to protect and preserve the agricultural area, forest and water bodies of Warangal region. The urbanization has increased very rapidly to 11.04% in Warangal urban area and in rural areas increased to 8.1% in the span of six years (Tables 6 and 7). Due to increase in the urbanization agricultural and forest areas have decreased drastically by 17.85% in urban areas and in rural areas by 15.3%. In Warangal Urban 231.09 km^2 of agricultural land cum forest area is decreased and in Warangal Rural 326.4 km^2 of agricultural land cum forest area is decreased. Urbanization area of 142.94 km^2 got increased in Warangal Urban and 173.25 km^2 of Urbanization area got increased in Warangal Rural.

4 Conclusions

The results indicate that the effect of urbanization on LULC of Warangal region is more which has to be planned efficiently. Urban development highly affected bear soil and agriculture cum forest both in rural and urban areas. Very little increase in water bodies is observed in the Warangal urban and rural areas, which will not be sufficient for the growing needs of the population. Increase in water bodies are observed due to increase in rainfall and implementation of Mission Kakatiya Scheme by the Telangana government. Proper planning in correlation with the growing population and LULC changes is to be made for both urban and rural areas of Warangal region. To know the effect of urbanization on LULC changes of Warangal using RS and GIS techniques is necessary. Therefore, measures shall be taken to overcome the problem of urbanization in the Warangal region. Effects of Urbanization on Land Use and Land Cover of Warangal Region Using RS and GIS techniques is important to understand the changes of Warangal land cover.

References

1. Mohan M, Pathan SK, Narendrareddy K, Kandya A, Pandey S (2011) Dynamics of urbanization and its impact on land-use/land-cover: a case study of megacity Delhi. J Environ Prot 2(09):1274
2. Moniruzzaman M, Roy A, Bhatt CM, Gupta A, An NTT, Hassan MR (2018) Impact analysis of urbanization on land use land cover change for Khulna City, Bangladesh using temporal landsat imagery. Int Arch Photogramm Remote Sens Spat Inf Sci 42(5):757–760
3. Jeykumar RKC, Chandran S (2019) Impact of urbanization on climate change and geographical analysis of physical land use land cover variation using RS-GIS. Global NEST J 21(2):141–152
4. Abd El-Kawy OR, Rød JK, Ismail HA, Suliman AS (2011) Land use and land cover change detection in the western Nile delta of Egypt using remote sensing data. Appl Geogr 31(2):483–494
5. Jaysawal D, Saha S (2014) Urbanization in India: an impact assessment. Int J Appl Sociol 4(2):60–65
6. Rawat JS, Kumar M (2015) Monitoring land use/cover change using remote sensing and GIS techniques: a case study of Hawalbagh block, district Almora, Uttarakhand, India. Egypt J Remote Sens Space Sci 18(1):77–84
7. Shalaby A, Tateishi R (2007) Remote sensing and GIS for mapping and monitoring land cover and land-use changes in the Northwestern coastal zone of Egypt. Appl Geogr 27(1):28–41

8. Treitz P, Rogan J (2004) Remote sensing for mapping and monitoring land-cover and land-use change-an introduction. Prog Plan 61(4):269–279
9. Mhawish YM, Saba M (2016) Impact of population growth on land use changes in Wadi Ziqlab of Jordan between 1952 and 2008. Int J Appl Sociol 6(1):7–14

Effect of LULC Changes on Land Surface Temperature

Rajashekar Kummari, Pavan Kumar Reddy Allu, Shashi Mesapam, Ayyappa Reddy Allu, Bhargavi Vinakallu, and Bhanu Prakash Ankam

Abstract The rapid urbanization of cities has led to significant human-induced alterations to the environment, characterized by the replacement of natural land cover with man-made structures such as concrete, bricks, asphalt, and metal. This transformation disrupts the natural processes of evapotranspiration, resulting in reduced cooling effects and increased heat storage within urban areas. This study involves the Land Surface Temperature (LST) variation over different classes of Land Use Land Cover (LULC) and the effect of change in LULC over the past two decades in and around Hyderabad City, Telangana, India. This is a Tier-1 Indian city experiencing notable LULC changes from the past two decades due to rapid Urbanization. This study examines how urbanization has affected the surface temperatures from the past twenty years, using Landsat 7 and Landsat 8. Spectral indices such a LULC, Radiance, Brightness temperatures and Land Surface Emissivity (LSE) have been generated and utilized for the generation of LST. A substantial change of vegetation increased by 28.92% from 2012 to 2022 period. And also, an urban area increasing by 108.8% but waterbodies have decreasing and increasing patterns in the period of 2012–2022. It is advised to implement efficient urban planning and design solutions to lessen the negative consequences of LST changes caused by LULC. Green space inclusion, vegetation restoration, cool pavement and roofing techniques, and sustainable urban design techniques are among those to be employed in order to reduce the effect of these temperature changes.

Keywords Urbanization · Land Surface Temperature (LST) · Urban Heat Island (UHI) · Normalized Difference Vegetation Index (NDVI)

R. Kummari · P. K. R. Allu (✉) · S. Mesapam · A. R. Allu · B. Vinakallu · B. P. Ankam
Department of Civil Engineering, National Institute of Technology, Warangal, India
e-mail: ap22cer2r05@student.nitw.ac.in

S. Mesapam et al. (eds.), *Developments and Applications of Geomatics*,
Lecture Notes in Civil Engineering 450, https://doi.org/10.1007/978-981-99-8568-5_12

1 Introduction

Rapid urbanization in the past decades has resulted in a great change in the Land Use Land Cover (LULC) and created a great impact on the environment and climate [1, 2]. One such important consequence of LULC change is changes in the Land Surface Temperature (LST) which has a major role in maintaining the local and regional climate patterns. Changes to the physical properties of the Earth's surface are referred to as LULC changes. These include modifications to the plant cover, urbanization, agricultural methods, and natural disturbances like deforestation or afforestation. These adjustments immediately alter the surface energy distribution, which in turn affects the thermal characteristics of the land surface and its temperature [3, 4].

1.1 Land Surface Temperature

The term Land Surface Temperature describes the temperature of the Earth's surface, specifically the temperature of the soil, plants, rocks, bodies of water and other land surface phenomenon. LST is essential to many environmental processes and applications, including urban planning, agriculture, hydrology and climate modeling. Numerous variables such as solar radiation, air conditions, vegetation cover, surface characteristics and land use land cover have an impact on LST. It can alter both geographically and temporally reflecting changes in surface temperature throughout the day and across the seasons [5].

1.2 Urban Heat Island

Urbanization also led to the formation of Urban Heat Island (UHI). Urban heat islands are formed when impermeable surfaces like buildings, roads, and pavements overtake naturally occurring landscapes, causing urban regions to experience greater LST than nearby rural or natural areas [6–8]. The local climate, human health, energy use, and general urban comfort may all be significantly impacted by its formation. On the other hand, modifications to the vegetation, including afforestation or deforestation, can also affect LST. Through mechanisms such as shade and evapotranspiration, vegetation has an impact on the surface energy balance. By limiting the quantity of solar energy received and encouraging water evaporation, which helps control temperature, vegetation cover has a cooling effect. As a result, changes in land cover that cause the loss or increase of vegetation can cause significant adjustments in LST patterns [9–11].

1.3 Objectives

Main aim of this paper is to analyze the Land Surface Temperatures of Hyderabad City

- To Analyze the variations of Land Use Land Cover Changes over the past two decades.
- To Estimate the change in surface temperatures due to LULC changes and to map the spatial patterns of LST.
- To Predict the future change in LULC for Hyderabad city for the year 2027.

2 Study Area

Hyderabad, a Metropolitan city in Telangana with a diverse and dynamic urban landscape is located between latitudes 17°18′ N to 17°35′ N and Longitudes 78°15′ E to 78°40′ E. The city originated on the banks of the river Musi. It is situated at an average elevation of 542 m above sea level and experiences a semi-arid type of climate. The study area for the current study includes in and around regions of Hyderabad till Outer Ring Road (ORR). Parts of Rangareddy, Sangareddy and Medchal-Malkajgiri districts are covered inside of ORR. HMDA region extended till ORR has experienced rapid urbanization. The population of these combined regions in 2002 was approximately 5,800,000 and in 2022 was approximately 10,000,000 constituting an increase of 72% population. This rapid increase in population increased the percentage of concrete sand bituminous structures in the city resulting in greater loss of natural land cover in and around the city (Fig. 1).

2.1 Data Used

In this present study, remotely sensed satellite images were used for the Land Use Land Cover classification and also for the generation and analysis of Land Surface Temperature (LST). For this Landsat 7 and Landsat 8 data have been used. Data sets for 2002, 2007 and 2012 have been used from Landsat 7 and for the years 2017 and 2022 Landsat 8 data is used (Table 1).

2.2 Methodology

The Methodology for the proposed study has shown in Fig. 2.

Fig. 1 Study Area Map

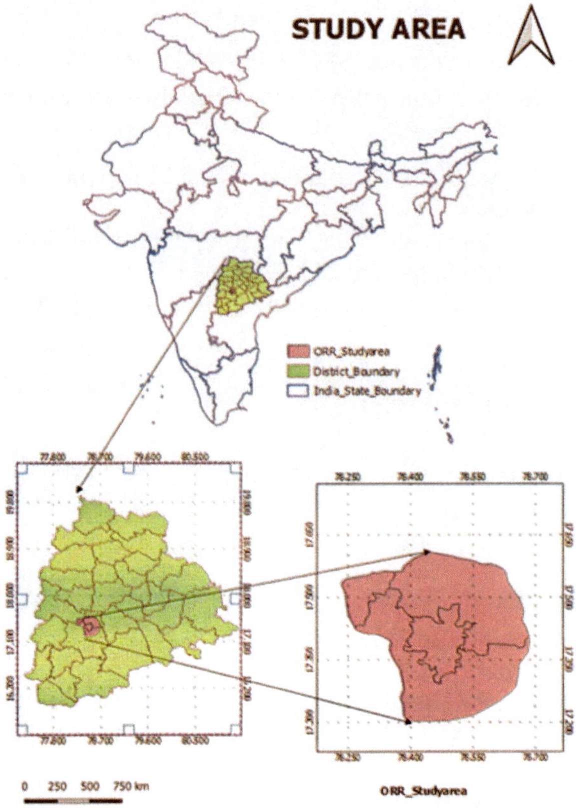

Table 1 Data collection from satellite imagery

Satellite	Type of sensor	Acquisition date	Path and row	Data source
Landsat 7	ETM+	07-04-2002	144, 048	USGS Earth Explorer
Landsat 7	ETM+	07-05-2007	144, 048	
Landsat 7	ETM+	20-05-2012	144, 048	
Landsat 8	OLI/TIRS	24-04-2017	144, 048	
Landsat 9	OLI/TIRS	24-05-2022	144, 048	

2.3 Image Pre-processing

The Raw data obtained from LANDSAT contains different types of errors and for that image pre-processing is required to take over the geometric, atmospheric and radiometric correction. Cloud free data of the winter season was used as it doesn't induce excess heat due to climate, only surface heat is dominant and would help in clearly analyzing the thermal properties and temperature changes.

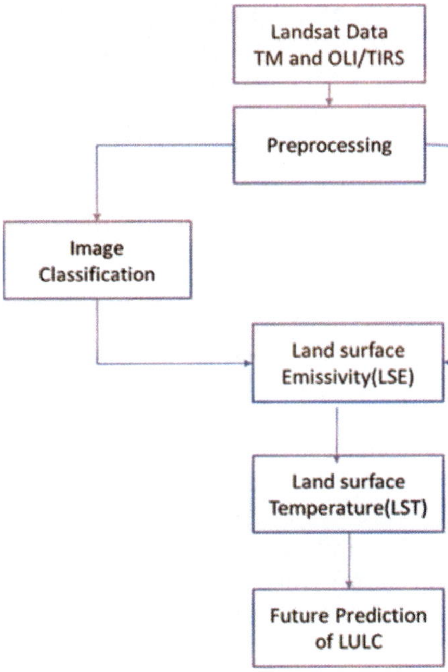

Fig. 2 Flow chart showing the methodology

The DN of images were converted into TO values which can be precisely used for calculating various indices error free. ERDAS imagine was used for the correction and the DOS 1 atmospheric correction was also done in QGIS Pre-processing. The first and vital step after data extraction helps in increasing the accuracy of the results.

2.4 *Image Classification*

A Typical method of remote sensing analysis is Image Classification. The goal of image classification is to classify the image's pixels into groups according to the types of land cover they represent by using a classifier algorithm, which groups pixels based on their reflectance value.

Two types of classification exist in general, which are unsupervised and supervised classification. In Unsupervised classification, the algorithm is self-defined and the software itself performs the classification without human intervention. In Supervised classification, the analyst uses samples of pixels from the image to teach the algorithm to recognize spectral classifications. For each land cover class, adequate samples must be used. Each class thought to differ from the others spectrally. To create samples, a polygon comprised of representative pixels from a class is created, and those pixels are subsequently loaded. Using the data from pixels as a signature. Following the collection of samples for each land cover class, the separability of the samples will be evaluated by looking at the signatures on a plot of wavelength versus reflectance value. Depending on the results, samples might need to be improved.

2.5 *Spectral Indices*

2.5.1 Normalized Difference Vegetation Index

Normalized Difference Vegetation Index (NDVI) is the mostly used Vegetation Index in the analysis of satellite images. It gives information about the density and health of vegetation cover based on the difference between the reflectance of Near-Infrared (NIR) and Red spectral bands. The health of plants depends on the electromagnetic spectrum. It is crucial to how NDVI functions and enables us to assess a plant's health based on how it reflects light and energy. A plant appears green to the human eye because of the chlorophyll pigment, which reflects green wavelengths and absorbs red ones. Plants' cell architecture reflects near-infrared (NIR) wavelengths. As a result, the plant develops, flourishes, and contains more cell structures when photosynthesis takes place. This implies that a healthy plant actively absorbs red light and reflects NIR—one with a lot of chlorophyll and cell structures.

NDVI can be calculated using the below formula:

$$\mathrm{NDVI} = \frac{\mathrm{NIR} - \mathrm{RED}}{\mathrm{NIR} + \mathrm{RED}}$$

NDVI values range from -1 to $+1$ with -1 indicating no vegetation and $+1$ indicating dense healthy vegetation and forest cover [12, 13].

2.5.2 Normalized Difference Built-Up Index

Normalized Difference Built-up Index (NDBI) is the mostly used Built-up or urban index in the analysis of satellite images. It quantifies the difference in reflectance between the shortwave infrared (SWIR) and near-infrared (NIR) spectral bands.

NDBI can be calculated using the below formula:

$$\mathrm{NDVI} = \frac{\mathrm{SWIR} - \mathrm{NIR}}{\mathrm{SWIR} + \mathrm{NIR}}$$

NDBI values range from -1 to $+1$. Greater positive values indicate a higher density of built-up areas. It highlights the contrast in reflectance between built-up structures (which have high reflectance in the SWIR band) and natural features like vegetation or bare soil (which have high reflectance in the NIR band).

2.6 LST Estimation from Landsat 7

The bands that are used for LST estimation using Landsat 7 data is Band 6 (Low gain or High gain). Firstly, the bands are to be rectified for scan line errors. This can be done by imposing the gap mask layers. Deriving LST from satellite images follows

1. Conversion of DN to Radiance
2. Conversion of Radiance to Brightness Temperature (BT)
3. Conversion of Kelvin to Degree Celsius

2.6.1 Conversion of DN to Radiance

DN values of the satellite images have to be converted to spectral radiance using the equation

$$L\lambda = \frac{L_{\max} - L_{\min}}{\mathrm{QCal}_{\max} - \mathrm{QCal}_{\min}} * (\mathrm{QCal} - \mathrm{QCal}_{\min}) + L_{\min}$$

where,

$L\lambda$	Spectral radiance at the sensor (W/m^2*sr*μm).
QCal	Quantized calibrated pixel value in DN.
$L_{\max}$	Spectral radiance scaled to $\mathrm{QCal}_{\max}$ (W/m^2*sr*μm).
$L_{\min}$	Spectral radiance scaled to $\mathrm{QCal}_{\min}$ (W/m^2*sr*μm).
$\mathrm{QCal}_{\min}$	The minimum Quantized calibrated pixel value in DN.
$\mathrm{QCal}_{\max}$	The maximum Quantized calibrated pixel value in DN.

2.6.2 Conversion of Radiance to Brightness Temperature (BT)

Here the converted spectral radiance of thermal band of ETM + sensor is further processed to temperature in Kelvin (K) by involving the following equation [12, 14–16].

$$\mathrm{BT} = \frac{K2}{\ln\left(\frac{K1}{L\lambda} + 1\right)}$$

where,

BT	At satellite brightness temperature (Kelvin).
$L\lambda$	Spectral radiance at the sensor (W/m^2*sr*μm).
$K1$ $K2$	are calibration constants.

2.6.3 Conversion of Kelvin to Degree Celsius

Further, the converted temperature scale in K of the thermal band can be again transformed into degree Celsius (°C) using the below equation

$$T(°\text{C}) = \text{BT} - 273.15$$

2.7 LST Estimation from Landsat 8

In this, Band 10 and Band 11 can be used for the generation of spectral radiance and brightness temperature. And NIR (Near-Infrared, band5) and Red (band4) bands are used for calculating Emissivity using NDVI method.

2.7.1 Conversion to TOA Radiance

Using the Radiance rescaling factor, Thermal Infra-Red Digital Numbers can be converted to TOA spectral radiance.

$$L\lambda = \text{ML} * \text{QCal} + \text{AL}$$

where,

$L\lambda$ TOA Spectral radiance (W/m^2*sr*μm).
ML Radiance multiplicative Band (No.)
AL Radiance Add Band (No.)
QCal Quantized and Calibrated standard product pixel values (DN).

2.7.2 Top of Atmosphere (TOA) Brightness Temperature

Spectral radiance data can be converted to top of atmosphere brightness temperature using the thermal constant values in metadata file

$$\text{BT} = \frac{K2}{\ln\left(\frac{K1}{L\lambda} + 1\right)} - 273.15$$

where,

BT TOA brightness temperature (°C).
$L\lambda$ Spectral radiance at the sensor (W/m^2*sr*μm).
K1 and K2 are calibration constants.

2.8 *NDVI Method for Emissivity Correction Calculation*

2.8.1 Calculate NDVI

NDVI can be calculated by using the following equation

$$\text{NDVI} = \frac{\text{NIR} - \text{RED}}{\text{NIR} + \text{RED}}$$

where, NIR is Near-Infrared which is band 5 and red is band 4 [12, 13].

2.8.2 Proportion of Vegetation

Proportion of Vegetation can be obtained by using the max and min values of the NDVI [17]

$$\text{PV} = \left(\frac{\text{NDVI} - \text{NDVI}_{\text{min}}}{\text{NDVI}_{\text{max}} - \text{NDVI}_{\text{min}}}\right)^2$$

where,

PV	Proportion of Vegetation.
NDVI	DN values from NDVI image.
NDVI_{min}	Minimum DN values from NDVI image.
NDVI_{max}	Maximum DN values from NDVI image.

2.8.3 Land Surface Emissivity

Land Surface Emissivity (LSE) is the average emissivity of an element on the surface of the Earth calculated from the PV [18]

$$E = 0.004 * \text{PV} + 0.986$$

E	Land surface emissivity.
PV	Proportion of Vegetation.

2.8.4 Land Surface Temperature

Land Surface Temperature is the radiative temperature which is calculated using Top of Atmosphere brightness temperature, wavelength of emitted radiance and Land surface emissivity [19–21]

$$\mathrm{LST} = \left(\frac{\mathrm{BT}}{\left(1 + \left(\frac{0.00115 * \mathrm{BT}}{1.4388}\right) * \ln(E)\right)} \right)$$

where,

BT Top of atmosphere brightness temperature (°C).
E Land Surface Emissivity.

2.9 *Future Prediction of LULC*

The Land use Land cover is the most important parameter for the LST analysis. The past LULC change dynamics would help in prediction the of Future prediction. Various methods like Random Forest, Artificial Neural Network (ANN), CA and multiple regression QGIS software were used for the LULC prediction over Hyderabad city and surrounding area up to ORR. The MOLUSCE (Modules for Land Use Change Evaluation) plugin tool was used for LULC simulation and prediction. The known algorithms CA, ANN, Logistic Regression and MCDA (Multiple criteria decision analysis) were used for the study area. The LULC data is validated with the predicted LULC map and later the future prediction is performed.

Pearson Correlation Constant was generated first and the area changes for different years and a transition matrix was generated. Transition potential modeling was used, the neural network was trained and the graph was generated between the trained data and the input data. Then simulation was under these spatial variables and after that validation was done for predicted 2022 and actual 2022. Using this relation, the future LULC for the years 2032 and 2042 are to be estimated.

The model that validates the simulated LULC map generates Kappa characteristics such as Overall kappa, Kappa histogram, Kappa location and % of correctness which are used for evaluating the accuracy of the model are given below,

$$K = \frac{P(A) - P(E)}{1 - P(E)}$$

Kappa location: $K_{\mathrm{loc}} = \frac{P(A) - P(E)}{P_{\max} - P(E)}$.
Kappa Histogram: $K_h = \frac{P_{\max} - P(E)}{1 - P(E)}$.
Where, $P(A) = P(A) = \sum_{i=1}^{c} p_{ii} = \sum_{i=1}^{c} p_{iT} p_{Ti}$, $P_{\max} = \sum_{l=i}^{c} \min(p_{iT} p_{Ti})$, p_{ij} is i, jth cell of contingency table, p_{iT} is the sum of all cells in the ith row, p_{Ti} is the sum of all cells in the jth column and c is the raster categories count.

3 Results and Discussions

3.1 *Land Use Land Cover Analysis*

In the present study, four land use land cover classes have been classified such as Vegetation, Waterbodies, Built-up and Barren Lands. It was discovered that the study area, which includes Hyderabad and its environs, had a distinct LC with the urban class dominating in the center. Urban class was confined to the study area's center part in 2001, when arid plains dominated the environment. And some parts of the north parts of the city were lined with dense vegetation, and there were also sections of it on the edges of the city. Its differentiation of the Land use for the years 2002, 2007, 2012, 2017 and 2022 can be viewed from the Land use land cover images (Fig. 3) and the area and percentage changes are listed in Table 2 and Fig. 3.

By analyzing the data from Table 2, there has been a significant decrease in barren lands over the years. In 2002, the Urban class covered an area of 307.23 km^2, which grew to 538.76 km^2 by 2022, indicating a substantial percentage increase of 75.36% over the two-decade period. Barren land, on the other hand, has declined by 49.6% during the same timeframe. Vegetation has exhibited an upward trend, particularly from 2012 onwards. There was a 19.99% increase from 2012 to 2017, followed by a further 7.14% increase from 2017 to 2022. Overall, the vegetation area expanded from 382.8 to 480.32 km^2 over the two-decade period. Furthermore, water bodies have shown an overall increase of 11% in the past two decades. These findings highlight the dynamic changes in land use and land cover in Hyderabad, with significant implications for urban development and environmental management.

3.2 *NDVI*

The Normalized Difference Vegetation Index (NDVI) is a spectral index used to represent the concentration of vegetation. In this study, NDVI values for the years 2002, 2007, 2012, 2017, and 2022 were generated using Band 4 (Red) and Band 5 (NIR) imagery. NDVI values ranging from -1 to $+1$ indicate the presence and density of vegetation. Positive values suggest areas with thick and dense vegetation, while negative values indicate sparse or no vegetation. In the study area, NDVI values range between -0.49 and $+0.53$, revealing a varied pattern of vegetation coverage. It is evident that certain parts of the city lack significant vegetation (Fig. 4).

Table 2 Area and percentage change over study area for different years

Land cover	Area (km^2)					Percentage changes for each class						
						Change for each 5 years				10 years		20 years
	2002	2007	2012	2017	2022	2002–2007	2007–2012	2012–2017	2017–2022	2002–2012	2012–2022	2002–2022
Vegetation	382.8	389.17	373.61	448.31	480.32	1.66	−4	19.99	7.14	−2.4	28.56	25.48
Waterbodies	118.47	142.72	128.32	121.51	131.62	20.47	−10.09	−5.31	8.32	8.31	2.57	11.1
Built-up	307.23	319.36	367.25	483.22	538.76	3.95	15	31.58	11.49	19.54	46.7	75.36
Barren	689.91	647.16	629.23	445.37	347.71	−6.2	−2.77	−29.22	−21.92	−8.8	−44.74	−49.6

Total extent of study area: 1498.41 km^2

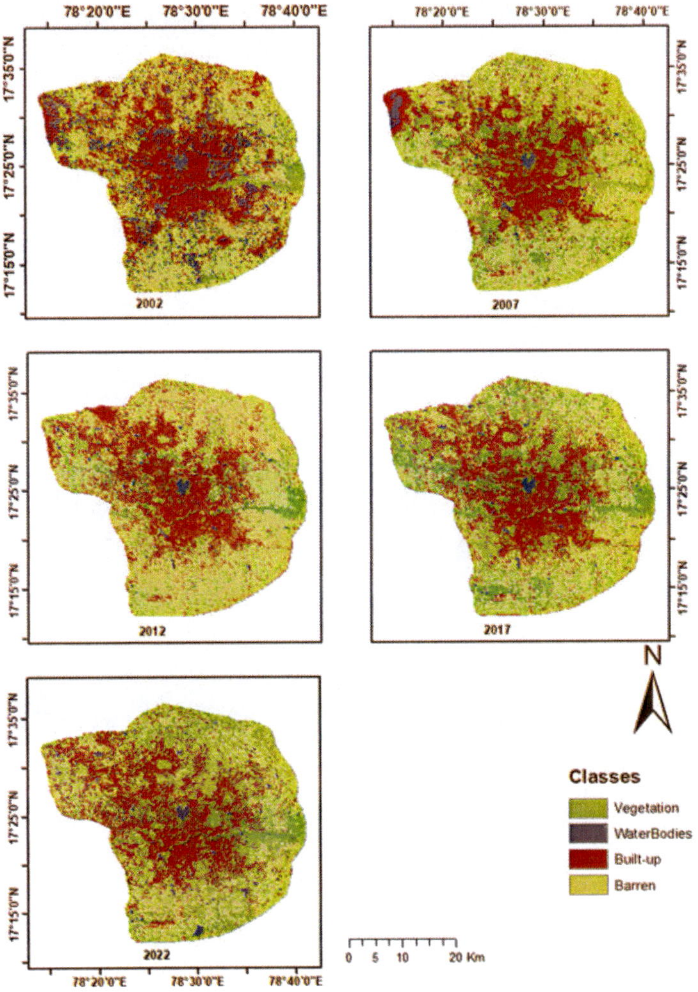

Fig. 3 LULC images

3.3 *Land Surface Temperature Analysis*

Land surface temperature is generated by using the spectral indices such as radiance, brightness temperature, proportion of vegetation and land surface emissivity. Hyderabad city is showing a mixed type of climate. The outer parts of the city excluding the central part have been experiencing greater temperatures. In the year 2002, only certain parts of the city experienced a temperature more than 35 °C and the max temperature was recorded as 42 °C. But for the year 2012, Hyderabad city has experienced a heat wave. Due to the lack of vegetation, the temperature has reached a maximum of 57.59 °C. And from the year 2017, the percentage of vegetation in and

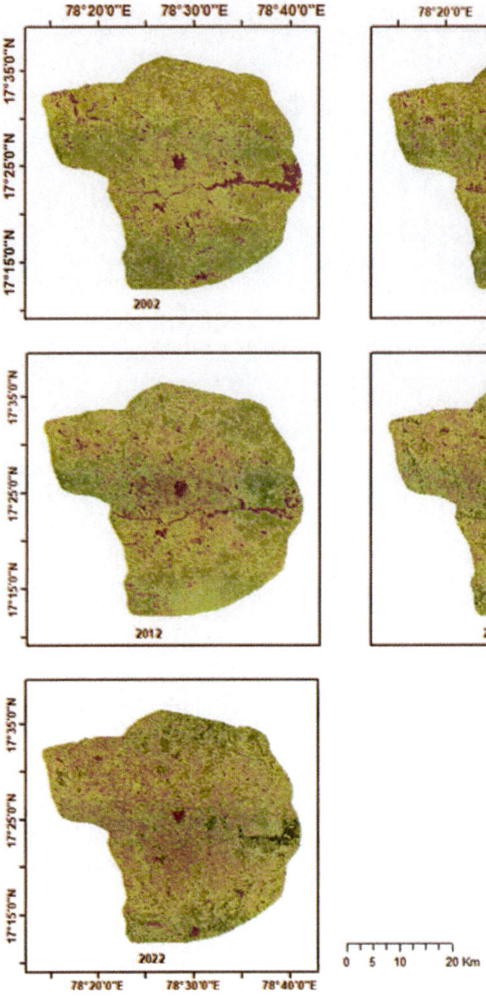

Fig. 4 NDVI images

around the parts of the city have been increasing resulting in a decrease in temperatures when compared with 2012. The maximum temperature recorded in 2017 was 50.2 °C and that of in 2022 was 41 °C. LULC have made a significant impact on the temperatures for the city as the vegetation percentage has increased over the years (Table 3 and Fig. 5).

Table 3 Max and min land surface temperatures

18.15	Max temp	Min temp	Mean temp
2002	42	21.41	31.705
2007	35.92	18.82	27.37
2012	57.69	18.15	37.92
2017	50.52	20.3	35.41
2022	40.55	19.85	30.2

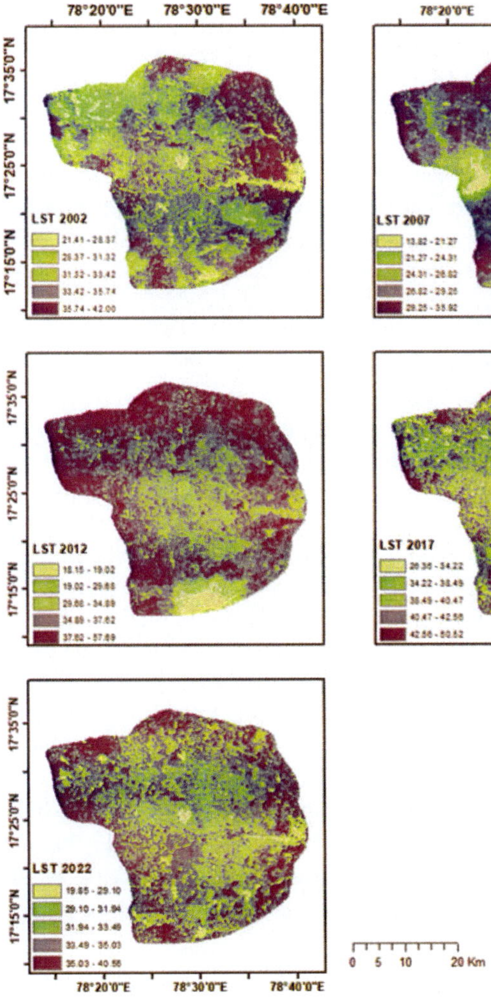

Fig. 5 Image showing LST variations

Table 4 Pearson's correlation

Initial	Final	Pearson's correlation
2002	2007	0.7912
2007	2012	0.8123
2012	2017	0.7879
2017	2022	0.76215

3.4 Future Prediction

3.4.1 Correlation Evaluation

Pearson correlation coefficient is the measure of the strength of linear relation between two variables. The value of this coefficient ranges between -1 to $+1$. Negative value indicates no correlation and positive values indicate a stronger linear relation. From the result, it is viewed that the years between 2002 and 2022 are more strongly related to linear correlation (Table 4).

3.4.2 Validation of the Model

The predicted LULC is to be validated with the existing datasets for its reliability. For this validate option in Molusce plugin is used. Two data sets are used for validation. The module generates different kappa coefficients such as kappa histogram, kappa overall, kappa location and % of correctness. Initial input data set is given the 2002 LULC and the final dataset is given the 2007 LULC map for the initial simulation of LULC for the year 2012. Kappa statistics with 0% indicates that there is no agreement and 100% indicates perfect agreement. Table 5 shows the result of the validated module to check the agreement between the simulated 2012 raster and the real classified raster of 2012.

The estimation of kappa (overall) is 0.7113, kappa (histogram) is 0.8127 and kappa (location) is 0.7095 while the % of correctness is 72.7421, which shows the consistency between the predicted 2012 LULC and the real 2012 LULC is good. And similarly, the model is used for simulating LULC for the years 2017 and 2022. The simulated maps are then validated with the reference 2017 and 2022 rasters respectively. On observing the % correction and other kappa indices, the model is

Table 5 Kappa index generated from model validation

	2002_07_12	2002_07_17	2002_07_22
% of correctness	72.7421	71.2379	69.9787
Kappa histogram	0.8127	0.7911	0.8871
Kappa overall	0.7113	0.7593	0.7261
Kappa location	0.7095	0.7293	0.7302

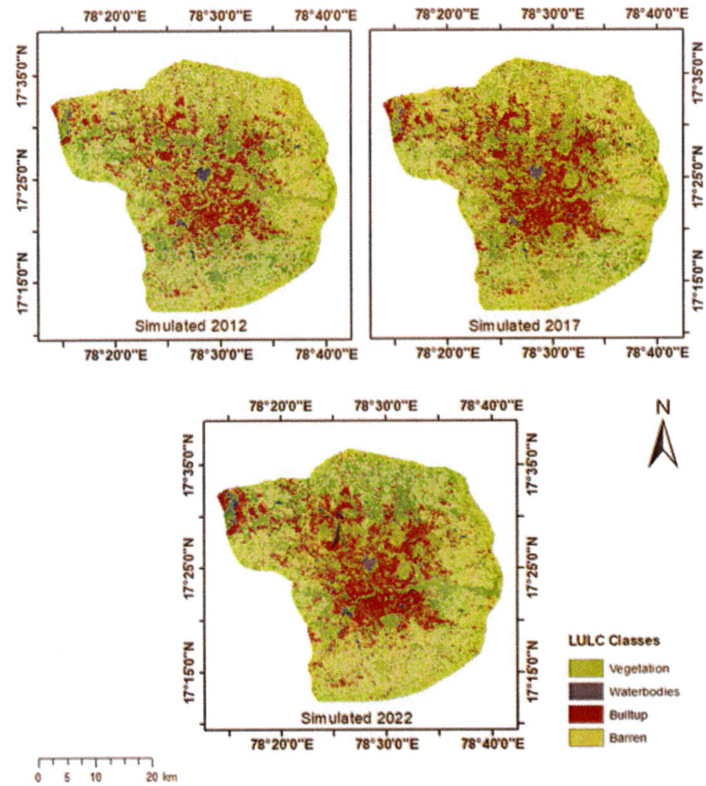

Fig. 6 Simulated LULC maps for the years 2012, 2017 and 2022

showing consistency for the simulated and real LULC. So this model suits best for Hyderabad city for the future prediction of LULC for the year 2027 (Fig. 6).

Using these inputs, the LULC for the year 2027 is predicted. From Fig. 7, it can be viewed that the built-up in the study area has been increased to a great extent by a decrease in the barren lands.

4 Conclusions

By using data from Landsat 7 ETM+ and Landsat 8 OLI/TIRS, this study has investigated about the Land Use Land Cover (LULC) changes over the past two decades, specifically from 2002 to 2022. It has also made an impact in studying about the Land Surface Temperature for the Hyderabad City with an extension up to Outer Ring Road (ORR). During this study, it is observed that Barren lands have experienced a major decrease of 49.6% from 2002 to 2022. Vegetation over the study area has witnessed the least increase in the period 2002–2012. But from 2012, there has

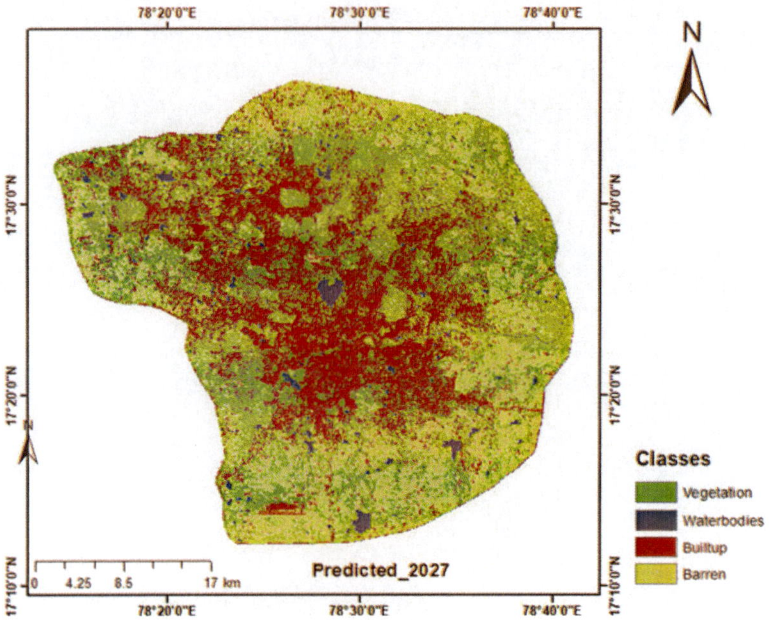

Fig. 7 LULC map predicted for the year 2027

been a better development of vegetation. There has been an increase of 25.48% from 2012 to 2022. Notably, the built-up in the city region has a growth in each period. It has experienced an overall increase of 75.36% from 2002 to 2022. Water bodies have experienced a least increase of 11.1% over the last two decades.

The surface temperature in the study region was evaluated using spectral indices such as radiance, brightness temperature, NDVI, proportion of vegetation, and emissivity. Land surface temperatures exhibited different trends over the past years. The lowest temperature recorded was 18.15 °C, while the maximum surface temperature of 57.69 °C was observed in 2012. Between 2012 and 2017, the city experienced higher temperatures, reaching 50 °C and above. However, due to an increase in vegetation from 2017 onwards, there has been a slight decrease in temperatures.

The aim of this research is to predict future Land Use and Land Cover (LULC) using QGIS and the Molusce plugin, employing a cellular automata-Markov chain-based simulation. LULC rasters were simulated for the years 2012, 2017, and 2022, which were then validated against reference LULC maps. The kappa indices indicate that the model provides the best fit for predicting LULC changes in the study area. A LULC map for the year 2027 was generated, revealing a significant increase in the built-up class and a considerable decrease in barren lands.

These findings contribute to a better understanding of the LULC dynamics and highlight the changing patterns of land cover in the study area. Further analysis and interpretation of these results can aid in urban planning, natural resource management, and decision-making processes to ensure sustainable development and efficient land use practices.

References

1. Dwivedi A, Khire MV (2018) Application of split-window algorithm to study urban heat island effect in Mumbai through land surface temperature approach. Sustain Cities Soc 41:865–877. https://doi.org/10.1016/j.scs.2018.02.030
2. John J, Bindu G, Srimuruganandam B, Wadhwa A, Rajan P (2020) Land use/land cover and land surface temperature analysis in Wayanad district, India, using satellite imagery. Ann GIS 26(4). https://doi.org/10.1080/19475683.2020.1733662
3. Singh P, Kikon N, Verma P (2017) Impact of land use change and urbanization on urban heat island in Lucknow city, Central India. A remote sensing-based estimate. Sustain Cities Soc 32:100–114. https://doi.org/10.1016/j.scs.2017.02.018
4. Sharma R, Pradhan L, Kumari M, Bhattacharya P (2021) Assessing urban heat islands and thermal comfort in Noida City using geospatial technology. Urban Clim 35:100751. https://doi.org/10.1016/j.uclim.2020.100751
5. Buscail C, Upegui E, Viel J-F (2012) Mapping heatwave health risk at the community level for public health action. Int J Health Geogr 11:38. http://www.ij-healthgeographics.com/content/11/1/38
6. Halder B, Bandyopadhyay J, Banik P (2021) Monitoring the effect of urban development on urban heat island based on remote sensing and geo-spatial approach in Kolkata and adjacent areas, India. Sustain Cities Soc 74:103186. https://doi.org/10.1016/j.scs.2021.103186
7. El-Hattab M, Amany SM, Lamia GE (2017) Monitoring and assessment of urban heat islands over the Southern region of Cairo Governorate, Egypt. Egypt J Remote Sens Space Sci 21:311–323. https://doi.org/10.1016/j.ejrs.2017.08.008
8. Kamboj S, Ali S (2020) Urban sprawl of Kota city: a case study of urban heat island linked with electric consumption. Mater Today: Proc 46(11). https://doi.org/10.1016/j.matpr.2020.08.783
9. Deilami K, Kamruzzaman M, Liu Y (2018) Urban heat island effect: a systematic review of spatio-temporal factors, data, methods, and mitigation measures. Int J Appl Earth Obs Geoinformation 67:30–42. https://doi.org/10.1016/j.jag.2017.12.009
10. Monteiro FF, Gonçalves W, Melo LD, Andrade B (2021) Assessment of urban heat islands in Brazil based on MODI'S remote sensing data. Urban Clim 35:100726. https://doi.org/10.1016/j.uclim.2020.100726
11. Equere V, Mirzaei PA, Riffat S, Wang Y (2021) Integration of topological aspect of city terrains to predict the spatial distribution of urban heat island using GIS and ANN. Sustain Cities Soc 69:10285. https://doi.org/10.1016/j.scs.2021.102825
12. Aniello C, Morgan K, Busbey A, Newland L (1995) Micro-urban heat islands using Landsat TM and a GIS. Comput Geosci 21(8):965–969. https://doi.org/10.1016/0098-3004(95)00033-5
13. Estoque RC, Murayama Y, Myint SW (2017) Effects of landscape composition and pattern on land surface temperature: an urban heat island study in the megacities of Southeast Asia. Sci Total Environ 577:349–359. https://doi.org/10.1016/j.scitotenv.2016.10.195
14. Mohan M, Kandya A, Battiprolu A (2011) Urban heat island effect over national capital region of India: a study using the temperature trends. J Environ Prot 2:465–472. https://doi.org/10.4236/jep.2011.24054
15. Shahmohamadi P, Che-Ani AI, Ramly A, Maulud KNA, Mohd-Nor MFI (2010) Reducing urban heat island effects: a systematic review to achieve energy consumption balance. Int J Phys Sci 5(14):2202–2207. https://doi.org/10.5897/IJPS.9000475

16. Weng Q, Lub D, Schubring J (2004) Estimation of land surface temperature-vegetation abundance relationship for urban heat island studies. Remote Sens Environ 89(4):467–483. https://doi.org/10.1016/j.rse.2003.11.005
17. Valor E, Caselles V (1996) Mapping land surface emissivity from NDVI: application to European, African and South American areas. Remote Sens Environ 57:167–184. https://doi.org/10.1016/0034-4257(96)00039-9
18. Van de Griend A, Owe M (1993) On the relationship between thermal emissivity and the normalized difference vegetation index for natural surfaces. Int J Remote Sens 14:1119–1131. https://doi.org/10.1080/01431169308904400
19. Avdan U, Jovanovska G (2016) Algorithm for automated mapping of land surface temperature using LANDSAT 8 Satellite Data. https://doi.org/10.1155/2016/1480307
20. Roy DP, Wulder MA, Loveland TR, Woodcock CE, Allen RG, Anderson MC, Helder D, Irons JR, Johnson DM, Kennedy R, Scambos TA, Schaaf CB, Schott JR, Sheng Y, Vermote EF, Belward AS, Bindschadler R, Cohen WB, Gao F, Hipple JD, Hostert P, Huntington J, Justice CO, Kilic A, Kovalskyy V, Lee ZP, Lymburner L, Masek JG, McCorkel J, Shuai Y, Trezza R, Vogelmann J, Wynne RH, Zhu Z (2014) Landsat-8: science and product vision for terrestrial global change research. Remote Sens Environ 145:154–172. https://doi.org/10.1016/j.rse.2014.02.001
21. Yu X, Guo X, Wu Z (2014) Land surface temperature retrieval from Landsat 8 TIRS—comparison between radiative transfer equation-based method, split window algorithm and single channel method. Remote Sens 6:9829–9852. https://doi.org/10.3390/rs6109829

Estimation of Aerosol Direct Radiative Forcing in Southern India

K. Tharani, Deva Pratap, Keesara Venkatareddy, and P. Teja Abhilash

Abstract The Aerosol Direct Radiative Forcing is a prominent parameter which is used to assess the effect of aerosols on temperature. Previous studies have shown the calculation of ADRF under clear sky conditions and model simulations to know the impact on changes in temperature. In the current study, a novel method is implemented to estimate ADRF under all-sky conditions using Modern-Era Retrospective analysis for Research and Applications (MERRA-2) data. The radiation diagnostics of MERRA-2 provide downward short wave and long wave fluxes under all-sky conditions but not the upwelling fraction of longwave radiation after interacting with the atmosphere. The upwelling fraction is estimated by factored method and difference method in this study and the time series is compared with the available downward flux data. The upwelling longwave flux data from the factored method is found to be appropriate in terms of matching the time series when compared to the data from the difference method. The estimation is carried out for five grid points, each from five climate zones of the study area and the times series plots have shown a better insight on the implemented factored method for the estimation of longwave flux. The ADRF is calculated using the available shortwave fluxes and estimated longwave fluxes at three different levels of atmosphere namely Top of Atmosphere (TOA), Surface and within the atmosphere for the year 2019. The ADRF at TOA and at surface level show that the warm semi-arid and sub-tropical oceanic highland climate regions have experienced higher ADRF when compared to the remaining climate zones. The ADRF in the atmosphere shows that positive ADRF was seen in the same climate regions. This illustrates that the rate of heating is more in the warm semi-arid and sub-tropical oceanic highland climate region in the year 2019.

Keywords Aerosol direct radiative forcing · Shortwave flux · Upwelling longwave flux

K. Tharani (✉) · D. Pratap · K. Venkatareddy
Research Scholar, Department of Civil Engineering, NIT Warangal, Warangal, India
e-mail: tharani@student.nitw.ac.in

P. T. Abhilash
Andhra University, Visakhapatnam, India

S. Mesapam et al. (eds.), *Developments and Applications of Geomatics*,
Lecture Notes in Civil Engineering 450, https://doi.org/10.1007/978-981-99-8568-5_13

1 Introduction

The major socio-economic issue in the current scenario is global and regional climate variability, which is due to an increase in urban emission concentrations. It affects the radiation budget of earth's surface and thus influences the climate parameters to a considerable extent. One of the components of urban emissions is aerosols. They are the minute particles suspended in the air or in any other gas. Aerosols have an impact on the climate system and weather in two primary ways. The first is through scattering and absorption of radiation that is considered as the direct effect of aerosols, which consequently, modifies the Earth's radiation budget [1, 3, 5, 7]. The absorbing aerosols (black carbon, mineral dust, etc.) decrease the Earth's albedo and heat the atmosphere which results in the greenhouse effect near the Earth's surface [13–15], whereas scattering aerosols (sulfate, nitrate, etc.) will increase the Earth's albedo and decrease the solar radiation reaching the Earth which results in a net cooling of the Earth-atmosphere system [2, 4, 6, 17]. Secondly, aerosols can alter cloud microphysical properties, increase cloud lifetime and enhance precipitation efficiency [9, 11, 16, 18].

Aerosol studies over Peninsular India focus primarily on aerosol loading, types and properties of aerosols and influence on the climate [10, 12]. The effect of aerosols on consecutive warming and cooling of the atmosphere is studied by using Aerosol Direct Radiative Forcing (ADRF). The main objective of the present study is to estimate ADRF under all-sky conditions.

2 Data

The present study concentrates on the direct effect of aerosols on the Earth's radiation budget. In order to understand the process, it is essential to have an insight on the forcing component of aerosols. Generally, radiative forcing is used to determine the effect of a particular component in the considered phenomenon. Hence, Aerosol Direct Radiative Forcing (ADRF) is defined as the component which describes the effect of aerosols on the radiative fluxes at top of atmosphere (TOA), at surface and absorption of radiation within the atmosphere. The study area is confined to peninsular India comprising of 9 states, which are divided into five climate zones as shown in Fig. 1. The datasets used are NetCDF files of Modern-Era Retrospective Analysis for Research and Applications (MERRA-2). The parameters from radiation diagnostics of MERRA-2 (M2T1NXRAD) are used to calculate ADRF at TOA, surface and in the atmosphere.

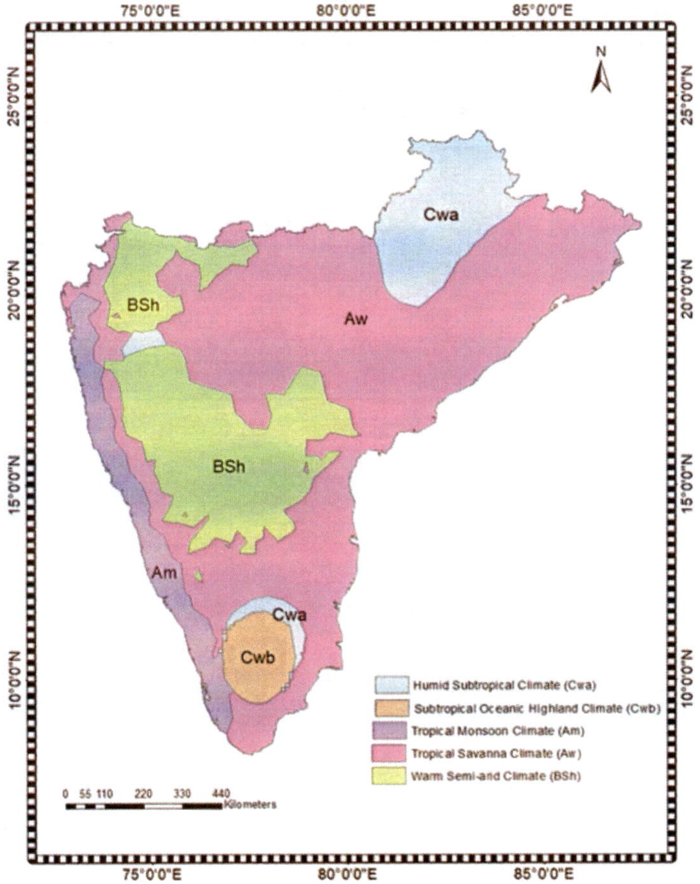

Fig. 1 Study area

3 Methodology

The methodology followed for the current study is shown in Fig. 2. Firstly, the parameters in NetCDF format are collected from MERRA-2 using GES DISK (https://disc.gsfc.nasa.gov/) platform. Then the parameters are averaged daily and written down in Excel using R statistical software. The missing parameters namely Surface Net Downward Longwave Flux assuming no aerosol (LWGNTCLN) and Upwelling Longwave Flux at TOA assuming no aerosol (LWTUPCLN) are calculated by the factorial method as given in Eq. 3. Secondly, ADRF at TOA and surface is calculated using the Eqs. 1 and 2 respectively. Finally, ADRF in the atmosphere $(ADRF_{ATM})$ is calculated by the difference of ADRF at TOA and at the surface.

Fig. 2 Methodology

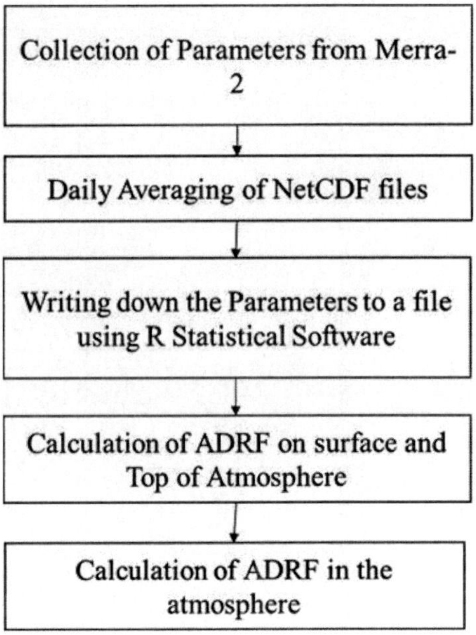

3.1 Calculation of ADRF

The parameters required for the calculation of ADRF are Surface Net Downward Shortwave Flux (SWGNT), Surface Net Downward Shortwave Flux assuming no aerosol (SWGNTCLN), Surface Net Downward Longwave Flux (LWGNT), Surface Net Downward Longwave Flux assuming no aerosol (LWGNTCLN), TOA Net Downward Shortwave Flux (SWTNT), TOA Net Downward Shortwave Flux assuming no aerosol (SWTNTCLN), Upwelling Longwave Flux at TOA (LWTUP), Upwelling Longwave Flux at TOA assuming no aerosol (LWTUPCLN) [8]. Out of all these parameters, six are directly available from MERRA-2. The LWGNTCLN and LWTUPCLN are not available from MERRA-2 which need to be calculated. Generally, the Aerosol Direct Radiative Forcing at surface Eq. (1) and at TOA Eq. (2) are calculated as follows:

$$ADRF_{SUR} = (SWGNT + LWGNT) - (SWGNTCLN + LWGNTCLN) \tag{1}$$

$$ADRF_{TOA} = (SWTNT + LWTUP) - (SWTNTCLN + LWTUPCLN) \tag{2}$$

The normal convention says that positive $ADRF_{ATM}$ warms up the atmosphere and vice versa. The current study aims to obtain the missing parameters by two methods namely factored method and difference method. The factored method uses the factor obtained by dividing SWGNTCLN with SWGNT and uses the value to calculate LWGNTCLN as shown in Eq. (3). The difference method uses the difference between SWGNTCLN with SWGNT and uses the value to calculate LWGNTCLN as shown in Eq. (4).

$$LWGNTCLN = \frac{SWGNTCLN}{SWGNT} * LWGNT \tag{3}$$

$$LWGNTCLN = SWGNTCLN - SWGNT + LWGNT \tag{4}$$

As a part of the analysis, five grid points each in a climate zone are considered and a time series plot for the data corresponding to 2019 of long wave flux calculated by factored and difference method is plotted to match the peaks and dips. The five grid points are Goa (Am), Nanded (Aw), Hyderabad (BSh), Raipur (Cwa) and Coimbatore (Cwb).

4 Results and Discussions

The time series plot for the calculated LWGNTCLN using factored and difference methods along with the available LWGNT are shown in Fig. 3. The parameter calculated by the difference method shows disparity in the values at peaks and dips and that calculated by the factored method matches with the existing data without aerosol. However, the occurrence of peaks and dips is same for both the methods. Hence, the data calculated by factored method is chosen to be the suitable one as it is in accordance with the parameter without aerosol.

4.1 ADRF at TOA, Surface and in the Atmosphere

The yearlong ADRF for the Peninsular India is calculated for the year 2019 at three different levels (Fig. 4). Most of the warm semi-arid, tropical monsoon, sub-tropical oceanic highland climate regions have witnessed a higher range of ADRF at TOA in the study area. The tropical savanna region has seen a lower ADRF at TOA in the year 2019. A similar pattern was visible at the surface except that a part of the tropical savanna towards the East coastal line has seen a reduction in ADRF.

The results show that the ADRF is high towards the southern part of the study area, moderate in the central region and low towards the northern part of the study area. Almost 159 lakh ha of the study area has experienced a positive ADRF whereas the remaining area has experienced negative ADRF in the atmosphere. Generally, positive ADRF indicates that there is a rise in temperature and vice versa. The positive ADRF in the range of 0–1 is distributed over warm semi-arid and sub-tropical oceanic highland climate regions. Further correlation with air temperature is required to assess

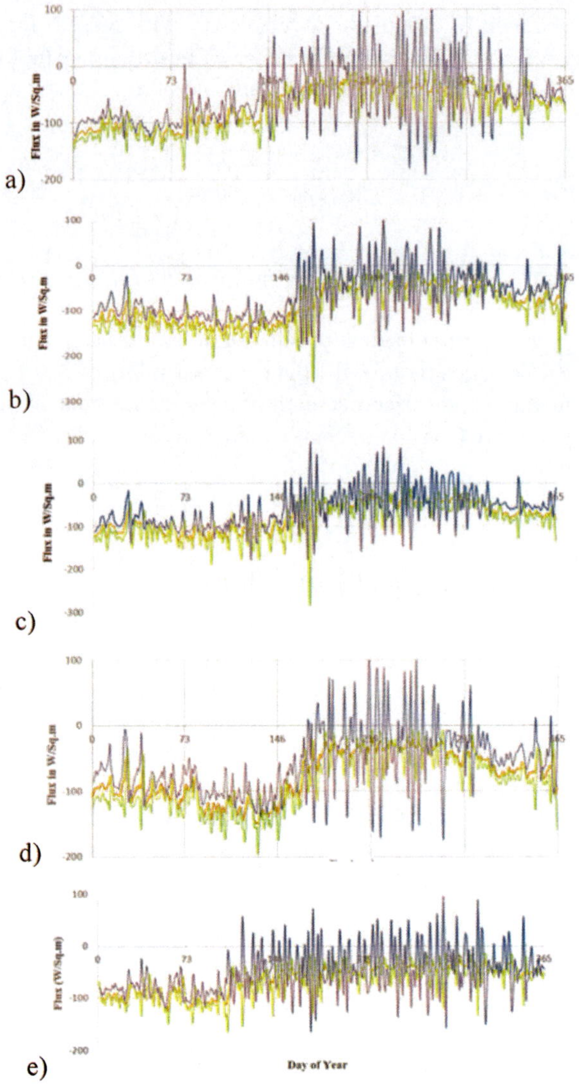

Fig. 3 Time series plot of LWGNT, LWGNTCLN_f, LWGNTCLN_d for **a** Goa **b** Nanded **c** Hyderabad **d** Raipur **e** Coimbatore for the year 2019

the effect of aerosols in the atmosphere. Also, the heating rate in each climate zone can help us to know the changes in global warming.

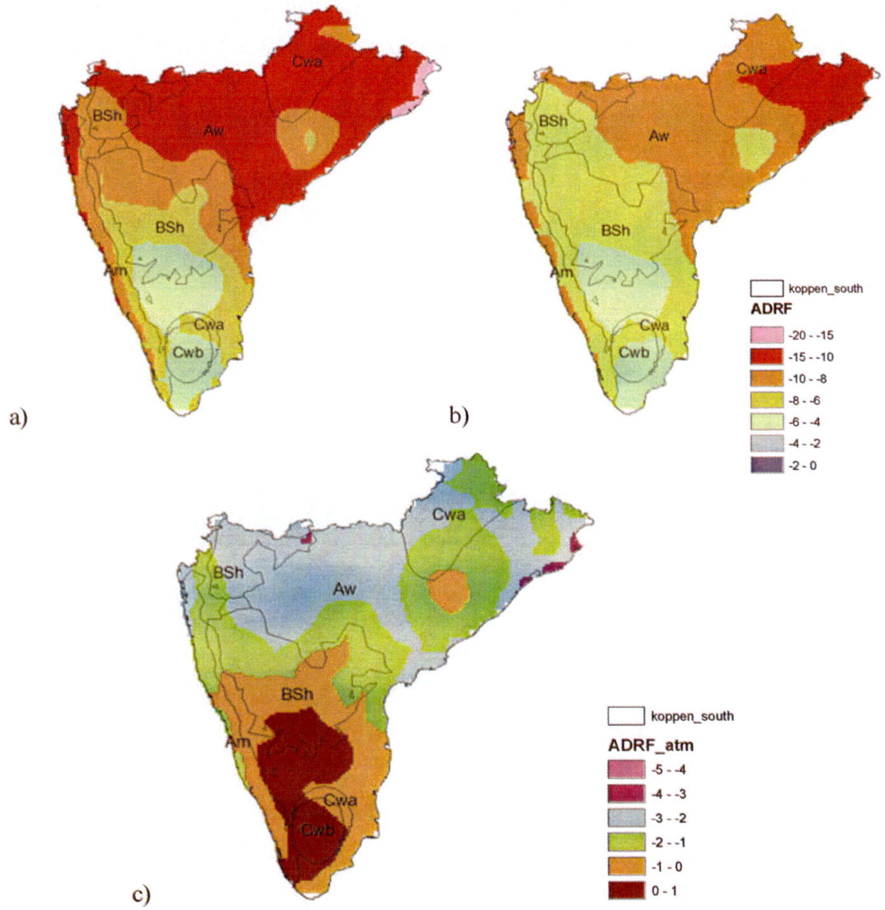

Fig. 4 ADRF at **a** Top of atmosphere **b** Surface **c** In the atmosphere

5 Conclusions

The current study has estimated the ADRF under all-sky conditions using MERRA-2 data. The upwelling longwave fluxes which were not available in the MERRA-2 radiation diagnostics data were estimated by factored and difference methods. The following conclusions are drawn from the analysis performed.

- The upwelling longwave flux estimated by the difference method was found to be inconsistent when compared to the available flux data. However, the peaks and dips of the time series curve were found to be in accordance.
- The upwelling longwave flux estimated by factored method was in agreement with the available flux data. Hence, the factored method was found to be the best one for the estimation of flux data.

- The estimated and available fluxes were used to calculate radiative forcing by aerosols. The ADRF was calculated at TOA, surface and within the atmosphere for the year 2019.
- The ADRF at TOA and surface was found to be negative in the study area whereas the ADRF within the atmosphere was positive only in the sub-tropical oceanic highland climate and a part of warm semi-arid climate regions.
- The positive ADRF might be attributed to the entrapment of radiation in the lower layers of the atmosphere, which might lead to atmospheric warming.

References

1. Adesina AJ, Kumar KR, Sivakumar V, Griffith D (2014) Direct radiative forcing of urban aerosols over Pretoria (25.75 S, 28.28 E) using AERONET sunphotometer data: first scientific results and environmental impact. J Environ Sci 26(12):2459–2474
2. Alam K, Trautmann T, Blaschke T (2011) Aerosol optical properties and radiative forcing over mega-city Karachi. Atmos Res 101(3):773–782
3. Aruna K, Kumar TL, Murthy BK, Babu SS, Ratnam MV, Rao DN (2016) Short wave aerosol radiative forcing estimates over a semi urban coastal environment in south-east India and validation with surface flux measurements. Atmos Environ 125:418–428
4. Dhar P, De BK, Banik T, Gogoi MM, Babu SS, Guha A (2017) Atmospheric aerosol radiative forcing over a semi-continental location Tripura in North-East India: model results and ground observations. Sci Total Environ 580:499–508
5. Feng H, Zou B (2019) Satellite-based estimation of the aerosol forcing contribution to the global land surface temperature in the recent decade. Remote Sens Environ 232:111299
6. García OE, Díaz JP, Expósito FJ, Díaz AM, Dubovik O, Derimian Y, Dubuisson P, Roger JC (2012) Shortwave radiative forcing and efficiency of key aerosol types using AERONET data. Atmos Chem Phys 12(11):5129–5145
7. Guleria RP, Kuniyal JC (2016) Characteristics of atmospheric aerosol particles and their role in aerosol radiative forcing over the northwestern Indian Himalaya in particular and over India in general. Air Qual Atmos Health 9(7):795–808
8. https://disc.gsfc.nasa.gov/
9. Jing X, Suzuki K, Michibata T (2019) The key role of warm rain parameterization in determining the aerosol indirect effect in a global climate model. J Clim 32(14):4409–4430
10. Johnson JS, Regayre LA, Yoshioka M, Pringle KJ, Lee LA, Sexton DM, Rostron JW, Booth BB, Carslaw KS (2018) The importance of comprehensive parameter sampling and multiple observations for robust constraint of aerosol radiative forcing. Atmos Chem Phys 18(17):13031–13053
11. Kant S, Panda J, Pani SK, Wang PK (2019) Long-term study of aerosol–cloud–precipitation interaction over the eastern part of India using satellite observations during pre-monsoon season. Theoret Appl Climatol 136(1):605–626
12. Kaskaoutis DG, Sinha PR, Vinoj V, Kosmopoulos PG, Tripathi SN, Misra A, Sharma M, Singh RP (2013) Aerosol properties and radiative forcing over Kanpur during severe aerosol loading conditions. Atmos Environ 79:7–19
13. Kumar KR, Kang N, Sivakumar V, Griffith D (2017) Temporal characteristics of columnar aerosol optical properties and radiative forcing (2011–2015) measured at AERONET's Pretoria_CSIR_DPSS site in South Africa. Atmos Environ 165:274–289
14. Lu Q, Liu C, Zhao D, Zeng C, Li J, Lu C, Wang J, Zhu B (2020) Atmospheric heating rate due to black carbon aerosols: Uncertainties and impact factors. Atmos Res 240:104891

15. Patel PN, Dumka UC, Kaskaoutis DG, Babu KN, Mathur AK (2017) Optical and radiative properties of aerosols over Desalpar, a remote site in western India: source identification, modification processes and aerosol type discrimination. Sci Total Environ 575:612–627
16. Qian YUN, Leung LR, Ghan SJ, Giorgi F (2011) Regional climate effects of aerosols over China: modeling and observation. Tellus B: Chem Phys Meteorol 55(4):914–934
17. Ramachandran S, Rupakheti M, Lawrence MG (2020) Aerosol-induced atmospheric heating rate decreases over South and East Asia as a result of changing content and composition. Sci Rep 10(1):1–17
18. Yu H, Dickinson RE, Chin M, Kaufman YJ, Zhou M, Zhou L, Tian Y, Dubovik O, Holben BN (2004) Direct radiative effect of aerosols as determined from a combination of MODIS retrievals and GOCART simulations. J Geophys Res: Atmos 109(D3)

Estimation of Groundwater Potential Zones in Southern Dry Agro-Climatic Area Using Geoinformatics and AHP Technique

A. B. Gireesh and M. C. Chandan

Abstract The world's demand for groundwater has been severely strained by overuse of groundwater and major climatic change over time. As the global need for drinking water for human consumption, agriculture, and industrial applications grows, so does necessity to assess groundwater. Due to the quick access to data, analysis, and knowledge, they give about resource for further improvement, GIS-based studies have grown in importance in groundwater research in recent years. In order to identify the groundwater potential zone of southern dry agro-climatic area of Mysuru and Mandya District, India, the current study has been carried out. A total of 15 theme layers were established and researched to help define groundwater potential zones. Depending on respective attributes and water potential capabilities, Analytic Hierarchy Process determines weights allocated to each class in all thematic layers. Utilizing data on groundwater prospects in area (CGWB), study's output was cross-validated, and total accuracy of methodology was 80.4%. The resulting groundwater potential zone was divided into three classifications: high, moderate, and low. According to the research, a moderate groundwater potential zone encompasses 65.8% of study area. There are zones with low and high groundwater potential in 6.75% and 27.43% of the area, respectively. The R^2 value of 0.8 further demonstrates that estimated groundwater potential index and groundwater level values in recommended AHP model are quite reliable in predicting the outcome. The demarcation of groundwater potential zones has been highlighted as a critical step towards achieving Sustainable Development Goals (SDG 6 and 13), contributing to the implementation of sustainable water and land management.

Keywords Groundwater potential zone · Geoinformatics · Analytical hierarchical process (AHP) · Sustainable Development Goals

A. B. Gireesh · M. C. Chandan (✉)
Department of Civil Engineering, The National Institute of Engineering, Mysuru 570008, India
e-mail: chandanmc@nie.ac.in

S. Mesapam et al. (eds.), *Developments and Applications of Geomatics*,
Lecture Notes in Civil Engineering 450, https://doi.org/10.1007/978-981-99-8568-5_14

1 Introduction

In addition to contributing to an area's economic growth and biodiversity, groundwater is one of the most valued sources of water for agricultural, urban, and industrial activities worldwide [16]. A significant source of freshwater, groundwater presently provides around 34% of the total yearly water supply. Therefore, evaluating this resource is essential for the long-term sustainability of groundwater management systems [17]. Water stress is regarded as one of the most crucial socio-environmental issues across several countries because of groundwater demand is increasing, and overexploitation of this vital resource is threatening future generations [24]. Increased groundwater withdrawal has manifested in a gradual decline in the water table. Groundwater occurrence and distribution are influenced by a variety of natural and anthropogenic causes such as topographical, hydrological, biological, geological, and atmospheric influences [6, 22, 27, 32, 34]. Novel strategies are required to produce accurate and helpful information for decision-makers since methods for determining the factors that affect the spatial distribution of groundwater and approaches for acquiring data are developing [1]. The first and most critical stage in groundwater management is groundwater potential mapping (GPM) [29]. Groundwater potential mapping has been described as a strategy for systematic water resource development and planning [9]. Accordingly, for the judicious use of water resources, a groundwater management strategy is recommended.

The primary focus of conventional techniques of identifying, characterizing, and mapping groundwater potential zones is ground surveys employing geophysical, geological, and hydrogeological equipment and a majority of which is expensive and time-consuming [5, 15, 24, 33]. Computer technology has been increasingly used in hydrological research in recent decades. The analysis, forecasting, and management of groundwater resources have all made substantial use of remote sensing (RS) and geographic information system (GIS) techniques [2, 8, 14].

Numerous research has been conducted using various multi-criteria decision-making approaches, geophysical methods, machine learning algorithms, and hybrid artificial intelligence. Analytical hierarchy process [3, 7, 15, 29], weight of evidence (WOE) [11, 34], frequency ratio (FR) [13, 26, 27], electrical resistivity, and vertical electrical sounding (VES) [18] (Fitteman and Stewart 1986) are some of the methods used by the researchers. The goal of current research is to estimate groundwater potential zones using geoinformatic and AHP method. The study objectives are to: (a) mapping groundwater zones using GIS methods; (b) establishing a robust model for estimating groundwater prospective zones, which will aid in agriculture production using GW and; (c) validate the obtained results with field data. As a result, findings of study provide vital recommendations and implementation measures to be taken by decision-makers, administrators, policymakers, and academics.

2 Materials and Method

The research area comprises Mysuru and Mandya district, which are situated in India's southern region (Fig. 1). The geographical extents of study area are latitude of 11°73'N to 13°04'N and Longitude 75°90'E to 77°32'E, which consist of an area of 11,815 km^2. It is located within the Kaveri River's watershed, which runs through northwestern and eastern parts of area. The Mysuru district's soil types are divided into three categories as red sandy, red loamy, and deep black soils. Except for a small portion of T. Narasipur taluk, almost the entire district is covered in red sandy soil. The Mandya District's soils range from red sandy loams to red clay loams, which are thin in hills and higher elevations but thick in valleys. The topographical elevation of study area ranges from 392 to 1175 m above mean sea level. According to the Koppen climatic classification, Mysore and Mandya have a tropical savanna climate that borders on a hot semi-arid climate. The district has moderate summer temperatures, up to 35 °C, and moderate winter temperatures, down to 20 °C. The average annual rainfall of study area ranges from 650 to 700 mm. In this study, 178 observation wells are considered monitored by Central Ground Water Board (CGWB). The groundwater level fluctuates from 0.4 to 19.6 m/bgl, according to CGWB data. The study area majorly consists of both urban as well as agricultural irrigated land.

The behavior and flow of groundwater sources are influenced by a number of variables. Their impact varies by region, therefore identifying the proper one is critical for determining the possible groundwater zone. Literature suggest the goal of

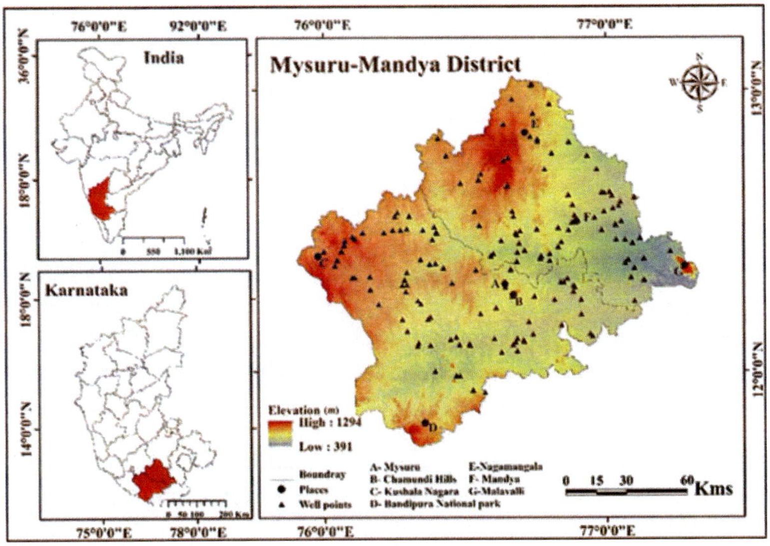

Fig. 1 Map of research area

the research, the nature of the region, and their availability influence their selection for investigation [21, 35].

The MCDM technique, i.e., the AHP utilized in this research for groundwater potential zone delineation, can be simplified down into five steps: thematic layer processing; pair-wise comparative matrix; weight calculation; consistency ratio; GWPZ map and validation.

2.1 Thematic Layer Processing

A total of 15 layers of information were collected for this study's groundwater potential zone delineation, including Geology, Geomorphology, Slope, Aspect, Elevation, Rainfall, Lineament Density, LULC, NDVI, Profile Curvature, Plan Curvature, Soil, TWI, Stream Power Index, and Slope length. Table 1 shows data used for estimating groundwater potential.

2.2 Analytic Hierarchy Process

In 1980, Saaty brought out the Analytic Hierarchy Process (AHP) approach. AHP is a problem-solving method that divides issues into groups and then arranges them in a hierarchical framework. To select priority criteria, this technique combines a comparison of criteria with a pre-determined measuring scale. The major input of the AHP technique is expert perception; hence, subjectivity plays a role in retrieval decisions (Putra et al. 2018). To identify a potential groundwater zone, multi-parameters have been examined. Each parameter's weights were computed based on their respective relevance in the decision-making process for determining groundwater zones. AHP's essential properties include hierarchical formulation, which is cost-effective and time-consuming, as well as exact outcomes [8].

AHP's adaptability for expert-based factor ordering and flexibility in reordering the rank in the case of decreased accuracy are its two main advantages. The weight of each parameter is computed by using the pairwise comparative matrix (PCM) base of judgment construction. In order to establish the weight of variables and give sub-classes a ranking in the current study, a mathematically based AHP method was employed. To remove biases in parameter ranking, a comprehensive research investigation was conducted. Together, the aforementioned elements can be used to identify a potential groundwater zone. PCM was created using a 1-to-9-point scale (Table 2). Based on each factor's significance and impact, AHP utilized PCM to evaluate each. Weights establish the circumstance or relevance degree of variables in a hierarchy. As a consequence, the comparison matrix was used to calculate the weights of the judgment-based parameters, which were then normalized. Normalized pairwise comparative matrix (NPCM) is made up of the quotients produced by dividing the sum and each judgment value in the same columns (Table 3). The result

Table 1 Description of database used for study

Data used	Source of data	Data type	Scale
Well location and groundwater level	India water resources information system (https://indiawris.gov.in)	Point	–
Rainfall	India meteorological department (https://mausam.imd.gov.in/)	Grid	0.25°*0.25°
Geomorphology	Geological survey of India-Bhukosh (https://bhukosh.gsi.gov.in)	Polygon	1:250,000
Geology	Geological survey of India-Bhukosh (https://bhukosh.gsi.gov.in)	Polygon	1:250,000
Elevation (DEM)	SRTM (https://dwtkns.com/srtm30m/)	Grid	30 m
Satellite image	Landsat-8 (https://earthexplorer.usgs.gov/)	Grid	30 m
Lineament	Geological survey of India-Bhukosh (https://bhukosh.gsi.gov.in)	Polygon	1:250,000
LULC	ESRI-ArcGIS (https://www.arcgis.com/)	Grid	10 m (resampled to 30 m)
Soil	National Bureau of Soil Survey (https://nbsslup.icar.gov.in/)	Polygon	1:500,000

was divided by total number of parameters in PCM after quotient values in each row were summed (Fig. 2).

A certain degree of consistency can exist during the paired evaluation of variables in AHP. As a result, it is important to assess the pairwise evaluation's formal consistency. Consistency level of weight of parameters is measured using consistency ratio (CR). Comparing the consistency index (CI) to the random consistency

Table 2 Rating scale of Satty's analytical hierarchical process

1/9	1/7	1/5	1/3	1	3	5	7	9
Extreme	**Very Strong**	**Strong**	**Moderate**	**Equal**	**Moderate**	**Strong**	**Very Strong**	**Extreme**
Less important ←				**Equal**	**More important** →			

Source Saaty [30]

Table 3 Random consistency index based on Saaty's scale

Number of variables	5	6	7	8	9	10	12	14	**15**
RI	1.11	1.25	1.35	1.40	1.45	1.49	1.51	1.57	**1.59**

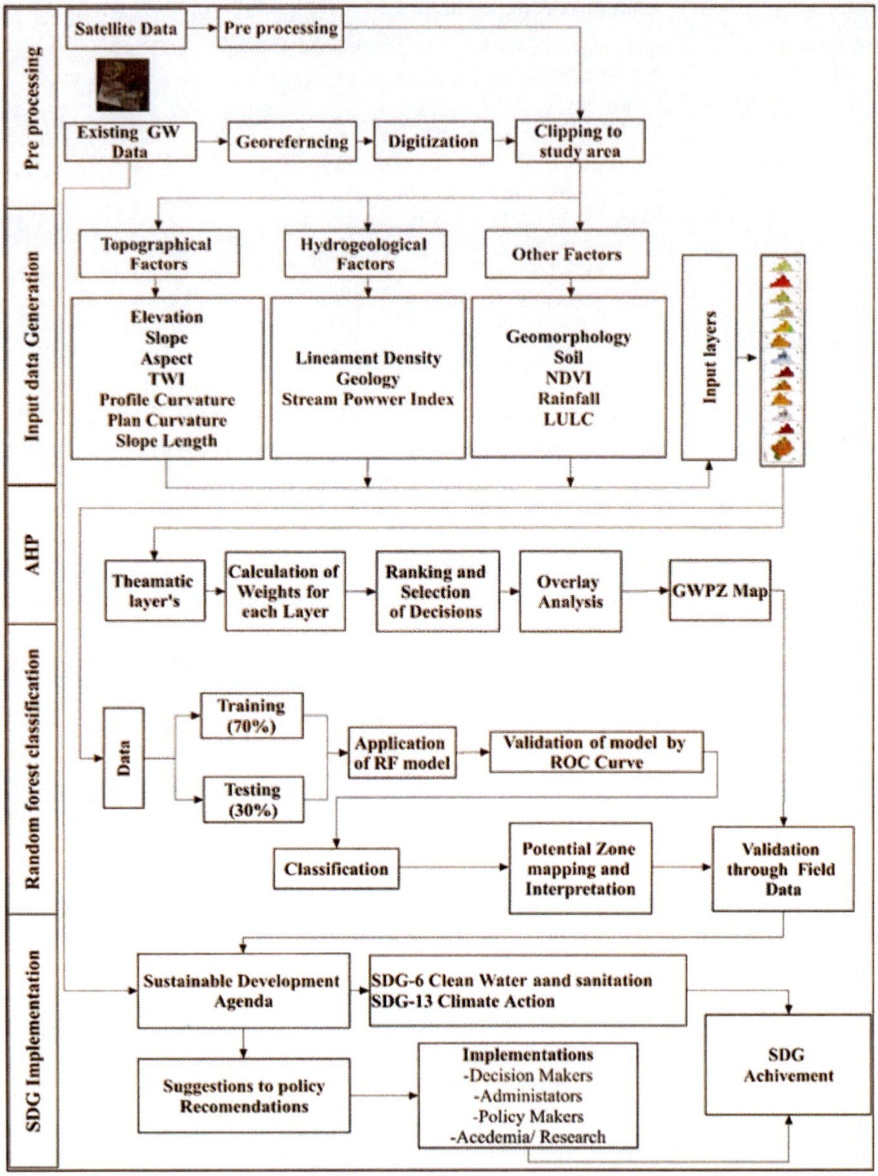

Fig. 2 Method adopted for the study

index is consistency ratio

$$\mathrm{CI} = \frac{(\lambda_{\max} - n)}{n - 1} \quad (1)$$

$$CI = \frac{(16.731 - 15)}{15 - 1} = 0.12$$

$$\mathrm{CR} = \frac{\mathrm{CI}}{\mathrm{RCI}} \quad (2)$$

$$\mathrm{CR} = \frac{0.12}{1.59} = 0.07 < 0.1$$

In the above equation, $\lambda_{\max}$ represents principal eigenvalue of matrix and n denotes number of factors used in estimation.

The consistency ratio is used to demonstrate the accuracy of weights obtained using NPCM. A judgment must be changed if the CR is larger than 0.10 even though it is considered more accurate if it is less than 0.10. The consistency ratio for this study was 0.07, which was judged suitable for further investigation. To create a groundwater potential zone map of the research region, all 15 thematic layers were integrated using weighted overlay analysis using a GIS platform.

$$\mathrm{GWPI} = \sum_{i}^{n} (Xa \times Yb) \quad (3)$$

where GWPI stands for Groundwater Potential Index, X is thematic layers' total weight, and Y is rank of the thematic layers' sub-class. The theme map is represented by a term (a = 1, 2, 3, ... X) and thematic map classes are represented by b term (b = 1, 2, 3, ... Y). The final groundwater potential zone map was divided into three categories: low, moderate, and high (Tables 4 and 5).

3 Results and Discussion

Geology, Geomorphology, Slope, Aspect, Elevation, Rainfall, Lineament Density, LULC, NDVI, Profile Curvature, Plan Curvature, Soil, TWI, Stream Power Index, and Slope length are among 15 criteria that are carefully evaluated to map distinct groundwater potential areas of study area. Figure 3 shows various thematic layers considered for the study. The following section is a detailed analysis of all of these factors (Table 6).

Table 4 Pairwise comparison matrix

Criteria	Aspect	Elevation	Geology	Geomorphology	LD	LS	LULC	NDVI	Plane curvature	Profile curvature	Slope	Soil	SPI	TWI	Rainfall
Aspect	1	0.33	0.33	0.33	3	3	3	3	3	5	7	7	9	5	3
Elevation	3	1	3	0.33	3	3	3	5	3	5	9	9	9	9	5
Geology	3	0.33	1	0.33	3	3	3	3	3	7	7	5	9	7	3
Geomorphology	3	3	3	1	3	3	3	5	3	9	9	9	9	9	5
LD	0.33	0.33	0.33	0.33	1	3	0.33	3	0.33	3	5	3	7	3	0.33
LS	0.33	0.33	0.33	0.33	0.33	1	0.33	3	0.33	3	3	3	3	3	0.33
LULC	0.33	0.33	0.33	0.33	3	3	1	3	5	7	5	7	9	3	3
NDVI	0.33	0.2	0.33	0.2	0.33	0.33	0.33	1	0.33	3	3	3	3	3	0.2
Plane curvature	0.33	0.33	0.33	0.2	3	3	0.2	3	1	5	7	3	3	3	3
Profile curvature	0.2	0.2	0.14	0.11	0.33	0.33	0.14	0.33	0.2	1	3	3	3	3	0.33
Slope	0.14	0.11	0.14	0.11	0.2	0.33	0.2	0.33	0.14	0.33	1	3	3	0.33	0.33
Soil	0.14	0.11	0.2	0.11	0.33	0.33	0.14	0.33	0.33	0.33	0.33	1	3	0.2	0.2
SPI	0.11	0.11	0.11	0.11	0.14	0.33	0.11	0.33	0.33	0.33	0.33	0.33	1	0.33	0.2
TWI	0.2	0.11	0.14	0.11	0.33	0.33	0.33	0.33	0.33	0.33	3	5	3	1	0.33
Rainfall	0.33	0.2	0.13	0.2	3	3	0.33	5	0.33	3	3	5	5	3	1

Table 5 Normalized pairwise comparison matrix

Criteria	Aspect	Elevation	Geology	Geomorphology	LD	LS	LULC	NDVI	Plane curvature	Profile curvature	Slope	Soil	SPI	TWI	Rainfall	Weights
Aspect	0.08	0.05	0.03	0.08	0.12	0.11	0.19	0.08	0.13	0.10	0.11	0.11	0.11	0.09	0.12	0.10
Elevation	0.23	0.14	0.30	0.08	0.12	0.11	0.19	0.14	0.13	0.10	0.14	0.14	0.11	0.17	0.20	0.15
Geology	0.23	0.05	0.10	0.08	0.12	0.11	0.19	0.08	0.13	0.13	0.11	0.08	0.11	0.13	0.12	0.12
Geomorphology	0.23	0.43	0.30	0.24	0.12	0.11	0.19	0.14	0.22	0.17	0.14	0.14	0.11	0.17	0.20	0.19
LD	0.03	0.05	0.03	0.08	0.04	0.11	0.02	0.08	0.01	0.06	0.08	0.05	0.09	0.06	0.01	0.05
LS	0.03	0.05	0.03	0.08	0.01	0.04	0.02	0.08	0.01	0.06	0.05	0.05	0.04	0.06	0.01	0.04
LULC	0.03	0.05	0.03	0.08	0.12	0.11	0.06	0.08	0.22	0.13	0.08	0.11	0.11	0.06	0.12	0.09
NDVI	0.03	0.03	0.03	0.05	0.01	0.01	0.02	0.03	0.01	0.06	0.05	0.05	0.04	0.06	0.01	0.03
Plane curvature	0.03	0.05	0.03	0.05	0.12	0.11	0.01	0.08	0.04	0.10	0.11	0.05	0.04	0.06	0.12	0.07
Profile curvature	0.02	0.03	0.01	0.03	0.01	0.01	0.01	0.01	0.01	0.02	0.05	0.05	0.04	0.06	0.01	0.02
Slope	0.01	0.02	0.01	0.03	0.01	0.01	0.01	0.01	0.01	0.01	0.02	0.05	0.04	0.01	0.01	0.02
Soil	0.01	0.02	0.02	0.03	0.01	0.01	0.01	0.01	0.01	0.01	0.01	0.02	0.04	0.00	0.01	0.01
SPI	0.01	0.02	0.01	0.03	0.01	0.01	0.01	0.01	0.01	0.01	0.01	0.00	0.01	0.01	0.01	0.01
TWI	0.02	0.02	0.01	0.03	0.01	0.01	0.02	0.01	0.01	0.01	0.05	0.08	0.04	0.02	0.01	0.02
Rainfall	0.03	0.03	0.03	0.05	0.12	0.11	0.02	0.14	0.01	0.06	0.05	0.08	0.06	0.06	0.04	0.06

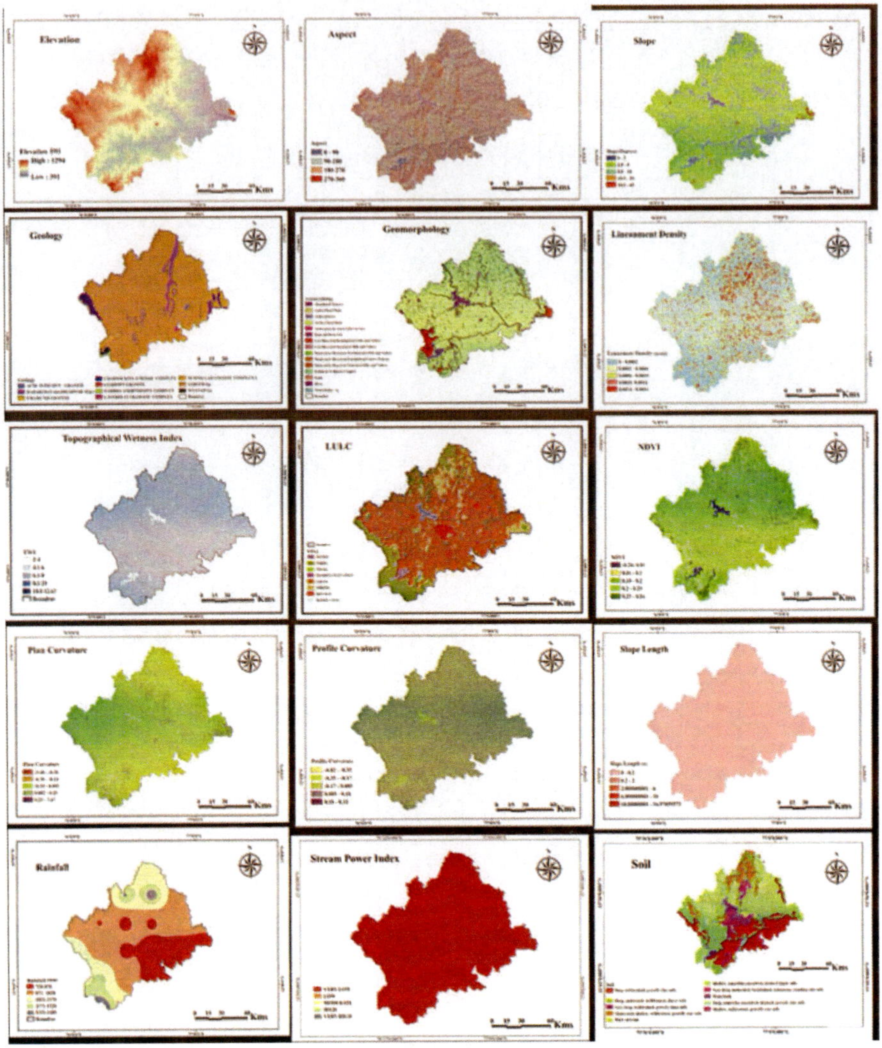

Fig. 3 Selected thematic layers of the groundwater affecting factors **a** Elevation, **b** Aspect, **c** Slope, **d** Geology, **e** Geomorphology, **f** Lineament density, **g** TWI, **h** LULC, **i** NDVI, **j** Plan curvature, **k** Profile curvature, **l** Slope length, **m** Rainfall, **n** Stream power index, **o** Soil

3.1 Assessment of Groundwater Potential Zones

Groundwater potential zone assessment and mapping using Geoinformatics is widely practiced in India and around the world [3]. The key to achieving reasonable results is precisely assigning weights [4]. The governing elements of groundwater potential are determined by study region and obviously, the available information. Thematic layers were generated from vector data based on these factors. The accuracy of

Table 6 Percentage of influencing factor based on Saaty's analytical hierarchical process

Influencing factor	% of influence
Geomorphology	19
Elevation	15
Geology	12
Aspect	10
LULC	9
Plan curvature	7
Rainfall	6
Lineament density	5
Slope length	3
NDVI	3
Profile curvature	3
Slope	3
TWI	2
Soil	2
Stream power index	1

Table 7 Classification of groundwater potential zone

Groundwater potential zone class	Area covered in km^2	Area in percentage
High	3240.85	27.43
Moderate	7775.45	65.81
Low	795.51	6.75

map improves as number of layers increases [15]. The prospective map's accuracy depends on how accurately each layer's weight is assigned [22].

To determine potential zone in research area, 15 criteria were integrated using a GIS and AHP-based approach. In order to produce geospatial mapping of groundwater potential in the WOA, the impact of these characteristics was initially assessed using PCM and AHP. Finally, the raster map output was produced. The groundwater potential zones have been reclassified into three classes based on the weight average: high groundwater potential zone, moderate groundwater potential zone, and low groundwater potential zone. The results are shown in Table 7 and Fig. 4 illustrates the groundwater potential spatial variability.

3.2 Validation of Groundwater Potential Zone

The groundwater potential map was created using a Geoinformatics and AHP-based approach using information that was accessible on several variables. However, the

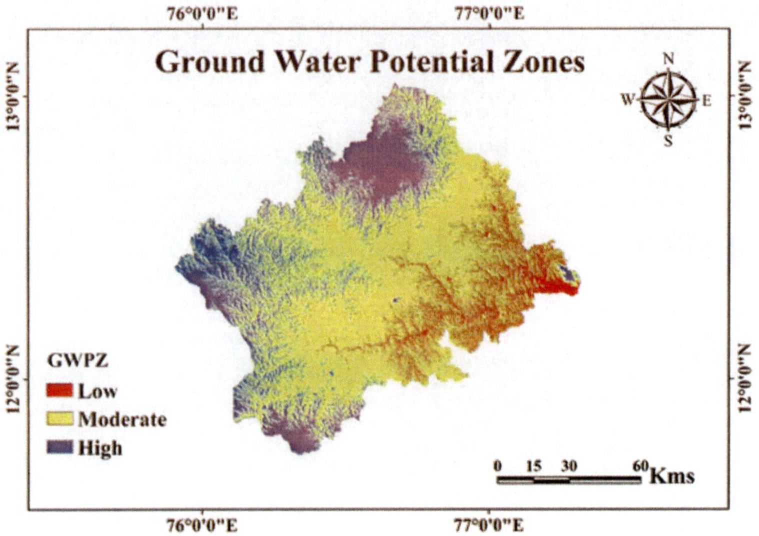

Fig. 4 Groundwater potential zone

outcomes must be verified utilizing real data that were gathered from ground. To validate generated groundwater potential maps, researchers used a number of different methods. To compare potential zone with field data, authors, [19] and [3], used depth to water level. In this investigation, which included 178 boreholes, the groundwater level was compared with groundwater potential mapped using a GIS-based approach. The groundwater data from field are divided into three categories: high level (0–5 mbgl), moderate level (5–10 mbgl), and low level (more than 20 mbgl). Table 8 shows the procedure in detail as well as comparison results.

Table 8 Validation of potential zones with CGWB actual information (Sample table showing 10 reference points)

Serial number	Latitude	Longitude	GWL	Zone	Evaluation
1	12.48333	76.68333	1.05	High	Positive
2	12.38889	76.68333	1.40	High	Positive
3	12.45833	76.84167	7.77	Moderate	Positive
4	12.18333	76.27500	1.45	High	Positive
5	12.81667	76.75833	2.12	High	Positive
6	12.40000	77.13333	4.80	High	Positive
7	12.45472	77.09306	4.99	High	Positive
8	12.28333	76.88333	2.54	High	Positive
9	12.33278	77.08750	7.14	High	Negative
10	12.33611	76.16111	8.13	High	Negative

Based on information about groundwater levels acquired from the Central Groundwater Board, the accuracy evaluation of groundwater potential zone map was carried out. There are 178 wells with a range of levels from 0 to 19.6 mbgl. These wells were categorized into three classes: high (0–5 mbgl), moderate (5–10 mbgl), and low (0–5 mbgl) (more than 10 mbgl). Model validation of predicted groundwater potential zones with groundwater level data indicates an accuracy of 80.4%, with around 142 wells meeting the predicted groundwater potential zones. It indicates that generated groundwater potential map agrees well with ground data.

The accuracy of correlation between study results and data from CGWB is acceptable. The location of groundwater-level observation point was overlain on groundwater potential zone map for validation. Following that, a scatter plot was produced between the groundwater level below ground and the related groundwater potential index. Figure 5 shows a regression plot depicting the association between GWPI and GWL. The high coefficient of determination (R^2) obtained indicates that two variables are well correlated. The R^2 value 0.8, further, implies that the suggested AHP model's estimated GWPI and GWL values are very reliable in predicting outcome. As a result, AHP-based groundwater potential zonation was successfully used, and it may be utilized as a realistic groundwater exploring strategy.

The utilization, management, and sustainability of groundwater are the three main aspects considered as essential factors to achieve SDGs. SDG 6 and 13 were included in the approach used to examine groundwater interlinkages with SDG objectives such as "Ensure availability and sustainable management of water and sanitation for everyone," sixth Sustainable Development Goal (SDG). The emphasis on data and governance is one of the most important aspects of the SDG agenda. SDG 13 implies

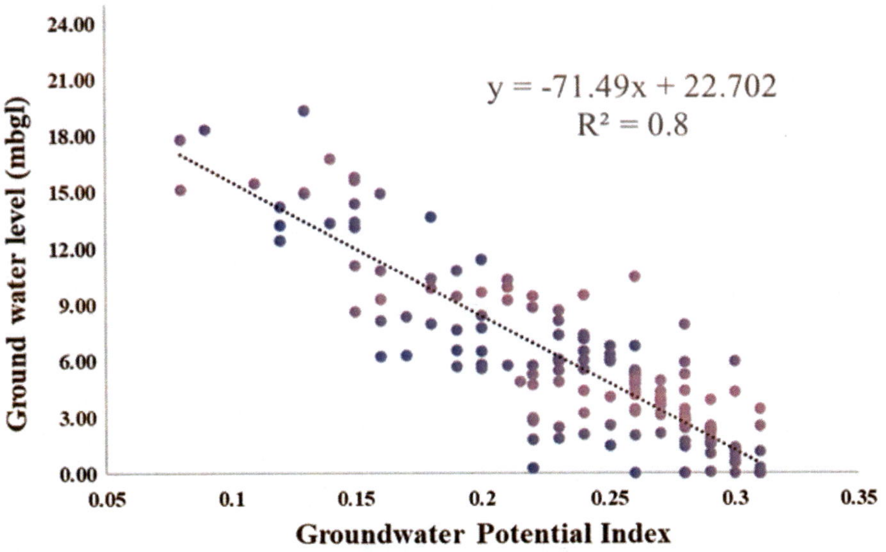

Fig. 5 Scatter plot between GWPI and GWL

to "Take immediate action to address climate change and its effects," which is treated to be the critical goals among others. These goals were successfully addressed in this research effort. The comprehensive AHP method may be taken into consideration as a well-established resource for groundwater strategy and management that will help to conserve water in line with SDGs.

4 Conclusion

In the southern dry agro-climatic region including Mysuru-Mandya, the comprehensive use of remote sensing and Geoinformatics for delineating groundwater potential zones has shown to be labor, time, and cost-effective. Based on the literature review, Geology, Geomorphology, Slope, Aspect, Elevation, Rainfall, Lineament Density, LULC, NDVI, Profile Curvature, Plan Curvature, Soil, TWI, Stream Power Index, and Slope Length are the input maps, created to estimate the groundwater potential zone for study area using weighted overlay in GIS. The 15 thematic layers were created using satellite data, geographical maps, and additional data sources. According to the results, the entire study area is divided into three groundwater potential zones: high (27.43%), moderate (65.81%), and low potential zone (6.75%). The majority of the study region lies in the moderate groundwater potential zone. The delineated groundwater potential zone map was validated using the groundwater level details about the research area given by the CGWB. The potential zone map from current study is provided to decision-makers for the optimal groundwater management. For agricultural and urban uses, the current study gives information to decision-makers for optimum groundwater management and planning. The study region is comprised of agricultural land, identified to cover larger land surface among other classes; therefore, this research will help to augment irrigation infrastructure and increase the region's agricultural production. This research can be extended by employing sophisticated machine learning techniques to improve and achieve enhanced results. In order to support the realization of the SDGs, this study will be helpful as a baseline for planning sustainable water resource management in the southern dry agro-climatic region in the coming years.

References

1. Agarwal R, Garg PK (2016) Remote sensing and GIS based groundwater potential and recharge zones mapping using multi-criteria decision making technique. Water Resour Manage 30(1):243–260
2. Al-Shabeeb A, et al (2018) Delineating groundwater potential zones within the Azraq Basin of Central Jordan using multi-criteria GIS analysis. Groundwater Sustain Dev 7:82–90
3. Arulbalaji P, Padmalal D, Sreelash K (2019) GIS and AHP techniques based delineation of groundwater potential zones: a case study from southern Western Ghats, India. Sci Rep 9(1):1–17

4. Arunbose S et al (2021) Remote sensing, GIS and AHP techniques based investigation of groundwater potential zones in the Karumeniyar river basin, Tamil Nadu, southern India. Groundw Sustain Dev 14:100586
5. Asoka A, et al 2018) Strong linkage between precipitation intensity and monsoon season groundwater recharge in India. Geophys Res Lett 45(11):5536–5544
6. Banks D, Robins N, Robins N (2002) An introduction to groundwater in crystalline bedrock. Norges geologiske undersøkelse, Trondheim
7. Chowdhury A et al (2009) Integrated remote sensing and GIS-based approach for assessing groundwater potential in West Medinipur district, West Bengal, India. Int J Remote Sens 30(1):231–250
8. Das S (2019) Comparison among influencing factor, frequency ratio, and analytical hierarchy process techniques for groundwater potential zonation in Vaitarna basin, Maharashtra, India. Groundw Sustain Dev 8:617–629
9. Díaz-Alcaide S, Martínez-Santos P (2019) Advances in groundwater potential mapping. Hydrogeol J 27(7):2307–2324
10. Fitterman DV, Stewart MT (1986) Transient electromagnetic sounding for groundwater. Geophysics 51(4):995–1005
11. Ghorbani Nejad S, et al (2017) Delineation of groundwater potential zones using remote sensing and GIS-based data-driven models. Geocarto Int 32(2):167–187
12. Guru B, Seshan K, Bera S (2017) Frequency ratio model for groundwater potential mapping and its sustainable management in cold desert, India. J King Saud Univ-Sci 29(3):333–347
13. Hammouri N, et al (2014) Groundwater recharge zones mapping using GIS: a case study in Southern part of Jordan Valley, Jordan. Arab J Geosci 7(7):2815–2829
14. Jha MK, et al (2007) Groundwater management and development by integrated remote sensing and geographic information systems: prospects and constraints. Water Resour Manage 21(2):427–467
15. Khoshand A, Kamalan H, Rezaei H (2018) Application of analytical hierarchy process (AHP) to assess options of energy recovery from municipal solid waste: a case study in Tehran, Iran. J Mater Cycles Waste Manage 20(3):1689–1700
16. Khosravi K, et al (2018) A comparison study of DRASTIC methods with various objective methods for groundwater vulnerability assessment. Sci Total Environ 642:1032–1049
17. Krishna Kumar S et al (2012) Hydrogeochemical study of shallow carbonate aquifers, Rameswaram Island, India. Environ Monit Assess 184(7):4127–4138
18. Kumar D, Ananda Rao V, Sarma VS (2014) Hydrogeological and geophysical study for deeper groundwater resource in quartzitic hard rock ridge region from 2D resistivity data. J Earth Syst Sci 123(3):531–543
19. Kumar S, Raizada A, Biswas H, Muralidhar W, Rao KS (2016) Groundwater extraction in Karnataka and its long term implications. Indian J Econo Dev 12(4):615–628
20. Lee S, Hong S-M, Jung H-S (2018) GIS-based groundwater potential mapping using artificial neural network and support vector machine models: the case of Boryeong city in Korea. Geocarto Int 33(8):847–861
21. Lee S, Kim Y-S, Oh H-J (2012) Application of a weights-of-evidence method and GIS to regional groundwater productivity potential mapping. J Environ Manage 96(1):91–105
22. Machiwal D, Jha MK, Mal BC (2011) Assessment of groundwater potential in a semi-arid region of India using remote sensing, GIS and MCDM techniques. Water Resour Manage 25(5):1359–1386
23. Mukherjee S, et al (2017) Increase in extreme precipitation events under anthropogenic warming in India. Weather and Climate 20:1–9
24. Naghibi SA, Ahmadi K, Daneshi A (2017) Application of support vector machine, random forest, and genetic algorithm optimized random forest models in groundwater potential mapping. Water Resour Manage 31(9):2761–2775
25. Naghibi SA, et al (2015) Groundwater qanat potential mapping using frequency ratio and Shannon's entropy models in the Moghan watershed, Iran. Earth Sci Inform 8(1):171–186

26. Oh H-J, et al (2011) GIS mapping of regional probabilistic groundwater potential in the area of Pohang City, Korea. J Hydrol 399(3–4):158–172
27. Patra S, Mishra P, Mahapatra SC (2018) Delineation of groundwater potential zone for sustainable development: a case study from Ganga Alluvial Plain covering Hooghly district of India using remote sensing, geographic information system and analytic hierarchy process. J Clean Pro 172:2485–2502
28. Pradhan B (2009) Groundwater potential zonation for basaltic watersheds using satellite remote sensing data and GIS techniques. Central Eur J Geosci 1(1):120–129
29. Rahmati O, Pourghasemi HR, Melesse AM (2016) Application of GIS-based data driven random forest and maximum entropy models for groundwater potential mapping: a case study at Mehran Region, Iran. Catena 137:360–372
30. Saaty TL (1980) The analytic hierarchy process: planning, priority setting, resources allocation. McGraw-Hill, New York
31. Saraf AK, Choudhury PR (1998) Integrated remote sensing and GIS for groundwater exploration and identification of artificial recharge sites. Int J Remote Sens 19(10):1825–1841
32. Senanayake IP et al (2016) An approach to delineate groundwater recharge potential sites in Ambalantota, Sri Lanka using GIS techniques. Geosci Front 7(1):115–124
33. Shahid S, Nath SK, Roy J (2000) Groundwater potential modelling in a soft rock area using a GIS. Int J Remote Sens 21(9):1919–1924
34. Tahmassebipoor N, et al (2016) Spatial analysis of groundwater potential using weights-of-evidence and evidential belief function models and remote sensing. Arab J Geosci 9(1):1–18
35. Verma N, Patel RK (2021) Delineation of groundwater potential zones in lower Rihand River Basin, India using geospatial techniques and AHP. The Egypt J Remote Sens Space Sci 24(3):559–570

Evaluation and Prediction of Land Use and Land Cover Changes in the Kumaradhara Basin, Western Ghats, India

N. Roopa, N. Namratha, H. Ramesh, and K. C. Manjunath

Abstract Land use land cover (LULC) is considered as the most significant and obvious indicator of changes in ecosystems. An understanding of current and potential future development opportunities is provided through analysis on the spatiotemporal shifting patterns of LULC and simulation of future scenarios. Kumaradhara river flows in the Western Ghats in southern peninsular, India. It is the major tributary of the Netravathi river, and the catchment has numerous perennial streams and is dominated by dense evergreen forests with high conservation value. In the present study, an integrated approach of remote sensing and geospatial techniques is used to assess LULC changes for the period of 2010–2020, and prediction of future LULC change was carried out by ANN model using MOLUSCE plugin of QGIS for the year 2025. The results have shown that the build-up land has increased considerably, and forest has decreased which is evident from the increase in cultivated land. The predicted LULC showed an increase in built-up land and a significant transformation of barren land. The results of this study indicate significant changes in the LULC pattern.

Keywords Land use/Land cover · Predicted LULC maps · MOLUSCE (ANN multilayer perception model) · Kumaradhara basin

N. Roopa · H. Ramesh
Department of Water Resources and Ocean Engineering, National Institute of Technology Karnataka, Surathkal, India

N. Namratha (✉)
Department of Hydraulics Engineering, The National Institute of Engineering, Mysuru, India
e-mail: namrathamys1998@gmail.com

K. C. Manjunath
Department of Civil Engineering, The National Institute of Engineering, Mysuru, India

S. Mesapam et al. (eds.), *Developments and Applications of Geomatics*,
Lecture Notes in Civil Engineering 450, https://doi.org/10.1007/978-981-99-8568-5_15

1 Introduction

The landscape structure or land cover of a river basin's watershed determines its hydrological yield. Anthropogenic and natural forces both contribute to the current changes in the earth's land surface [1]. These factors have an impact LULC change directly or indirectly. Anthropogenic activities include deforestation, urbanization, overgrazing, and conversion to agricultural land. These are the primary proximal reasons for LULC change [2]. Earthquakes, landslides, droughts, and floods are the natural factors that affect the land use and land cover [3].

LULC changes affect an area's climate, which in turn impacts on natural resources such as water, wetlands, and biodiversity [4]. It is also known to have an impact on the hydrology of any catchment area. The LULC patterns are greatly influenced by climatic factors such as rainfall, temperature, and these factors need long-term research to fully comprehend the trend.

LULC change detection is a complex function and requires different methods of analysis based on spatial and social data in order to understand the drivers and impacts of change with time. The LANDSAT imagery archive runs from 1973 to the present (LANDSAT 8) and contains the most recent satellite data that are reasonable with prior collective years [5, 6]. Remote Sensing and Geographic Information Systems (GIS) when compared to ground-based surveys are the most cost-effective techniques for mapping and assessing LULC changes [7, 8]. Modules of land-use change evaluation (MOLUSCE), a Quantum GIS (QGIS) plugin is used by most of the researchers to assess prospective LULC changes using CA model and also contains a probability of a transition matrix [9].

Western Ghats (WG) of India is one of the 34 global biodiversity hotspots and habitat loss is a wreaking havoc for the past few decades [10]. Along the west side of the Western Ghats, the Kumaradhara river is a significant feature of rugged forested terrain. It is the major tributary of the Netravathi river and the catchment has numerous perennial streams and is dominated by dense evergreen forests with high conservative value [11]. The availability of relevant historical satellite datasets makes quantifying LULC change very simple, however, identifying the causes for such change is difficult. The climate, magnitude, time, ecological, social, political institutions, economies, and international protocols all influence the drivers [12]. Studying the land change pattern and its mechanism will help one to further analyze the effects of changes in the hydrology of a catchment, which affects various hydrological cycle metrics.

The aim of the present study is to adopt geospatial and remote sensing techniques to evaluate and track LULC shifts for three different years in the Kumaradhara river basin. The present paper attempts to evaluate LULC changes between various classes from 2010 to 2020, forecast LULC change, and prepare a prediction map for 2025. The study also highlights the importance of evaluating LULC change and its applications for further studies.

2 Materials and Methodology

2.1 Study Area

Kumaradhara river is the tributary of the Netravathi river, which has its origin in Western Ghats of Karnataka, flowing to a length of 126 km and joins the Arabian Sea on the west. Kumaradhara river basin extends from 12°29′4″ N to 12°58′33″ N latitude and 75°95′8″ E to 75°47′48″ E longitude with the catchment area of 1776 km^2, originates in Central Western Ghats at an elevation of 1480AMSL [10]. Kumaradhara basin is rich in biodiversity and the freshwater ecosystems are extremely diverse, distinctive, and vital to livelihoods and economies. It spreads across three districts, Hassan, Kodagu, and Dakshina Kannada. The catchment is shown in Fig. 1.

Kumaradhara river has tributaries, namely, Hongadahalla and Kadumanehalla rivers, and sub-catchments covering these parts of the Mookanamane and Marenahalli rivers are chosen for the present study. Mookanamane catchment is flanked by Bisle and Kaginele forests on either side and covers an area of 42.24 km^2, which extends from 12°45′0″ N to 12°49′30″ N latitude and 75°43′30″ E to 75°46′30″ E longitude with an elevation of 1179 m AMSL, the annual average rainfall in this area is 2978 mm, and annual average maximum and minimum temperatures are 28 °C and 19 °C, respectively. Marenahalli catchment covers an area of 64.19 km^2 extending from 12°52′30″ N to 12°57′30″ N latitude and 75°40′0″ E to 75°46′30″ E longitude with an elevation of 1293 m AMSL, the annual average rainfall, maximum, and

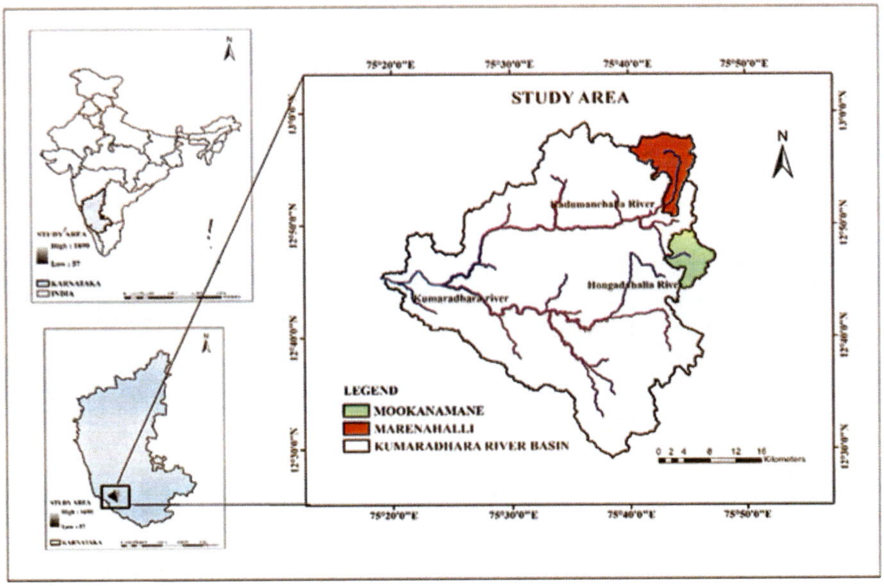

Fig. 1 Study area

minimum temperatures are 3439.4 mm, 29 °C, and 26 °C. The study area mainly covers forests and landscapes associated with Kumaradhara river.

2.2 *Data Used in the Study*

Remotely sensed LANDSAT data were used to discriminate between the various LULC cover categories. The fluctuations in the Kumaradhara river catchment were monitored using the pictures acquired from the USGS Earth Explorer website. The ETM+ sensor of LANDSAT 7 can produce images for 2010 and 2015, and LANDSAT 8 Operational Land Imager (OLI) and TIRS CI-Level 1 sensor can produce images for 2020. For the months of February and March in 2010, 2015, and 2020, all these images were downloaded. Due to the region's harsh weather conditions and significant seasonal variations in LULC, especially vegetation, these months were chosen for data collection.

The following characteristic data were picked for the study, digital elevation model (DEM), slope maps, aspect maps, and distance from road maps. The road map was downloaded from the website of open street maps. Table 1(a) and 1(b) provides specifics of the original source of datasets.

Table 1 a Source of datasets. b LANDSAT data used for the study

Data	Criteria	Description	Source
(a): Source of datasets			
Digital elevation model	DEM Gradient Aspect	SRTM with 30 m spatial resolution	National remote sensing centre/ Bhuvan/DEM/ elevation
Road map	Distance from road map	Road map of Kumaradhara river basin	Open street map

Year	Satellite ID	Sensor ID	Path	Row	Acquisition date
(b): LANDSAT data used for the study					
2010	LANDSAT 7	ETM+	145	051	18/02/2010
2015	LANDSAT 7	ETM+	145	051	27/03/2015
2020	LANDSAT 8	OLI and TIRS	145	051	09/03/2020

2.3 Methodology

2.3.1 Preparation of LULC Maps

The LULC classification method depends on the nature of the study area, spatial, and temporal resolution of the data [13]. The LANDSAT images downloaded from USGS website are georeferenced to Universal Traverse Mercator (UTM) projection system, 43° N WGS 84, which was used throughout the study. The satellite images are imported to QGIS and False-color composite (FCC) was created using the bands 4, 3, and 2 for LANDSAT 7 and bands 5, 4, and 3 for LANDSAT 8. Radiometric corrections are applied to reduce the inconsistencies of the image and to enhance the quality of the satellite images.

2.3.2 Classification and Accuracy Assessment

The LULC maps are classified into five different classes, water bodies, forest, cultivated land, built-up land, and barren land, using the LANDSAT level 1 classification in the present study. A maximum likelihood algorithm is used to determine the number of land cover types and the training pixels for each of the desired classes using land cover information. The description of the land use classes and land cover statistics is given in Table 2. The adopted equations for the calculation of user's accuracy, producer's accuracy, overall accuracy, and kappa coefficient are as follows,

$$\text{Users' accuracy} = \left(\frac{xi}{xj}\right) * 100$$

where,

xi = Number of correctly classified samples in each category

xj = Total number of reference samples in that category (Row table)

$$\text{Producers' accuracy} = \left(\frac{xi}{xk}\right) * 100$$

where,

xk = Column total of the reference samples

$$\text{Overall accuracy} = \left(\frac{xd}{xT}\right) * 100$$

where,

xd = Diagonal total of correctly classified samples

xT = Total number of reference points

$$\text{Kappa coefficient} = \left(\frac{(xd * xT) - B}{(xT * xT) - B}\right)$$

where,

B = Sum of products of row total and column total for each LULC type in confusion matrix.

In the present study, a confusion matrix was produced to calculate the overall accuracy and kappa coefficient for each land use land cover category.

2.3.3 Prediction Map of 2025

The cellular automaton model describes the LULC modifications using the probabilities obtained from the transitions through ANN learning process [14]. MOLUSCE is computing land use change analysis effectively [15], and future LULC maps will be created using the plugin QGIS 2.18.24. The correlations among the spatial variable parameters are calculated using the Pearson's correlation, and then the change in LULC between the initial year (2010) and the final year (2020) is calculated. The methodology is shown in Fig. 2.

3 Results and Discussion

3.1 Comparison of Landcover Datasets

The study area was classified according to the requirements of the research work and its applications. Marenahalli catchment spreads over 64.19 km^2. The classification maps for the years 2010, 2015, and 2020 are shown in Fig. 3. Table 3 explains the percentage of area covered by each class in the Marenahalli catchment. According to 2010 results, the maximum area of the catchment, i.e., 93.79% (60.18 km^2) of the area is covered by forest. In the year 2015, it covers 92.94% (59.67 km^2) and in 2020 92.61% (59.43 km^2). Mookanamane catchment spreads over 42.24 km^2 in which 89.25% (36.80 km^2) of the total area is covered by forest in the year 2010, and in 2015 and 2020, the area covered is 87.64% (35.39 km^2), and 87.22% (35.22km^2). The classification maps of Mookanamane are shown in Fig. 4, and Table 4 explains the percentage of area covered by each class. The forest covers in these two catchments are the lands with tree canopy, evergreen, semi-evergreen vegetation types producing other forest products. Built-up land is the area developed due to non-agricultural uses and barren lands are the degraded lands devoid of vegetation.

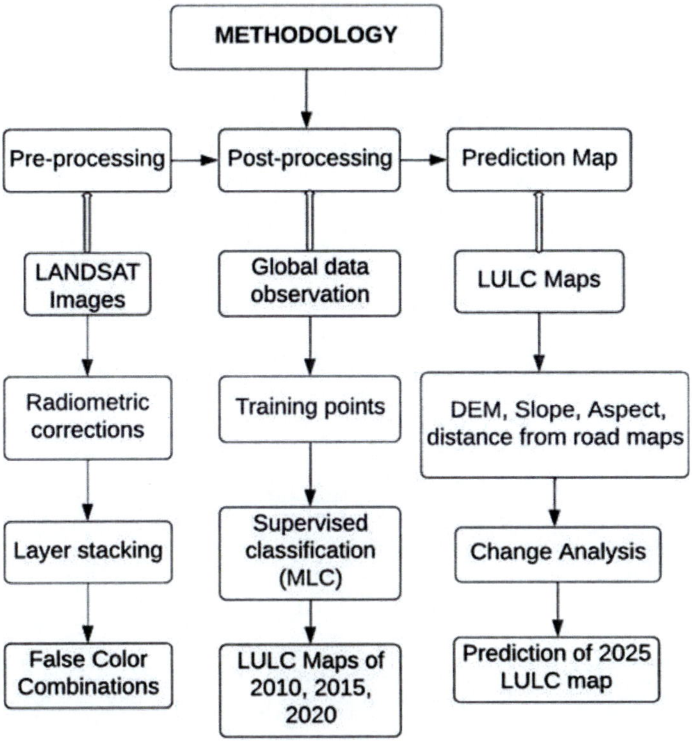

Fig. 2 Flowchart of methodology

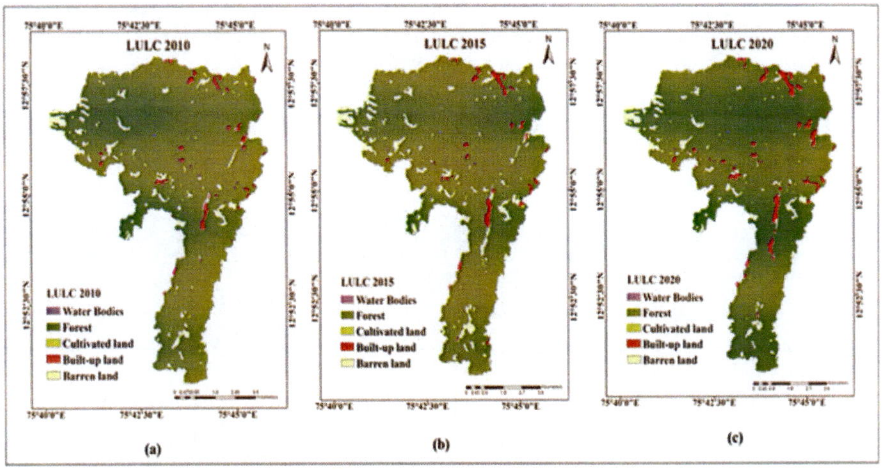

Fig. 3 LULC maps of Marenahalli, 2010 (**a**), 2015 (**b**), 2020 (**c**)

Table 2 Land use land cover categorization scheme for the study area

Serial number	Class	Description
1	Water bodies	Surface water that is retained as ponds, rivers, flowing as streams and other bodies of water
2	Forest	Land covering dense or evergreen forests, deciduous forest, tall grass, and scrublands
3	Cultivated land	A cycle of a cultivated area, involving a brief period of farming, harvest and then a return to bare soil
4	Built-up land	The land occupied with buildings and other man—made constructions and also settlements such as roads, industries etc.,
5	Barren land	Land having lack of water, soil management, and sparsely vegetated terrain and Bare rock exhibits evidence of erosion, and ground deformation

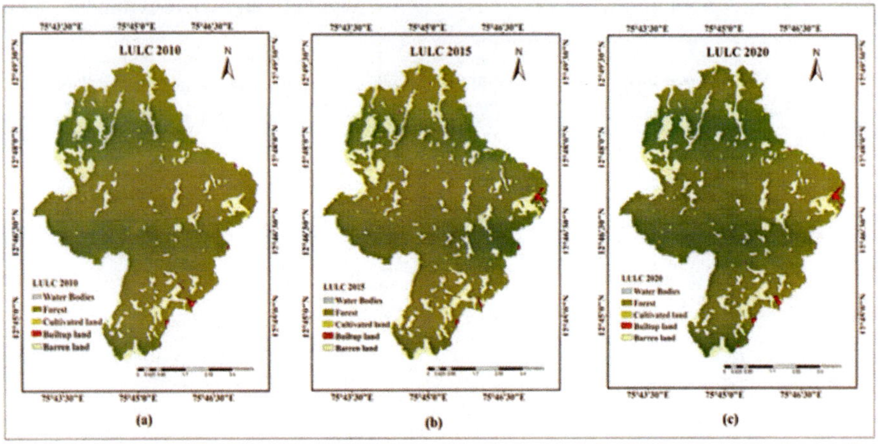

Fig. 4 LULC maps of Mookanamane, **a** 2010, **b** 2015, **c** 2020

Table 3 LULC classification of Marenahalli

LULC class	2010		2015		2020	
	km^2	%	km^2	%	km^2	%
Water bodies	0.019	0.030	0.027	0.042	0.015	0.030
Forest	60.185	93.779	59.672	92.940	59.43	92.57
Cultivated land	0.006	0.009	0.012	0.018	0.019	0.030
Built-up land	0.931	1.451	1.208	1.780	1.469	2.290
Barren land	3.035	4.730	3.257	5.207	3.235	5.090

Table 4 LULC classification of Mookanamane

LULC class	2010		2015		2020	
	km^2	%	km^2	%	km^2	%
Water bodies	0.015	0.036	0.012	0.020	0.008	0.010
Forest	36.808	89.251	35.396	87.640	35.277	87.220
Cultivated land	0.034	0.083	0.053	0.129	0.039	0.100
Built-up land	0.086	0.210	0.136	0.322	0.197	0.480
Barren land	4.296	10.418	5.246	11.870	5.627	12.190

3.2 *Accuracy Assessment and Change Analysis of Classified Images*

Accuracy assessment and change detection of land use land cover are very important for understanding the interaction and relationship between humans and nature [16]. The accuracy of LULC is calculated using various measures such as overall accuracy, user's accuracy, producer's accuracy, and Cohen's kappa coefficient [17]. In this study, a confusion matrix was produced to calculate the overall accuracy and kappa coefficient for each land use and land cover category. The value of overall accuracy and kappa coefficient is given in Table 5. According to Landis and Koch [18], Cohen's Kappa shows substantial and perfect agreement if the values fall between 0.6 and 1. In the present study, the kappa coefficients of both the catchments lie in the above range and hence it is considered as substantial agreement.

3.3 *LULC Change Detection*

The LULC classes and its changes of Marenahalli catchment in the year 2010–2015 and 2015–2020 are shown in Table 6. Water bodies and barren land exhibited both negative and positive changes whereas the cultivated land and built-up land showed positive changes, but it is observed that forest area is decreasing with negative changes. Minimal changes were observed in all other classes and built-up land is having the most substantial changes. The forest covers the maximum area of the catchment and it showed a decreasing trend. The decrease in forest is attributed to conversion of forest lands into developed area being used for various types of developmental works. The area under built-up land increased due to gradual increase in demand for shelter by inhabitants in and around Marenahalli. The area of barren land is rising due to farmer's disregard of agricultural operations and clearing of land, which suggests a clear plan for new structures and communities.

The changes in LULC classes for the years 2010–2015 and 2015–2020 are presented in Table 7 for Mookanamane catchment built-up land, barren land, and cultivated land shows an increasing trend. Similar to Marenahalli, the forest area

Table 5 Accuracy assessment

Catchment	Year	Overall accuracy	Kappa coefficient
Marenahalli	2010	91.00	0.88
	2015	91.00	0.90
	2020	89.04	0.86
Mookanamane	2010	91.02	0.88
	2015	88.05	0.85
	2020	89.00	0.86

Table 6 Change analysis in Marenahalli

Year	2010–2015			2015–2020		
	Area in km^2		Changes in %	Area in km^2		Changes in %
	2010	2015		2015	2020	
Water bodies	0.019	0.027	+0.008	0.027	0.015	−0.012
Forest	60.185	59.672	−0.513	59.672	59.43	−0.242
Cultivated land	0.006	0.012	+0.006	0.012	0.019	+0.007
Built-up land	0.931	1.208	+0.277	1.208	1.469	+0.261
Barren land	3.035	3.257	+0.222	3.257	3.235	−0.022

is decreasing. Built-up land shows an increasing trend due to gradual increase in communities of people. It has been recognized as a tourist spot due to which there is an increase in buildings, roads, and hotels. Forest areas are reducing due to human intrusion at forest areas and other settlement activities.

3.4 Forecasting Future LULC Changes

Forecasting LULC changes is important for recognizing and emphasizing probable future adjustments and modifications to the landscape [19]. The prediction is supported by the previous LULC patterns seen in the study area. In the present study, a prediction of LULC change for the year 2025 has been attempted. The variables used for the prediction are LULC maps, slope maps, aspect maps, and the distance from the road maps. The trends of the historical changes have been used for the prediction of future LULC maps. Tables 8 and 9 show the predicted land area for the year 2025 (Fig. 5).

Table 7 Change analysis in Mookanamane

Year	2010–2015			2015–2020		
	Area in km^2		Changes in %	Area in km^2		Changes in %
	2010	2015		2015	2020	
Water bodies	0.015	0.012	−0.003	0.012	0.008	−0.004
Forest	36.808	35.396	−1.412	35.396	35.277	−0.119
Cultivated land	0.034	0.053	+0.019	0.053	0.039	−0.014
Built-up land	0.086	0.136	+0.050	0.136	0.197	+0.061
Barren land	4.296	5.246	+0.950	5.246	5.627	+0.381

Table 8 Distribution of LULC classes in 2025 at Marenahalli

LULC class	2010		2015		2020		2025	
	km^2	%	km^2	%	km^2	%	km^2	%
Water bodies	0.019	0.030	0.027	0.042	0.015	0.030	0.014	0.022
Forest	60.18	93.77	59.67	92.94	59.43	92.57	59.30	92.40
Cultivated land	0.006	0.009	0.012	0.018	0.019	0.030	0.021	0.033
Built-up land	0.931	1.451	1.208	1.780	1.469	2.290	1.571	2.448
Barren land	3.035	4.730	3.257	5.207	3.235	5.090	3.305	5.150

Table 9 Distribution of LULC classes in 2025 at Mookanamane

LULC class	2010		2015		2020		2025	
	km^2	%	km^2	%	km^2	%	km^2	%
Water bodies	0.015	0.036	0.012	0.020	0.008	0.010	0.015	0.036
Forest	36.80	89.25	35.39	87.64	35.27	87.22	35.14	85.22
Cultivated land	0.034	0.083	0.053	0.129	0.039	0.100	0.0776	0.19
Built-up land	0.086	0.210	0.136	0.322	0.197	0.480	0.2569	0.62
Barren land	4.296	10.418	5.246	11.870	5.627	12.190	5.7609	13.97

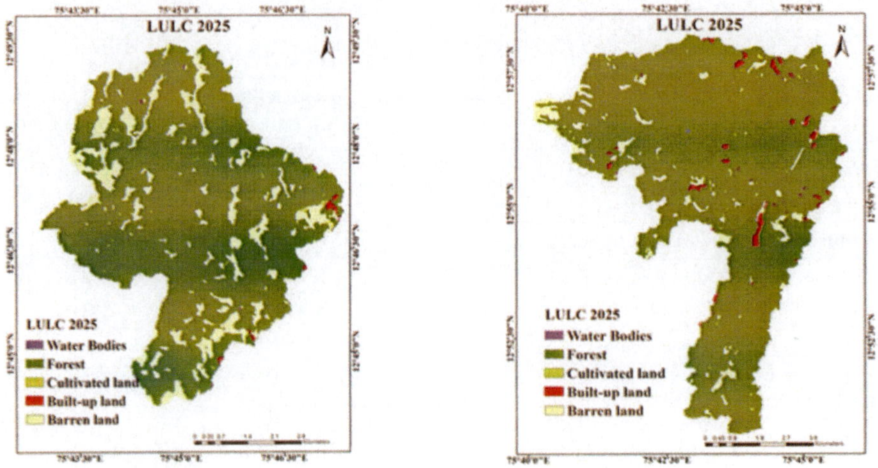

Fig. 5 Prediction map of (1) Mookanamane and (2) Marenahalli

4 Conclusion

Kumaradhara catchment is rich in biodiversity and is a major source of varieties of species, flora, and fauna. The present study investigates LULC changes for sustainable development through an object-oriented image classification approach. This study has revealed that there are changes in LULC categories over the year (2010–2020), which is presented in Tables 6 and 7.

The results of the change detection in Marenahalli catchment showed that the largest change was recorded in the forest land cover class −0.853% for a span of 10 years from 2010 to 2020. The barren land and built-up land were the next dynamic land use categories. The area under barren land increased to 0.22% and again decreased 0.022% from 2015 to 2020. Built-up land increased 0.48% from the year 2010 to 2020. Cultivated land also increased considerably but at a very lower rate. The predictions of future LULC change also revealed that the trends in decreasing forests are going to continue in future, and there will be minimal increase in built-up and cultivated lands. A substantial amount of preservation and conservation is required for the forest area in and around Marenahalli catchment, which is a crucial area. Marenahalli LULC classified images show an overall accuracy of 91% in 2010, 91% in 2015, and 89% in 2020.

The change detection results of the Mookanamane catchment revealed that the forest land cover class saw the higher level of change −1.58% from the year 2010–2020. The other land use classes that increased were barren land at the rate of 0.95% and built-up land at the rate of 0.06%. Although at much slower rate, the cultivated land expanded from the year 2010 to 2015 but gradually it showed a decreasing trend −0.014% from the year 2015 to 2020. This catchment showed a considerable drop in land use classes such as forest and the water bodies and an increase in barren and built-up land. The projections of future LULC change also showed that there will be a significant increase in built-up and cultivated regions, as a continuation of trends towards decreasing woods. The forest region in and around the Mookanamane watershed decreases significantly. Also, Mookanamane LULC classified images show an overall accuracy of 91% in 2010, 88% in 2015, and 89% in 2020. Forecasted LULC classes are shown in Tables 8 and 9.

Changes in land use and cover become a key part of modern resource management and environmental change monitoring systems. Natural resources such as water, soil, nutrients, plants, and animals are significantly impacted by land use and land management strategies. Information on land usage may be utilized to provide solutions for problems with managing natural resources like salinity and water quality. The study of historical LULC change in Marenahalli and Mookanamane catchment will be helpful to evaluate the relation between LULC change, and the results of this study can be used for further watershed hydrology studies through advanced modeling tools. In conclusion, the present study highlights the scarcity of research conducted in the aforementioned study area concerning land use land cover (LULC) analysis. By assessing and predicting LULC changes, this paper contributes valuable insights for future investigations, aimed at establishing correlations with hydrological parameters, thereby paving the way for enhanced resource management and environmental change monitoring in the region.

References

1. Aspinall RJ (2008) Land use change, science, policy and management. CRC Press, Taylor and Francis group
2. Lambin EF (2008) Land- use and land-cover change local processes and global impacts. Springer, Berlin Hyderberg New York
3. Gupta S, Roy M (2012) Land use/ land cover classification of an urban area- a case study of Burdwan municipality, India. Int J Geomatics Geosci 2:1024–1036
4. Pachauri R (2007) Climate change, 2007, synthesis report. IPCC
5. Hakim BS, Arif S (2019) Spatial dynamic prediction of land use/land cover change (Case study: Tamalanrea sub district, Makassar city. IOP conference Earth Environmental science. Institute of Physics
6. Rawat JS, Biswas V (2013) Changes in land use /cover using geospatial techniques: a case study of Ramnagar town area, district Nainital, Uttarakhand, India. Egypt J Remote Sens Space Sci 16(1):111–117
7. Subhashini S, Thirumaran K (2016) A comparative analysis of land surface retrieval methods using Landsat 7 and 8 data to study urban heat island effect in Madurai. Int J Earth Sci Eng 09(04):1397–1404
8. Srivastava R (2020) Changes in vegetation cover using GIS and remote sensing: a case study of south campus BHU, Mirzapur, India. J Sci Res 64:135–141
9. Saputra MH, Lee H (2019) Prediction of land use and land cover changes for North Sumatra, Indonesia, using an artificial-neural-network-based cellular automaton. Sustainability 11
10. Ramachandra TV (2013) Kumaradhara River Basin, Karnataka Western Ghats: need for conservation and sustainable use. Environmental Information System [ENVIS]

11. Rakesh Kumar Sinha TE (2018) Effects of historical and projected land use/cover change on runoff and sediment yield in the Netravati river basin, Western Ghats, India. Environ Earth Sci 77(3):1–19
12. Mangueira JR, Vieira LT, Azevedo TN (2021) Plant diversity conservation in highly deforested landscapes of the Brazilian Atlantic Forest. Perspect Ecol Conserv 19(1):69–80
13. Kaufmann R, Seto KC (2001) Change detection, accuracy, and bias in a sequential analysis of Landsat imagery in the Pearl River Delta, China: econometric techniques. Agric, Ecosyst Environ 85:95–105
14. NextGIS (2017) MOLUSCE—quick and convenient analysis of land cover changes. https://nextgis.com/blog/molusce/
15. Sathya BA, Shashi M (2020) Future land use land coverscenario simulation using open source GIS for the city of Warangal, Telangana, India. Appl Geomatics 12:281–290
16. Chughtai AH, Abbasi H (2021) A review on change detection method and accuracy assessment for land use land cover. Remote Sens Appl: Soc Environ 22:100482
17. Congalton RG (2019) Assessing the accuracy of remotely sensed data: principles and practices, 3rd Edition. CRC Press, India
18. Landis JR, Koch GG (1977) An application of Hierarchical kappa-type statistics in the assessment of majority agreement among multiple observers. Biometrics 33(2):363–374
19. Marwa Waseem A, Halmy PE (2015) Land use/land cover change detection and prediction in the north-western coastal desert of Egypt using Markov-CA. Appl Geogr 63:101–112
20. Bangalore D (2013) S Land use and land cover mapping by using remote sensing and GIS techniques—a case study of Kasaba Hobli, Hoskote Taluk, Bangalore Rural District, Karnataka, India. Int J 2(1):1–6

Evaluation of Surface Soil Moisture Using Remote Sensing and Field Studies

T. N. Santhosh Kumar and Abhishek A. Pathak

Abstract Soil moisture (SM) is an important quantity to examine in terms of agriculture, meteorology, and hydrology to understand the evaporation cycle and drought mechanisms. This study aims to estimate surface soil moisture in arid areas using Sentinel-1A SAR data. In order to collect soil samples from sampling grids that are synchronized with Sentinel-1A passes, study area is divided into 80 grids, each measuring 10 m by 10 m. Six SAR images were collected from Copernicus Open Access Hub website. The vegetation index (NDVI) was calculated using a Sentinel-2A image. The SNAP software was used to process the SAR images, and R studio was used to extract NDVI values and backscattered energy of each sample grid. In this study, an empirical equation was developed to model surface soil moisture using the dielectric constant and backscattering coefficients. The performance of the model was assessed using statistical indicators such as the coefficient of correlation, Nash–Sutcliffe efficiency, and root mean square error, which yielded results of 0.85, 1.46, and 0.75, respectively.

Keywords SAR data · Soil mapping · Semi-empirical model · Bare field

1 Introduction

Surface soil moisture (SSM) is a crucial element that influences hydrological cycle, ecology, and energy fluxes between ground and atmosphere [16]. SSM data are critical in many industries, including agriculture, meteorology, and hydrology [19]. Soil moisture is one of the least monitored of all the hydrologic variables. It is greatly influenced by unpredictable and intermittent precipitation, varying evapotranspiration rates, heterogeneous soils, land cover, and topography and is extremely changeable in both space and time [15]. Monitoring soil moisture, thus, becomes quite challenging.

T. N. S. Kumar · A. A. Pathak (✉)
Department of Civil Engineering, The National Institute of Engineering, Mysuru, India
e-mail: abhishekpathak@nie.ac.in; abhipathak2013@gmail.com

S. Mesapam et al. (eds.), *Developments and Applications of Geomatics*,
Lecture Notes in Civil Engineering 450, https://doi.org/10.1007/978-981-99-8568-5_16

In order to study global behavior and earth changes, remote sensing is a valuable instrument. Moreover, for determining distribution and amount of soil moisture at various scales, remote sensing provides an affordable, efficient, and time-consuming alternative to in situ measurements [10, 17]. Different remote sensing techniques have demonstrated their capacity to retrieve soil moisture from the earth's surface, including optical, thermal, and microwave. However, microwave remote sensing has a number of advantages, including high penetration and all-weather capabilities [12, 22]. Numerous studies have been published to assess soil moisture content using X-band [1, 24], L-band [2, 7, 16, 17] and C-band [3, 11, 14, 16, 20, 21, 23], SAR images. From the available literature, it is known that soil moisture can be obtained with reasonable accuracy using C-band SAR over barren and grasslands.

Several approaches for estimating soil moisture across bare soil surfaces using SAR measurements have been developed, such as physical model, advanced integral equation model, integral equation model for multiple scattering, and empirical and semi-empirical models [13], and various approaches assumed that the connection between surface soil moisture and SAR backscattering energy is linear [5, 6, 18]. The main objective of the study is to develop an empirical model for surface soil moisture using the dielectric constant and backscattering coefficients. This study uses Sentinel-1A data to estimate and map the surface soil moisture in the study region on a plot scale.

2 Study Area

An agricultural plot in Mysuru district was chosen as the study area. To identify whether the study area is grassland or barren, a reconnaissance study was first conducted. The study area consists of two plots, which consist of two different soil types, i.e., sandy loam and silty clayey. Location of study area (Fig. 1a) along with Google Earth image of study area and the sample grids are shown in Fig. 1b. In these sites, a drip irrigation is practiced to irrigate the agricultural lands. Therefore, to regulate irrigation water in this area, investigations of soil moisture are essential.

3 Data

3.1 Sentinel-1A

The European Space Agency (ESA) launched Sentinel-1A satellite on 3 April 2014. The SAR payloads utilize a C-band frequency of 5.405 GHz and have various operational modes, such as Interferometric wide swath, stripe map, extra wide swath, and Wave. These payloads have a 12-day repetitive cycle and capture data in dual-polarization mode (VV + VH). To obtain Sentinel-1A data for this study, the

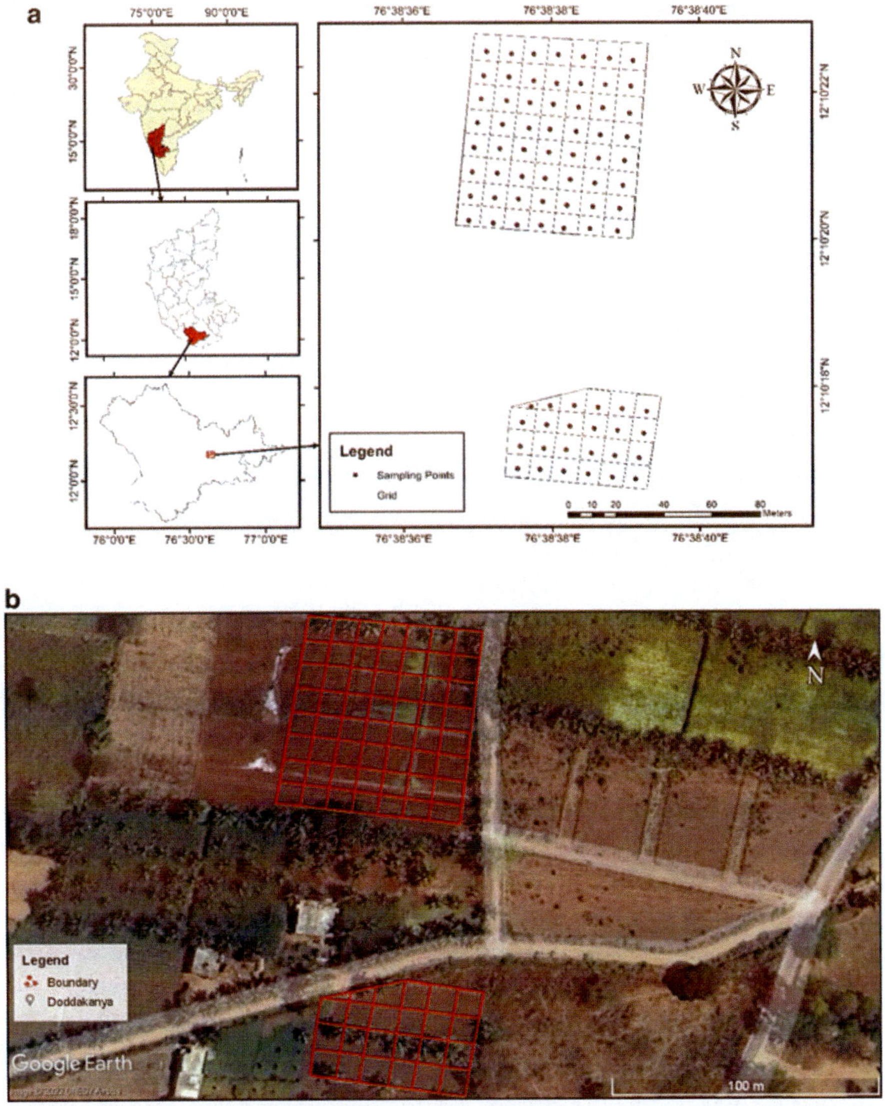

Fig. 1 **a** Study area map. **b** Google Earth image of study area along with sampling grids

Copernicus Open Access Hub is utilized (https://scihub.copernicus.eu/dhus/#/home). Table 1 provides an overview of Sentinel-1A data features, and Table 2 lists the dates on which data were collected.

Table 1 Sentinel-1A features

Frequency band	C-band
Resolution	(10*10 m)
Frequency	5.405 Ghz
Satellite revisit period	12 days
Polarization	VV and VH (dual-pol)
Period	2014–present
Orbit direction	Descending
Swath (km)	100–150

Table 2 Data acquisition dates

Number of pass	1	2	3	4	5	6
Acquisition dates	21/3/2022	02/04/2022	26/04/2022	8/05/2022	20/05/2022	01/06/2022

3.2 Sentinel-2A

The Copernicus Open Access Hub is used to acquire Sentinel-2A L1C data. Sentinel-2A is an optical data collecting mission with a 10-day return time that intends to document changes in land surface conditions and assist surface modifications. It has a resolution of 10*10 m.

3.3 Field Data

Two sets of soil samples were gathered from the field. The first set of samples, taken from a depth of 0–5 cm, was intended to determine the gravimetric moisture content. The second set of samples, collected using a core cutter, was taken to interpret the gravimetric moisture content as volumetric moisture content and to assess the soil texture in accordance with IS 2720–4, 1985 [9]. Soil samples were collected during each satellite pass over study region from 21 March 2022 to 1 June 2022, with the temporal resolution of 12 days (Fig. 2). The surface roughness of the soil is measured using a roller chain.

4 Method

Figure 3 illustrates the overall methodology. The following sections describe the image pre-processing, field data analysis, and the semi-empirical model used in this study.

Fig. 2 Collecting soil samples on-site and drying them in an oven

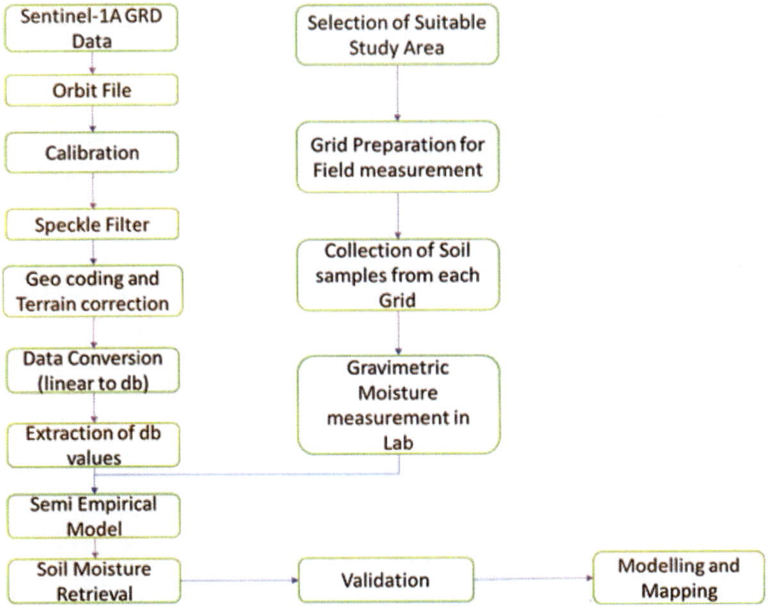

Fig. 3 Flowchart of methodology

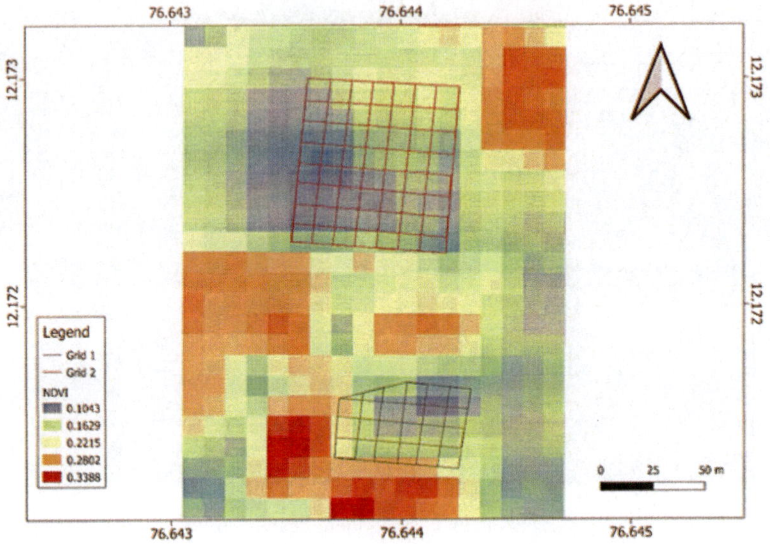

Fig. 4 NDVI map

4.1 Image Processing

Sentinel Application Platform (SNAP) software was used to pre-process Sentinel-1A images. It contains standard procedures such as importing data, applying orbit file, radiometric calibration, multi-looking, speckle filter, geometric terrain correction, and data conversion (linear to dB). Sentinel-2A L1C data were used to develop NDVI map. In this study, single image that was downloaded on the date 22 March 2022 was used since plant growth is a variable that changes slowly. For each sample grid, the normalized difference vegetation index (NDVI) data are generated using NDVI processor tool in SNAP. NDVI map of study site is shown in Fig. 4. Using R software, backscattered value and resultant NDVI values were extracted from each sample grid.

4.2 Data Collection from Field

Soil samples are taken throughout all sampling grids at 6:30 a.m. ± 1/2 hour, synchronized with satellite passes. During a single satellite visit, 80 samples were collected from the study site. To calculate the gravimetric moisture content, topsoil (0–5 cm) is collected and labeled in zip-lock packets. As per IS 2720–4, 1985 [9], the process of collecting soil samples involves using a core cutter and subsequently subjecting them to laboratory analysis for texture and bulk density. The samples are transported to the laboratory, weighed to record their initial weight, and then dried in an oven at a temperature of 105 °C to determine the gravimetric soil moisture. After the samples have been oven-dried, their dry weights are measured, and an equation is used to calculate soil moisture is:

$$\text{Soil moisture} = \left(\frac{w_\omega - w_d}{w_d}\right) \times 100 \quad (1)$$

where, w_ω= wet weight of soil, w_d = dry weight of soil.

A roller chain approach was used to measure the Surface Soil Roughness (SSR) at the location. The statistical metric RMS (vertical variation) was used to explain SSR [16], which was obtained using Eq. (2).

$$\text{RMS} = \sqrt{\left(\frac{\sum (h_i - h)^2}{n - 1}\right)} \quad (2)$$

where, h = mean height of the soil surface, h_i = soil surface height, and n = number of samples

5 Analysis

5.1 Field Data Analysis

The barren area was chosen to study how soil moisture affects backscattered energy in the absence of vegetation. Figure 5 shows the spatiotemporal change of soil moisture within the plot. The box plot shows that soil moisture on 20 May 2022 has increased due to the high rainfall during that period. The soil moisture on 2 April 2022, 26 April 2022, and 1 June 2022 has decreased due to less or no rainfall at that period.

5.2 Soil Moisture Retrieval Model

A semi-empirical model was developed using multi-linear regression analysis. The main factors affecting the backscattered value are used to predict the surface soil moisture. Since it is a barren field, NDVI indicates no correlation, hence it is ignored. Many studies have proposed empirical models to estimate soil moisture, and these models demonstrated a basic relationship between surface soil moisture and backscattering energy [7, 14, 16]. The backscattering energy and dielectric constant were considered to model the soil moisture, and these are presented in Eqs. (3) and (4).

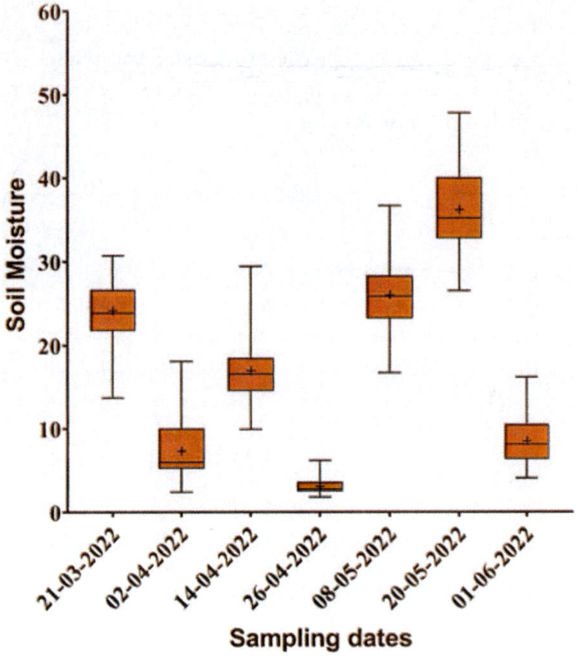

Fig. 5 Variation of soil moisture within the plot over the study period

$$\mathrm{Mv} = \mathrm{f}\left(\sigma^{o}_{vv}, \sigma^{o}_{vh}, \varepsilon\right) \quad (3)$$

$$\mathrm{Mv} = \mathrm{a} + \mathrm{b} * \sigma^{o}_{vh} + \mathrm{c} * \sigma^{o}_{vv} + \mathrm{d} * \varepsilon \quad (4)$$

where, σ^{o}_{vh} and σ^{o}_{vv} are the backscattering coefficient, ε = dielectric constant.

6 Results and Discussion

6.1 Surface Soil Roughness and Backscattering Energy

As described in the prior section, the surface roughness of the study area was established. Figures 6 and 7 examine the connection between surface roughness and the backscattered coefficients. The VH and VV polarizations showed minimal correlation with SSR, with R^2 values of 0.01 and 0.02, respectively, indicating that backscattered energy is only marginally affected by SSR. Hence, it is neglected.

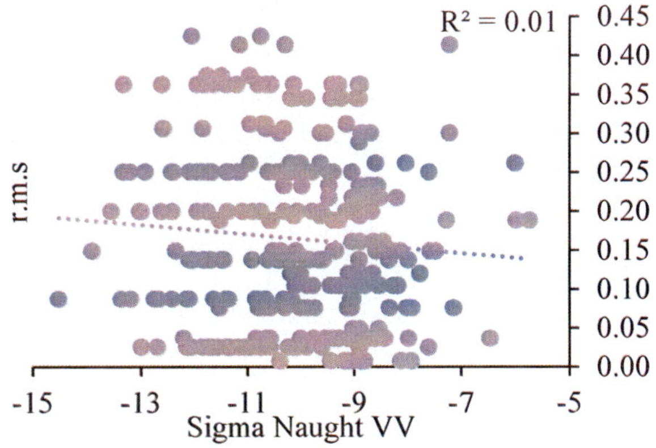

Fig. 6 Relationship between R.M.S and VV polarizations

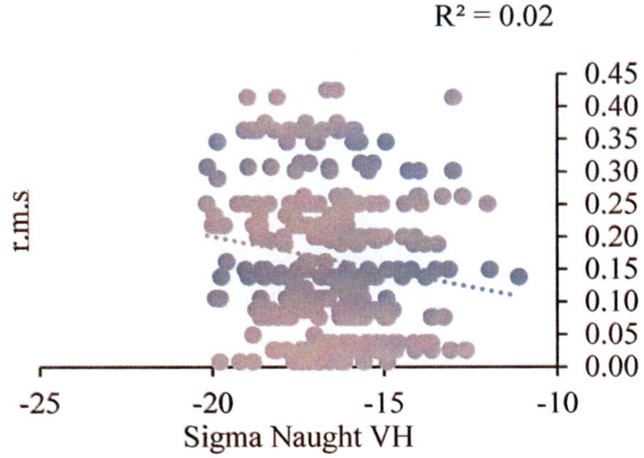

Fig. 7 Relationship between R.M.S and VH polarizations

6.2 Vegetation and Backscattering Energy

Sentinel-2A data were used to generate the NDVI values of each sampling grid. NDVI and backscattered coefficients of VV and VH polarizations are analyzed in Figs. 8 and 9. Correlation study shows that there is a weaker association between NDVI and backscattered energy, with $R^2 = 0.01$ for $\sigma°_{vv}$ and $R^2 = 0.15$ for $\sigma°_{vh}$. NDVI values are within the range, i.e., NDVI < 0.3. As a result, it is not used to model soil moisture.

6.3 Dielectric Constant and Backscattering Energy

Field soil moisture and texture data were used to compute the dielectric constants of each sample. Dry soil has a dielectric constant of roughly 3, while water has a

Fig. 8 Relationship between NDVI and VV polarizations

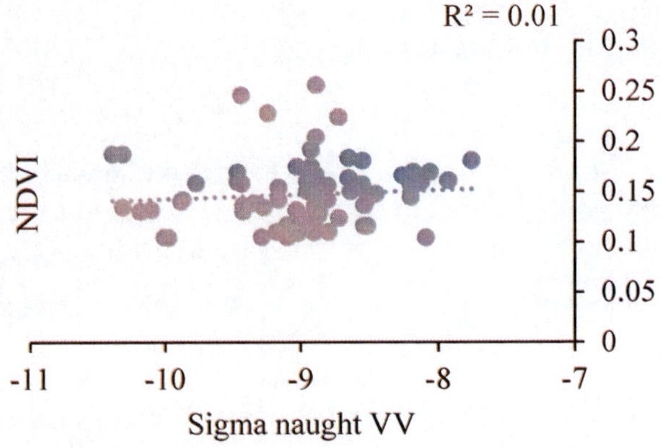

Fig. 9 Relationship between NDVI and VH polarizations

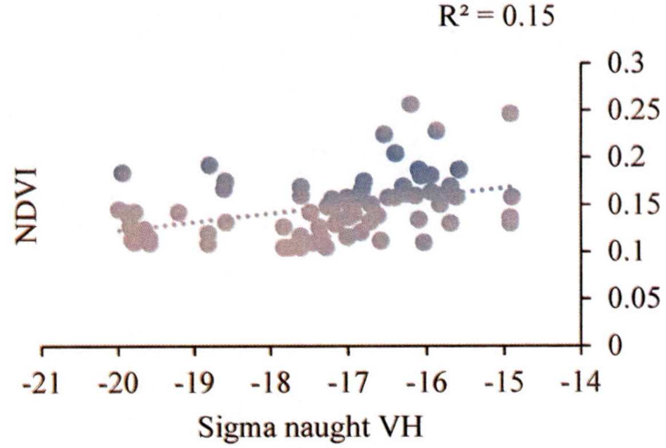

dielectric constant of 80 [4]. As a result, the water-to-soil ratio fluctuates between these ranges. According to [8], for a certain microwave frequency, dielectric constant may be determined using Eq. (5).

$$\varepsilon = (x_0 + x_1 s + x_2 c) + \left(y_0 + y_1 s + y_2 c\right) M_v + (z_0 + z_1 s + z_2 c) M_v^2 \quad (5)$$

where, ε is dielectric constant, *s* denotes sand percentage, *c* stands for clay percentage, M_v represents volumetric moisture content and xi, yi, zi are frequency-dependent coefficients adopted from [8]. Backscattered energy is linearly dependent on dielectric constant, as shown in Figs. 10 and 11. R^2 for VH polarization is 0.27 and 0.40 for VV polarization.

Fig. 10 Relationship between dielectric constant and VV polarization

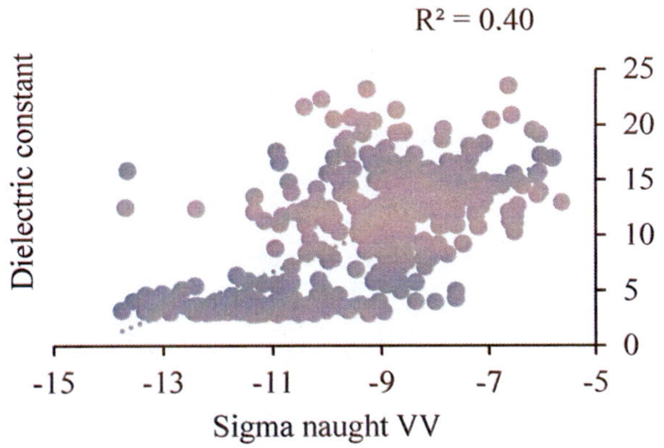

Fig. 11 Relationship between dielectric constant and VH polarization

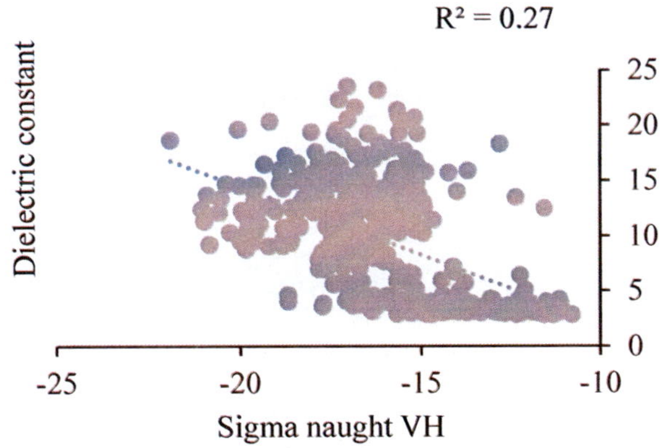

6.4 Soil Moisture and Backscattering Energy

For data obtained at 12-day intervals from 21 March 2022 to 1 June 2022. The correlation between volumetric moisture content and backscattered energy is portrayed in Figs. 12 and 13. With an R^2 of 0.45 for VV polarization and 0.34 for VH polarization, backscattered energy is strongly dependent on soil moisture. The findings indicate that the backscattering coefficient of VV polarization has the strongest correlation with soil moisture as observed in the field. This is primarily due to the fact that co-polarization has superior penetration capabilities compared to cross-polarization (Rawat et al. 2017).

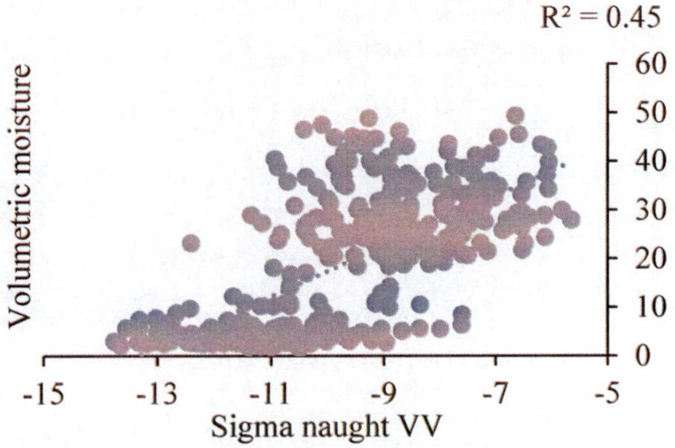

Fig. 12 Relationship between volumetric moisture content and VV polarization

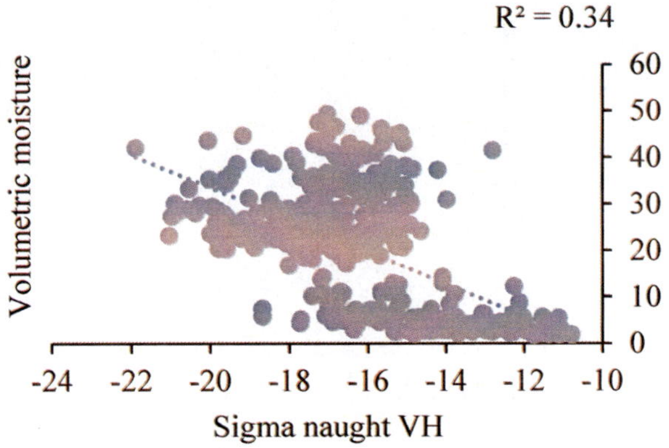

Fig. 13 Relationship between volumetric moisture content and VH polarization

6.5 Model Development

Major factors that influence the sensitivity of backscattered energy are used to model the surface soil moisture. Multilinear regression analysis is used to develop a model, with volumetric moisture content (Mv) as the dependent variable and backscattered energy, dielectric constant as the independent variables. The proposed model has a 95% confidence level with an adjusted R^2 of 0.85 and RMSE of 2.11. Equation 6 describes the empirical model that was developed. The individual association between Mv and other dependent variables is moderate; which manifests the significance of considering the multivariable in modeling SSM. Similar results can be observed in Gururaj et al. [7].

$$Mv = -1.8417 - 0.71564 * \sigma^o vh + 0.9228 * \sigma^o vv + 1.987255 * \varepsilon \quad (6)$$

6.6 Validation

The developed model was validated using filed data collected on 1 June 2022. Figure 14 shows the relationship between actual soil moisture and soil moisture predicted by a model. Statistical metrics like R^2, RMSE, and NSE were used to evaluate the model. Soil map that was developed for the date 1 June 2022 is shown in Fig. 15.

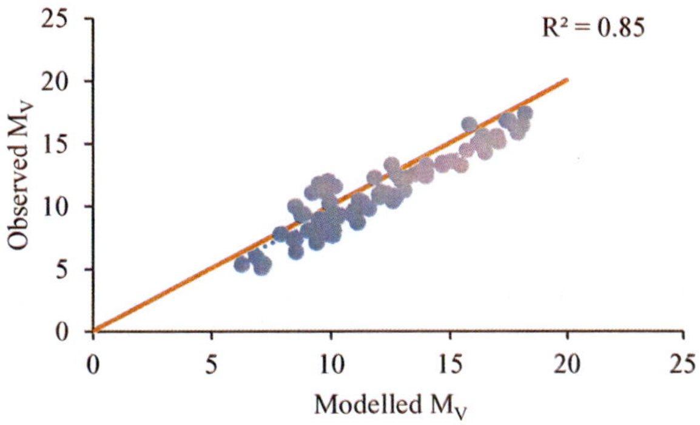

Fig. 14 Relation between observed and modeled soil moisture

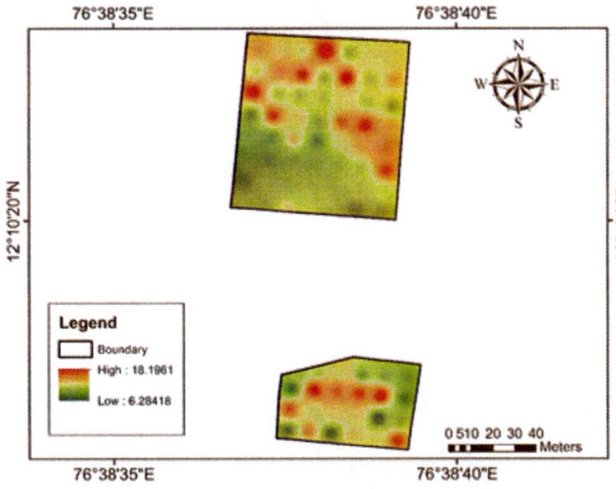

Fig. 15 Soil moisture map

7 Conclusion

In this work, surface soil moisture was modeled using Sentinel-1A data using an empirical equation. Sentinel-1A has demonstrated its ability to detect surface soil moisture at field scale. Semi-empirical model was developed using multiple regression analysis, with soil moisture as the dependent variable and the backscattering coefficient and dielectric constant as independent variables. On bare agricultural land, the proposed model can predict soil moisture content with an acceptable degree of accuracy, with an adjusted R^2 of 0.90 at a 95% confidence level. The suggested empirical model estimates surface soil moisture content with $R^2 = 0.85$, RMSE $= 1.46$, and NSE $= 0.75$. This study will serve as a foundation for designing irrigation water management strategies and to simulate regional hydrology based on field soil moisture. The results of the study aid local farmers in their irrigation water management. However, researchers are encouraged and recommended to do further hydrological modeling studies using field soil data and generated soil moisture maps.

Declaration Myself and my team declare that the photo of soil sample collection from the agricultural field is no objection to use in the publication.

References

1. Baghdadi N, Dubois-Fernandez P, Dupuis X, Zribi M (2012) Sensitivity of main polari-metric parameters of multifrequency polarimetric SAR data to soil moisture and surface roughness over bare agricultural soils. IEEE Geosci Remote Sens Lett 10(4):731–735
2. Balenzano A, Satalino G, Lovergine F, Rinaldi M, Iacobellis V, Mastronardi N, Mattia F (2013) On the use of temporal series of L- and X-band SAR data for soil moisture retrieval. Capitanata plain case study. Eur J Remote Sens 46. https://doi.org/10.5721/EuJRS20134643
3. Beale J, Waine T, Corstanje R, Evans J (2021) Improved soil moisture estimation with Sentinel-1 for arable land at the field scale
4. Dubois PC, Van Zyl J, Engman T (1995) Measuring soil moisture with imaging radars. IEEE Trans Geosci Remote Sens 33(4):915–926
5. Esetlili MT, Kurucu Y (2016) Determination of main soil properties using synthetic aperture radar. Fresenius Environ Bull 25(1):23–36
6. Gao Q, Zribi M, Escorihuela MJ, Baghdadi N (2017) Soil moisture retrieval from Sentinel-1 and Modis synergy. In: EGU General Assembly Conference Abstracts, p 8087
7. Gururaj P, Umesh P, Shetty A (2021) Assessment of surface soil moisture from ALOS PALSAR-2 in small-scale maize fields using polarimetric decomposition technique. Acta Geophys 69(2):579–588
8. Hallikainen MT, Ulaby FT, Dobson MC, El-Rayes MA, Wu LK (1985) Microwave dielectric behavior of wet soil-part 1: Empirical models and experimental observations. IEEE Trans Geosci Remote Sens 1:25–34
9. IS 2720–4 (1985) Methods for test soil. Bureau of Indian Standards. 1–39
10. Kornelsen KC, Coulibaly P (2013) Advances in soil moisture retrieval from synthetic aperture radar and hydrological applications. J Hydrol 476:460–489
11. Liang J, Liang G, Zhao Y, Zhang Y (2021) A synergic method of Sentinel-1 and Sen-tinel-2 images for retrieving soil moisture content in agricultural regions. Comput Electron Agric 190:106485

12. Ma C, Li X, McCabe MF (2020) Retrieval of high-resolution soil moisture through combination of Sentinel-1 and Sentinel-2 data. Remote Sens 12(14):2303
13. Montaldo N, Fois L, Corona R (2021) Soil moisture estimates in a grass field using Sentinel-1 radar data and an assimilation approach. Remote Sens 13(16):3293
14. Parida BR, Pandey AC, Kumar R, Kumar S (2022) Surface soil moisture retrieval using Sentinel-1 SAR data for crop planning in Kosi River basin of North Bihar. Agronomy 12(5):1045
15. Santos WJR, Silva BM, Oliveira GC, Volpato MML, Lima JM, Curi N, Marques JJ (2014) Soil moisture in the root zone and its relation to plant vigor assessed by remote sensing at management scale. Geoderma 221:91–95
16. Sekertekin A, Marangoz AM, Abdikan S (2020) ALOS-2 and Sentinel-1 SAR data sensitivity analysis to surface soil moisture over bare and vegetated agricultural fields. Comput Electron Agric 171:105303
17. Sekertekin A, Marangoz AM, Abdikan SAYGIN, Esetlili MT (2016) Prelim-inary results of estimating soil moisture over bare soil using full-polarimetric ALOS-2 data. The Int Arch Photogrammetry, Remote Sens Spatial Inf Sci 42:173
18. Şekertekin A, Marangoz AM, Abdikan S (2018) Soil moisture mapping using Senti-nel-1A synthetic aperture radar data. Int J Environ Geoinform 5(2):178–188
19. Srivastava PK, O'Neill P, Cosh M, Kurum M, Lang R, Joseph A (2014) Evalua-tion of dielectric mixing models for passive microwave soil moisture retrieval using data from ComRAD ground-based SMAP simulator. IEEE J Selected Topics Appl Earth Observations Remote Sens 8(9):4345–4354
20. Sutariya S, Hirapara A, Meherbanali M, Tiwari MK, Singh V, Kalubarme M (2021) Soil moisture estimation using Sentinel-1 SAR data and land sur-face temperature in Panchmahal district, Gujarat State. Int J Environ Geoinform 8(1):65–77
21. Thanabalan P, Vidhya R, Kankara RS (2021) Soil moisture estimation using RISAT-1 and SENTINEL-1 data using modified Dubois model in comparison with averaged NDVI. Geocarto Int, 1–21
22. Yang M, Wang H, Tong C, Zhu L, Deng X, Deng J, Wang K (2021) Soil moisture retrievals using multi-temporal sentinel-1 data over nagqu region of Tibetan plateau. Remote Sens 13(10):1913
23. Zayani H, Zribi M, Baghdadi N, Ayari E, Kassouk Z, Lili-Chabaane Z, Michot D, Walter C, Fouad Y (2022) Potential of C-Band Sentinel-1 data for estimating soil moisture and surface roughness in a watershed in western France
24. Zribi M, Kotti F, Lili-Chabaane Z, Baghdadi N (2012) Soil texture mapping over a semi-arid area using TERRASAR-X radar data over a semi-arid area. IEEE Trans Geosci Remote Sens Lett 9:353–357

Evaluation of the Influence of Land Use and Climate Changes in Runoff Simulation Using Semi-Distributed Hydrological Model

M. S. Saranya and Vinish V. Nair

Abstract Water resources must be managed effectively to meet current and future demands, ensure sustainability, and meet the needs of a growing population. Identification of the characteristics of resources in the basin, such as the land use land cover (LULC), climatic parameters, and runoff, makes it possible to manage water resources for a long time. In this study, using the hydrological model, Soil and Water Assessment Tool (SWAT), a separation strategy, was implemented in the Meenachil River basin in the Kottayam district of Kerala to differentiate the effects of change in climate and LULC on runoff. The calibration and validation of the SWAT model were facilitated by runoff data available at gauging station Kidangoor within the study area from 1987 to 2010 for the calibration (1987–2004) and validation periods (2005–2010), respectively. Four distinct scenarios were examined to determine the relative impact of LULC and climate on runoff. The findings showed that the variation in streamflow over the study area is significantly affected by climate change (84.86%), with land use having a 15.13% influence.

Keywords SWAT · Climate change · LULC change · Meenachil river basin

1 Introduction

Water is a very important natural resource that is necessary for all living things to survive. Water resources must be managed effectively to meet current and future demands, ensure sustainability, and meet the needs of a growing population. With more warming in the twenty-first century, it is expected that a larger share of the world's population will be affected by water shortages and major river floods. The hydrologic response of a watershed is strongly influenced by the changes in climatic conditions and LULC changes. These changes are accelerated by the rise in regional population, increase in greenhouse gas (GHG) emission, urbanisation, and other

M. S. Saranya (✉) · V. V. Nair
R.I.T. Govt. Engineering College, APJ Abdul Kalam Technological University, Kerala, India
e-mail: saranyams@rit.ac.in

S. Mesapam et al. (eds.), *Developments and Applications of Geomatics*,
Lecture Notes in Civil Engineering 450, https://doi.org/10.1007/978-981-99-8568-5_17

anthropogenic activities within the region [1–3]. It is vital to understand how the changing conditions influence the availability of water at watershed scale. Developing countries like India are potentially at risk from the negative impacts of global climate change since they lack the resources to respond effectively to the problem. A thorough understanding of the major factors affecting runoff makes it easier to manage and plan for water resources over a long period of time. However, reliable prediction of hydrological components is a significant challenge for the research community due to the uncertainty and variability of hydrological processes in terms of location and time. Generally, experimental investigation, statistical evaluation of region-specific hydrologic parameters, and hydrological modelling are utilised to evaluate the influence of change in climate and LULC on runoff [4]. Hydrological modelling is the most widely used method for simulating hydrological processes due to its ability to depict complex water cycles and spatial variability of runoff.

It is broadly acknowledged that variation in climatic parameters and LULC has the greatest influence on streamflow, which is a critical element of the hydrological cycle. Several studies have been carried out globally to assess current and future variations in streamflow by taking into account the influence of change in climate and LULC independently and together [5–9]. For the purpose of creating practical adaptation strategies and aiding decision-makers in accomplishing water security, it is essential to investigate how change in climate and LULC affect hydrological processes in a given region under different scenarios [9]. Making better planning decisions requires a quantitative estimation of how change in climate and LULC will affect regional or watershed hydrology and the identification of potential issues with water resources. This study attempted to separate the individual contributions of change in climate and LULC on runoff variation in Meenachil river basin, Kerala, which would aid in planning and management of water resources for long period in the area under investigation.

2 Materials and Methods

2.1 *Study Area*

Meenachil river basin is a 78-km-long humid tropical river basin that begins in Vagamon hills, flows through major towns in Kottayam district, and empties into Vembanad lake. It is bordered by the basins of Muvattupuzha in the north, Periyar in the east, and Manimala in the south. It has a catchment area of 1272 km^2 and is situated between within 9°25'–9°55N latitude and 76°30'–77° E longitude. The Meenachil river is threatened by extensive illegal sand mining. Historically, the riverbanks were heavily enriched with sand. Now the riverbanks are overgrown with grass and mud. Due to this illegal sand mining, the river is becoming deeper every day. Another threat to the river is sewage contamination. This river is not an outlier when it comes

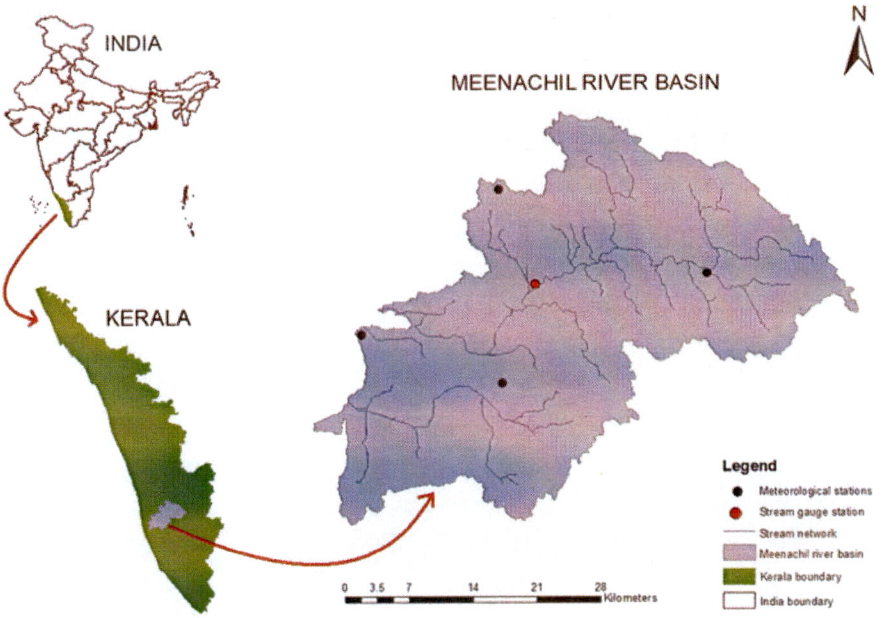

Fig. 1 Map of the Meenachil river basin

to monsoon floods and resulting public destruction. The location map Meenachil river basin is illustrated in Fig. 1.

2.2 *Input Data*

Inputs for the hydrological model SWAT include climate parameters such as precipitation, maximum and minimum temperatures, relative humidity, wind speed, and solar radiation, as well as spatial data such as digital elevation model (DEM), LULC map, and soil map. The DEM of the Meenachil river basin was downloaded at 30 m resolution from NASA's Earth Science Data System (ESDS) [10]. The digital soil map of the world at a scale of 1:5,000,000 was downloaded from the Food and Agricultural Organisation (FAO) website and used to extract the soil map for the study area. Land use and land cover maps for the study area were generated using LANDSAT satellite images with a spatial resolution of 30 m, which are freely available in the USGS's Earth Explorer interface [11]. The LANDSAT images available for the period 2000 and 2008 that were cloud-free were chosen for this study. Landsat 7 and 5 satellite images were obtained as a tile from the USGS website using path 144 and row 53, correspondingly for 2000 and 2008. Images were then individually clipped based on the study area boundary using the ArcGIS software. With the aid of ArcGIS's supervised classification tool and the maximum likelihood algorithm, the clipped satellite images were divided into nine different land use land cover categories (Fig. 2). These nine categories were water, grassland, tea plantation, urban area, forest, paddy field, mixed vegetation, barren land and rubber plantation. The percentage change in LULC categories for the years 2000 and 2008 is shown in Table 1. The meteorological data from 1980 to 2010 were collected from the government agencies IMD (India Meteorological Department) and IDRB (Irrigation Design and Research Board of Kerala). Data were gathered for four stations within the study area. Since the calibration and validation of a hydrological model are essential for its effective application, streamflow data from the Kidangoor station within the Meenachil river basin were collected for the purpose of calibration and validation. It was obtained from the Central Water Commission (CWC) [12] for the years 1987–2010.

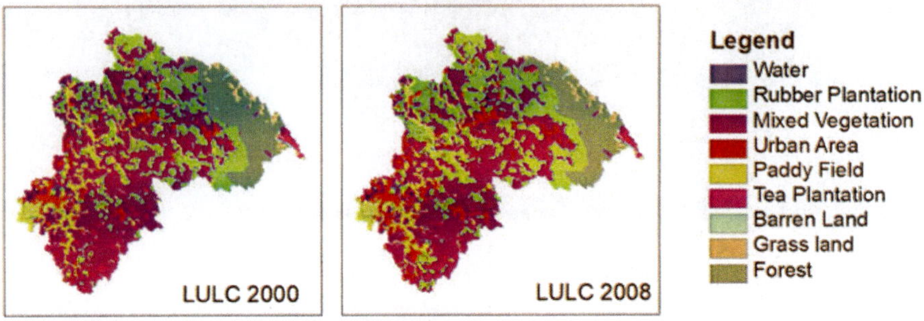

Fig. 2 LULC map for the years 2000 and 2008

Table 1 Percentage area covered by LULC types between 2000 and 2008

Land type	Percentage area covered (%)	
	2000	2008
Rubber plantation	20.897	25.265
Urban area	5.690	8.370
Paddy field	10.242	10.109
Mixed vegetation	48.238	41.337
Barren land	0.271	0.288
Grassland	1.427	1.407
Tea plantation	0.378	0.383
Water	0.630	0.627
Forest	12.226	12.214

2.3 Description of the SWAT Model

One of the hydrological models used by numerous researchers and agencies for simulating hydrologic components of a watershed is the SWAT model, developed in the United States [13]. It has made a significant contribution to many scientific fields worldwide. The SWAT model is a continuous-time model with a daily time step, at watershed scale [14]. The framework of SWAT model structure is shown in Fig. 3. The water balance equation governs hydrology in SWAT. It is capable of generating long-term yields in order to determine the effect of land-management practices [15]. We can simulate streamflow, sediment transport, and nutrient runoff at watershed scale using this model's user-friendly interface. It can model the hydrologic cycle at three distinct scales, including the whole watershed, sub-watersheds, and hydrologic response units (HRUs). HRU is the model's smallest unit, characterised by homogeneous land use, soil, and slope characteristics in a given sub-watershed based on the thresholds established by the user for each of these categories. In the current study, a threshold of 20%, 20%, and 10%, respectively, for land use, slope, and soil on area, was assigned. Depending on the user's preference, SWAT will use either a weather generator or observed weather data for meteorological inputs.

In hydrologic modelling with SWAT, the various steps include watershed delineation, HRU generation, weather definition, input table writing, hydrological simulation, calibration, and validation. A DEM-based method available in SWAT was utilised to automatically define the watershed and sub-watershed boundaries. Figure 4 displays the DEM of the study area, and Fig. 5 depicts the delineated watershed with 35 sub-basins. In order to conduct the HRU analysis, the prepared land use map was reclassified in accordance with the SWAT land use classification table. Additionally, user-defined soil types were added into the model and reclassified correspondingly.

Then HRUs were obtained by overlaying the reclassified land use, soil, and slope maps of user-defined thresholds. The threshold values are established to eliminate insubstantial land use, soil type, and slope in each sub-basin, and thus preventing the formation of an excessive number of HRUs [1]. In this study area, 123 HRUs are being produced. Following that, the model was fed with all the necessary climatic variables. In the event that station data were not available, the weather generator within the SWAT can simulate weather parameters for which there are no observed data and to fill in any missing weather data. The model uses the SCS curve number method [16] to calculate runoff, and the most accurate Penman–Monteith equation was chosen to compute the potential evapotranspiration. The runoff is estimated independently for each HRU, and then it is routed in order to get the overall runoff for the watershed. The simulation included a 3-year warm-up period. The simulation was scheduled to run from 1980 to 2010, inclusive of the warm-up period.

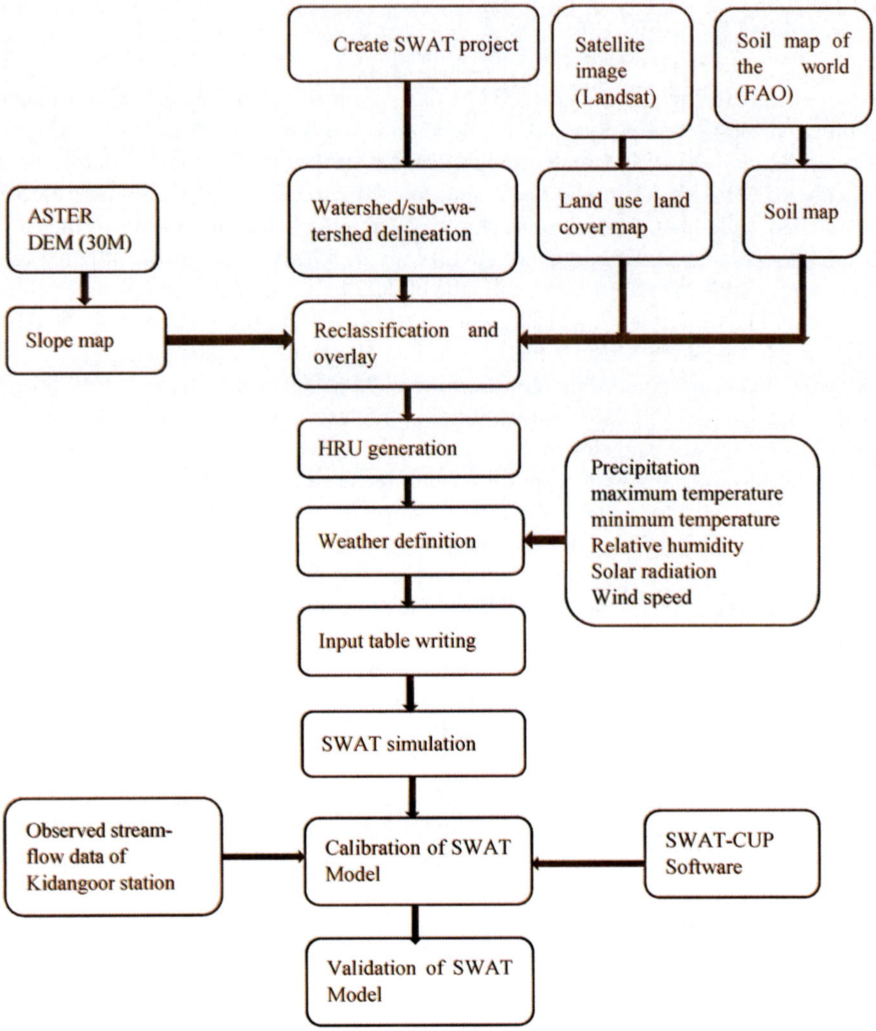

Fig. 3 Framework of SWAT model structure

2.4 *Calibration and Validation*

Calibration is the process of selecting the best parameters for running a model, whereas validation is the process of proving that a model can make predictions that are sufficiently accurate for the intended purpose of the project. The successful application of hydrologic models relies on the calibration and sensitivity analysis of their parameters. Observed data are necessary for the calibration and validation process to be carried out effectively. The runoff data available at the Kidangoor station within the study area for the period 1897–2010 were used for the same. The SWAT model

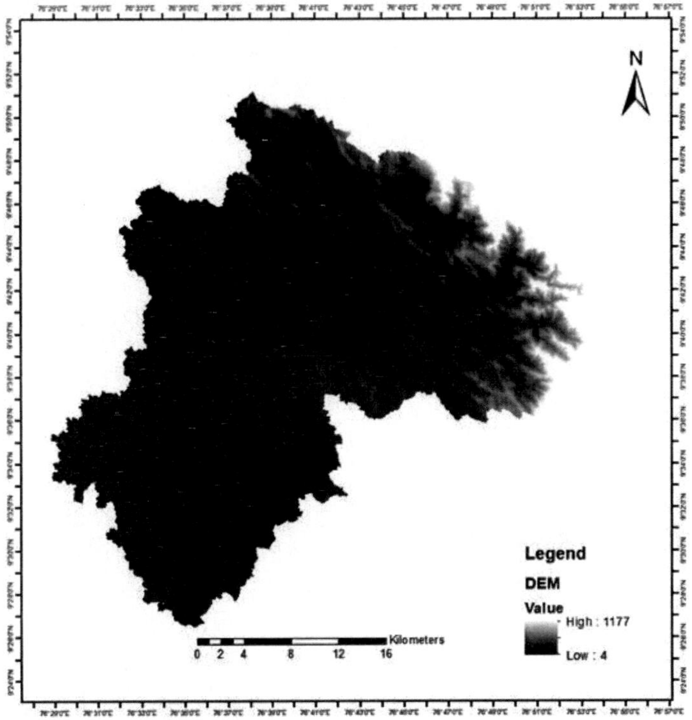

Fig. 4 DEM of Meenachil river basin

can be calibrated using the software SWAT-CUP (SWAT Calibration and Uncertainty Programs). SWAT-CUP uses the SWAT input files to change the parameter sets and run SWAT simulations. The Sequential Uncertainty Fitting algorithm (SUFI-2) was chosen from the four calibration techniques found in SWAT-CUP because it can run with multiple parameters in the fewest number of model runs to produce good prediction uncertainty ranges. 18 parameters were chosen for sensitivity analysis based on the relevant literature review and the SWAT-CUP software use manual. Two indicators, such as p-value and t-stat, govern sensitivity analysis primarily. The global sensitivity approach ranks the parameters based on their respective t-statistics and p-values after examining the sensitivity of one parameter in relation to another [14]. In the case of t-statistics, a parameter is more sensitive if its absolute value is greater, while for p-values, it is more sensitive if it is closer to zero. Calibration was performed by comparing simulated monthly runoff to observed monthly runoff data for the historical period from 1987 to 2004, and validation was performed using data from 2005 to 2010. Statistical indicators such as root mean square error to standard deviation ratio (RSR), percent bias (PBIAS), and Nash–Sutcliffe efficiency (NSE) were used to evaluate the performance of the model during the calibration and validation periods using the evaluation criteria proposed by Moriasi et al. [17] for the monthly time step.

Fig. 5 Sub-basins of Meenachil river basin

2.5 Separation of the Impact of Climate and LULC Change

Four scenarios were used to separate the percentage contribution of climate and LULC change on runoff. In order to ascertain the individual as well as combined contribution of climate and LULC change on runoff, simulations were carried out for the period 1983–2010 under four scenarios (scenario 1 (SR_1), scenario 2 (SR_2), scenario 3 (SR_3) and scenario 4 (SR_4)) using the validated SWAT model [4]. As shown in Table 2, streamflow simulations under SR_1 and SR_2 were run for period 1, and simulations under SR_3 and SR_4 were run for period 2. The difference between SR_1 and SR_2 represents runoff variation caused by LULC change, whereas the difference of SR_1 and SR_3 indicates runoff variation caused by climate change. The difference between SR_1 and SR_4 shows how much LULC and climate have contributed together. The following equations were used to assess the above-mentioned analysis:

$$\Delta O_L = O_2 - O_1 \tag{1}$$

$$\Delta O_C = O_3 - O_1 \tag{2}$$

Table 2 Scenarios considered for separation of the impact

Scenario	Time period	Land use map used	Streamflow
SR_1	1983–1996	2000	O_1
SR_2	1983–1996	2008	O_2
SR_3	1997–2010	2000	O_3
SR_4	1997–2010	2008	O_4

$$\Delta O = O_4 - O_1 \tag{3}$$

$$\Delta O_m = O_C + O_L \tag{4}$$

Theoretically,

$$\Delta O = \Delta O_m \tag{5}$$

where O_1, O_2, O_3, and O_4 correspond to the average streamflow simulated under SR_1, SR_2, SR_3, and SR_4, respectively. The impact of climate change on streamflow is given by $\Delta O_C/\Delta O_m \times 100\%$ and the impact of change in LULC is given by $\Delta O_L/\Delta O_m \times 100\%$.

3 Results and Discussion

3.1 *Sensitivity Analysis, Calibration, and Validation*

Eliminating the parameters that were found to be insensitive is made easier by sensitivity analysis. The global sensitivity analysis of SUFI-2 employs the t-test and p-value, to see the effects of each parameter on the objective function [6]. We used 1000 iterations to distinguish the sensitive parameters and selected the 15 most sensitive parameters with t-statistics greater than 1 [2]. The selected parameters are SOL_AWC, CN2, ALPHA_BNK, REVAPMN, CH_K2, ESCO, RECHARGE_DP, GW_REVAP, SOL_BD, GWQMN, GW_DELAY, ALPHA_BF, EPCO, CH_N2 and SOL_K, which are listed in order of sensitivity in Table 3.

The table also contains the best-fitted value, p-value, and t-statistic for each of the above parameters. SOL_AWC is found to be the most sensitive parameter for the study basin. Figure 6a and b depicts the comparison of the monthly observed and simulated runoff hydrographs for the calibration (1987–2004) and validation (2005–2010) periods, respectively. Based on the values obtained in Table 4, SWAT has a performance rating of "very good" for the calibration (NSE = 0.78 and $R^2 = 0.8$) and validation (NSE = 0.86 and $R^2 = 0.9$) of monthly runoff simulation. The peak flows during rainy months are under estimated, despite the fact that simulated monthly

Table 3 The characteristic values of chosen sensitive parameters

Serial number	Parameter and its description	p-value	t-stat	Fitted value
1	SOL_AWC (Available water capacity of the soil layer)	0.00	3.941	0.175
2	CN2 (SCS runoff curve number)	0.036	2.420	96.42
3	ALPHA_BNK (Baseflow alpha factor for bank storage)	0.062	2.121	0.325
4	REVAPMN (Threshold depth of water for revap or percolation to occur)	0.069	2.101	87.5
5	CH_K2 (Effective hydraulic conductivity in main channel alluvium)	0.073	2.007	487.5
6	ESCO (Soil evaporation compensation factor)	0.101	−1.984	0.475
7	RECHARGE_DP (Deep aquifer percolation factor)	0.102	−1.962	0.225
8	GW_REVAP (Groundwater "revap" coefficient)	0.119	1.865	0.168
9	SOL_BD (Moist bulk density (g/cm^3))	0.172	1.591	1.90
10	GWQMN (Threshold depth of water in the shallow aquifer required for return flow (mm))	0.185	1.437	2875
11	GW_DELAY (Groundwater delay (days))	0.198	1.248	212.50
12	ALPHA_BF (Baseflow alpha factor (days))	0.234	1.086	0.875
13	EPCO (Plant uptake compensation factor)	0.289	1.054	0.325
14	CH_N2 (Manning's n value for the main channel)	0.312	−1.043	0.0172
15	SOL_K (Saturated hydraulic conductivity)	0.326	−1.015	1.450

hydrographs performed satisfactorily during the calibration and validation period. This might be as a result of the streamflow simulation method's (SCS CN method) limitations. The underestimation of extreme events can be disregarded though, as the intention of the current work is to comprehend how LULC and climate change affect runoff.

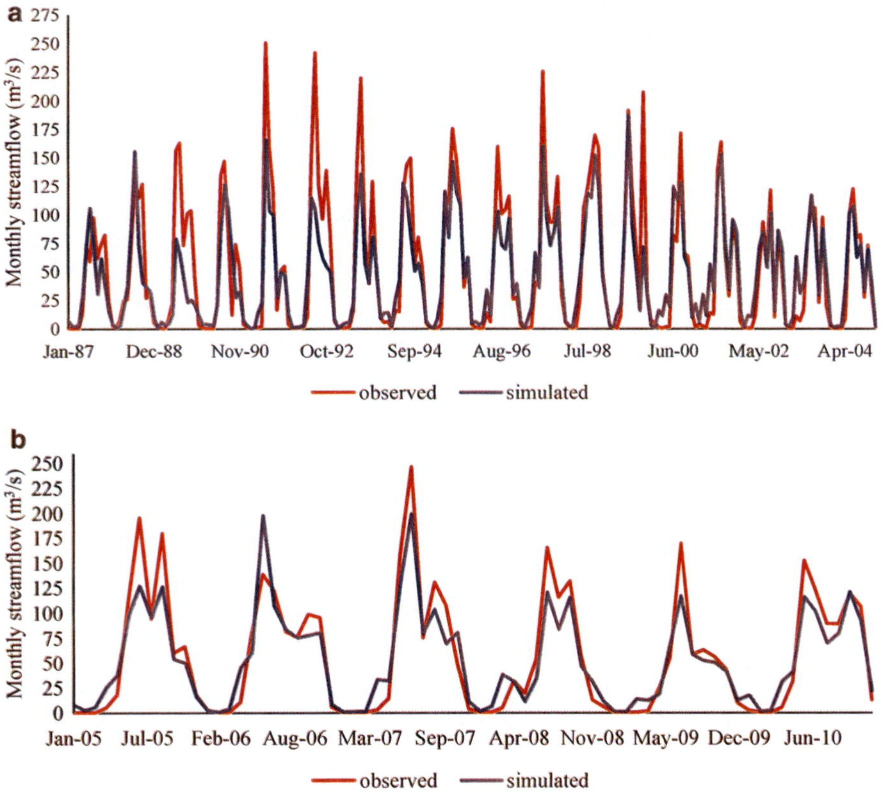

Fig. 6 **a** Comparison of streamflow during calibration period 1987–2004. **b** Comparison of streamflow during validation period 2005–2010

Table 4 Model performance during calibration and validation

	NSE	RSR	PBIAS (%)	R^2
Calibration	0.78	0.47	4.7	0.8
Validation	0.86	0.363	5.84	0.9

3.2 Relative Contribution Assessment

The annual runoff simulated under the four scenarios SR_1, SR_2, SR_3, and SR_4 was utilised to assess the effect climate and LULC variation. Table 5 depicts the mean annual streamflow corresponding to the four scenarios. This separation strategy was performed for the historical period 1980–2010. Due to the 3-year warm-up period considered during SWAT simulation, the actual simulation began in 1983. During period 1 (1983–1996), the simulation for scenarios SR_1 and SR_2 differed only by the land use map considered (see Table 2), and the simulated streamflow was 46.20 $m^3/$

Table 5 Relative contribution rate analysis

Scenario	Mean annual runoff (m^3/s)	Variation in runoff (m^3/s)	Climate contribution (%)	Land use contribution (%)
SR_1	46.20	–	84.86	15.13
SR_2	48.08	2.04		
SR_3	51.66	11.44		
SR_4	53.30			

s and 48.08 m^3/s, respectively. The variation in streamflow caused by a change in land use was 2.04 m^3/s, while the total variation in streamflow (sum of O_L and O_C) was 13.48 m^3/s. Using Eq. 4, it was determined that 15.13% of the runoff variation was due to land use change. Likewise, climate change accounted for 84.86% of the variation in runoff. Even though climate change is the most influential factor on runoff of the Meenachil basin, land use change cannot be ignored.

4 Conclusions

This study employed SWAT for hydrological simulation in the Meenachil river basin. Using observed data from a stream gauge station in the basin, the simulations are effectively validated. The applicability of the model for this river basin has been proven in terms of statistical indicators such as NSE, PBIAS, RSR, and R^2 during both calibration and validation. The separation of the effects of climate and LULC variation on runoff was demonstrated by simulating runoff under four distinct scenarios. When climate data from 1980 to 2010 and land use maps of 2000 and 2008 were used to separate the individual contributions on runoff simulation, it was discovered that climate change contributed 84.86% of the streamflow and LULC change contributed 15.13%. The simulations under various scenarios showed that runoff changes in the study area are primarily caused by climate change. Land use change impacts must also be considered for efficient watershed management. The management of water resources may be impacted by the various land use activities that change the hydrological cycle and the hydrological processes. The sustainable management of water resources depends on the ability to foresee the consequences of change in climate and land use on freshwater supplies [18]. The only way to identify and implement methods to improve water management actions is to examine the likely effects on a regional/watershed level.

References

1. Chen Y, Niu J, Sun Y, Liu Q, Li S, Li P, Sun L, Li Q (2020) Study on streamflow response to land use change over the upper reaches of Zhanghe Reservoir in the Yangtze River basin. Geosci Lett 7:1–12. https://doi.org/10.1186/s40562-020-00155-7
2. Sinha RK, Eldho TI, Subimal G (2020) Assessing the impacts of land use/land cover and climate change on surface runoff of a humid tropical river basin in Western Ghats, India. Int J River Basin Manage. https://doi.org/10.1080/15715124.2020.1809434
3. Loukika KN, Keesara VR, Buri ES, Sridhar V (2022) Predicting the effects of land use land cover and climate change on Munneru river basin using CA-Markov and soil and water assessment tool. Sustainability 14(9):5000. https://doi.org/10.3390/su14095000
4. Guo J, Su X, Singh VP, Jin J (2016) Impacts of climate and land use/cover change on streamflow using SWAT and a separation method for the Xiying River Basin in northwestern China. Water 8:192
5. Zhang X, Srinivasan R, Hao F (2007) Predicting hydrologic response to climate change in the Luohe river basin using the SWAT model. Trans ASABE 50(3):901–910. https://doi.org/10.13031/2013.23154
6. Zuo D, Xu Z, Yao W, Jin S, Xiao P, Ran D (2016) Assessing the effects of changes in land use and climate on runoff and sediment yields from a watershed in the Loess Plateau of China. Sci Total Environ 544:238–250. https://doi.org/10.1016/j.scitotenv.2015.11.060
7. Ahiablame L, Sinha T, Paul M, Ji J, Rajib A (2017) Streamflow response to potential land use and climate changes in the James River watershed, upper Midwest United States. J Hydrol: Regional Stud 14(1):150–166. https://doi.org/10.1016/j.ejrh.2017.11.004
8. Quansah JE, Naliaka AB, Fall S, Ankumah R, Afandi GE (2021) Assessing future impacts of climate change on streamflow within the Alabama river basin. Climate 9(55). https://doi.org/10.3390/cli9040055
9. Saranya MS, Vinish VN (2022) Impact evaluation and analysis at a river basin scale under projected climate and land use change. Water Supply, ws2022417. https://doi.org/10.2166/ws.2022.417
10. Nasa EARTH DATA Homepage. https://search.earthdata.nasa.gov/search. Accessed 18 July 2021
11. USGS Earth Explorer Homepage. https://earthexplorer.usgs.gov/. Accessed 19 July 2021
12. India WRIS Homepage. https://indiawris.gov.in/wris/#/. Accessed 11 May 2020
13. Zhao K, Li J, Ma X, Deng C (2022) The effects of land-use and climatic changes on the hydrological environment in the Qinling mountains of Shaanxi province. Forests 13:1776. https://doi.org/10.3390/f13111776
14. Abbaspour KC (2015) SWAT-CUP: SWAT calibration and uncertainty programs—a user manual. Eawag: Swiss Federal Institute of Aquatic Science and Technology, p 100
15. Arnold JG, Moriasi DN, Gassman PW, Abbaspour KC, White MJ, Srinivasan R, Santhi C, Harmel RD, Van Griensven A, Van Liew MW, Kannan N, Jha MK (2012) SWAT:model use, calibration, and validation. Trans ASABE 55(4):1491–1508. https://doi.org/10.13031/2013.42259
16. Neitsch SL, Arnold JG, Kiniry JR, William JR (2009) Soil and water assessment tool theoretical documentation. AgriLIFE Research & Extension
17. Moriasi DN, Arnold JG, Van Liew MW, Bingner RL, Harmel RD, Veith TL (2007) Model evaluation guidelines for systematic quantification of accuracy in watershed simulations. Trans ASABE 50(3): 885–900. https://doi.org/10.13031/2013.23153
18. Raihan F, Ondrasek G, Islam MS, Maina JM, Beaumont LJ (2021) Combined impacts of climate and land use changes on long-term streamflow in the Upper Halda Basin, Bangladesh. Sustainability 13(21):12067. https://doi.org/10.3390/su132112067

Flood Damage Assessment of a River Basin Using HEC-GeoRAS

K. C. Amal Vishnu and Vinish V. Nair

Abstract Flood is one of the most catastrophic events among the many different types of natural hazards. It seriously harms people, property and places used for industry and agriculture. A GIS extension called HEC-GeoRAS provides a series of steps, tools and options for preparing river geometry grid data for input into HEC-RAS, which is utilised to create the final inundation map. The DEM and a Basin Landuse Map must be provided as input data for the preparation of river geometry using HEC-GeoRAS model. The flood hydrograph is derived from the synthetic unit hydrograph and probable maximum precipitation, and the PMP value obtained was about 157.3 cm. The total area of the basin was found to be 787 km^2 in which the inundated area was 274.279 km^2, which is about 35.851% of the total area.

Keywords DEM · Landuse map · HEC-GeoRAS

1 Introduction

1.1 General Background

Among the many different types of natural hazards, flooding is one of the most catastrophic occurrences. It seriously harms people, infrastructure and places used for industry and agriculture. The most frequent natural calamities in a tropical nation like India are floods. Flooding is defined by the European Union Floods Directive (EUFD) as the water flooding of territory that is not normally covered by water. Thus, the land that is normally dry is submerged by an overflow of water. When water overflows from a river, lake, or ocean, the levees may be breached or overtopped.

K. C. A. Vishnu (✉) · V. V. Nair
Department of Civil Engineering, Rajiv Gandhi Institute of Technology Kottayam, APJ Abdul Kalam Technological University, Kerala, India
e-mail: amalvishnu217@rit.ac.in

V. V. Nair
e-mail: vinish@rit.ac.in

S. Mesapam et al. (eds.), *Developments and Applications of Geomatics*,
Lecture Notes in Civil Engineering 450, https://doi.org/10.1007/978-981-99-8568-5_18

Flooding can also happen when precipitation collects on soggy terrain. Thus, the primary forces of flood are heavy rainfall, drainage cramming, impervious soil and erosion of banks. Although the size of lakes, rivers and other water bodies will alter due to seasonal differences in rainfall and snowmelt, these changes are unlikely to be significant unless they result in property flooding or domestic animals drowning. Some floods form gradually, while others, like flash floods, can form quickly and without any outward signs of rain. Also, floods can either be very massive, affecting entire river basins, or very minor, affecting a specific area or town.

Due to the continuous developmental activities, it is difficult or rather impossible to prevent flooding of the urban areas. Hence, flood management aiming at reduction of loss of life and property must be the prime concern. Many studies in the area of developing models for the forecasting of rural and urban flooding have been conducted in recent years. Every year, in most parts of India, a repeated occurrence of flood is found with states in southern part of India the most affected by its ferocity. To identify the flood-prone areas and to have knowledge on probability of occurrence or return period of floods has become inevitable for planning and mitigation of damage caused by this natural phenomenon. To understand the effect of flooding on life and property, flood inundation mapping (FIM) is done [8]. FIM gives the significant data such as spatial extent and depth of the areas affected by flood, which are required to adopt appropriate flood management strategies.

Spatial analysis tools such as geographic information system (GIS) and remote sensing (RS) are effectively used for assessment of the flood risk. In spite of the fact that floods cannot be prevented, studies done to date show, the inundation caused due to flood can be reduced if the flood-prone areas were known in advance this can be achieved with the help of a flood inundation map developed for the study area using available technological advances. It also helps in effective and complete flood management system development.

1.2 Objectives

The objectives of the study are

- To obtain the water delineation of the digital elevation model of Vamanapuram basin using ArcHydro tool.
- To determine the synthetic unit hydrograph and hence to develop a flow hydrograph for Vamanapuram basin.
- To prepare the probable maximum precipitation.
- To prepare flood inundation map using HEC-RAS software.
- To conduct a flood damage assessment of Vamanapuram basin using HEC-GeoRAS software.

1.3 Study Area

Study area comprises the region from Chellangi (Panavoor) to Thuruthu (Chirayinkeezhu) of Vamanapuram River Basin. It is spread over Thiruvananthapuram and Kollam districts of Kerala state. The basin is bounded by Kottarakara Taluk of Kollam district in the north and by the Arabian Sea in the west. The Vamanapuram river rises in the Western Ghats' Chemmunji mottai at a height of roughly 1706 m above sea level and empties into Anjengo Lake. The river has a drainage area of 787 sq. km and is 80 km long. It lies between 51°18′59″ E and 51°26′58″ E longitude and 33°38′6″ N and 33°46′19″ N latitude.

2 Literature Review

2.1 General

Flood, being a disastrous natural phenomenon, has to be assessed and managed so as to reduce its dreadful effects. Proper and timely warning if provided can mitigate the flood damages to a greater extent. Developmental activities are continuous in urban regions and thus to prevent urban flooding it is almost become impossible. In the past few decades, various research has been published in connection with the development of models for the forecasting and prediction of floods in both urban and rural settings. Floods are a common occurrence in most of India each year, but they are most severe in the southern states of India.

2.2 Flood Damage Assessment

To understand the effects mainly on an important structure or on any particular region flood inundation mapping or FIM is required. FIM gives the significant data such as spatial extent and depth of the areas affected by flood, which are required to adopt appropriate flood management strategies. Sahoo and Sreeja [8] conducted a study in an urban catchment in northeast India to quantify the flood risk and to use flood inundation mapping (FIM) as an effective tool in flood management. Storm water management model (SWMM) was used to estimate the surface runoff. The flooding pattern for peak rainfall intensities was then acquired using the local drainage network and correlating to various return periods. Based on the potential flood hazard brought on by the 100-year return period flood, a flood hazard map was created. Based on the hazard rankings developed using the shared impact of vulnerability parameters such as flood depth, inundated area, land use, affected population, and roadways, flood risk zones have been defined. The effectiveness of HR for locating the worst flood-affected area is discussed in the study.

Getahun and Gebre [3] carried out a flood hazard assessment and mapping of the flood inundation area in Awash River basin, Ethiopia. Factors causing floods such as slope, elevation, rainfall, drainage density, land use and soil type were rated and the flood hazard regions are delineated using a multi-criteria evaluation technique in a GIS. To obtain the flood danger map, a final weighted overlay analysis of all elements was obtained by calculating the weight of each flood-generating factor by pair-wise comparison. All return periods had high flood levels, especially from Dubti to Lake Abe. According to the study, there were less inundated areas in the upper and intermediate reaches of the Awash River Basin than there were in the lower reaches. The Awash River Basin's northwest, southwest and western escarpments all fall under the low to very low flood hazard danger category. In particular in low-lying flood-prone areas, afforestation and good land use management are crucial to reducing the detrimental consequences of floods. The study is very beneficial for the relevant bodies in helping them create measures based on the potential for flooding in the area. The findings of this paper are crucial for planning land use, formulating policies, making investment decisions and for security considerations.

Kourgialas and Karatzas [6] presented a method to evaluate the benefits of a flood warning system, as well as a method to determine the hazardous areas. A sound flood control strategy and an evaluation of the flood-prone areas make up a trustworthy flood management plan. They analysed six factors in order to determine the spatial distribution of the dangerous locations—flow accumulation, rainfall intensity, slope, geology, land use and elevation. Five regions were chosen for the study that ranged in flood threat from extremely low to very high. The locations with a high risk of flooding are delineated on the created map of flood-hazard areas. The Koiliaris River basin was examined using the suggested methodology to identify the settlements and areas at risk of floods. The results were then verified using information from previous floods in the basin.

Chithra and Sumam [2] used hydraulic model HEC-RAS for flood inundation mapping of Kurumali river basin for probable maximum flood event. The model's inputs were created with the help of ArcGIS, this is a software product developed by ESRI, which is a compatible programme Hydrologic Engineering Center's Geographical River Analysis System (HEC-GeoRAS). The flow and stage hydrographs for the research region could be reasonably predicted by the hydraulic model. Hydrologic Engineering Center's River Analysis System (HEC-RAS) was used to route the flood for the greatest likely flood occurrence. The mapping of flood inundation was done using the results that were collected. For studies on flood risk assessment, the created flood inundation maps could be used. Accuracy of model predictions can be increased by using high-resolution Digital Elevation Model (DEM) or field observed data.

3 Methodology

3.1 General

The generalised methodology includes watershed delineation from the DEM developing the cross-section and streamline in HEC-GeoRAS, development of flood inundation map and assessment of the flood impact of Vamanapuram river basin using HEC-GeoRAS software.

3.2 Description of the HEC-GeoRAS Model

The final inundation map is produced using the GIS extension HEC-GeoRAS, which has a number of steps, tools and options for preparing river geometry GIS data for import into HEC-RAS. Triangular Irregular Network (TIN), DEM and land use are the input data needed to prepare the river geometry for the HEC-GeoRAS model. The river geometry file and stream flow data are also utilised as input files for the HEC-RAS model, which generates the river's water surface level. HEC-GeoRAS serves as an interface for data management between ArcGIS and HEC-RAS. The river streamline, centreline, banklines, flow route centrelines and XS cut lines must be digitally recovered from an earlier river file, from aerial photographs or from topographical data using the HEC-GeoRAS interface. The HEC-GeoRAS geodatabase file contains information on river reach, cross-section and other relevant topics. With the exception of the river and reach names, which are manually filled in for the cross-section and river data layers, all other properties are automatically computed by the HEC-GeoRAS. A GIS environment will be used to store the outcomes of the HEC-RAS model simulation, and the HEC-GeoRAS tool will be used to conduct additional studies. The GIS data transmitted between HEC-RAS and ArcGIS are in.sdf file format. The HEC-RAS editing tools can be used to modify the exported GIS geometric data in the model. The Geometric Editor, a Graphical User Interface (GUI) used to handle the geographic data, imports the.xml file exported from the HEC-GeoRAS [3]. For the cross-sections of each reach, the Manning friction values are entered in this editor. The steady flow data editor receives the stream flow data. The data for the reaches are taken out of the geometric editor by this editor. The model needs to be aware of the initial water level at the beginning and finish of unconnected reaches as well as at junctions with other reaches in order to calculate the water surface level (boundary conditions). Water surface profile changes for various flow rates with variable recurrence intervals in desired river lengths, current velocity values, normal depth, critical depth, hydraulic features and parameters in the river are the output data of the HEC-RAS model [1].

3.3 *Watershed Delineation of Vamanapuram River Basin*

The region upstream from a given outflow point serves as the physical boundary for watersheds or catchments. Both physically using paper maps and digitally using a GIS environment are acceptable methods. In this study, watershed delineation is done using the DEM obtained from Bhuvan (National Remote Sensing Centre, Cartosat-1 A image).

3.4 *Preparation of Flood Inundation Map*

The flood inundation mapping of Vamanapuram River basin is done using HEC-RAS, for which a set of data such as cross-section, streamline and centreline of the river and flood hydrograph are required. The cross-section, streamline and centreline are extracted with the help of HEC-GeoRAS. The flood hydrograph is developed as per the procedure proposed by Central Water Commission (CWC) [10].

3.4.1 Estimation of Flood Hydrograph

Flood hydrograph estimation requires a Synthetic Unit Hydrograph (SUH) and the Probable Maximum Precipitation (PMP) for duration equal to the base period of SUH [4]. The step wise procedure of estimation of SUH and PMP is as follows:

(i)(i) Development of SUH

A catchment's SUH is a unit hydrograph with a unit duration. It was created using relationships between the physiographic and unit hydrograph parameters of sample catchments in a hydro-meteorologically uniform region. Physiographic characteristics of the region include catchment area, length of main stream and equivalent stream slope. Catchment area (A) is considered as the area enclosed within the watershed boundary. Length of main stream (L) as well as the length from centre of gravity to the gauging site (L_c) is obtained from the DEM. Equivalent stream slope (m/km) is obtained from the Eq. 1 [10].

$$s = \frac{\Sigma(L_i(D_{i-1} + D_i)}{L^2} \tag{1}$$

where,

L_i = length of the ith segment (km)

D_{i-1}, D_i = depth of the river at the point of intersection of (i-1) and ith contours from the datum at the level of point of study (m)

L = length of longest stream (km)

Catchment plan, showing the foresaid measurement, is shown in Fig. 1. The various parameters of the 1-h SUH are derived using the following Eqs. (2–10) [10].

$$t_p = 0.553\left(\frac{L * L_c}{\sqrt{S}}\right)^{0.405} \quad (2)$$

$$q_p = \frac{2.043}{t_p^{0.872}} \quad (3)$$

$$Q_p = q_p * A \quad (4)$$

$$T_B = 5.083 * t_p^{0.733} \quad (5)$$

$$T_m = t_p + 0.5t_r \quad (6)$$

where,

t_p = basin lag
q_p = peak discharge per sq.km area of catchment
Q_p = peak discharge of the unit hydrograph for the catchment area (A)
T_B = time base of SUH
T_m = peak time of SUH

Width of SUH at 50 and 75% of peak discharge ordinate Q_p in hours is computed using Eqs. 7–10.

$$W_{50} = 2.197/q_n^{1.067} \quad (7)$$

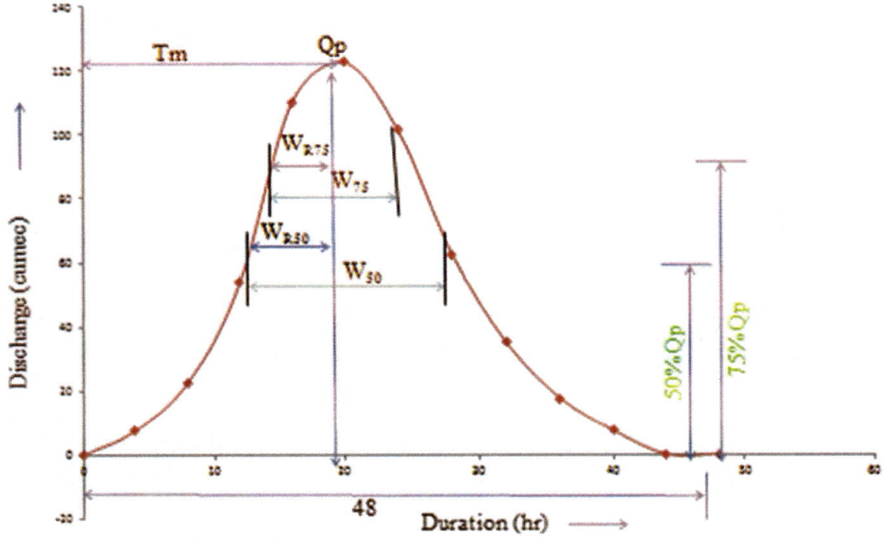

Fig. 1 1-h SUH

$$W_{75} = 1.325/q_n^{1.088} \quad (8)$$

$$W_{R50} = 0.799/q_n^{1.138} \quad (9)$$

$$W_{R75} = 0.536/q_n^{1.109} \quad (10)$$

A pictorial representation of a sample SUH is shown in Fig. 1.

(ii)(ii) Estimation of PMP

It is defined as the greatest depth of precipitation, which is possible for a given time and duration over a given storm area under known meteorological conditions [6]. There are in general two approaches for the estimation of PMP:

- Traditional frequency analysis method based on Chow's general frequency equation.
- Approach specified by World Meteorological Organisation (WMO), 1986, which is basically frequency analysis method.

Since the latter involves graphical methods, the approximation of values varies from person to person. Hence, the traditional frequency analysis is used. It was proposed by Chow [6]. Chow's equation is given in Eq. 11.

$$PMP = X_n + K_m \sigma_n \quad (11)$$

where, X_n = mean of n annual maximum values
σ_n = standard deviation
K_m = frequency factor

$$K_m = \frac{K_m - K_{n-m}}{\sum K_{n-m}} \quad (12)$$

X_m = highest value in the series

$X_{n\text{-}m}$ = mean of the series excluding the highest value

$\Sigma X_{n\text{-}m}$ = standard deviation excluding the highest value

The PMP, thus, obtained is divided into corresponding interval by multiplying its value with the 48-h time duration coefficient and hence the storm rainfall is obtained. The 48-h time duration coefficient is obtained from the Central Water Commission (CWC), Flood Estimation Report for Kaveri basin [10]. The rainfall increment values are then computed. Then the effective rainfall values are obtained by reducing the loss rate, i.e., 0.1 cm/hr (CWC).

(iii) Derivation of flow hydrograph

The rainfall excess increments are arranged against the hourly SUH ordinates such that the maximum value of rainfall excess comes against the peak discharge of SUH, the next lower value of rainfall excess increment comes against the next lower discharge ordinate and so on. The summation of the product of unit hydrograph ordinate and 1-h rainfall excess produces the direct runoff [5]. The design base flow rate is 0.05 cumec per sq.km [10], and hence the total base flow for area of the catchment is obtained by multiplying the catchment area with the design base flow rate and the base flow is finally derived [9]. The base flow is multiplied to the total direct runoff values to get the maximum discharge. The maximum discharge ordinates hence obtained is plotted against the duration to obtain the flood hydrograph.

3.4.2 Extraction of Cross-Section, Streamline and Centreline of the River

The watershed delineation done using ArcHydro tool is used to extract the cross-section, streamline and centreline of the river. The extraction is done using HEC-GeoRAS, which is a customised toolbar in ArcGIS; the RAS geometry menu being mainly used [3]. The polylines (XS cutlines) are created at the points were the cross-section are needed. The cross-section, centreline, bank line are generated by developing corresponding feature class [3]. This acts as a pre-processing input for HEC-RAS one-dimensional model used for comparison with the field data.

3.4.3 Development of Flood Inundation Map

The flood hydrograph and the river geometrical data derived are used to develop the flood inundation details about that particular river cross-section by running the unsteady flow.

condition and hence the rise in water level corresponding to the cross section is obtained. The RAS GIS import file (final river geometry file), which can be used as input for HEC-RAS, is produced by the HEC-GeoRAS model following the creation of the RAS geometry data. The cross-section is then verified, the river geometry is edited and the river geometry file in the HEC-RAS model is finalised and corrected. Boundary conditions are used following the creation of the final river geometry file [3]. The boundary conditions for the upstream generally have a flow or stage hydrograph, and the downstream end has a typical rating or normal depth, internal boundary conditions can be added to define a gate operation or to add a flow within a river system. The unsteady flow simulation is done and the water level for different return periods is generated in HEC-RAS for a particular cross-section. In order to get the flood inundation map of the basin, 2D modelling is to be done [1]. In 2D modelling, a 2D mesh is created over the terrain of the basin, the terrain is the float file of the DEM, which is imported into HEC-RAS from the ArcGIS, the boundary conditions with the main river, tributaries and lower boundary are also set. Unsteady conditions are applied to the boundary conditions with the respective flood hydrograph, which was obtained from the synthetic unit hydrograph that was generated for the basin using the PMP. The inundation map for each return period is then exported into HEC-GeoRAS for final inundation area mapping along the river.

3.5 Flood Damage Assessment

The inundated area is exported in TIF file format and is imported on the HEC-GeoRAS. The land use land cover map of the corresponding river basin is prepared using Arc GIS 10.1 software. It is then used as a layer on which the inundated map is to be overlaid. The land use land cover map may contain attributes such as plantation, builtup area, forest, water body, etc. of the corresponding basin and hence the percentage of inundation over the attributes is known from the attribute table. This helps one to assess the possible damage caused by a flood.

4 Results and Discussion

4.1 General

Daily rainfall data of the past 21 years were collected from Indian Meteorological Department (IMD), Trivandrum. Toposheets of the Vamanapuram basin were collected from Survey of India Department. Watershed was delineated from DEM using HEC-GeoRAS, for which the DEM was obtained from Bhuvan (National Remote Sensing Centre). The process of developing the cross-section and streamline was carried out in HEC-GeoRAS, whereas the development of flood inundation map was done in HEC-RAS. The assessment of the flood impact of Vamanapuram river basin using HEC-GeoRAS software was then carried out.

4.2 Watershed Delineation of Vamanapuram River Basin

Using the DEM acquired from Bhuvan, watersheds or catchments were defined as the region upstream from a specific outlet point, (National Remote Sensing Centre, Cartosat-1 A image). Watershed delineation was done in Arc GIS, and the resultant delineated map is shown in Fig. 2.

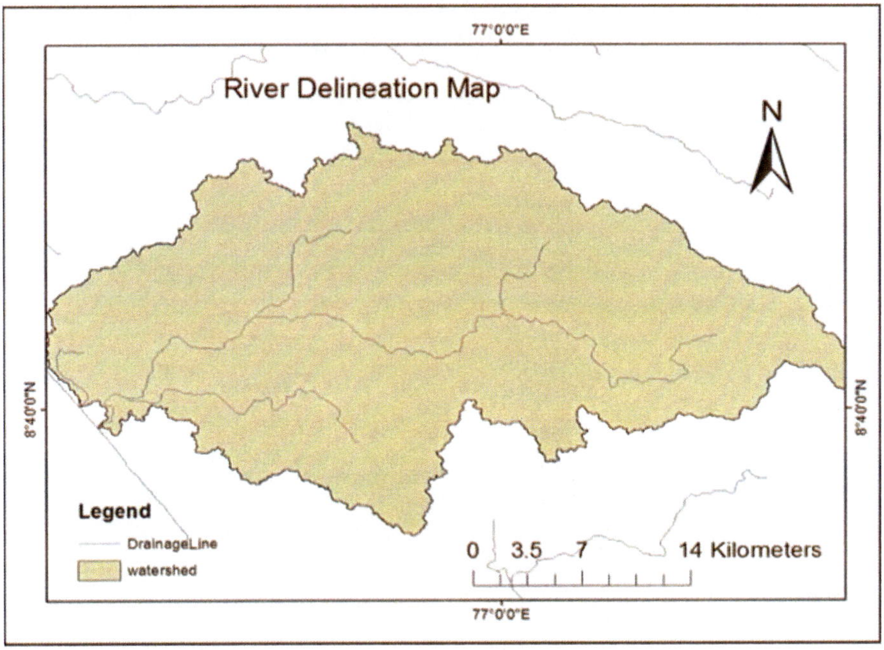

Fig. 2 River delineation map

4.3 *Preparation of Flood Inundation Map*

The flood inundation mapping of Vamanapuram River basin was done using HEC-RAS. The required cross-section, streamline and centreline of the river were extracted using HEC-GeoRAS. The flood hydrograph was developed as per the procedure proposed by Central Water Commission (CWC) [10].

4.3.1 Estimation of Flood Hydrograph

(i) Development of SUH

Catchment area (A) was taken as the area enclosed within the watershed boundary and is equal to 787 km^2. Length of main stream (L) is about 88 km, and the length from centre of gravity to the gauging site (L_c) was obtained from the DEM as 45 km. Equivalent stream slope (m/km) was obtained using the Eq. 1 as 0.411 m/km. The various parameters of the 1-h SUH derived using the Eqs. 2–10 are as follows:

$$t_p = 18.966\ hr$$

$$q_p = 0.1569\ cumecs$$

$$Q_p = 123.655\ cumecs$$

Table 1 Synthetic unit hydrograph ordinates

Duration (hr)	Discharge (cumec)
0	0
4	7.5
8	22.5
12	53.7
16	110
20	122.5
24	101.3
28	62.5
32	35
36	17.5
40	7.5
44	0
48	0

$T_B = 43.942\ hr$

$T_m = 19.466\ hr$

$W_{50} = 15.852\ h\ \mathrm{r}$

$W_{75} = 9.939\ hr$

$W_{R50} = 6.575\ hr\ \mathrm{r}$

$W_{R75} = 4.18\ hr$

The developed SUH coordinates are given in Table 1, and SUH graph is shown in Fig. 4.5.

(i) Estimation of PMP

The PMP is referred to as the maximum depth of precipitation that is feasible across a specific storm area for a given amount of time and duration under established meteorological circumstances. The traditional frequency analysis as proposed by Chow (Eq. 11) is used, and the corresponding parameters are obtained using Eqs. 11 and 12 and are as follows

$X_m = 1671$ cm
$X_{n-m} = 1518$ cm
$\Sigma X_{n-m} = 219$ cm
$X_n = 1408$ cm
$s_n = 239$ cm
$K_m = 0.69$

$PMP = 157.3\ cm$

Table 2 Computed effective incremental rainfall

Duration	TD coefficients	Storm rainfall (cm)	Rainfall increment (cm)	Effective rainfall (cm)
0	0	0	0	0
4	0.3	50	50	50
8	0.5	74	24	23
12	0.6	88	14	14
16	0.7	102	14	14
20	0.7	110	7.9	7.8
24	0.8	120	9.4	9.3
28	0.8	126	6.3	6.2
32	0.9	137	11	11
36	0.9	142	4.7	4.6
40	0.9	146	4.7	4.6
44	1	153	6.3	6.2
48	1	157	4.7	4.6

The Probable maximum precipitation thus obtained is equal to 157.3 cm. It was then divided into corresponding interval by multiplying the PMP value with the 48-h time duration coefficient and hence the storm rainfall was obtained. The computed rainfall increment values and the effective rainfall values are given in Table 2.

(ii) Derivation of flow hydrograph

The base flow of a catchment is obtained by multiplying the design base flow rate with the area of the catchment. The design flow rate as per CWC from flood estimation report Kaveri basin report is 0.005 cumec per sq.km, and hence the total base flow for the catchment of area 788 km^2 was obtained as 39.4 cumec. To obtain the maximum discharge, the entire direct runoff values were added to the base flow. The maximum discharge ordinates as given in Table 3 are plotted against the duration to obtain the flood hydrograph and are shown in Fig. 3.

4.3.2 2D Modelling in HEC-RAS for River Catchment in Vamanapuram Basin

The 2D modelling is done by creating a 2D mesh of the required space in the loaded terrain of Vamanapuram basin. Mesh created had 134,048 cells. The main river, the tributary within the mesh were marked using the tools and a lower boundary is also given, to mark the final flood area.

The flood hydrograph generated is given as an input for the main river and for tributaries, and the normal depth value of 0.01 is given as the input for the lower boundary. The unsteady flow analysis is done respective to the above conditions and

Table 3 Flood hydrograph ordinates

Direct runoff (cumec)	Base flow (cumec)	Flood hydrograph ordinate (cumec)
0	39.4	39.4
34.6425	39.4	74.0425
174.7125	39.4	214.1125
585.9207	39.4	625.3207
2584.45	39.4	2623.85
6153.91	39.4	6193.31
1423.9741	39.4	1463.3741
878.5625	39.4	917.9625
326.83	39.4	366.23
108.36	39.4	147.76
46.44	39.4	85.84
0	39.4	39.4
0	39.4	39.4

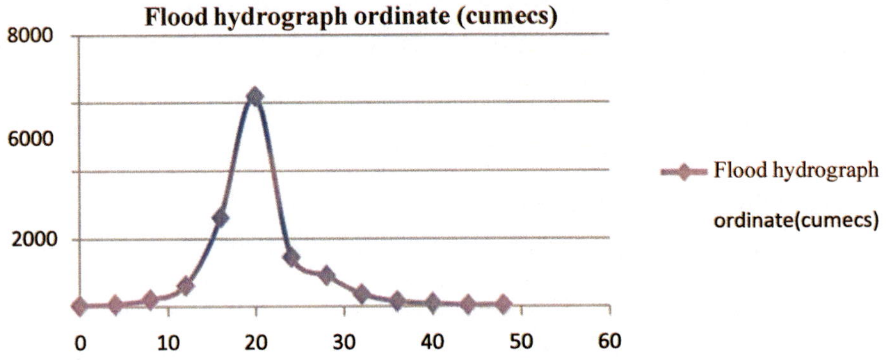

Fig. 3 Flood hydrograph

hence the flood inundation map over the basin is obtained, which is shown in Fig. 6. Apart from the Inundation map, the arrival time and the water surface elevation were also obtained as shown in Figs. 4 and 5.

4.3.3 Flood Damage Assessment in HEC-GeoRAS

The landuse map was obtained from the Landuse Board with major classifications. The flood inundation map obtained was overlayed on the landuse map in order to assess the damage caused over the classification of the landuse. Figure 7 shows the image with the landuse map overlayed with inundation map.

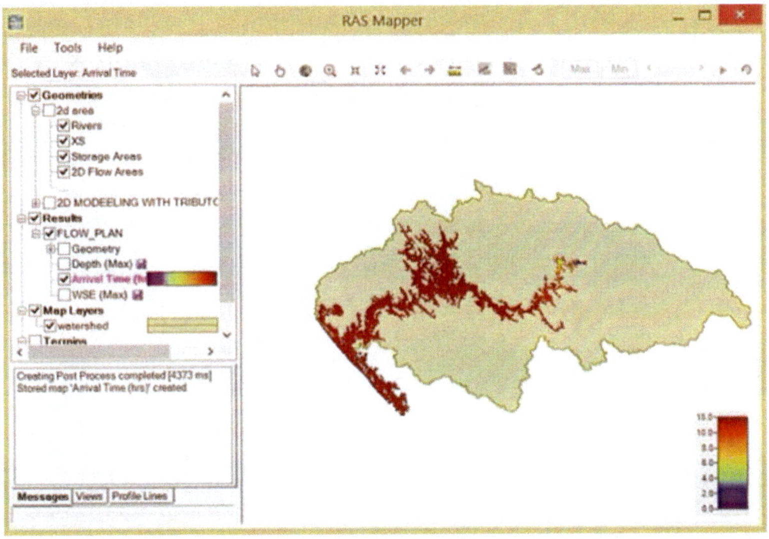

Fig. 4 Arrival time of the flood over the terrain

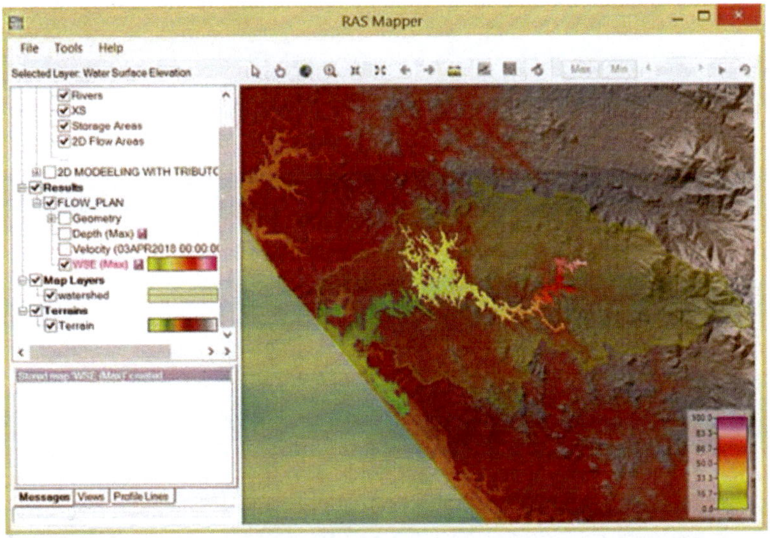

Fig. 5 Maximum water surface elevation of Vamanapuram basin

The main classification contains the water body in that region, Plantation 1 which represent the plantations done on high elevated area example Rubber, Plantation 2 that represent the plantations done on the low-lying areas such as wheat, paddy etc., the other classifications are mixed vegetation, forest, builtup area and barren land.

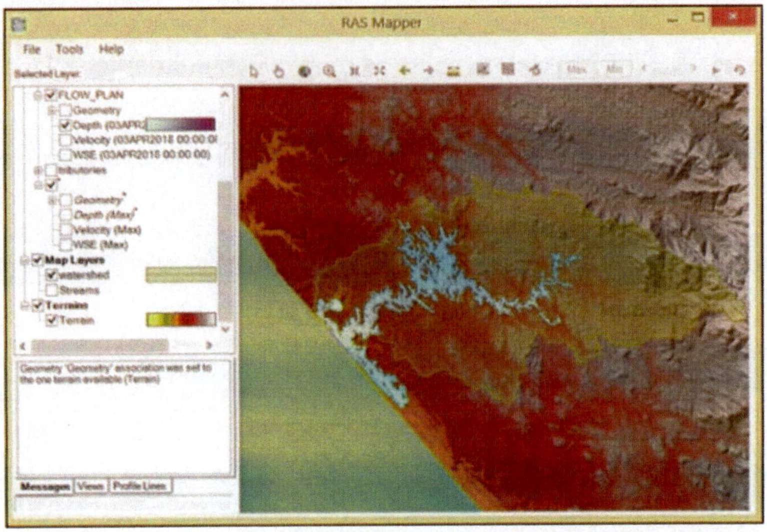

Fig. 6 The flood inundation map of Vamanapuram basin

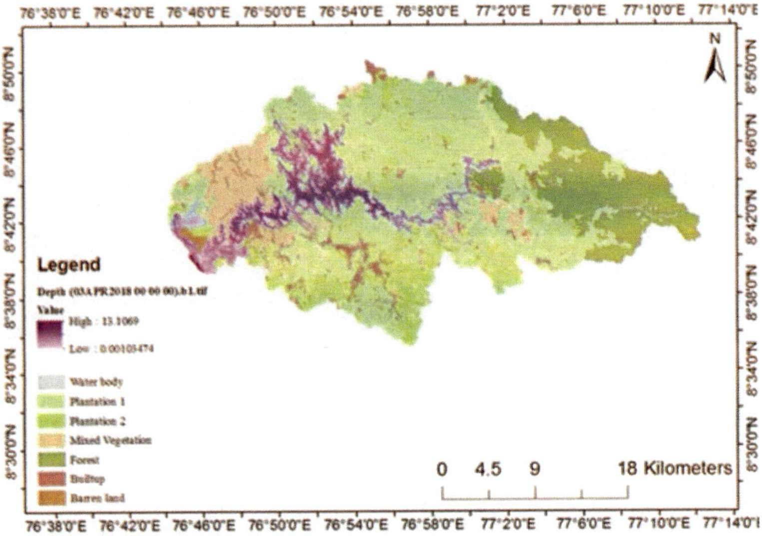

Fig. 7 Flood inundation map over landuse map

The damage area affected on each classification is individually calculated by clipping the flood inundation over the landuse classification (Table 4).

The area of the classification water body is 32.12 km^2, and hence the total area of the catchment becomes 787 km^2. Therefore, the total flood damage of the basin thus obtained is 260.383 km^2 and is 33.085% of the total area.

Table 4 Classification-wise damage assessment

Classifications of the landuse map	Total area of the classification on the land use map (km^2)	Clipped area excluding the flood-inundated region (km^2)	Total damage area of the classification (km^2)	Percentage of damage in each classification (%)
Plantation 1	321.802	174.8031	146.998	45.679
Plantation 2	94.662	65.2983	29.3637	31.019
Mixed vegetation	205.994	128.17004	77.82396	37.779
Forest	122.249	119.7361	2.5129	2.055
Builtup	4.6201	3.5279	1.0922	23.640
Barren land	5.5532	2.9610	2.5922	46.679

5 Conclusion

The water delineation for the Vamanapuram basin was obtained using ArcHydro tool from the digital elevation model. 21 years of daily rainfall value of the Vamanapuram region was collected from Indian Meteorological Department. Using 1-h synthetic unit hydrograph and the probable maximum precipitation values, the flow hydrograph was developed. The Vamanapuram basin terrain was then used to create the flood inundation map using the HEC-RAS programme. The flood damage assessment of Vamanpuram basin was done in HEC-GeoRAS and ArcGIS.

The PMP value obtained was 157.3 cm and a 48-h duration flood hydrograph was derived. The total area was found to be 787 km^2 in which the inundated area was about 260.383 km^2 and was 33.085% of the total area. The landuse classification that affected the most was barren land, about 46.679% of the classification was affected due to the flood. The remaining classifications such as Plantation 1, Plantation 2, mixed vegetation, forest and builtup area in which the area affected are 45.679%, 31.019%, 37.779%, 13.423%, 2.055% and 46.679%, respectively.

References

1. Ali H, Lee Teang S, Majid M, Hadi M (2012) Incorporation of GIS based program into hydraulic model for water level modeling on river basin. J Water Resour Protect
2. Chithra M, Sumam KS (2014) Flood risk assessment by GIS and hydraulic model. In: Proceedings of international conference on materials for the future—innovative materials, processes, products and applications—ICMF, pp 315–318
3. Getahun AA, Gebre SL (2015) Flood hazard assessment and mapping of flood inundation area of the Awash River Basin in Ethiopia using GIS and HEC-GeoRAS/HEC-RAS. J Civil Environ Eng 5(4):1–12
4. Singh KJ, Sarkar S (2012) Development of GIUH for the catchment contributing to Loktak lake, North East India. J Ind Soc Remote Sens
5. Reddy NA, Seelam JK, Rao S, Nagaraj MK (2018) Flood estimation at ungauged catchments of western catchments of Karnataka, West coast of India. ISH J Hydraul Eng

6. Kourgialas NN, Karatzas GP (2011) Flood management and a GIS modelling method to assess flood-hazard areas—a case study. Hydrol Sci J 56(2):211–225
7. Lan P, Lin B, Zhang Y, Chen H (2017) Probable maximum precipitation estimation using the revised km-value method in Hong Kong. J Hydrol Eng
8. Sanat Nalini Sahoo and, Pekkat, Sreeja (2017) Development of flood inundation maps and quantification of flood risk in an urban catchment of Brahmaputra River. J Risk Uncertainity Eng Syst 20:1–11
9. Tazyeen S, Nyamathi SJ (2015) Flood routing in the catchment of urbanized lakes. Aquatic Procedia
10. Flood Estimation Report for Kavery Basin Subzone – 3 (i), (1985) Central Water Commission

Flood Hazard Mapping for Amaravati Region Using Geospatial Techniques

Sampath Kumar, Talari Reshma, Savitha Chirasmayee, Kasa Priyanka, Kokku Priyanka, and Gokla Ram

Abstract Floods pose a significant natural hazard that can cause extensive damage to infrastructure, property, and human life. The use of geospatial techniques, such as remote sensing, Global Positioning Systems (GPS), and Geographic Information System (GIS), has become increasingly important for flood hazard mapping. The research paper aims to develop flood hazard maps for the Amaravati region, the new capital city of Andhra Pradesh, India, using geospatial techniques, including an Analytic Hierarchy Process (AHP). The region is susceptible to floods due to its location and topography, which includes low-lying areas and water bodies. In the present study, remote sensing techniques were employed to extract information on land use, land cover, topography, and drainage patterns, and GIS to integrate this information with other relevant data, such as rainfall and river flow data, to develop flood hazard maps. Additionally, the study utilized the AHP process, a multi-criteria decision-making method, to weigh and rank various factors involved in the development of the d hazard maps. The AHP process provided a structured and systematic approach to prioritize the importance of different variables and factors in the flood hazard mapping process. The developed maps provide valuable information for decision-makers, urban planners, and emergency management agencies to plan for and mitigate the impact of potential flood events in the region. The AHP process, in combination with geospatial techniques, contributed to the accuracy and reliability of the flood hazard maps. By utilizing geospatial techniques and the AHP, this research paper contributes to the existing knowledge on flood hazard mapping. The findings provide valuable insights that can be applied in flood risk management and disaster preparedness in the Amaravati region.

Keywords Analytic hierarchy process · Amaravati · Flood hazard maps · Geographic information systems · Remote sensing

S. Kumar · T. Reshma · S. Chirasmayee (✉) · K. Priyanka · K. Priyanka · G. Ram
National Institute of Technology Andhra Pradesh, Tadepalligudem, India
e-mail: pj032101@student.nitandhra.ac.in

T. Reshma
e-mail: reshma@nitandhra.ac.in

S. Mesapam et al. (eds.), *Developments and Applications of Geomatics*,
Lecture Notes in Civil Engineering 450, https://doi.org/10.1007/978-981-99-8568-5_19

1 Introduction

Floods are one of the most devastating natural disasters that cause widespread damage to infrastructure, property, and human life worldwide. With the increasing frequency and severity of floods, accurate flood hazard mapping has become crucial for mitigating and managing the impact of these disasters. Flood hazard mapping involves the identification and assessment of areas at risk of flooding based on factors such as topography, land use, and hydrological data. Geospatial has emerged as powerful tools for flood hazard mapping. Remote sensing images provide high-resolution data on land use, land cover, and topography, while Geographic Information systems (GIS) enables the integration and analysis of multiple datasets, including rainfall and river flow data. The integration of these technologies has resulted in more accurate and reliable flood hazard maps.

Several studies have highlighted the significance of geospatial techniques for flood hazard mapping. Flood hazard zonation mapping using GIS emphasized its effectiveness in identifying flood-prone areas is reviewed [1]. Few studies [2, 3] assessed flood hazard using remote sensing and GIS and concluded that remote sensing techniques provide accurate data for flood hazard mapping. Reference [4] conducted a case study on flood hazard mapping using remote sensing and GIS and machine learning in Yuyao, China and demonstrated its usefulness in identifying high-risk areas. Few studies [5] developed novel flood hazard maps using remote sensing and GIS in the Mahanadi River Basin, India and found that it was effective in predicting potential flood-prone areas. Sharma et al. [6] developed flood hazard maps using remote sensing and GIS in the Alaknanda River Basin, India and showed that the technology can provide valuable insights for flood risk management.

Moreover, several studies have demonstrated the effectiveness of the Analytic Hierarchy Process (AHP) for flood hazard mapping. The AHP process was used to develop a flood hazard map for the River Basin in Philippines [7]. The authors found that AHP helped to identify the most critical factors affecting flood risk and enabled the creation of a more accurate flood hazard map. References [8, 9] employed AHP to develop a flood hazard map for the different river basins, India. The study demonstrated that AHP can be used to assess and rank different factors affecting flood risk, resulting in a more reliable flood hazard map. Furthermore, the integration of AHP with other geospatial techniques, such as remote sensing and GIS, has resulted in more accurate and reliable flood hazard maps. A recent study by Gupta and Dixit [10] used AHP to develop a flood hazard map for the Assam region, India and found that the integration of AHP with geospatial techniques resulted in a more comprehensive understanding of flood risk.

Flood hazard mapping is essential in areas susceptible to floods, such as the Amaravati region of Andhra Pradesh, India which is the capital of the state. Geospatial techniques can aid in flood hazard mapping by providing high-resolution data on land use, topography, and hydrological characteristics. The integration of these geospatial techniques with AHP can enhance the accuracy of flood hazard maps by ranking and assessing the relative importance of various factors affecting flood risk. In this paper, we aim to develop a flood hazard map for the Amaravati region using geospatial techniques and AHP. The map will serve as a valuable tool for flood risk management, disaster preparedness, and infrastructure planning.

2 Study Area

The study area of this research is the Amaravati region, located in the state of Andhra Pradesh in India, between the coordinates of 16.58° N, 16.29°S, 80.67° E, 80.16°W, as depicted in Fig. 1. It covers an area of 1020.632 square kilometers, encompassing the regions of Tadikonda, Pedakurapadu, Thullur, Vaddeswaram in Guntur district, and the outskirts of Amaravati, Satenapalle, Guntur, Mangalagiri in Guntur district, and Vijayawada in Krishna district, as shown in Fig. 1. Amaravati is situated between the Krishna River and the low hills of the Deccan Plateau, and it is the point where the Kondaveeti Vagu meets the Krishna River, causing inundation of 13,500 acres of land during the monsoon season each year. Amaravati is known for its rich cultural heritage as an ancient Buddhist center, and it is a popular destination for sightseeing and pilgrimage. From various studies, it is observed that 70% of the Amaravati area is vulnerable to flooding, and the region experiences floods three times on average every year, with severe flooding occurring in 1977, 1998, 2009, and 2019 in the Krishna River. The region also experiences frequent water shortages and droughts. These observations demonstrate the need for effective flood management and mitigation strategies to minimize the adverse impacts of floods on the region's socio-economic and environmental well-being.

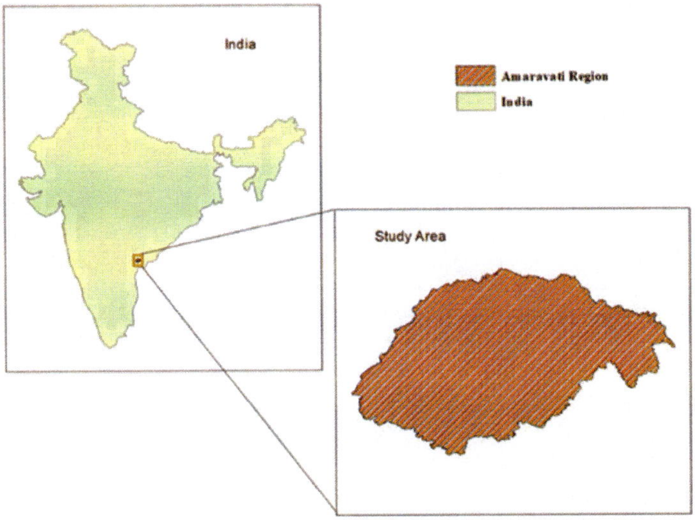

Fig. 1 Location of Amaravati study area

3 Methods and Methodology

3.1 Data Used

In this study, we obtained essential data from various sources to analyze the watershed characteristics and develop effective management strategies for water resources. Specifically, we collected and downloaded Digital Elevation Model (DEM) data from the USGS Earth Explorer website, Landsat 8 data for generating the Land Use and Land Cover (LULC) map, and soil data from the FAO website as shown in Table 1 for the year 2019. We utilized flow direction data as input for preparing the Flow Accumulation Map, and we generated the slope map and rainfall intensity map using geospatial technologies. The seven maps produced using these geospatial technologies are essential inputs for understanding the hydrological processes and identifying the areas that are prone to floods and water scarcity in the region are shown in Fig. 2a–i. These maps serve as a critical resource for policymakers and stakeholders to develop sustainable water resource management strategies for the region.

3.2 Methodology

In the present study, AHP a widely used multi-criteria decision-making method for analyzing complex decisions is used for preparing flood hazard maps. It is used for ranking a set of alternatives or selecting the best alternative among the given alternatives. In this study, using AHP, the flood hazard map of the Krishna River basin is obtained by choosing six different parameters that influence flood occurrence and assigning weightage to each parameter as shown in Table 2 and Fig. 3. The AHP method involves several steps. Firstly, the problem and criteria are defined. In this study, the problem is to estimate the flood hazard map of the Krishna River basin, and the criteria include six parameters, namely, slope, elevation, flow accumulation, land use and land cover, rainfall intensity, and distance from the drainage

Table 1 Sources of data collection of different process

Factor	Source	Download URL
Elevation (ASTER DEM)	USGS earth explorer	https://earthexplorer.usgs.gov/
Land use and landcover (Landsat 8)	USGS earth explorer	https://earthexplorer.usgs.gov/
Rainfall intensity	IMD Pune	https://www.imdpune.gov.in/
Soil (global soil maps)	FAO website	https://www.fao.org/home/en
Slope, flow accumulation, distance from drainage network	Generated using DEM in ArcMap 10.7.1 (licensed version)	–

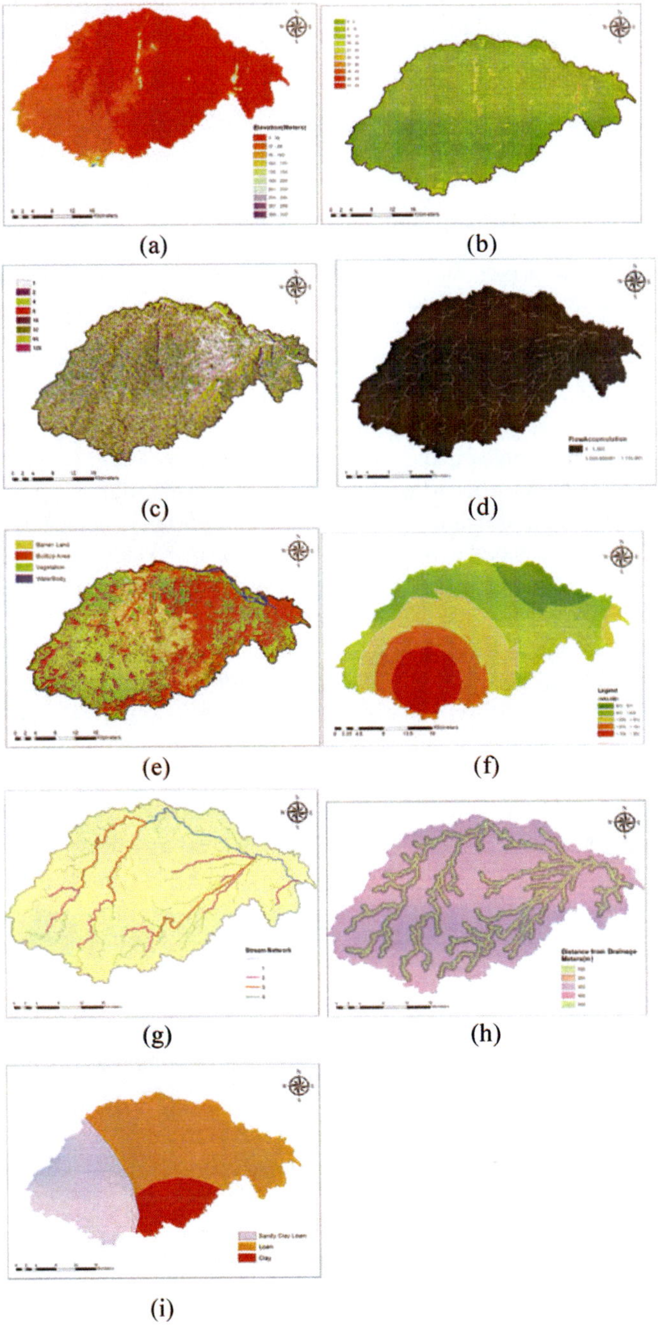

Fig. 2 **a** Digital elevation model, **b** slope map, **c** flow direction map, **d** flow accumulation map, **e** land use and land cover map, **f** rainfall intensity map, **g** stream order map, **h** distance from drainage map, **i** soil map

Table 2 Relation between factors that influence flood risk

Factor changing	Major effect	Minor effect	Factor rate (FR)
Flow accumulation (F)	LULC	S	1.5
Slope (S)	LULC, F	–	2
Land use and landcover (LULC)	F, R	S	2.5
Rainfall intensity (R)	F	LULC	1.5
Elevation (E)	LULC, F, R	S	3.5
Distance from drainage network (D)	LULC, F, R	S	3.5
Soil (A)	LULC, F, S	–	3

network. Secondly, the alternatives are defined. In this study, the alternatives refer to the hazardous areas within the study area. Next, priority is established amongst the criteria and alternatives using pairwise comparison. The pairwise comparison involves comparing each criterion or alternative with every other criterion or alternative and assigning a score that reflects the relative importance or preference between them. Later, the consistency amongst the pairwise comparison is checked to ensure the validity of the comparisons. Next, the relative weights are evaluated from the pairwise comparisons, and the overall priorities for the alternatives are calculated based on these weights. By employing the methodology presented by Kourgialas and Karatzas (2016), the weightage of each parameter is determined. The last step involves conducting a sensitivity analysis to evaluate how variations in the weightage allocated to each parameter affect the robustness of the outcomes. Overall, the AHP method provides a step-by-step approach to decision-making, and it can be applied to various complex problems, including flood hazard mapping. The method is effective in identifying the most important factors that influence the decision and assigning appropriate weights to these factors, thereby providing a systematic and transparent decision-making process.

In this study, a GIS-based weighted linear combination approach was used to combine the seven maps that were generated as shown in Eq. 1. The technique involved assigning a percentage weight to each factor and multiplying it by the corresponding factor. The resulting products were summed up to create the final map of the hazardous area. Figure 3 illustrates the detailed methodology that was employed in this process.

$$\textbf{Flood Risk Map} = \sum X_i W_i = FW_r + SW_s + LW_l + RW_r + EW_e + DW_d + AW_a \tag{1}$$

where X_i = the map of each parameter i, W_i = the weight of each parameter i, F = flow accumulation factor, S = slope factor, L = land use factor, R = rainfall intensity factor, E = elevation factor, D = distance from drainage network. A = slope map and the respective weights is shown in Table 3 and total flood hazard area is shown in Table 4.

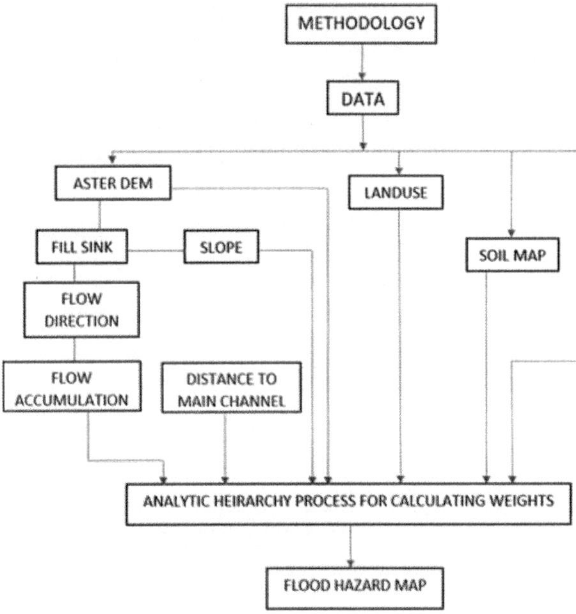

Fig. 3 Flowchart of methodology

Table 3 Weightage of factors causing flood risk

Factors	Range of effect	Description	Weight of effect (A)	Rate (B)	Weighted rating (A × B)	Total weight	Percentage (%)
Elevation	153–332	Very high	5	1.5	7.5	22.5	8.62
	86–152	High	4		6		
	47–85	Moderate	3		4.5		
	27–46	Low	2		3		
	3–26	Very low	1		1.5		
Slope	22–57	Very high	5	2	2	30	11.49
	12–21	High	4		4		
	7–11	Moderate	3		6		
	4–6	Low	2		8		
	0–3	Very low	1		10		
Flow accumulation	601,227–1,110,961	Very high	5	1.5	7.5	22.5	8.62
	304,971–601,226	High	4		6		
	161,199–304,970	Moderate	3		4.5		
	52,282–161,198	Low	2		3		
	0–52,281	Very low	1		1.5		

(continued)

Table 3 (continued)

Factors	Range of effect	Description	Weight of effect (A)	Rate (B)	Weighted rating (A × B)	Total weight	Percentage (%)
Landuse and landcover	Water Body	Very high	5	3	15	33	12.64
	Vegetation	Moderate	3		9		
	Built-up area	Low	2		6		
	Barren land	Very low	1		3		
Rainfall intensity	1134–1202	Very high	5	1.5	7.5	22.5	8.62
	1075–1133	High	4		6		
	1026–1074	Moderate	3		4.5		
	978–1025	Low	2		3		
	876–977	Very low	1		1.5		
Distance from drainage		Very high	5	4.5	22.5	67.5	25.86
		High	4		18		
		Moderate	3		13.5		
		Low	2		9		
		Very low	1		4.5		
Soil	Loam	Moderate	3	3	9	18	6.9
	Sandy clay loam	Low	2		6		
	Clay	Very low	1		3		

Table 4 Flood hazard area

Flood impact	Area (in Km^2)	Percentage of area (%)
High	4.603	0.4
Moderate	299.9	29.63
Low	616.206	60.87
Very low	91.312	9.02
Total	1012.24	100

4 Results and Discussion

The generation of results in this study involved the preparation of seven input maps: Digital Elevation Model (DEM), Slope Map, Flow Accumulation Map, Land Use and Land Cover Map, Rainfall Intensity Map, Soil Map, and Distance from Drainage Map as shown from Figs. 4, 5, 6, 7, 8, 9 and 10. To ensure an equal class distribution for all maps and account for their relative importance, each input map was reclassified into five classes using the reclassify tool in ArcGIS software. This allowed us to spread the weightage values and obtain a more accurate output of the final hazard map shown in Fig. 11. The reclassified maps are listed below. The reclassified maps are as follows.

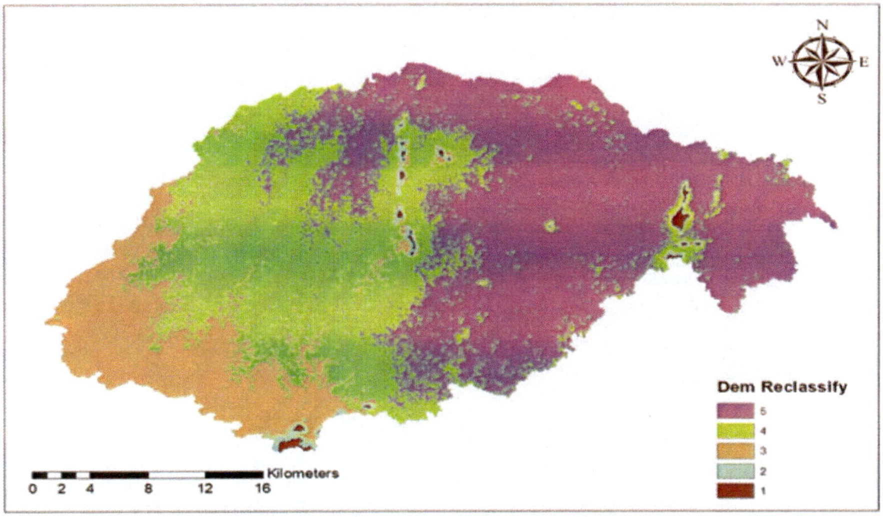

Fig. 4 Reclassified DEM map

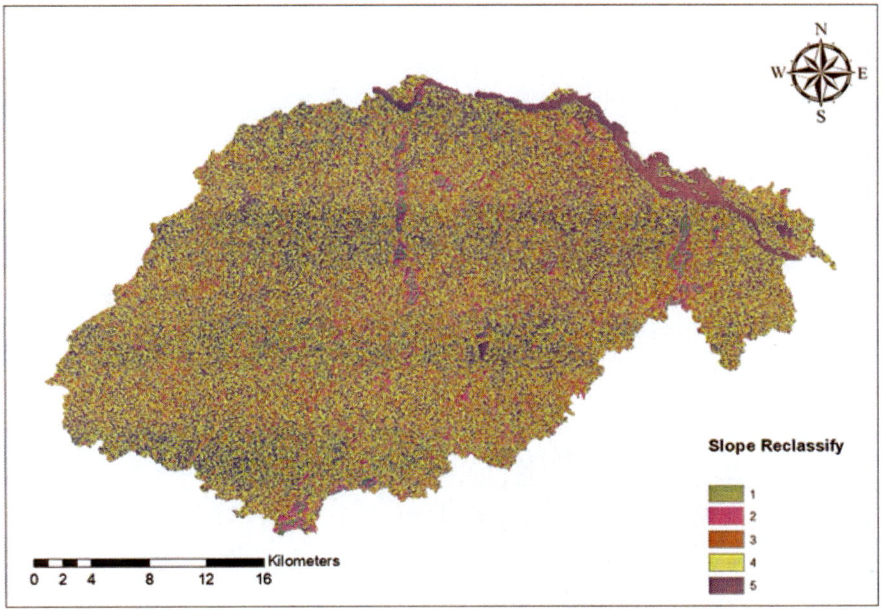

Fig. 5 Reclassified slope map

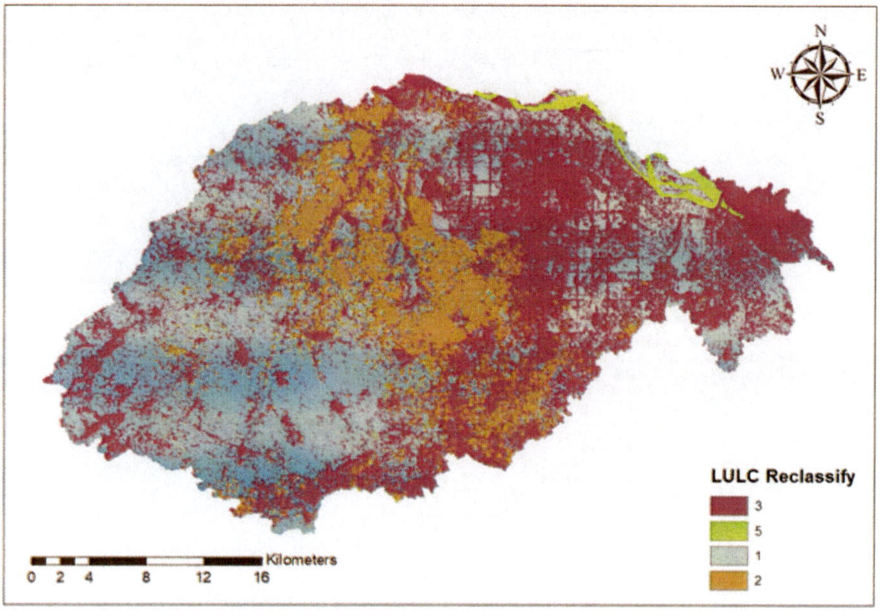

Fig. 6 Reclassified LULC map

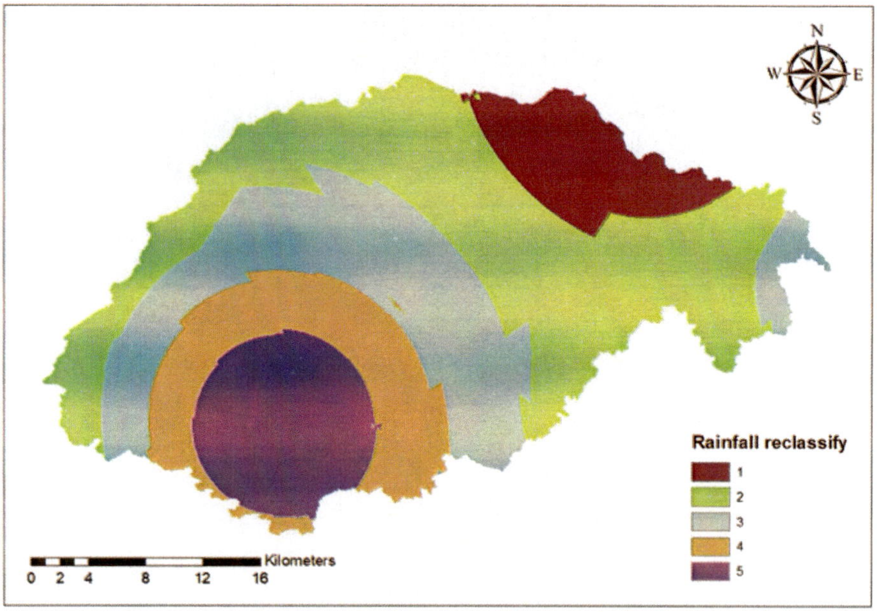

Fig. 7 Reclassified rainfall intensity map

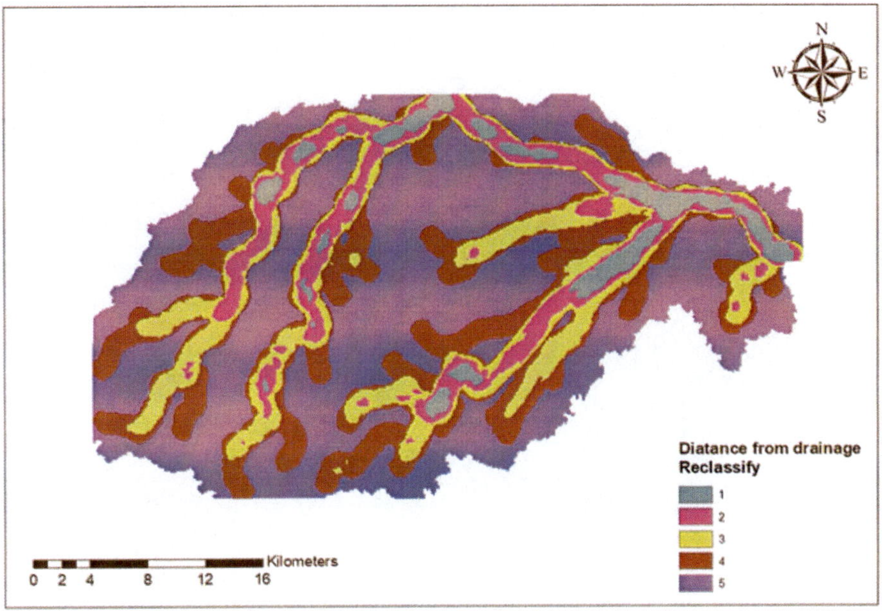

Fig. 8 Reclassified distance from drainage map

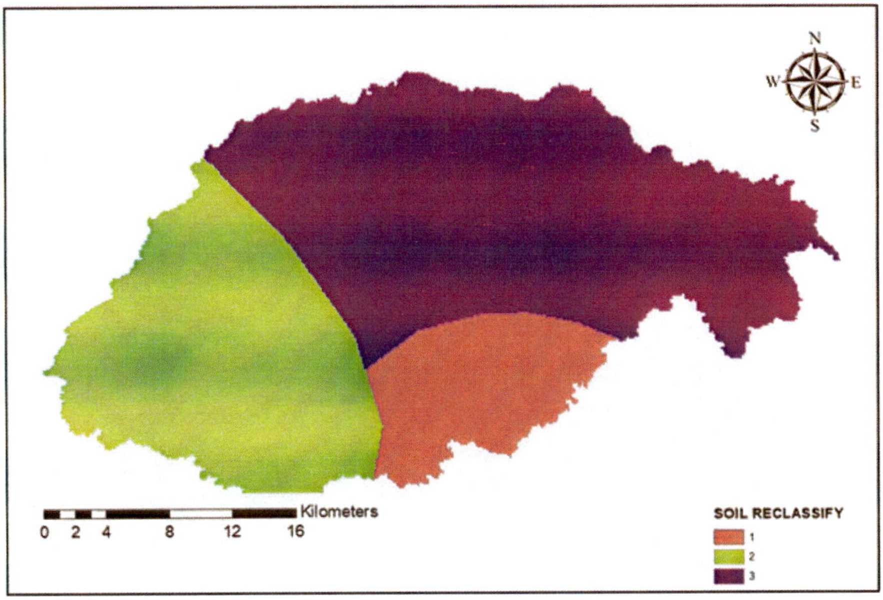

Fig. 9 Reclassified soil map

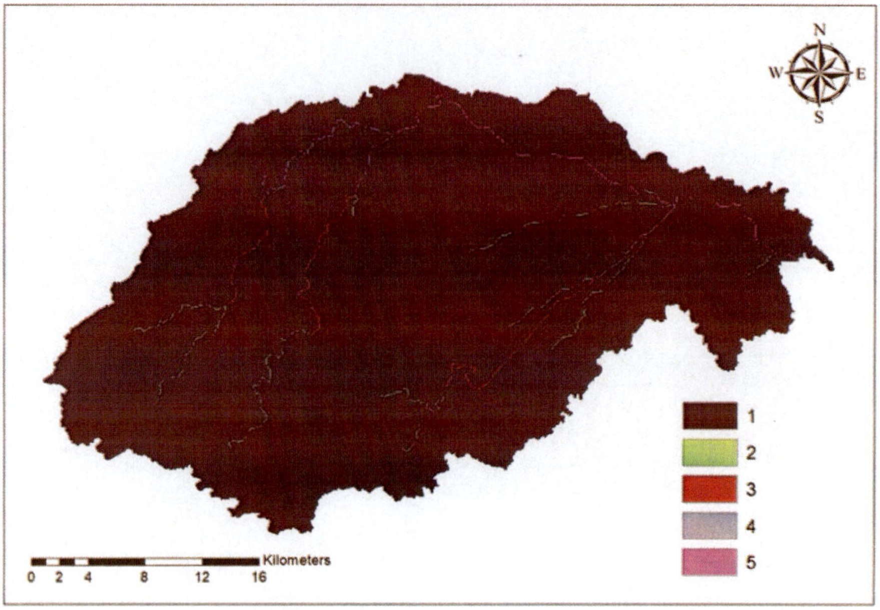

Fig. 10 Reclassified flow accumulation map

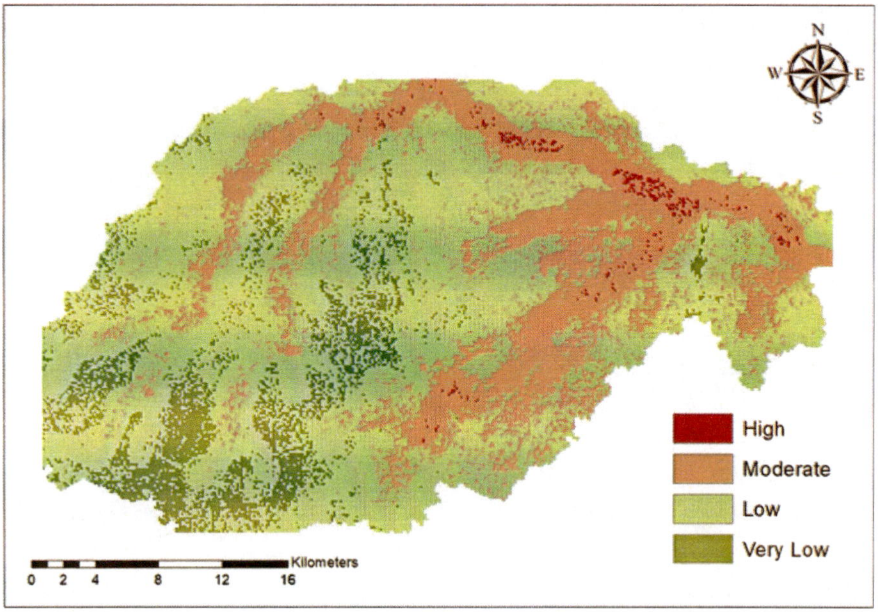

Fig. 11 Final flood hazard map

After the reclassification of these maps into the required number of classes (say 5 here), all these maps are given as input to the Weighted Overlay Table tool in ArcGIS software with the respective weights of each map. This tool makes all these maps overlay on each other with respect to the weightages given (sum of all the given weightages should be 100) and give the output of the final hazard map based on all the given inputs, which are a part of AHP. Finally, the final hazard map is shown in Fig. 11.

The study aimed to assess flood risk in a selected study area of 1012.24 Km^2 and identify areas vulnerable to flooding. Our modeling results revealed a high-hazard flood zone covering 4.603 Km^2, indicating the significant vulnerability of this region. Conversely, a vast area of 616.206 Km^2 was observed to be less prone to flooding, reflecting the impact of human interventions, such as infrastructure development and flood management measures. These findings have significant implications for flood risk reduction and mitigation strategies in the study area. The identification of high-risk flood zones can inform targeted investments in flood protection measures, reducing the potential impact of future flood events on people's lives and livelihoods. Furthermore, our study highlights the need for continued investment in flood risk reduction measures to minimize the impact of natural disasters on vulnerable communities. Overall, our results provide valuable insights into flood risk assessment and management and can inform future policy decisions aimed at improving the resilience of communities in flood-prone regions. However, there are lack of monitoring datasets for calibration and validation is one of the limitations of the present study; However, demonstrates the power of geospatial technology in flood risk analysis and decision-making. The comprehensive flood hazard and flood shelters model developed in ArcGIS can be easily used by novice GIS users, allowing for effective flood management and disaster response. With the increase in rainfall due to climate changes leading to more floods, the use of geospatial data and AHP process will provide a valuable source of analysis.

5 Conclusions

Floods are natural phenomena that cannot be prevented, but human activity is exacerbating their impact. This study aimed to create a flood hazard map of the Amaravati region using remote sensing and GIS techniques. By analyzing data from seven different factors, we identified low-lying areas near the stream network as particularly susceptible to flooding during heavy rains. However, our focus area, Amaravati, experiences less rainfall and is less vulnerable to floods. This study demonstrates the power of geospatial technology in flood risk analysis and decision-making. The comprehensive flood hazard and flood shelters model developed in ArcGIS can be easily used by novice GIS users, allowing for effective flood management and disaster response. The final flood hazard map generated from our analysis can inform preventive measures and evacuation plans in flood-prone areas while building designs can be modified to reduce human and financial losses. The methodology adopted in the present study using the AHP approach can be widely applied to generate flood hazard maps for other regions. This study supports the conclusion that remote sensing and GIS technology can provide valuable information to operational users for planning flood-related emergency responses. Ultimately, this study highlights the importance of utilizing technology to improve disaster preparedness and response to natural disasters.

References

1. Mudashiru RB, Sabtu N, Abustan I, Balogun W (2021) Flood hazard mapping methods: a review. J Hydrol 603:126846. https://doi.org/10.1016/j.jhydrol.2021.126846
2. Alarifi SS, Abdelkareem M, Abdalla F, Alotaibi M (2022) Flash flood hazard mapping using remote sensing and GIS techniques in Southwestern Saudi Arabia. Sustainability 14(21). https://doi.org/10.3390/su142114145
3. Elkhrachy I (2015) Flash flood hazard mapping using satellite images and GIS tools: a case study of Najran City, Kingdom of Saudi Arabia (KSA). Egypt J Remote Sens Space Sci 18(2):261–278. https://doi.org/10.1016/j.ejrs.2015.06.007
4. Feng Q, Gong J, Liu J, Li Y (2015) Flood mapping based on multiple endmember spectral mixture analysis and random forest classifier—the case of Yuyao, China. Remote Sens 7(9). https://doi.org/10.3390/rs70912539
5. Surwase T, Manjusree P, Nagamani PV, Jaisankar G (2019) Novel technique for developing flood hazard map by using AHP: a study on part of Mahanadi River in Odisha. SN Appl. Sci. 1(10):1196. https://doi.org/10.1007/s42452-019-1233-6
6. Sharma VS, Naithani BP, Singh M (2014) Remote sensing and GIS approach for hazard vulnerability assessment of upper Alaknanda Basin, Garhwal Himalaya (Uttarakhand), India. In: Landscape ecology and water management. Tokyo, pp 273–286. https://doi.org/10.1007/978-4-431-54871-3_20
7. Morales FJ, De Vries W (2021) Establishment of natural hazards mapping criteria using analytic hierarchy process (AHP). Front Sustain 2:667105. https://doi.org/10.3389/frsus.2021.667105
8. Ghosh A, Kar SK (2018) Application of analytical hierarchy process (AHP) for flood risk assessment: a case study in Malda district of West Bengal, India. Nat Hazards 94(1):349–368. https://doi.org/10.1007/s11069-018-3392-y
9. Mahapatra M, Ramakrishnan R, Rajawat AS (2015) Coastal vulnerability assessment using analytical hierarchical process for South Gujarat coast, India. Nat Hazards 76(1):139–159. https://doi.org/10.1007/s11069-014-1491-y
10. Gupta L, Dixit J (2022) A GIS-based flood risk mapping of Assam, India, using the MCDA-AHP approach at the regional and administrative level. Geocarto Int 37(26):11867–11899. https://doi.org/10.1080/10106049.2022.2060329

GIS and RS-Based Soil Erosion and Sediment Yield Modelling in Manikpur, Chhattisgarh, India

B. Himajwala and A. D. Prasad

Abstract The Manikpur coalfield is 300 km^2 and is in the Korba district of the Indian state of Chhattisgarh. In India, one of the most severe issues is soil erosion. Calculating exact soil erosion over a specific time is extremely difficult. Huge quantities of mining waste are typically released as over burden dump (OBD) materials by opencast mines. They impact agriculture since they are prone to soil erosion, sedimentation, and poor water quality. An empirical equation of the Revised Universal Soil Loss Equation (RUSLE) and the Sediment Delivery Distributed Model (SEDD) were employed to quantify soil erosion and sediment yield. The findings of these equations were contrasted to those of direct field measurements obtained around the opencast region utilizing an appropriate suspended sediment sampler. The maximum soil erosion value obtained is 79.2 tons/ha/year, and the maximum sediment yield is 57.92 tons/ha/year, according to the findings of this study. The sediment load values from these two models follow the same trend. These models were evaluated for performance and found to be effective, with values of simulated and observed sediment yields $R^2 = 0.81$ and RRMSE = 0.66. The observations revealed that most of the area has a low slope gradient. The opencast mine and overburden dump area have a slight erosion potential due to the higher slope inclination. According to the study, GIS is an effective tool for modelling soil erosion potential and sediment output.

Keywords GIS · OBD · SEDD

1 Introduction

Soil erosion is "the removal of soil from the exposed surface of land due to the impact of raindrops and running water, including rainfall-runoff, melting snow and ice, wind or gravity, and human activities." Soil erosion is mainly caused by deforestation, cattle overgrazing, building, and mining operations. The soil sheet is removed during opencast mining, and over burden dumps (OBD) are built up from the fractured

B. Himajwala (✉) · A. D. Prasad
Department of Civil Engineering, NIT Raipur, Raipur, CG 492010, India
e-mail: bandhuhimajwala123@gmail.com

S. Mesapam et al. (eds.), *Developments and Applications of Geomatics*,
Lecture Notes in Civil Engineering 450, https://doi.org/10.1007/978-981-99-8568-5_20

rock [1]. The dumping materials are dispersed on the ground as OBD; they cover a significant amount of land, preventing it from serving its intended purpose and, as a result, degrading soil quality [2]. On the other hand, the mining sector has a bad reputation since it can endanger public health and safety and harm a local, regional, and global scale; the environment includes land, soil, water, and trees [3]. They cover a vast region of ground that is no longer used for its intended purpose and has poorer soil quality [4]. Surface mining areas can produce 100–2,000 times more sediment than forested areas and 10 times more sediment than grazing areas [5]. Physical and human factors influence sediment generation and delivery [6]. RUSLE utilizes a combination of Geospatial Information System (GIS) and Remote Sensing (RS) to more precisely estimate annual soil loss in the catchment (GIS).

2 Study Area

Manikpur opencast mine is situated in the Korba Coalfield, in the Korba district of Chhattisgarh, India. The study area depicted in (Fig. 1) is 115.4 km^2 area. The region's terrain is plain, with open mines, forests, and streams, among other things. The average yearly temperature in the area is between 45 and 55 °C.

Fig. 1 Study area with sample collection locations and soil erosion observed in the studied region: **a** Rill and inter-rill erosion at location no.17, **b** stream bank erosion at location no. 3

2.1 Data Used

From 2011 to 2021, daily rainfall data were obtained from (SECL, Manikpur) for six rain gauge stations in the study area: Manikpur, Aurai, Nagoi, Samarkana, and Pondi. The velocity of a stream was determined using current meter revolutions at various cross-sections using surface water samples taken in the Manikpur coalfield region at 13 locations. Digital Elevation Model (DEM) data from the Shuttler Radar Topographic Mission produced a model with a cell size resolution of 30 m × 30 m (SRTM, USGS). The data on land use, land cover, and soil texture were provided by the Chhattisgarh Council of Science and Technology (CCOST), Raipur, Chhattisgarh.

3 Methodology

This describes how to use the direct approach as well as the RUSLE model to calculate the yearly average soil loss rate and sediment yield. The graphs illustrate sediment production and soil degradation for the chosen parameters in (Fig. 2).

3.1 Soil Erosion and Sediment Yield Modelling Using Empirical Equations

In this study, the RUSLE model is used to determine how much soil is lost to the watershed. Climate, soil qualities, topography, and land cover characteristics are the primary elements impacting soil erosion, and the RUSLE model established an equation to reflect these factors [7]. The RUSLE is an empirical soil loss forecasting

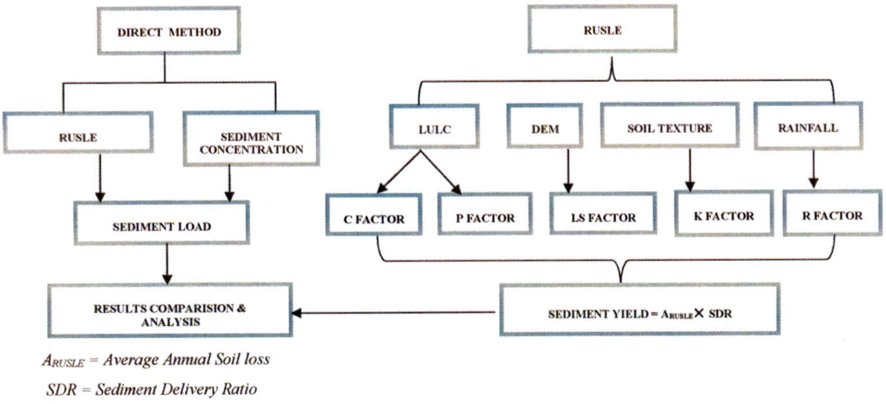

Fig. 2 Methodology

model which predicts mean yearly soil loss and sediment yield [8]. quantitatively forecasts soil erosion. It evaluates the yearly averages of inter-rill and rill erosion-related soil loss and sediment [9]. This is how the formula is written:

$$A = R \times K \times LS \times C \times P \tag{1}$$

R is the rainfall-runoff erosivity factor (MJ mm/ha*h*y); K is the soil erodibility factor) (t* ha* h/ha*MJ*mm); A_{RUSLE} is estimated average soil erosion (t/ha/year); LS is an expression for slope length and gradient, whereas crop management is denoted by the letter C. P is a consideration of support practice (0–1).

Rainfall Erosivity Factor (R): The R factor is an MJ.mm/(ha.hr.year) erosion index [10]. The R factor estimates the volume and rate of flow connected to events causing rainfall [11]. The KE of a storm in the metric unit is calculated by equation developed [7].

$$R = 79 + 0.363 \times P\left(MJ\frac{\text{mm}}{\text{ha}} * \text{h} * \text{y}\right) \tag{2}$$

Soil Erodibility Factor(K): K Factor is controlled by organic material, soil conditions, and soil structure all affect the K factor. The soil's percolation ability and high stability of the soil impact the K factor's value. The soil's organic matter content and soil distribution of particle sizes are the primary soil variables needed to compute the soil erodibility K-value (sand, silt, and clay) [9]. The main types of soil in the research area are Loamy, clayey, and loamy skeletal. The K variable values are obtained from literature [12], and the geographic area under each soil class has been given.

Slope Length and Steepness Factor (LS): RUSLE studies an advanced method for finding the slope length factor. DEM's equation for calculating the LS factor is as follows [13]:

$$L = (\lambda/22.1)^{0.5} \tag{3}$$

$$S = 10 \times \sin\theta + 0.03 \text{ if slope} < 9\% \tag{4}$$

$$S = 16.8 \times \sin\theta - 0.05 \text{ if slope} > 9\% \tag{5}$$

Cover Management Factor(C): The cover management component considering the most practicably achievable calculated yearly soil loss of the catchment came within a range varied from classes in the study region in comparison to earlier studies, C is arguably the far more significant of the RUSLE variables [14]. Other land use categories for allocating C factor values are listed in Table 1.

Support Practice Factor (P): These methods have a negative effect on soil loss by managing the runoff's flow pattern and decreasing its volume and velocity [15]. According to previous research [16], the values 0.28 and 1 were attributed to paddy and non-paddy fields, respectively, on the P factor map.

Table 1 LULC of the studied region with C values

LULC	C	Area (ha)	% Area
Cropland	0.28	8384.77	36.7
Built up	0.23	969.63	4.5
Forest	0.13	8040.62	36.4
Water bodies	0	311.4	1.4
Wastelands	0.14	270.26	20.3

Sediment Delivery Ratio by GIS-Based Sediment Delivery Distributed Model. It is the proportion between a region's sediment production and overall erosion. The ratio of sediments distribution for the catchment, SDR, is approximated using the following equation [17] as a function of travel time.

$$\mathrm{SDR_i} = \mathrm{e}^{(-\beta \times \mathrm{ti})} \tag{6}$$

$$t_i = \sum\nolimits_{i=1}^{k} (\lambda \mathrm{n}/Vi) \tag{7}$$

The stream velocity is determined from Manning's equation, which is a combination of inclination of the ground surface and characteristics of the forest cover [18].

$$V_i = a_i \times (S_i)^{0.5} \tag{8}$$

where: it is the travel time (h), β is a parameter particular to a watershed. λ_n = length of a flow route segment n, m, a_i = surface roughness coefficient for cell n, m/s, S_i = slope for cell n, m/m.

3.2 *Direct Method for the Estimation of Sediment Yield*

In direct method analysis, surface water samples were taken with a bottle-type sampler at thirteen locations across the Manikpur coalfield area to determine the suspended sediment yield. The stream discharge is measured using the area velocity approach. A filtration test is performed on each sample to measure the suspended sediment concentration.

$$\text{Sediment load} = \text{Discharge(Q)} \times \text{Suspended sediment concentration(C)} \tag{9}$$

4 Results and Discussion

The erosion map is obtained by multiplying the raster images utilizing the ArcGIS raster converter. The calculated yearly soil degradation of the catchment fell within a span of 0–79.2 tons per hectare per year, as shown in (Fig. 7) having an average value of 0.30. The soil loss map is categorized in accordance with the erosion risk groups proposed by [16] for Indian circumstances, as described in Table 2. Rainfall-runoff erosivity of the catchment was found high at 1 and 3 point around Korokoma shown in Fig. 2 and low at point 16 around Bhilaikurd.

In the study area, Loamy soil is present at 1, 3 points around Korokoma and 8, 12, 13, 16 points around Manikpur, Nakhtikar, Kudri, and Bhilaikhurd, respectively, which is undergone moderate to severe soil erosion. K factor map as shown in Fig. 4 with an average value of 0. Steep and rough slopes encompass the opencast mine and overburden dumps (Fig. 5) . Cover management is dimensionless, as demonstrated in Fig. 6. Around 73.1% of the studied area was comprised of vegetative covers, such as woods and croplands. In comparison, only roughly 26.9% comprised built-up regions, water bodies, and wastelands.

The study area's P-factor image, as displayed in Fig. 7, was created by assigning P-values to various land use classifications. The sediment arrival rate shown in Fig. 9 has a high spatial distribution at the dam site and outlet, especially in areas with high slopes. This was due to a nearby outlet time of concentration being low and flow velocity being high at sloppy locations. Slight spatial variability in sediment yield throughout the whole catchment region (Fig. 8) ranges from 0 to 57.93 t/ha/year. Due to enhanced flow accumulation downslope from the source to the catchment outlet, high sediment production occurs in river valleys and tends to grow downstream (Fig. 10).

Table 2 Distribution of various erosion classes

Range	Soil erosion class	Area (ha)	% Area
<5	Slight	11,407.18	99
5–10	Very slight	83.22	0.72
10–15	Moderate	12.44	0.11
15–20	Moderately severe	7.11	0.06
20–40	Severe	11.53	0.1
40–80	Very severe	2.43	0.02

Soil erosion rates in India, 1992

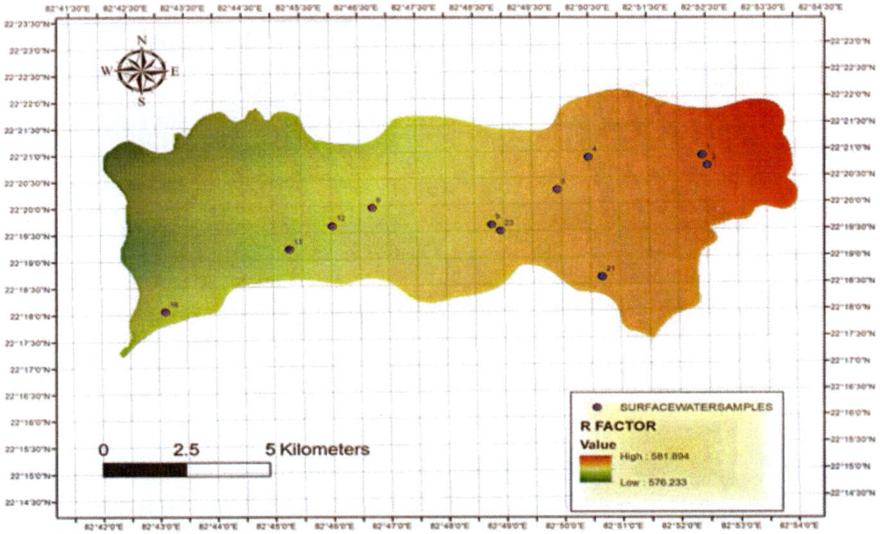

Fig. 3 Rainfall erosivity factor map of the studied region

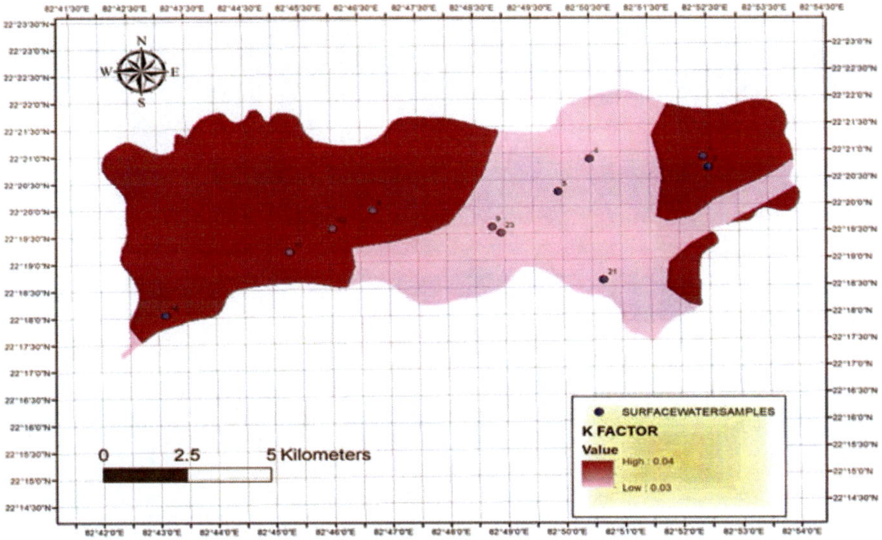

Fig. 4 Soil erodibility factor map of the studied region

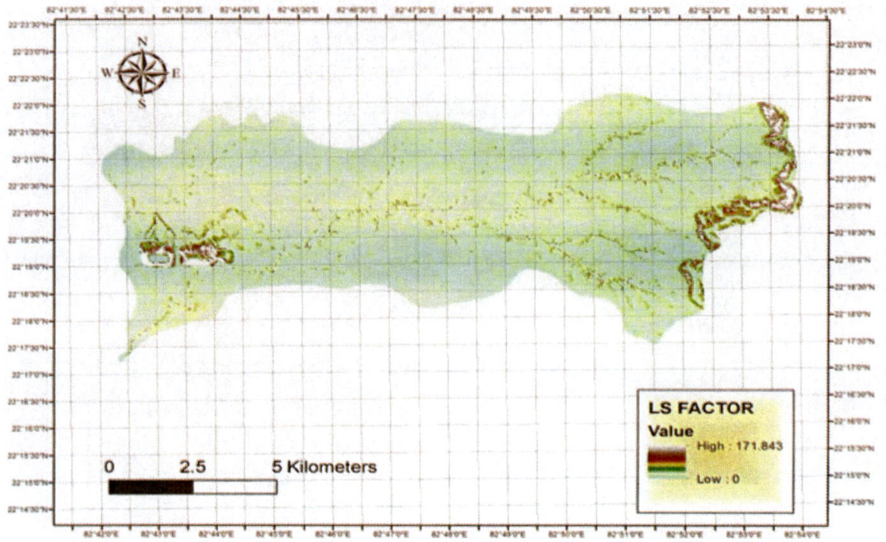

Fig. 5 Slope length and steepness factor (LS) map of the studied region

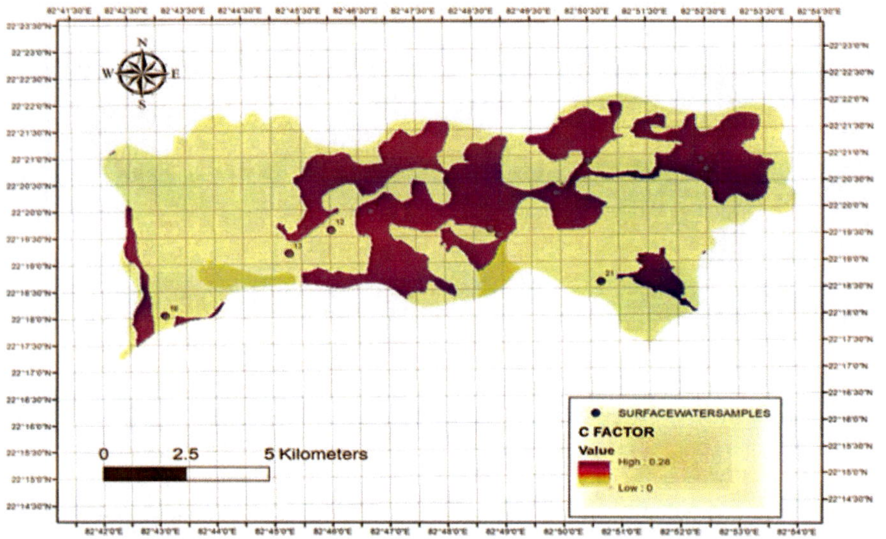

Fig. 6 Cover management factor (C) map of the studied region

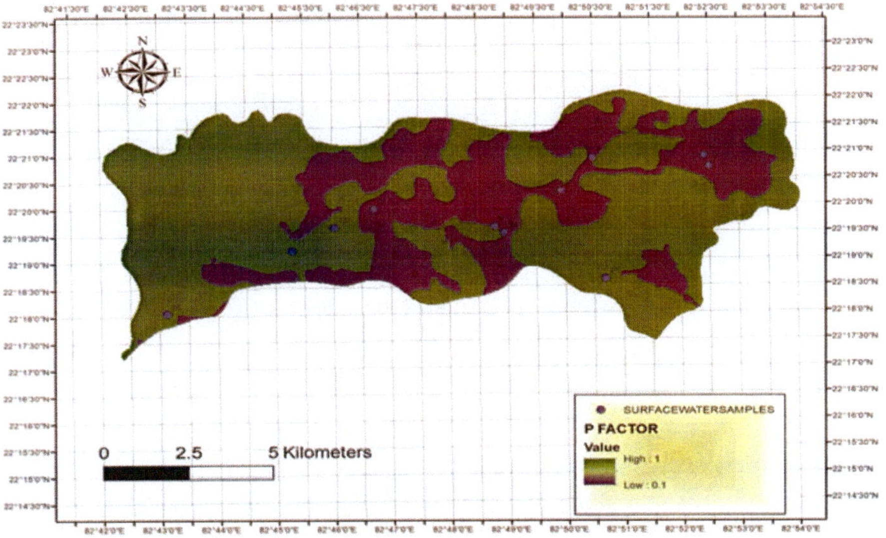

Fig. 7 Support practice factor (P) map of the studied region

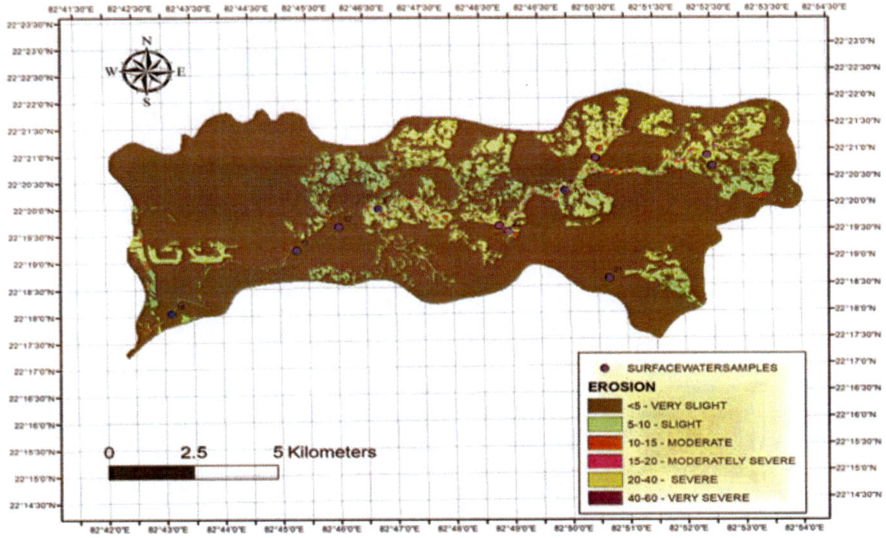

Fig. 8 Soil erosion map of the studied region

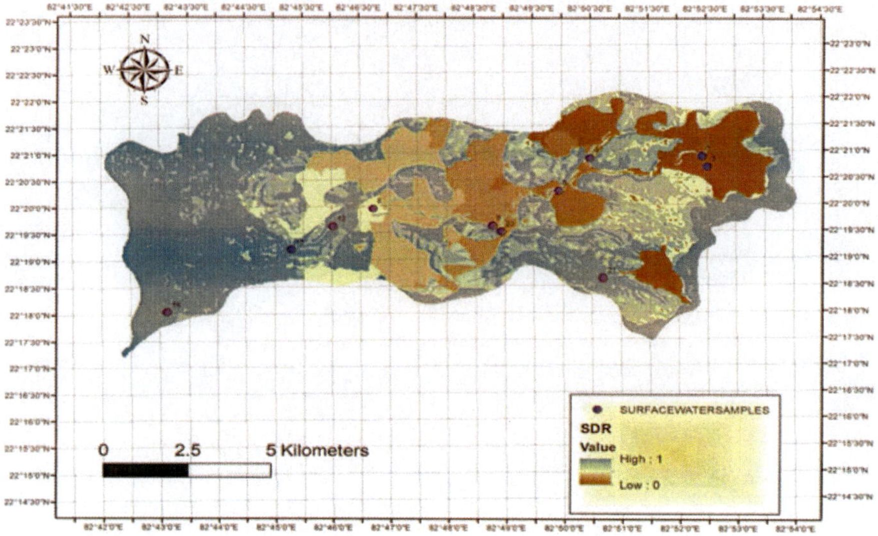

Fig. 9 Sediment delivery ratio map of the studied region

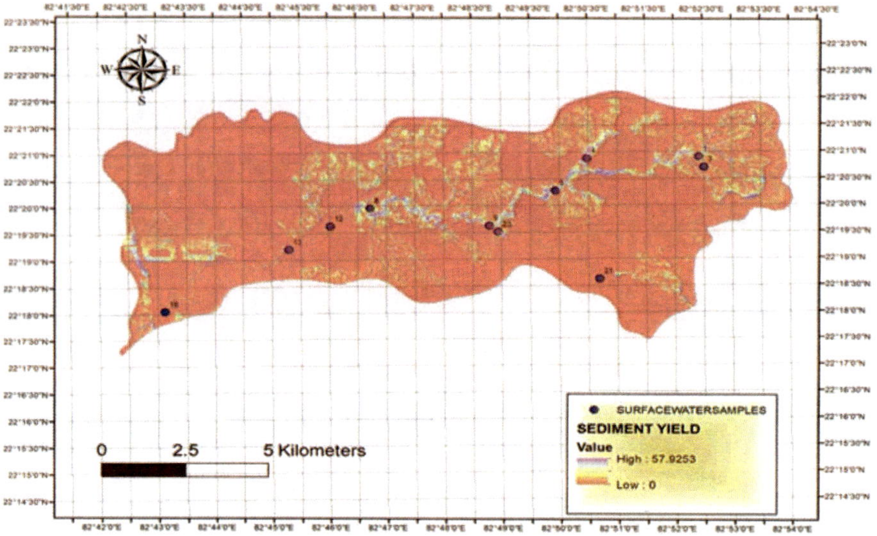

Fig. 10 Sediment yield map of the studied region

5 Conclusions

Incorporating RUSLE into GIS yields sedimentation and soil loss produce outcomes, which offer dynamic spatial dimension to soil degrading and sediment production in region of study. Considering the good vegetative cover and low slope gradient, surface soil erosion and sediment output were very low in the studied region, given the principal elements and processes. Very few zones are subject to moderate erosion risk, and these are built-up areas surrounding the opencast mine and overburden dumps, as well as built-up areas along steep slopes. Under the slight erosion class, the final output of soil erosion covered 99% of total land use. The model accuracy of the result obtained by this study was estimated using a relative mean square error (R^2) and the relative mean square error square error (RRMSE). R^2 and RRMSE are 0.81 and 0.66, and the result for this study was good and satisfactory. The sediment yield values calculated from the field observation are compared with the simulated sediment yield by an empirical equation of the RUSLE model. The sediment yield values at locations 4 and 13 are nearly equal; for others.

Acknowledgements The author would like to thank the Project Director of Chhattisgarh Council of Science and Technology (CCOST), Raipur, for allowing herself to work with them at the Manikpur coal mine area and providing the data for this study.

References

1. Ghose AK (2009) Technology vision 2050 for sustainable mining. Procedia Earth Planet Sci 1(1):2–6. https://doi.org/10.1016/j.proeps.2009.09.003
2. Kothyari UC (1996) Erosion and sedimentation problems in India. IAHS-AISH Publ 236(236):531–540
3. Choi Y, Song J (2016) Sustainable development of abandoned mine areas using renewable energy systems: a case study of the potential photovoltaic assessment at the tailings dam of abandoned Sangdong mine, Korea. Sustainability 8(12):1320
4. Gupta AK, Paul B (2015) Ecorestoration of coal mine overburden dump to prevent environmental degradation: a review. Res J Environ Sci 9(7):307–319. https://doi.org/10.3923/rjes.2015.307.319
5. Ward A, Smith A, Caldwell J (1984) Surface erosion and sediment control at opencast mines in southern Africa. Challenges African Hydrol Water Resour Proc Harare Symp (144):517–527
6. De Boer DH, Crosby G (1996) Specific sediment yield and drainage basin scale. IAHS AISH Publ 236(236):333–338
7. Wischmeier WH, Mannering (1962) Division S-6—soil and water management. (15)
8. Smith DD, Wischmeier WH (1962) Rainfall erosion. In: Advances in agronomy, vol 14. Elsevier, pp 109–148
9. Sharpley AN, Williams JR (1990) EPIC. Erosion/productivity impact calculator: 1. Model documentation. 2. User manual
10. Devatha CP, Deshpande V, Renukaprasad MS (2015) Estimation of soil loss using USLE model for Kulhan Watershed, Chattisgarh—a case study. Aquatic Procedia 4:1429–1436
11. Wang Y et al (2017) Increasing trends in rainfall-runoff erosivity in the Source Region of the Three Rivers, 1961–2012. Sci Total Environ 592:639–648

12. Kumar T, Jhariya DC, Pandey HK (2020) Comparative study of different models for soil erosion and sediment yield in Pairi watershed, Chhattisgarh, India. Geocarto Int 35(11):1245–1266. https://doi.org/10.1080/10106049.2019.1576779
13. Moore ID, Burch GJ (1986) Physical basis of the length-slope factor in the universal soil loss equation. Soil Sci Soc Am J 50(5):1294–1298
14. AEM using R and G on Sahu DC, Baghel T, Sinha MK, and Ahmad I (2017) Soil Erosion Modeling using Rusle and GIS on Dudhawa Catchment. Int J Appl Environ Sci 12(6):1147–1158
15. Renard KG, Foster GR, Weesies GA, McCool DK, Yoder DC (1996) Predicting soil erosion by water: a guide to conservation planning with the Revised Universal Soil Loss Equation (RUSLE). Agric Handb 703:25–28
16. Singh G, Babu R, Narain P, Bhushan LS, Abrol IP (1992) Soil erosion rates in India. J Soil Water Conserv 97–99
17. Ferro V, Porto P, Tusa G (1998) Testing a distributed approach for modelling sediment delivery. Hydrol Sci J 43(3):425–442
18. Fu G, Chen S, McCool DK (2006) Modeling the impacts of no-till practice on soil erosion and sediment yield with RUSLE, SEDD, and ArcView GIS. Soil tillage Res 85(1–2):38–49

Groundwater Level Trends Over Southern India

Kotapati Narayana Loukika, Keesara Venkatareddy, and Eswar Sai Buri

Abstract Groundwater is essential for obtaining freshwater and plays a crucial role in promoting sustainable development in agriculture, industry, and the socioeconomic status of an area. However, due to the extensive use of groundwater, many areas in India are facing a decline in their groundwater levels. Groundwater level time series analysis aids in detecting trends, analyzing behavior, and determining the causes of water level decline. We considered four seasons for identifying the trends in groundwater levels i.e. Post-Monsoon Rabi, Monsoon, Pre-Monsoon and Post-Monsoon Kharif seasons. Using Mann–Kendall test, analysis of trends for Southern India which consists of 8 states was performed. Data on the piezometric level of 181 observation wells from 1996 to 2018 has been taken into consideration. The significant declining trend was observed in 47 observation wells in monsoon season, 61 wells in post-monsoon Rabi, 27 wells in post-monsoon Kharif and 45 wells in pre-monsoon from the results. Out of all wells only one well exhibited increasing trend over 8 states. The outcomes of the study are going to help the groundwater development authorities in the respective states for resource management, water supply planning, groundwater recharge management, environmental protection and policy formulation and regulation.

Keywords Mann–Kendall · Groundwater · Trend analysis

1 Introduction

Groundwater is one of the major and finite sources of freshwater, which is under continuous pressure [1]. A better understanding of groundwater helps in understanding the hydrological cycle [2]. Observing and understanding the groundwater change in the long term, are crucial for the maintenance of ecosystems and to develop the economy. The rate at which groundwater resources are being depleted is higher

K. N. Loukika (✉) · K. Venkatareddy · E. S. Buri
Department of Civil Engineering, National Institute of Technology Warangal, Hanamkonda, Telangana, India
e-mail: loukikak@student.nitw.ac.in

S. Mesapam et al. (eds.), *Developments and Applications of Geomatics*,
Lecture Notes in Civil Engineering 450, https://doi.org/10.1007/978-981-99-8568-5_21

than their natural recharge rate, primarily because of the excessive use of groundwater for anthropogenic activities like irrigation. The population growth and expanding irrigation areas made an increase in the usage of groundwater. Improper planning in the development of groundwater in some regions has further complicated the problem, as a consequence of this, there has been a swift reduction in groundwater levels resulting in a vast number of shallow wells being dried up. Groundwater availability predictions are often problematic because of insufficient data quantity, quality and a limited number of monitoring sites. Groundwater depletion on the large spatial and temporal scale has impacts on salt-water intrusion, natural streamflow and land subsidence in deltaic areas, which causes groundwater quality deterioration and ecosystems degradation. Groundwater resources have become increasingly relied upon in arid and semiarid regions due to the uncertainty and decreased surface flow and uneven distribution of rainfall. Managing groundwater is crucial in semiarid regions, as excessive use can lead to a significant decline in groundwater levels. The observations of the groundwater levels should be reliable and frequent to monitor the storage of groundwater and its changes. The reduction in groundwater levels has resulted in environmental problems, such as drought, land subsidence and groundwater depletion. Continuous monitoring of water levels is crucial to detecting trends in groundwater levels. Due to the spatial and temporal fluctuations in groundwater, sustainable water resource management has become a global priority [3].

The Mann–Kendall test is the most commonly used nonparametric method for analyzing trends in time series data [4]. The Mann–Kendall test is commonly used to analyze trends in various climate-related variables, such as climatic parameters, stream flows, reference evapotranspiration, evaporation and groundwater levels [5, 6]. The analysis of water level patterns is crucial for determining the future scenario of accessible groundwater resources of the state.

2 Study Area

The study area considered for current study is southern India which consists of 8 states and they are Andhra Pradesh, Tamil Nadu, Orissa, Maharashtra, Chattisgarh, Karnataka, Kerala and Telangana. The study area is shown in the Fig. 1. The climate in southern India is semi-arid and agriculture is the primary source of income for most of the population. In the majority of the region, the southwest monsoon season is regarded as the primary source of rainfall. Rapid population growth has resulted in a significant increase in utilization of water resources.

The study area with IMD grid points and groundwater observation wells is shown in Fig. 2.

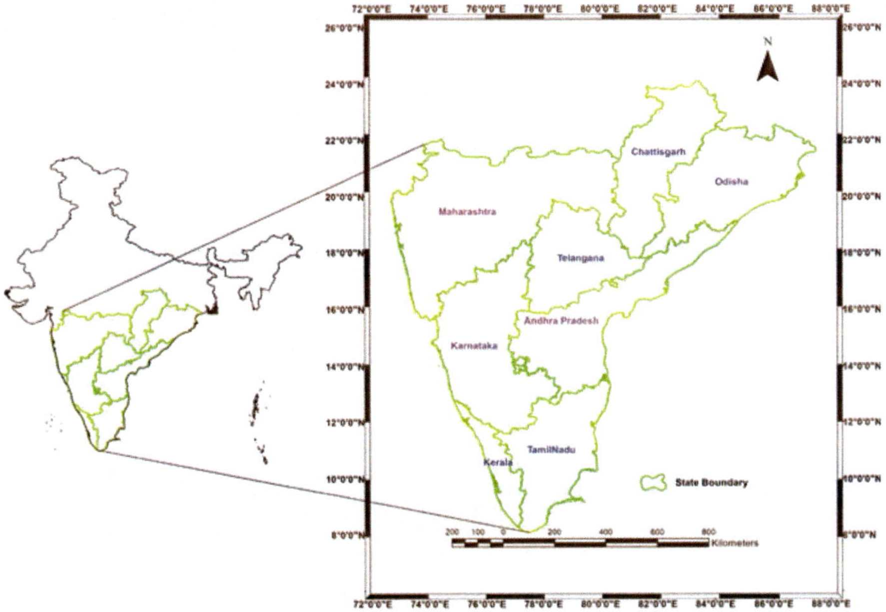

Fig. 1 The study area map location

3 Data and Methods

The Data on groundwater levels was obtained from the India-Water Resources Information System (WRIS) and comprised of groundwater level readings for the Monsoon, Post-Monsoon Kharif, Post-Monsoon Rabi, and Pre-Monsoon periods from 1996 to 2018 used for analysis. This data is available online in at Water Resource Information system of India (WRIS) website (Source: http://indiawris.gov.in/wris/#/groundWater) which is an open source data. The methodology followed in the present study is shown in the Fig. 3. Groundwater level time series analysis aids in detecting trends, analyzing behavior, and determining the reason behind the reduction in groundwater levels.

3.1 Mann–Kendall Test

The non-parametric test which is a well-known MK-test [7, 8] was utilized to analyze long-term trend patterns. If a series of dimensions n includes annual values x_j and x_i from years j and i respectively (with $j > i$), the resulting statistics for the MK-test are as follows.

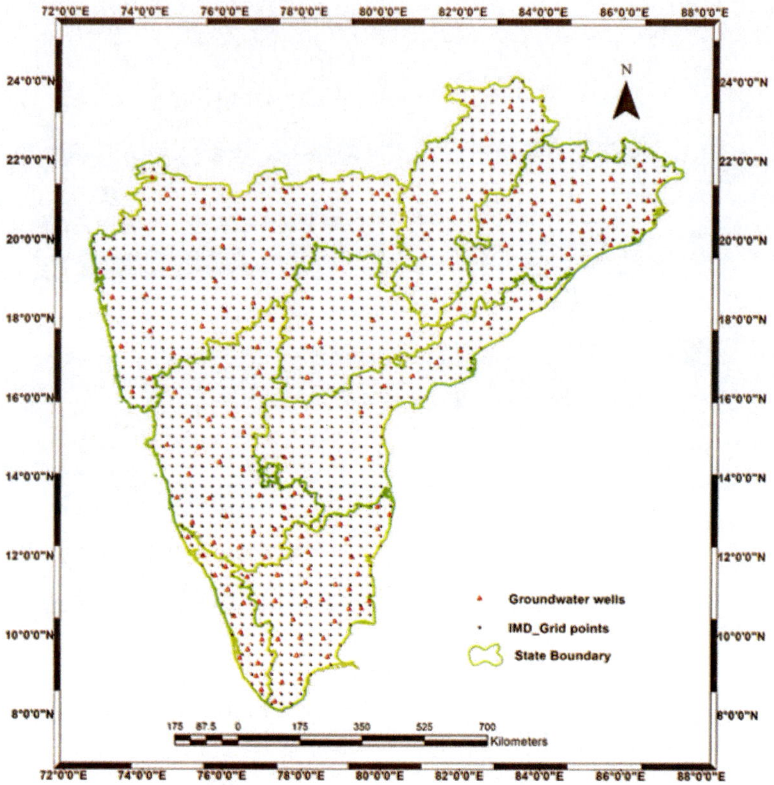

Fig. 2 IMD grid points and groundwater observation wells

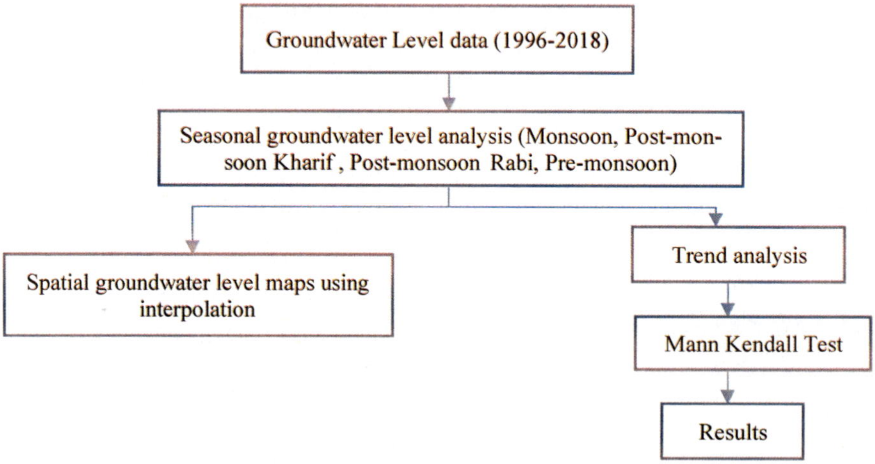

Fig. 3 Methodology flowchart

$$S = \sum_{i=1}^{n-1} \sum_{j=i+1}^{n} sgn(x_j - x_i) \tag{1}$$

With

$$sgn(x_j - x_i) = \left\{ \begin{array}{c} 1, if\left(x_j - x_i\right) > 0 \\ 0, if\left(x_j - x_i\right) = 0 \\ -1, if\left(x_j - x_i\right) < 0 \end{array} \right\}. \tag{2}$$

If the x_i values are independent and randomly ordered and $n > 10$, the statistic S follows a normal distribution with a mean of zero and a variance given by the following equation.

$$Var(S) = \left[n(n-1)(2n+5) - \sum_{i=1}^{m} t_i i(i-1)(2i+5) \right] / 18 \tag{3}$$

In this context, m represents the number of tied groups in the entire data set, while t_i represents the total number of data points in the ith tied group.

The equation used to calculate the standardized test statistic Z_{MK} is as follows:

$$Z_{MK} = \left\{ \begin{array}{c} \frac{S-1}{\sqrt{Var(S)}} for S > 0 \\ 0 for S = 0 \\ \frac{S+1}{\sqrt{Var(S)}} for S < 0 \end{array} \right. \tag{4}$$

A two-tailed test is used to determine if the null hypothesis should be rejected for a specific significance level α (90, 95), where Z_{MK} must be greater than $Z_{1-\alpha/2}$. .The trend will be identified as increasing or decreasing based on the sign of the test statistic. The magnitude of the trend is determined using the Theil-Sen approach (TSA) [9].

$$\beta = \text{median}\left[\frac{x_j - x_i}{j - i}\right] \text{for allj} > \text{i}. \tag{5}$$

3.2 *Results*

Figure 4 shows the spatial variation of mean monthly seasonal groundwater levels for the period 1996–2018. The groundwater level is high in the pre-monsoon season compared to other three seasons. During the monsoon period from 1996 to 2018, Coimbatore in Tamil Nadu had a mean monthly groundwater level of 12.17 m. In the

Post-Monsoon Rabi period, the mean monthly groundwater level was 10.7 m, while in the Post-Monsoon Kharif period it was 11.36 m. The Pre-monsoon mean monthly groundwater level was 12.4 m, which was higher compared to other observation wells. Depth to below groundwater level in Post-Monsoon Kharif is less in Jajapur which is in Odisa state.

Trends in groundwater level for the Monsoon, Post-monsoon Kharif, Post-monsoon Rabi and Pre-monsoon seasons were analyzed using the MK test is shown in the Fig. 5. The study utilized data on piezometric levels from 181 observation wells during the period from 1996 to 2018.

Monsoon season

In the Telangana state, Nalgonda has exhibited a decreasing trend, while all other observation wells showed no trend in the Monsoon season. Similarly, in Andhra Pradesh, five wells located in Vizianagaram, East Godavari, Srikakulam, Nellore, and Visakhapatnam showed a decreasing trend, while the remaining wells showed no trend during the Monsoon season. In Chattisgarh state, Bastar, Dantewada, Raigarh, Uttar Bastar Kanker, Narayanpur, and Raipur regions exhibited a decreasing trend, while other wells showed no trend. In Karnataka, Bijapur, Davanagere, Dharwad,

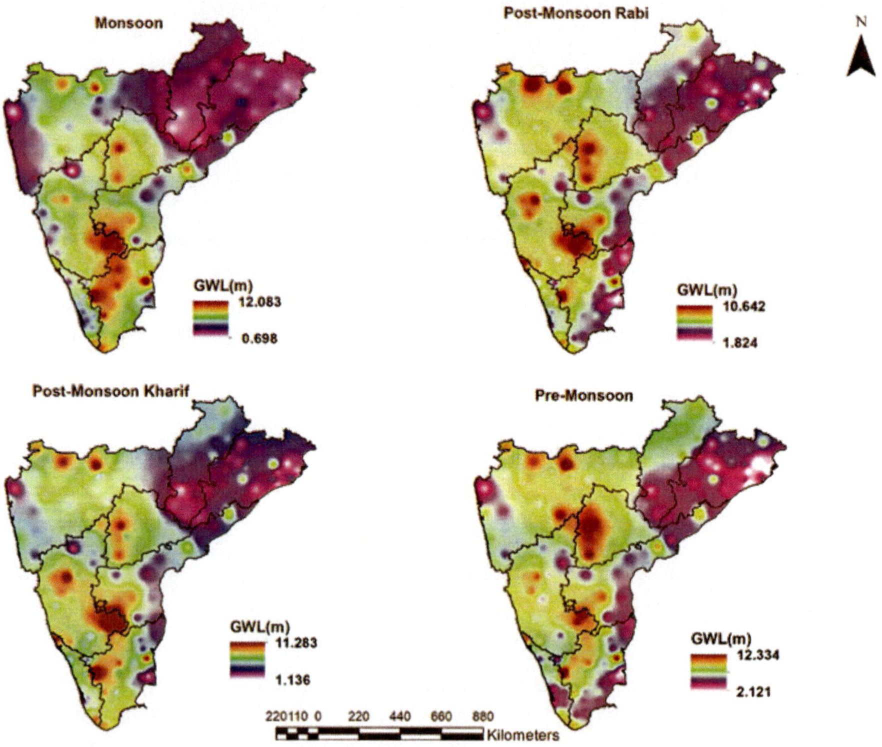

Fig. 4 Spatial variation of Groundwater levels for the period 1996–2018

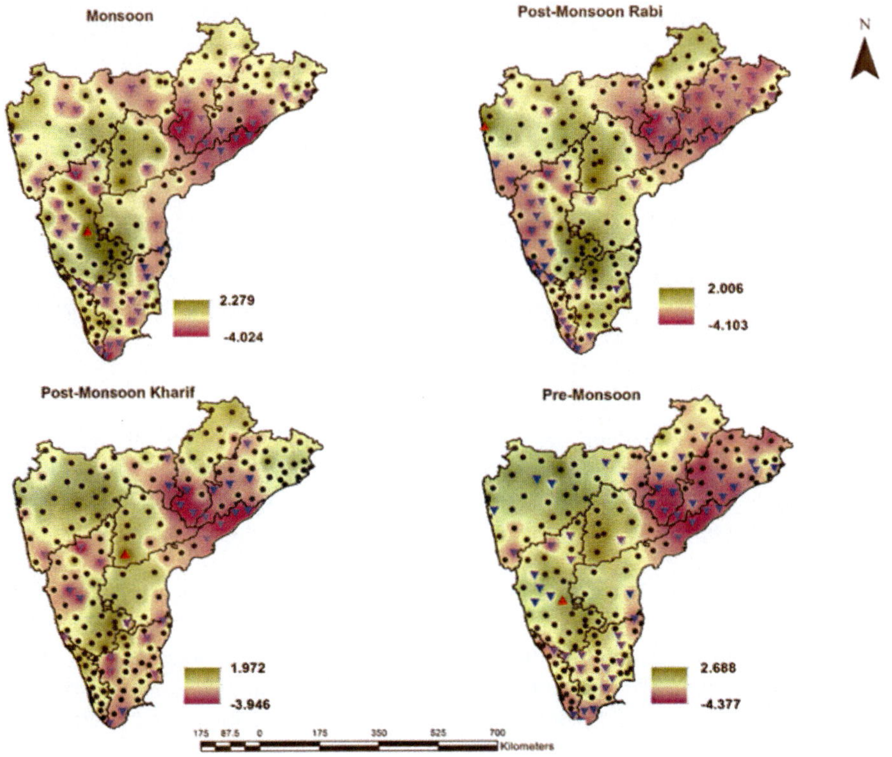

Fig. 5 The spatial distribution of Z-values and trends of grid points in groundwater levels across southern India during the period 1996–2018

Gulbarga, Haveri, Raichur, Shimoga exhibited a decreasing trend, while Chitradurga showed an increasing trend. From Kerala state, Wayanad showed a decreasing trend. Maharashtra state had a decreasing trend in the Aurangabad, Chandrapur, Jalna, Kolhapur, and Wardha regions. Similarly, in Tamil Nadu, 13 wells exhibited a decreasing trend. Additionally, the decreasing trend was observed in Cut tack, Dhenkanal, Gajapati, Ganjam, Kandhamal, Kendrapara, Koraput, and Malkangiri from Odisha state. Finally, Thiruvananthapuram in Kerala exhibited a decreasing trend.

Post-monsoon Rabi

Observations from several states in India indicate that some wells are showing a decreasing trend. For instance, Andhra Pradesh had four such wells, Chattisgarh and Kerala had three and five wells, respectively, while Karnataka had 13 wells with a decreasing trend. Maharashtra had one well in Greater Bombay that exhibited an increasing trend, but eight other wells had a decreasing trend. Odisha had the highest number of wells with 21 showing a decreasing trend, while Tamil Nadu had five wells with the same pattern. Only one well from Telangana (Khammam) displayed a decreasing trend.

Post-monsoon kharif

In Andhra Pradesh, four wells located in Vizianagram, East Godavari, Srikakulam, and Visakhapatnam have shown a decreasing trend. Similarly, the only well located in Chattisgarh, Dantewada, has also exhibited a de creasing trend. In Karnataka, five wells and two wells in Kerala, namely Thiruvarur and Kasaragod respectively, have also shown a decreasing trend pattern. Additionally, four wells in Maharashtra, five wells in Odisha, and six wells in Tamil Nadu have exhibited decreasing trend patterns. However, in Telangana, Mahabubnagar has shown an increasing trend pattern.

Pre-monsoon

There has been a decreasing trend observed in several districts of Andhra Pradesh, including Vizianagaram, East Godavari, West Godavari, Srikakulam, and Visakhapatnam. Bijapur in Chattisgarh has experienced an increasing trend, while Dantewada, Bastar, Narayanpur, and Uttar Bastar Kanker have shown a decreasing trend. Only one well located in Bijapur, Karnataka, has exhibited a decreasing trend. Similarly, 20 wells in Odisha and 4 wells in Kerala have exhibited a decreasing trend pattern. The trend has been increasing in Greater Bombay and Buldana in Maharashtra, while Kolhapur, Garhchiroli, and Gondiya have exhibited decreasing trend patterns. Additionally, 20 wells in Odisha and 7 wells in Tamil Nadu have shown a decreasing trend. Mahabubnagar and Hyderabad have shown an increasing trend pattern, while Khammam has experienced a decreasing trend.

4 Conclusions

The study found that there was a notable decrease in water levels in several observation wells across India during different seasons. Specifically, 47 wells experienced a declining trend during the monsoon season, 61 wells during the Post-Monsoon Rabi season, 27 wells during the Post-Monsoon Kharif season, and 45 wells during the Pre-Monsoon season. Only one well showed an increasing trend across the eight states studied. The results showed that there were more declining trends during the Post-Monsoon Rabi and Pre-Monsoon seasons. The insights from this present study were useful for the policy makers and stakeholders especially in the irrigation and agricultural sector in an inevitable manner for getting maximum benefits.

References

1. Bhanja SN, Mukherjee A, Rodell M, Wada Y, Chattopadhyay S, Velicogna I, Panglauru K, Famiglietti JS (2017) Groundwater rejuvenation in parts of India influenced by water-policy change implementation. Sci Rep 7(1). https://doi.org/10.1038/s41598-017-07058-2
2. Miro EM, Famiglietti JS (2018) Downscaling GRACE remote sensing datasets to high—resolution groundwater storage change maps of California's Central Valley. Remote Sensing 10(1)
3. Patle GT, Singh DK, Sarangi A, Rai A, Khanna M, Sahoo RN (2015) Time series analysis of groundwater levels and projection of future trend. J Geol Soc India 85:232–242. https://doi.org/10.1007/s12594-015-0209-4
4. Kendall MG (1955) Rank correlation methods. Charles Griffin, London
5. Patle GT, Singh DK, Sarangi A, Rai A, Khanna M, Sahoo RN (2013) Temporal variability of climatic parameters and potential evapotranspiration. Indian J Agric Sci 83(4):518–524
6. Jagadeesh P, Anupama C (2013) Statistical and trend analyses of rainfall: A case study of Bharathapuzha river basin, Kerala, India. ISH J. Hydraulic Eng. 20(2):119–132. https://doi.org/10.1080/09715010.2013.843280
7. Mann HB (1945) Non-parametric tests against trend. Econ-metrica 13:245–259
8. Kendall MG (1962) Rank correlation methods. HafnerPublishing Company, New York
9. Kumar KS, Rathnam EV (2009) Analysis and prediction of groundwater level trends using four variations of Mann Kendall tests and ARIMA modelling. J Geol Soc India 94:281–289. https://doi.org/10.1007/s12594-019-1308-4

Impact of Climate Change on Streamflow Over Nagavali Basin, India

Nageswara Reddy Nagireddy and Keesara Venkatareddy

Abstract Natural disasters such as cyclones pose a serious threat to the Indian coast. The Nagavali basin is an interstate east-flowing river that supports agricultural and domestic water demands in Koraput, Kalahandi, and Rayagada districts in Odisha, as well as Vizianagaram and Srikakulam in Andhra Pradesh. Furthermore, this basin is especially prone to cyclones generated by low-pressure depressions in the Bay of Bengal. The present research aims to simulate the streamflow of the Nagavali basin using the SWAT model and to provide historical information as well as future changes within streamflow in response to climate change. Calibration (1991–2005) and validation (2006–2014) of the SWAT model showed a satisfactory for monthly streamflow. The downscaled bias-corrected geographically separated the NASA NEX-GDDP dataset was employed for simulating the future streamflow under two RCP scenarios, 4.5 and 8.5. The analysis periods were divided into 27-year blocks that included a historical period (1980–2005) as well as three future periods, namely near term (2022–2047), the middle future (2048–2073), and the far future (2074–2099). Under both scenarios, the CNRM-CM5 model predicted a rising trend in precipitation and streamflow. Under RCP 4.5, the CNRM-CM5 model showed the greatest percentage change of 22.06% in the far future, with a corresponding streamflow change of 36.26%. Under RCP 8.5, the BNU-ESM, CNRM-CM5, and IPSL-CM5A-MR models represent percentage change in precipitation ranging from 22.09 to 26.28% and corresponding streamflow ranging from 33.81 to 45.11%, respectively, in the far future. Peak discharges are estimated at various return periods with Log Pearson Type III probability distribution. The estimated peak discharge (5626 m^3/s) of the 100-year return period was observed in 2006. This study findings are useful to water resource managers.

Keywords Nagavali basin · Climate change · Flood frequency analysis · SWAT

N. R. Nagireddy (✉) · K. Venkatareddy
Department of Civil Engineering, NIT Warangal, Hanamkonda, Telangana, India
e-mail: nnagireddy@student.nitw.ac.in

S. Mesapam et al. (eds.), *Developments and Applications of Geomatics*,
Lecture Notes in Civil Engineering 450, https://doi.org/10.1007/978-981-99-8568-5_22

1 Introduction

Climatic situations over the twenty first century forecasted by Global Climate Models (GCMs) revealed a warming trend around the world. However, the severity and nature of these effects vary greatly across regions. Floods and droughts will become more common in the majority of the globe, resulting in lower crop productivity and an increase in the risk of hunger and poverty [1]. The consequences of warming temperatures on hydrological processes and the management of water resources have become a major cause of attention in recent years. Numerous simulations, studies, and experiments based on general circulation models (GCMs) predicted an increase in temperature as greenhouse gas levels increased. Climate processes are expected to worsen as a result, with serious effects for agricultural and energy output, ecosystems, flooding, and drought. For long-term strategic management of a nation's water assets in the context of evolving changing climate consequences, it is essential that the consequences of climate change be assessed using high temporal and spatial resolution across the basin scale. Wagner et al. [2] employed SWAT to examine the consequences of climate and land use changes in the Western Ghats. India is an important economically developing nation, with almost sixty percent of its population relying directly on climate-sensitive sectors such as farming and fisheries [3]. The spatial patterns of Indian rainfall suggested that the west coast and northeast India would receive the most rain under both the A2 and B2 IPCC scenarios [4]. There will be plenty of water in the Godavari, Brahmani, and Mahanadi basins, but catastrophic flooding is possible [5]. The Indian summer monsoon rainfall, which occurs from June to September accounting for more than 75% of total precipitation in Indian river basins [6]. So, the accurate projections of the summer monsoon is important, because it is the most important drivers for climate change policies and adaptation for Indian river basins [7]. Raghavan et al. [8] investigated the NEX-GDDP dataset, which includes statistically downscaled CMIP5 past and future climate forecasts at daily and 25 km spatial resolution. They concluded, surface temperatures in Southeast Asia will rise by greater than 3.5 °C in the future. Rao et al. [9] forecasted future shifts in extreme rainfall over south India with high-resolution NEX-GDDP datasets. This NEX-GDDP dataset replicates mean rainfall fairly well. Rainfall may increase up to 11% in the near and up to 38% in the far future, under RCP 4.5 and 8.5 scenarios. According to future climate predictions, the severity, as well as the timing of rainfall extremes in most parts of south India, may increase throughout the monsoon season under both scenarios. Nagireddy et al. [10] evaluated the climate change impact on streamflow and sediment in the Nagavali and Vamsadhara watersheds using CMIP6 climate models data and SWAT model. They found the range of projected changes in the percentage of average annual precipitation and average temperature ranged from 0 to 41.7% and 0.7 to 2.7 °C, respectively, indicating a warmer and wetter climate in the Nagavali and Vamsadhara watersheds.

The occurrence and severity of peak discharges connected with individual events are used to give significant details on the probability of floods. The flood frequency assessment depends on existing data from each year's independent flood discharge maxima. Dams, bridges, and flood prevention structures can all benefit from understanding frequency and magnitude relationships. Griffis and Stendinger [11] estimated maximum yearly precipitation and discharges using the Log Pearson III distribution. In the case of Indian rivers, the Log Pearson-III distribution produces reliable outcomes for calculating peak discharges up to a 100-year return interval [12, 13]. The primary goal of this research is to evaluate the effects of climate change on streamflow in the Nagavali basin using the SWAT model and to calculate the peak discharges at distinct return periods using Log Pearson type III probability distribution.

2 Materials and Methods

2.1 Study Area

The Nagavali basin is an intrastate east-flowing river basin residing adjacent to the Godavari and Mahanadi basins (Fig. 1). The Nagavali river originate in the vicinity of Lakhbahal village in Odisha's Kalahandi district. It flows 256 km before joining the Bay of Bengal. The drainage area of the Nagavali basin is 9200 km^2.

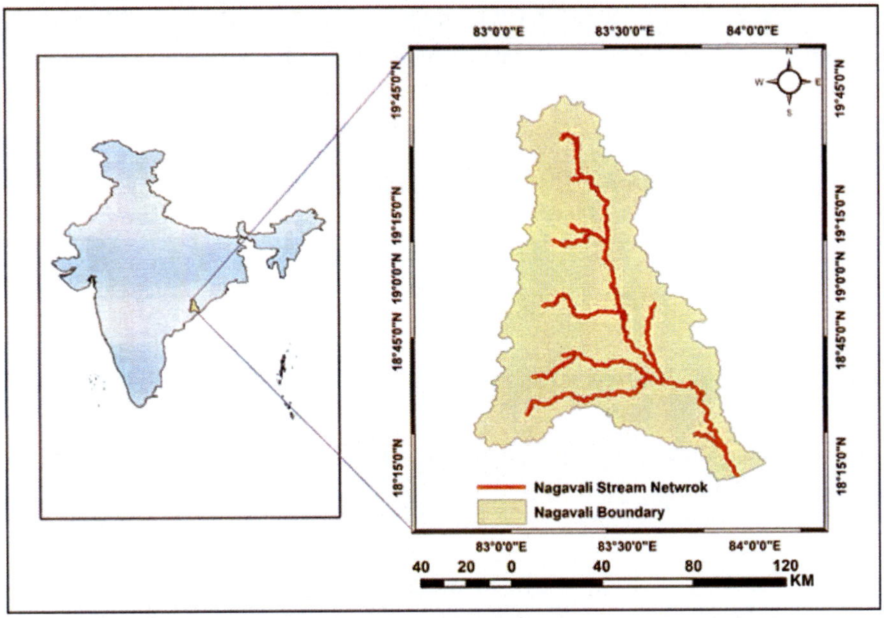

Fig. 1 Study area map

2.2 Datasets

Spatial datasets DEM, LULC and Soil map used in the present study found in [14]. NEX-GDDP dataset obtained from Coupled Model Inter-comparison Project Phase 5 (CMIP5) simulations under two RCP emission scenarios, 4.5 and 8.5 were used. It has a spatial resolution of 0.25°. These datasets include a set of global, high-resolution, bias-corrected climate change forecasts which may be employed for evaluating the consequences of climate change at finer scales.

2.3 Model Setup and Methodology

The SWAT model is a broad, ongoing system constructed by the U.S. Department of Agriculture to analyse and predict hydrologic processes over long time periods. It is intended to simulate daily surface runoff, sediment yield, nutrient and chemical loads in farming operations. The model has been exhaustively described by [15]. To begin, the SWAT model was created by projecting a DEM, LULC map, and soil map into the WGS 1984 UTM 44N projection. Using the QSWAT tool on the QGIS interface, the Nagavali basin was split into 34 subbasins and 2153 hydrological response units (HRUs) depending on components like soil uniformity, landuse, slope, and the threshold area of 100 ha. The SWAT model was initially calibrated and validated with streamflow. To evaluate the model performance, the objective function Nash Sutcliffe efficiency (NSE) [16] was used. It was determined that the model performance was satisfactory if the NSE was greater than 0.5 [17]. Based on previous reported research, we have selected three CMIP5 models in this study such as CNRM-CM5, BNU-ESM and IPSL-CM5A-MR models.

3 Results and Discussions

3.1 Flood Frequency Analysis

Annual Maximum Series (AMS) data from the Srikakulam station over a 29-year period (1990–2018) were used for the analysis of flood magnitude and frequency in the Nagavali basin. The observed yearly Maximum Series in Table 1 were used to calculate a designed flood. Peak flows for 2-year, 5-year, 10-year, 25-year, 50-year, and 100-year return periods were calculated employing the Log Pearson type III probability distribution, and the calculated discharges are presented in Table 2. The highest (5626 m^3/s) calculated discharge of 100 year return period was observed in the year of 2006 over Nagavali basin, the flood frequency analysis for 29 years of annual maximum data showed the 13 events fall within the 2-year return period,

12 events fall under the 5-year return period, 2 events fall under the 10-year return period, and one event falls under the 50- and 100-year return periods.

Table 1 Annual maximum series of streamflow and their rank

Year of occurrence of event	Ranked maximum streamflow (m^3/s)	Rank
2006	5624.739	1
2014	4224.446	2
2013	2141.784	3
1992	2012.6	4
1995	1917	5
2012	1900	6
2008	1703.286	7
1994	1640.2	8
1991	1497	9
2011	1442.89	10
2018	1389.468	11
2009	1374.639	12
2010	1338.666	13
1996	1284	14
2015	1200	15
2017	1191	16
2003	1086.72	17
2007	1014.324	18
1990	1009.5	19
2000	852.94	20
2005	796.293	21
2016	772.102	22
2001	758.841	23
2004	658.59	24
1997	577.8	25
1998	451.05	26
1999	370.34	27
1993	337.52	28
2002	301.176	29

Table 2 Estimated streamflow using log pearson type III distribution

Return period	Skew coefficient K(0.0805)	Streamflow Q(m^3/s)
2-year	−0.014	1114
5-year	0.837	1979
10-year	1.290	2686
25-year	1.778	3735
50-year	2.097	4630
100-year	2.386	5627

3.2 Calibration and Validation

The observed streamflow data at the Srikakulam station was used to calibrate the SWAT model over the Nagavali basin. It should be emphasized that for a complicated model like SWAT, where there are many parameters to estimate, manually achieving the ideal parameter set is challenging. As a result, auto calibration with parameter optimization was performed to obtain the best SWAT model parameters. 15 parameters influencing the streamflow were considered for streamflow calibration. Table 3 presents the calibrated parameters. The calibrated parameters were within the ranges and details of the parameters that were described in [18] and in the SWAT user manuals. A measure of Nash Sutcliffe Efficiency (NSE) was chosen as an objective function. The NSE statistics for the monthly streamflow of the calibration and validation period were 0.84 and 0.71. More information about the calibration and validation of the SWAT model is available in [14].

3.3 Climate Change Impact on Precipitation

The average yearly precipitation of Nagavali basin is 1251 mm, over the baseline period of 26 years (i.e., 1980–2005). The percentage bias between the IMD gridded precipitation and climate model precipitation for the baseline period is presented in Table 4.

As shown in Table 4, the precipitation bias between the observed IMD and climate models ranged from 5.15 to 8.6%. The selected climate models represents the slightly overestimated precipitation over the basin. The percentage change between the precipitation predicted by the climate models in the near term (2022–2047), middle future (2048–2073) and far future (2074–2099) in comparison to historical data (1980–2005) is presented in Table 5.

According to the table, the BNU-ESM and IPSL-CM5A-MR models revealed a decreasing precipitation in the near term and a rise in the middle and far future under the RCP 4.5 scenario. The CNRM-CM5 model predicted an increasing trend of precipitation throughout the future. For the RCP 4.5 scenario, the CNRM-CM5

Table 3 Calibrated parameters, range and fitted values

S.No	Parameter_Name	Method	Range	Fitted values
1	CN2.mgt	Relative	±0.1	−0.088
2	ALPHA_BF.gw	Replace	0 to 1	0.642
3	GW_DELAY.gw	Absolute	-30 to 90	84.300
4	GWQMN.gw	Absolute	±1000	5
5	GW_REVAP.gw	Replace	0.02 to 0.2	0.053
6	REVAPMN.gw	Absolute	±750	−498.75
7	ALPHA_BF_D.gw	Replace	0–1	0.45
8	RCHRG_DP.gw	Absolute	−0.05 to 0.05	−0.019
9	SOL_AWC.sol	Relative	±0.1	0.04
10	ESCO.hru	Replace	0.3 to 0.6	0.53
11	CANMX.hru	Replace	0–20	0.45
12	CH_N2.rte	Replace	0.01–0.1	0.033
13	CH_K2.rte	Replace	0–100	74.75
14	CH_K1.sub	Replace	0–100	73.25
15	CH_N1.sub	Replace	0.01–0.3	0.19

Source Nagireddy et al. [14]

Table 4 Percentage bias between the IMD and historical models precipitation

Model name	% bias in precipitation
BNU-ESM	8.6
CNRM-CM5	5.15
IPSL-CM5A-MR	6.18

model showed the highest percentage change of 22.06% in the far future, and the IPSL-CM5A-MR model showed the lowest percentage change of −1.87% in the near term. Under the RCP 8.5 scenario, the IPSL-CM5A-MR model showed decreasing precipitation in the near term and increasing precipitation in the middle and far future. The BNU-ESM and CNRM-CM5 models showed an increasing trend in precipitation throughout the future. Under the RCP 8.5 scenario, the IPSL-CM5A-MR model showed the highest percentage change of 26.28% in the far future, and the lowest percentage change of −4.89% in the near future. The mean monthly precipitation and streamflow over the historical period of 1980–2005 are presented in Fig. 2.

According to Fig. 2, the IMD observed precipitation peaks occurred in August with an intensity of 254.65 mm for the basin. The BNU-ESM and IPSL-CM5A-MR models' historical precipitation peaks were observed in July, but the CNRM-CM5 model showed its peak precipitation in August. The projected mean monthly precipitation under two RCP scenarios is shown in Fig. 3.

The BNU-ESM model projected an identical historical peak in near and mid futures, with a peak shift to August in the far future under the RCP 4.5 scenario. The

Table 5 Percentage change in precipitation and streamflow with respect to historical data

RCP scenario	Period	Model name	% change in precipitation	% change in streamflow
4.5	Near term (2022–2047)	BNU-ESM	−0.65	−4.92
		CNRM-CM5	+8.54	+16.67
		IPSL-CM5A-MR	−1.87	−9.12
	Middle future (2048–2073)	BNU-ESM	+7.03	+10.13
		CNRM-CM5	+6.57	+7.82
		IPSL-CM5A-MR	+17.12	+25.62
	Far future (2074–2099)	BNU-ESM	+10.78	+18.40
		CNRM-CM5	+22.06	+36.26
		IPSL-CM5A-MR	+13.58	+10.38
8.5	Near term (2022–2047)	BNU-ESM	+0.85	−3.27
		CNRM-CM5	+5.21	+ 7.46
		IPSL-CM5A-MR	−4.89	−15.69
	Middle future (2048–2073)	BNU-ESM	+21.4	+39.78
		CNRM-CM5	+11.54	+11.92
		IPSL-CM5A-MR	+12.54	+14.12
	Far future (2074–2099)	BNU-ESM	+22.58	+45.11
		CNRM-CM5	+22.09	+33.81
		IPSL-CM5A-MR	+26.28	+42.43

Note '+' sign indicates increasing in future, '−'sign indicates decreasing in future

CNRM-CM5 model predicted peak precipitation in August at a similar historical peak. The IPSL-CM5A-MR model predicted that the peak would shift to August. The BNU-ESM model predicted a near-historical peak in near and mid futures, with a peak shift to August in the far future under the RCP 8.5 scenario. The CNRM-CM5 model predicted peak precipitation in July for the near and mid future, and August for the far future. The IPSL-CM5A-MR model predicted that the peak would shift to August.

3.4 Climate Change Impact on Streamflow

According to the table, the BNU-ESM and IPSL-CM5A-MR models revealed a negative streamflow in the near term and an increase in the middle and far future under the RCP 4.5 scenario. The CNRM-CM5 model predicted an increasing trend of streamflow throughout the future. For the RCP 4.5 scenario, the CNRM-CM5 model showed the highest percentage change of 36.26% in the far future, and the IPSL-CM5A-MR

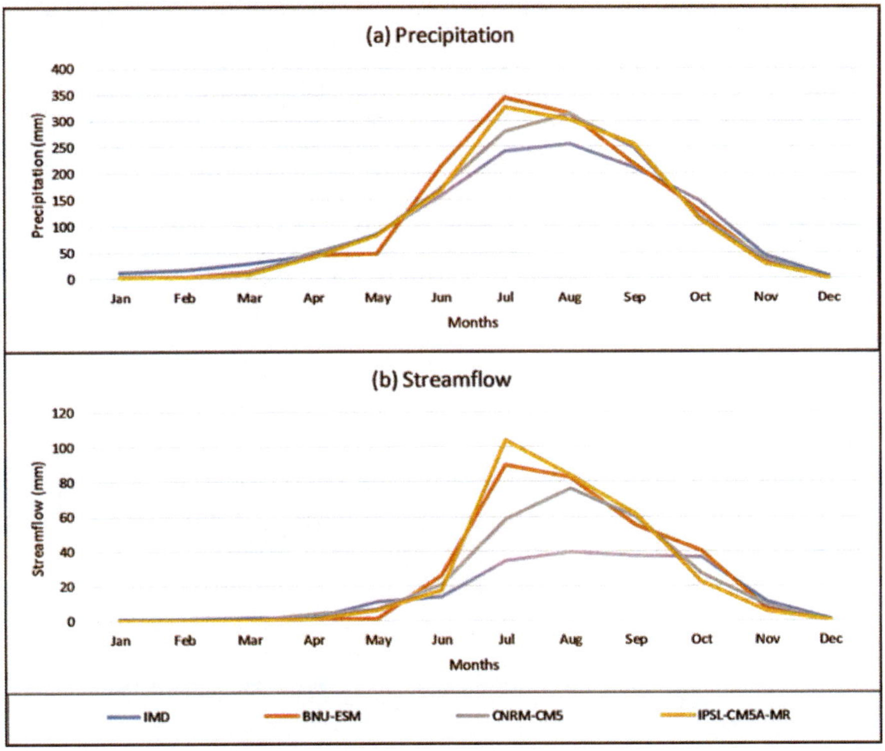

Fig. 2 Mean monthly precipitation and streamflow for the historical period (1980–2005)

model predicted the lowest percentage change of −9.12% in the near term. The BNU-ESM and IPSL-CM5A-MR model predicted decreasing streamflow in the near term and increasing streamflow in the middle and far future, under the RCP 8.5 scenario. The CNRM-CM5 model revealed an increasing trend in streamflow throughout the future. The BNU-ESM model showed the highest percentage change of 45.11% in the far future, and the IPSL-CM5A-MR model revealed the lowest percentage change of −15.69% in the near term, under the RCP 8.5 scenario. The findings are consistent with the results reported by Nagireddy et al. [10]. According to Fig. 2, the simulated streamflow peaks with IMD data occurred in August with an intensity of 39.31 mm for the basin. The BNU-ESM and IPSL-CM5A-MR models' historical streamflow peaks were observed in July, but the CNRM-CM5 model showed its peak streamflow in August. The projected mean monthly streamflow under two RCP scenarios is shown in Fig. 4.

The BNU-ESM model projected a similar historical peak in the near and middle future, but a peak shift to August in the far future, under the RCP 4.5 scenario. The CNRM-CM5 model predicted the peak streamflow in August at a similar historical peak. The IPSL-CM5A-MR model predicted that the peak would shift to August. Under the RCP 8.5 scenario, the BNU-ESM model predicted a similar historical

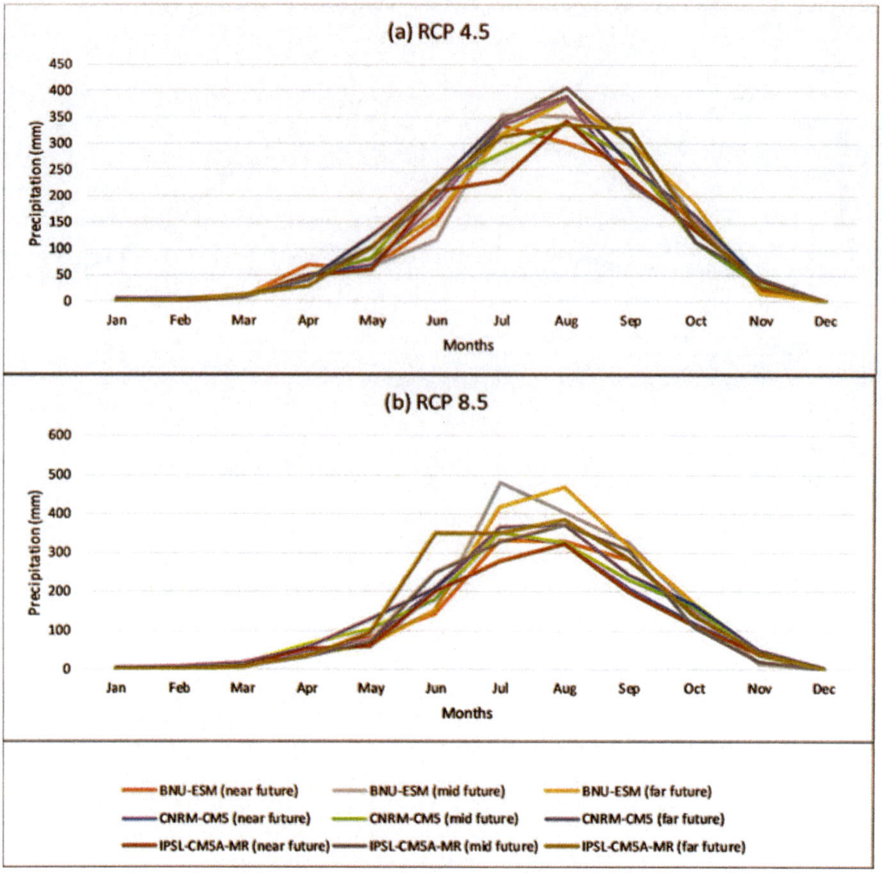

Fig. 3 Projected mean monthly precipitation under two RCP scenarios

peak in the near and mid future, but a peak shift to August in the far future. The CNRM-CM5 model predicted the peak precipitation in July for the near and mid future, and August for the far future. The IPSL-CM5A-MR model predicted that the peak would shift to August.

4 Conclusions

Flood frequency assessment using log Pearson type III distribution shows more frequent low magnitude floods and longer period of record will be critical for a complete evaluation for designing structures such dam, bridges, culverts etc. The percentage bias between the IMD precipitation and climate models precipitation over the basin represents slightly overestimation. The percentage change between

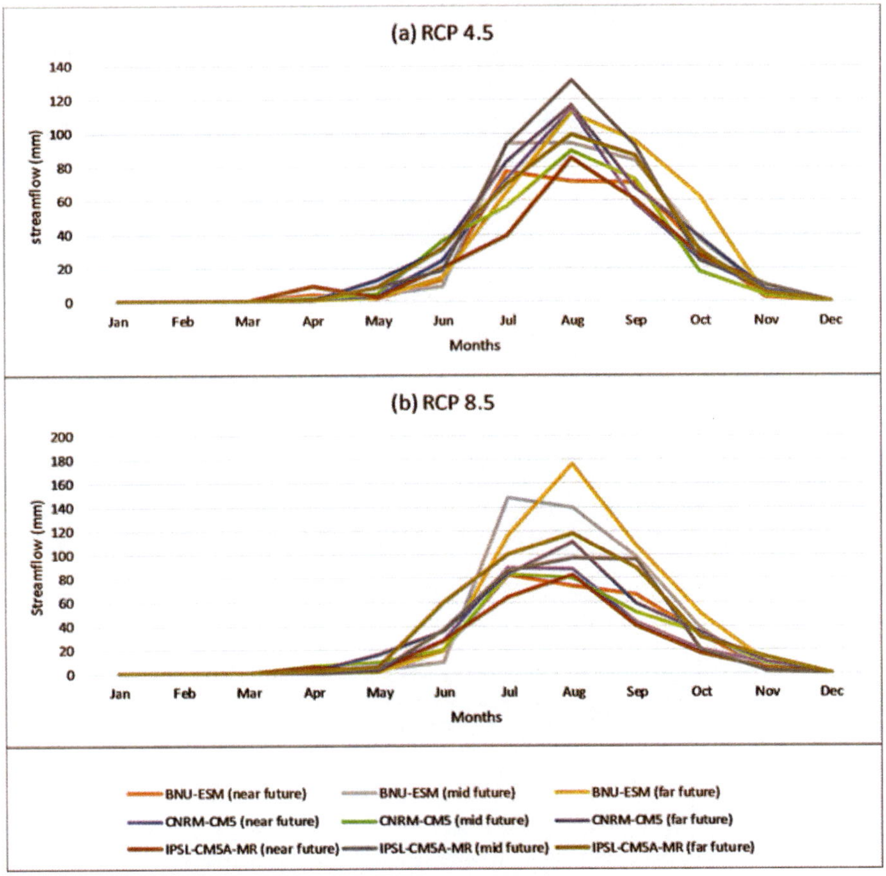

Fig. 4 Projected mean monthly streamflow under two RCP scenarios

the historical and future simulated streamflow patterns followed the climate model precipitation patterns. The BNU-ESM and IPSL-CM5A-MR models revealed a negative precipitation and streamflow over near term and a positive over the middle and far future. The CNRM-CM5 model predicted an increasing trend of precipitation and streamflow throughout the future under both RCP scenarios. The IMD observed precipitation and simulated streamflow peaks occurred in August. The BNU-ESM and IPSL-CM5A-MR models' historical precipitation and streamflow peaks were observed in July, but the CNRM-CM5 model showed its peak precipitation and streamflow in August. The BNU-ESM model precipitation and streamflow showed a similar historical peak in the near and mid future, but a peak shift to August in the far future, under the RCP 4.5 scenario. The CNRM-CM5 model predicted the peak precipitation and streamflow in August similar to historical peak. The IPSL-CM5A-MR model precipitation and streamflow predicted that the peak would shift to August. The BNU-ESM model projected a similar historical peak in the near and mid future, but a peak shift to August in the far future, under the RCP 8.5 scenario. The CNRM-CM5 model predicted the peak precipitation and streamflow in July for the near and mid future, and August for the far future. The IPSL-CM5A-MR model predicted that the peak would shift to August.

References

1. Abbaspour KC, Faramarzi M, Ghasemi SS, Yang H (2009) Assessing the impact of climate change on water resources in Iran. Water Resour Res 45(10)
2. Wagner PD, Bhallamudi SM, Narasimhan B, Kantakumar LN, Sudheer KP, Kumar S, Schneider K, Fiener P (2016) Dynamic integration of land use changes in a hydrologic assessment of a rapidly developing Indian catchment. Sci Total Environ 539:153–164
3. Dhar S, Mazumdar A (2009) Hydrological modelling of the Kangsabati river under changed climate scenario: case study in India. Hydrol Process: Int J 23(16):2394–2406
4. Gosain AK, Rao S, Basuray D (2006) Climate change impact assessment on hydrology of Indian river basins. Curr Sci 346–353
5. Gosain, A. K., Sandhya Rao., Anamika Arora.: Climate change impact assessment of water resources of India. *Current Science* (2011): 356–371.
6. Menon A, Levermann A, Schewe J (2013) Enhanced future variability during India's rainy season. Geophys Res Lett 40(12):3242–3247
7. Choudhury BA, Rajesh PV, Zahan Y, Goswami BN (2021) Evolution of the Indian summer monsoon rainfall simulations from CMIP3 to CMIP6 models. Clim Dyn 1–26
8. Raghavan SV, Hur J, Liong SY (2018) Evaluations of NASA NEX-GDDP data over Southeast Asia: present and future climates. Clim Change 148: 503–518
9. Rao KK, Kulkarni A, Patwardhan S, Kumar BV, Kumar TL (2020) Future changes in precipitation extremes during northeast monsoon over south peninsular India. Theoret Appl Climatol 142:205–217
10. Nagireddy NR, Keesara VR, Venkata Rao G, Sridhar V, Srinivasan R (2023) Assessment of the impact of climate change on streamflow and sediment in the Nagavali and Vamsadhara Watersheds in India. Appl Sci 13(13):7554
11. Griffis VW, Stedinger JR (2007) Log-Pearson type 3 distribution and its application in flood frequency analysis. I: distribution characteristics. J Hydrol Eng 12(5):482–491
12. Pawar U, Hire P (2018) Flood frequency analysis of the Mahi Basin by using Log Pearson Type III probability distribution. Hydrospatial Anal 2(2):102–112
13. Bhat MS, Alam A, Ahmad B, Kotlia BS, Farooq H, Taloor AK, Ahmad S (2019) Flood frequency analysis of river Jhelum in Kashmir basin. Quatern Int 507:288–294
14. Nagireddy NR, Keesara VR, Sridhar V, Srinivasan R (2022) Streamflow and sediment yield analysis of two medium-sized east-flowing river basins of India. Water 14(19):2960
15. Neitsch SL, Arnold JG, Kiniry JR, Williams JR (2011) Soil and water assessment tool theoretical documentation version 2009. Texas Water Resources Institute
16. Nash JE, Sutcliffe JV (1970) River flow forecasting through conceptual models part I—a discussion of principles. J Hydrol 10(3):282–290
17. Moriasi DN, Arnold JG, Van Liew MW, Bingner RL, Harmel RD, Veith TL (2007) Model evaluation guidelines for systematic quantification of accuracy in watershed simulations. Trans ASABE 50(3):885–900
18. Arnold JG, Moriasi DN, Gassman PW, Abbaspour KC, White MJ, Srinivasan R, Santhi C et al (2012) SWAT: model use, calibration, and validation. Trans ASABE 55(4):1491–1508

Impervious Surface Area Prediction Using Landsat Satellite Imagery and Open Source GIS Plugin

Ayyappa Reddy Allu and Shashi Mesapam

Abstract In India, population growth is rapidly increasing and the rural people are moving to urban areas for improving their socio-economic activity of life. For sustaining the human needs, most of the land use land cover features are migrating to impervious surfaces, which may lead to decreasing the infiltration capacity of the soil and increasing the flood frequency. Land Use Land Cover (LULC) maps are helpful to monitor and predict the impervious surface area using the Remote Sensing techniques. The proposed work aims to simulate the impervious surface area of Jagtial, Telangana, India, in the year 2050. This is accomplished by using Landsat satellite images from the years 2000, 2005, 2010, 2015, and 2020, and applying the Random Forest classification algorithm to generate LULC maps. The maximum tree depth is set at 30, the maximum number of trees is 250, and the maximum number of samples per class is 1000. A land use simulation model, based on Cellular Automata and Markov Chains, is employed to calibrate and optimize the LULC images. The model predicts the LULC map for the years 2020 and 2050, which are validated using existing classified LULC images. The imperviousness index of the LULC classes is used to estimate the impervious surface area of the location. The analysis of the multi-temporal LULC images shows that biophysical and socioeconomic factors have a significant impact on the increase in built-up areas and the decline in water bodies by the year 2050.

Keywords Cellular automata · Impervious surface area · Land use land cover · Markov chain

1 Introduction

In urban areas, population growth is rapidly growing due to the increase in population and migration of the people from rural places. LULC classes are transforming to the built up and industrial areas for meeting the requirements of the increased population,

A. R. Allu (✉) · S. Mesapam
Department of Civil Engineering, National Institute of Technology, Warangal, India
e-mail: aa721015@student.nitw.ac.in

S. Mesapam et al. (eds.), *Developments and Applications of Geomatics*,
Lecture Notes in Civil Engineering 450, https://doi.org/10.1007/978-981-99-8568-5_23

which may lead to increase in the impervious surface. Impervious surface is one of the land cover form which may prevents the infiltration of water into the soil and it may also impact the ground water table and increasing the runoff [1, 2]. Human made surfaces such as pavements, buildings, parking lots, are constructed with the concrete, whereas the asphalt material are considered as an impervious surface [1, 3]. Impervious surface area will be playing a major role in urban planning and development, and resource management [1, 3, 4]. So, there is a need to monitor and predict the impervious surfaces to regularize the development.

Remote Sensing is one of the technology which uses the satellite images to continuous assessment and monitoring of impervious surfaces [5]. Several techniques were successfully applied to extract the impervious surface location using the satellite images in the previous research works, including Artificial Neural Network (ANN), sub pixel classification, regression tree analysis, multiple regression analysis, spectral mixture analysis, LULC classification [3, 5–9] etc. In that, the LULC classification method is one of the easiest method to solving the environmental problem [10, 11]. LULC images are playing a vital role in the identification of impervious surfaces by classifying the land use features on the ground into number of classes [12]. Each class has different impervious surface characteristics, that will be helpful for extracting the impervious surface area [10, 13]. Classifying the features into classes in the satellite images might be a challenging task due to their spatial and spectral information. Training dataset will be helpful for classifying the satellite images by training the random sample of each class from the respective satellite image [12].

Many classification algorithms, such as Maximum Likelihood, Support Vector Machine, and Random Forest, have been shown to be useful for classifying the features to prepare the Land Use and Land Cover (LULC) maps using training datasets [4, 9, 11, 14–16]. Each classification algorithm has its own strengths and limitations. The Random Forest classification algorithm, in particular, uses an ensemble of multiple decision trees to make decisions, resulting in high accuracy in classifying the features into the relevant classes [3, 6, 8, 15, 17–20]. This algorithm builds a decision tree for each sample to train the dataset.

The combination of impervious nature of LULC classes provides the impervious surface location [1, 2, 8, 21] of the respective satellite image but for prediction, it needs to simulate the future image of the LULC by using the existing dataset. Modelling is an effective method to provide the opportunity to explore multiple pathways to analyses and predict the future from temporal data [10, 19]. Several studies worked in the modelling of LULC change to simulate land use changes and develop the interaction between land cover and environmental changes [9, 14, 15, 19]. LULC changes mainly affect the biodiversity, vegetation, water resources, and human life, and are linked to sustainable economic development [4]. LULC change analysis in most of the studies is worked with the regression model or transition model to simulate and predict the future datasets. Spatial transition models simulate the future development with possible estimations using the Monte Carlo's based Cellular Automata (CA) and Markov Chain (MC) or other methods [11, 16, 17].

The CA-MC is an integrated model which is suitable to detect the land use changes and simulates to predict the land cover changes by considering the spatial and temporal components of the dynamic change in land cover [14]. This CA-MC model is included in various software such as TerraSet, QGIS [14]. In that, QGIS is an open source software which uses the MOLUSCE plugin [4, 12, 14, 16, 17, 22] (Modules for Land Use Change Evaluation) (https://plugins.qgis.org/plugins/molusce/) to simulate the future images from the past LULC images using the ANN, Logistic Regression (LR), Weight of Evidence and Multi Criteria Evaluation techniques [16, 17, 19]. For these techniques, transitional area change of LULC classes plays a major role for modelling and simulating the future developments and changes [22]. The periodic gap between the LULC images is also an impacting factor for predicting the LULC changes but it depends on the growth of the study area location.

The primary objective of this study is to predict the extent of impervious surfaces in the year 2050 using Landsat satellite images from the years 2000, 2005, 2010, 2015, and 2020 on the QGIS platform. LULC classification has been prepared from the Landsat satellite images to extract the impervious surface area of the study area location in QGIS software. The future LULC image will be predicted using the MOLUSCE plugin in the QGIS software. The novelty of this approach is to predict the impervious surface of the study area by utilizing different periodic change combination of the initial and final year satellite images.

2 Study Area

Jagtial district is located in the state of Telangana, India (refer to Fig. 1). It is situated between the latitude 18°42' and longitude 78°54' and has an average elevation of +293 m above the sea level. According to the 2011 census, the district had a population of 9,85,417 and a total area of 2,419 KM^2. Jagtial district is divided into 5 municipalities, 18 mandals, and 380 villages. Industries, educational institutions, and research institutes are located in the Jagtial and Korutla municipalities, which have led to significant changes in these regions.

3 Methodology

The methodology flowchart (refer to Fig. 2) outlines the steps involved in data collection, preparation, processing, analysis, and validation to estimate and simulate the impervious surface area for the year 2050 using the Landsat images.

3.1 Data Collection

In this study, Landsat satellite imagery from the years 2000, 2005, 2010, 2015, and 2020 were utilized to estimate and predict the impervious surface area of the study

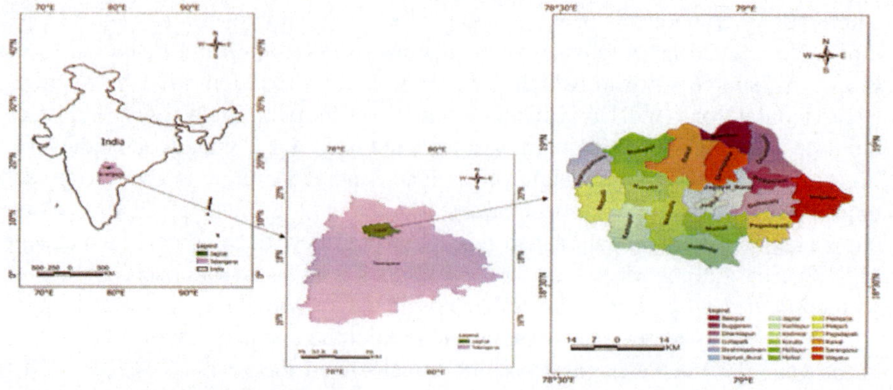

Fig. 1 Jagtial location in India map

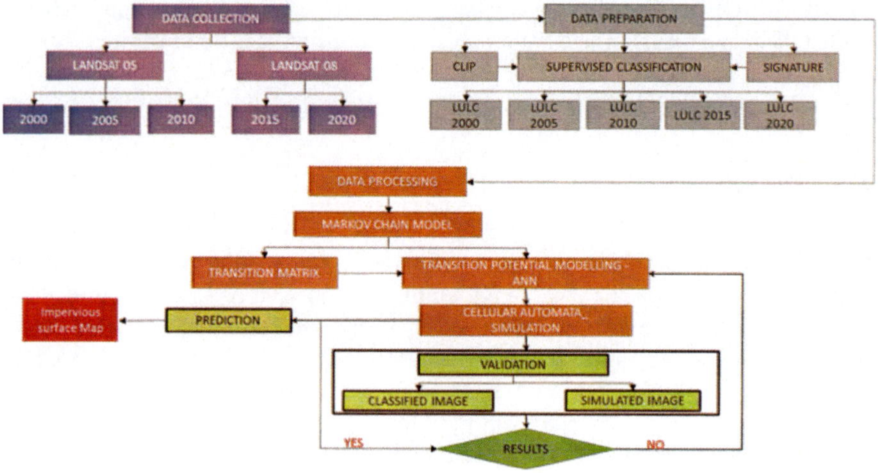

Fig. 2 Methodology flowchart of impervious surface map using Landsat satellite images

location. The satellite imagery from the years 2000, 2005, and 2010 was obtained from the Landsat 05 TM (Thematic Mapper), while the imagery from 2015 and 2020 was obtained from the Landsat 8 OLI (Operational Land Imager) through GloVis USGS (https://glovis.usgs.gov/). The blue, green, red, and Near Infra-Red (NIR) spectral bands of the Landsat satellite imagery were utilized with a spatial resolution of 30 m. The satellite images acquired from Landsat were all in the same reference system (WGS 1984 UTM Zone 44N), and the information regarding the collected data is presented in Table 1. For validation purposes, Google Earth, field validation, and socio-economic activities were considered as ground truth data.

Table 1 Data acquisition information about the Landsat satellite imagery

Data	Path	Row	Date of collection	Bands
Landsat 05 TM	144	47	02 October 2000	1, 2, 3, 4
Landsat 05 TM	144	47	07 April 2005	1, 2, 3, 4
Landsat 05 TM	144	47	23 May 2010	1, 2, 3, 4
Landsat 08 OLI	144	47	19 April 2015	2, 3, 4, 5
Landsat 08 OLI	144	47	11 January 2020	2, 3, 4, 5

3.2 Data Preparation

Preprocessing techniques were applied to the Landsat images to perform atmospheric correction. The preprocessed bands of the satellite imagery for the years 2000, 2005, 2010, 2015, and 2020 were layer stacked, and the study area location was extracted from the layer-stacked image for the respective years by using its administrative boundary (shown in Fig. 3). Subsequently, image classification techniques were performed for further processing and analysis.

Image Classification

The Random Forest (RF) algorithm was employed to classify the layer-stacked images. The samples of each class for every year were used to train the dataset. The area was classified into four (LULC) classes: barren, built-up, vegetation, and water bodies, which represented the features in the satellite image (Table 2). For

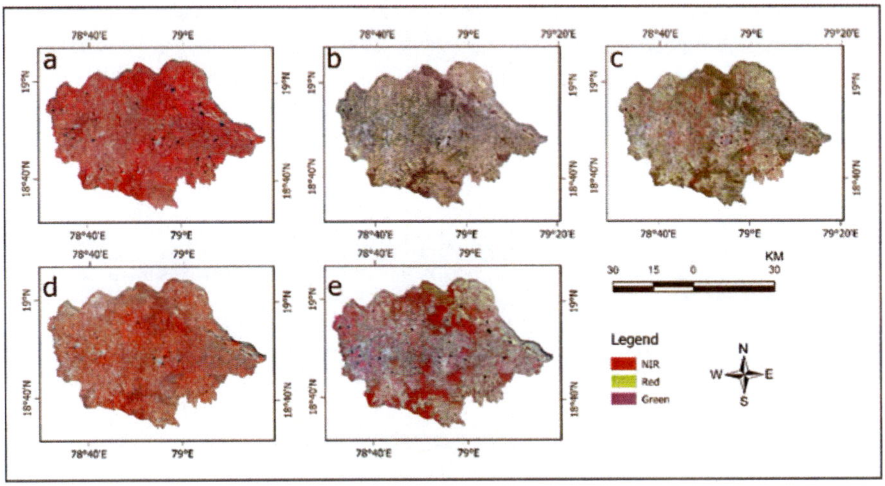

Fig. 3 Layer stacked images of Jagtial location for the year **a** 2000, **b** 2005, **c** 2010, **d** 2015 and **e** 2020

each class, 30 training samples were selected, and the satellite images were classified using the Random Forest algorithm with a maximum tree depth of 30, a maximum number of trees of 250, and a maximum number of samples per class of 1000. The raster images were generated in the open-source GIS software, QGIS (https://www.qgis.org/en/site/), and the satellite images were classified using the Semi-Automatic Classification (https://github.com/semiautomaticgit/SemiAutomaticClassificationPlugin) plugin.

The LULC classification was performed for the years of 2000, 2005, 2010, 2015 and 2020 using the RF supervised machine learning algorithm. The results of the classification are presented in Fig. 4.

Table 2 Land cover feature of the LULC class

S. No	Class	Feature
1	Barren	Bare lands, sediment deposits, exposed rocks, mixed grasslands
2	Builtup	Residential, commercial, industrial, roads, railway network, and other uncategorized surface areas
3	Vegetation	Agriculture, forest grass land, shrubs and bush areas
4	Water Body	Open water bodies like ponds, lakes, rivers, reservoirs

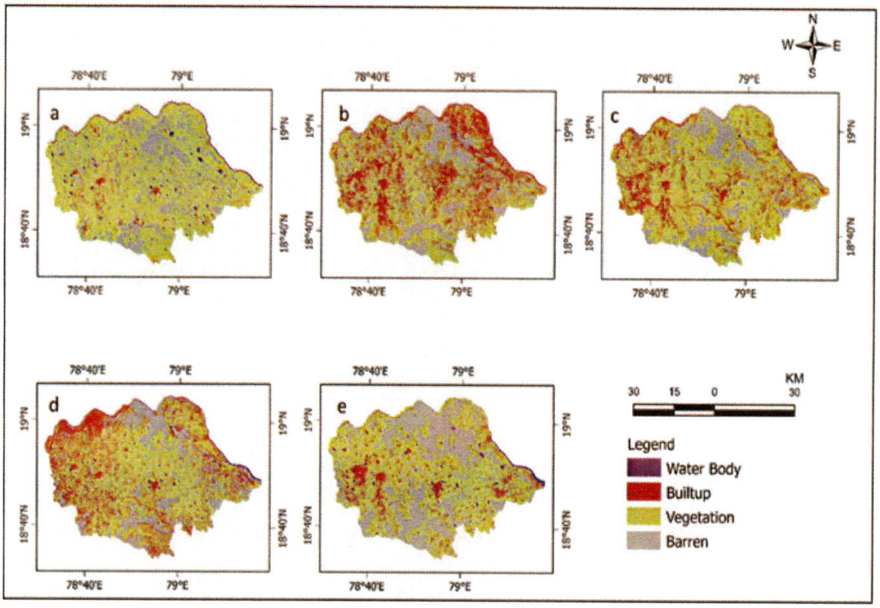

Fig. 4 LULC images of the year: **a** 2000, **b**2005, **c** 2010, **d** 2015 and **e** 2050

Table 3 Accuracy of the classified images for the years of 2000, 2005, 2010, 2015 and 2020

Year of LULC image	2000	2005	2010	2015	2020
Overall accuracy (%)	86.05	78.17	82.15	81.73	80.54
Kappa coefficient (K_c)	0.795	0.744	0.781	0.797	0.760

Accuracy Assessment

The accuracy of the classified LULC images was validated using the ground truth information collected from Google Earth and field. A confusion matrix was created to evaluate the accuracy of the classified images, which was measured through the overall accuracy and Kappa coefficient. The equations used for this evaluation are as follows:

$$\text{Overall Accuracy}(OA) = \frac{\text{Number of pixels classified correctly}}{\text{Total number of pixels}} \times 100$$

$$\text{Kappa Coefficient}(K_c) = \frac{N \sum_{i=1}^{r} x_{ij} - \sum_{i=1}^{r} (x_{i+} \times x_{+j})}{N^2 - \sum_{i=1}^{r} (x_{i+} \times x_{+j})}$$

where, x_{ij} is the number of observations in row i and column j, x_{i+} is the marginal total of row i, x_{+j} is the marginal total of column j, r is the total number of rows and columns in the error matrix and N is the total number of observations. The range of the Kappa coefficient (K_c) typically falls between -1 and $+1$, but it is usually between 0 and 1. A value of 1 indicates complete agreement between the classified and reference data but zero represents the no agreement. The accuracy of the classified LULC images (Table 3) is useful for calculating the impervious surface area and predicting future data.

3.3 Data Processing

Impervious Surface Ratio

The Impervious Surface Ratio (ISR) of the study location indicates the amount of land covered with hard materials that resist water infiltration. The LULC images were used to evaluate and predict the impervious surface by monitoring changes in the natural landscape. In this study, the features were classified into four classes, with the vegetation, barren, and water body classes considered as pervious surfaces and the built-up class considered as the impervious surface. The ISR was calculated using the following equation:

$$\text{Impervious Surface Ratio} = \frac{\text{Total Impervious Surface Area}}{\text{Total(Pervious + Impervious)Area}}$$

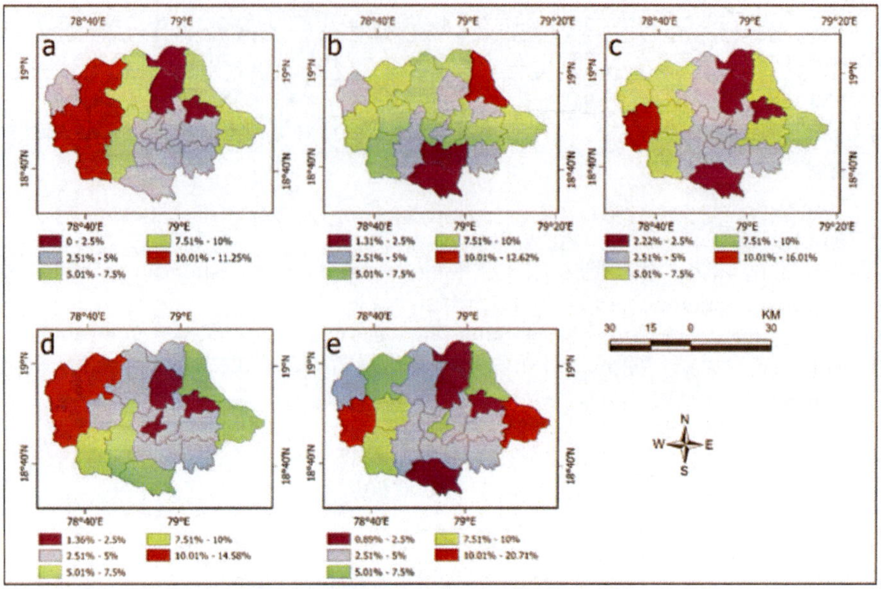

Fig. 5 Impervious surface map for the year **a** 2000, **b** 2005, **c** 2010, **d** 2015 and **e** 2020

ISR maps (Fig. 5) for the years 2000, 2005, 2010, 2015 and 2020 were prepared using the classified LULC images from those respective years. Simulated LULC images will be useful for predicting the ISR for the year of 2050.

Simulation of LULC using MOLUSCE

This study predicts the LULC for the year of 2050 through simulation of three different combinations of the initial and final year images: (i) the LULC of 2000 and 2005, (ii) the LULC of 2000 and 2010, and (iii) LULC of 2010 and 2015. CA based MC calibration and simulation model was used to predict the future LULC image from past LULC maps. The model predicts the transition from one state at T + t to T + n(t) state based on the state T.

$$L_{(T+t)} = P_{ij} \times L_T$$

$$P_{ij} = \begin{bmatrix} P_{11} & P_{12} & P_{13} \\ P_{21} & P_{22} & P_{23} \\ P_{31} & P_{32} & P_{33} \end{bmatrix}$$

where $L_{(T)}$ is the land use status at the year of T, $L_{(T+t)}$ is the land use at the year of $T + t$, T is the initial year, t is the change in time, n is number of times of simulation and P_{ij} is the i, jth cell of contingency table.

The MOLUSCE plugin was utilized to predict the LULC of the year 2050 through the simulation of LULC changes between different periodic combinations. The functional procedure of the MOLUSCE plugin is described as follows:

Correlation Evaluation

A correlation analysis was conducted to evaluate the linear relationship between the initial and final year LULC images. This was accomplished by assessing the statistical measures of the spatial variable factors using the Pearson's correlation coefficient.

Area Change

Transition area and transition probability matrices were created between the initial and final year LULC images to analyze the trends in the transformation of LULC features. The change in area (KM^2) and probability of area change (%) from one class to another classes were measured in the transition matrix.

Transitional Potential Modelling

Modeling the variation of LULC features and recognizing the patterns is a challenging task for human brains, but algorithms can be used to model complex patterns. Fuzzy logic requirements were incorporated into an ANN, to allow for a continuous range between 0 and 1, determined by the usability of the terrain. The core elements of the ANN inter-connected between the neurons, with modifications to the weight connections were used to learn the patterns of transforming direction of LULC features.

Cellular Automata Simulation

CA is used to simulate the LULC by taking the transition probabilities from the transition potential modelling. Transition simulator maps, certaincy and simulation maps were generated for the each simulation.

Validation

The predicted LULC images were validated using existing datasets to predict future LULC images. The validation module used four parameters of kappa statistics to calculate the accuracy of the model: kappa location (K_{loc}), kappa histogram (K_{hist}), overall kappa (K) and percentage of correctness.

$$K_{\text{loc}} = \frac{P(A) - P(E)}{P_{\max} - P(E)}$$

$$K_{\text{hist}} = \frac{P_{\max} - P(E)}{1 - P(E)}$$

$$K = K_{\text{loc}} \times K_{\text{hist}} = \frac{P(A) - P(E)}{1 - P(E)}$$

where

$P(A) = \sum_{i=1}^{c} p_{ii}$, $P(E) = \sum_{i=1}^{c} (p_{ia} \times p_{aj})$, $P_{\max} = \sum_{i=1}^{c} \min(p_{ia}, p_{aj})$, p_{ij} is the i, jth cell of contingency table, p_{ia} is the sum of all cells in ith row, p_{aj} is the sum of all cells in jth column, and c is count of raster categories.

4 Results and Discussion

The changes in LULC features are not constant over time and may be influenced by various factors, such as socio-economic activities, traffic, etc. The analysis of the classified LULC images shows that barren land areas have changed to vegetation and built-up areas from 2005 to 2010 and 2015, while vegetation has changed to built-up and barren land. However, the built-up area has increased from 2000 to 2020. Table 5 shows the changes in the area of classified features for the years 2000 to 2020, which can be used to predict future LULC images.

In this study, simulation techniques are used to predict the LULC for the year 2050 by validating the predicted LULC of 2020 with the classified LULC of 2020. The Pearson's correlation coefficient between the input images is high (as shown in Table 4), indicating a strong relationship between the LULC images.

A transition probability matrix was created between the initial and final year LULC images, representing the probability of each feature class changing from one class to another. Table 5 shows that the transformation of water body features to built-up is high compared to the other classes. Vegetation features are mostly transforming to barren and built-up classes in every combination, while barren classes are transforming to built-up at a much lower rate from 2010 to 2015 compared to other combinations. On the other hand, most of the built-up features are transforming to vegetation in the period between 2000 and 2010, and between 2010 and 2015, compared to the period between 2000 and 2005. This suggests that there was an increase in vegetation and a decrease in the rate of change of built-up area from 2010 to 2015, with most of the built-up area being constructed in 2005.

The MC-CA technique used an Artificial Neural Network (ANN) algorithm to predict the future LULC images for 2020 and validated the results using the classified LULC image of that year. If the accuracy of the model was validated, it could then be used to predict the LULC image for 2050. Figure 6 shows the simulated LULC images for the years 2020 and 2050, created using a combination of initial and final images from 2000 and 2005, 2000 and 2010, and 2010 and 2015. The accuracy of the simulated images is presented in Table 6.

Table 4 Pearson's correlation coefficient between initial and final year images

Initial image	2000	2000	2010
Final image	2005	2010	2015
Perason's coefficient	0.91524	0.93101	0.94427

Table 5 Transition probability matrix between initial and final year

2000–2005	Water body	Builtup	Vegetation	Barren
Water body	0.182212	0.433089	0.188618	0.196082
Builtup	0.001585	0.51499	0.296325	0.1871
Vegetation	0.000463	0.173612	0.519008	0.306916
Barren	0.000161	0.178949	0.345664	0.475226
2000–2010				
Water body	0.194328	0.305635	0.375121	0.124916
Builtup	0.001131	0.444525	0.426566	0.127779
Vegetation	0.000911	0.112287	0.679013	0.207789
Barren	0.000186	0.110661	0.474718	0.414435
2010–2015				
Water body	0.650648	0.230543	0.056886	0.061923
Builtup	0.021064	0.395258	0.485827	0.097851
Vegetation	0.002802	0.193037	0.655715	0.148446
Barren	0.005275	0.056041	0.249508	0.689176

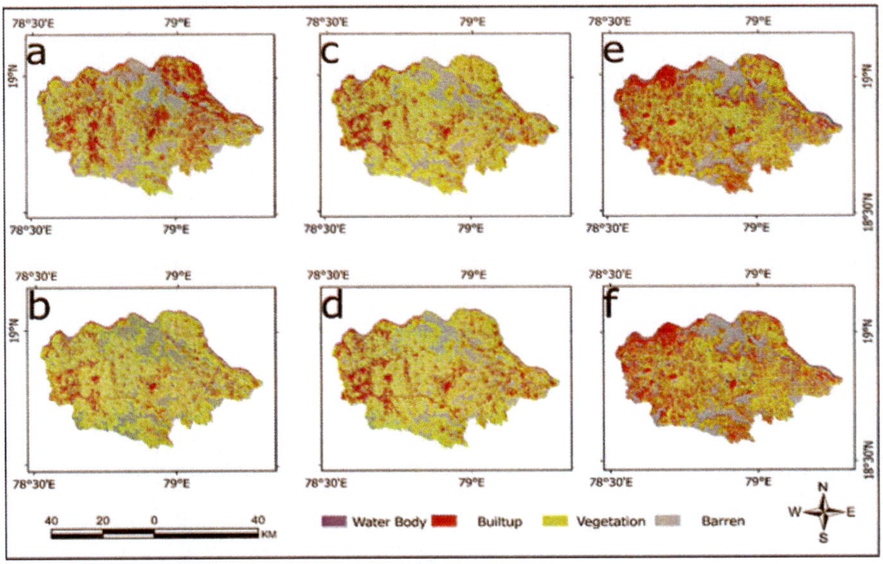

Fig. 6 Simulated LULC images: **a**, **c**, **e** represent the simulated year of 2020, created using the initial and final images of 2000, 2005; 2000, 2010; and 2010, 2015 respectively. **b**, **d**, **f** represent the simulated year of 2050, created using the same combinations of initial and final images

Table 6 Accuracy of simulated LULC images of the year 2020 and 2050 using the initial and final year images

Initial	Final	Simulated	Kappa Histo	Kappa local	Kappa	% of correctness
2000	2005	2020	0.62369	0.80497	0.77479	73.19752
2000	2005	2050	0.63475	0.75241	0.72386	71.25894
2000	2010	2020	0.84649	0.91443	0.77406	85.38166
2000	2010	2050	0.76889	0.77242	0.69391	78.64269
2010	2015	2020	0.80482	0.71247	0.57341	69.25772
2010	2015	2050	0.85098	0.69286	0.58961	70.68681

The simulated LULC images for the year 2020 and 2050 are accurately predicted using the initial and final images of 2000 and 2010. These images depict an increase in vegetation and built-up areas and a decrease in barren and water body areas when compared to the latest year LULC image (2020). The most significant increase in built-up area was observed in the west side (Ibrahimpatnam, Metpalli, Kathlapur, Mallapur, and Korutla mandals) and the north-east side (Dharmapuri and Velgatur mandals) of the study area. However, the percentage area of these feature classes in both the classified and simulated LULC images is shown in Table 7. These results demonstrate the potential of the simulation techniques and the power of the ANN algorithm in predicting future LULC images and understanding the impact of socio-economic activities and environmental changes on the study area.

Figure 5 indicates that, there has been a limited increase in the impervious surface ratio in the study area. The results of the study show that the increase in built-up areas is concentrated in the west and north-east regions of Jagtial district, while the central region of the district has remained largely unaffected (shown in Fig. 7). This suggests that the impact of socio-economic activities and environmental changes on the study area has been relatively limited, at least with regards to the impervious surface ratio. The results of the study may be useful for decision-makers, planners and other stakeholders in the area to develop strategies for managing and mitigating the impact of future growth and development in the region.

Table 7 Percentage area of the various classes in classified and simulated LULC images

Class/year	Classified area (%)					Simulated area from 2000 and 2005 (%)		Simulated area from 2000 and 2010 (%)		Simulated area from 2010 and 2015 (%)	
	2000	2005	2010	2015	2020	2020	2050	2020	2050	2020	2050
Barren	34.90	35.78	27.44	28.97	40.71	35.94	56.63	22.84	21.91	31.31	35.24
Builtup	4.24	19.65	13.07	18.21	13.57	19.95	7.70	16.48	18.83	23.37	24.28
Vegetation	58.33	44.07	58.93	51.87	43.42	43.74	35.23	60.27	59.07	44.39	39.53
Water Body	2.53	0.50	0.56	0.95	2.29	0.37	0.44	0.41	0.19	0.94	0.95

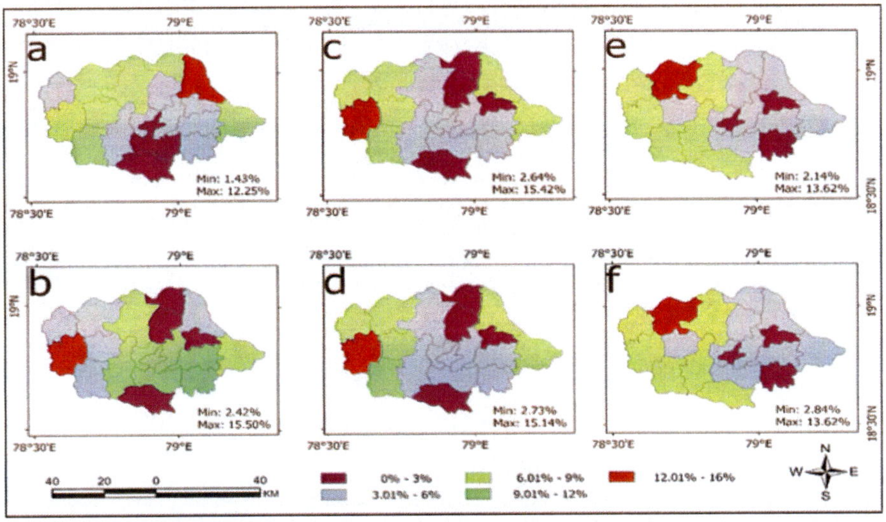

Fig. 7 Impervious Surface Ratio images of the simulated LULC images: **a**, **c**, **e** are simulated year of 2020 from the initial and final images of 2000, 2005; 2000, 2010; and 2010, 2015 respectively, **b**, **d**, **f** are simulated year of 2050 from the initial and final images of 2000, 2005; 2000

5 Conclusion

The impervious surface map for 2050 was predicted using open-source GIS software. The prediction was based on the classified LULC images from the years 2000, 2005, 2010, 2015, and 2020, which were processed using the MOLUSCE plugin in QGIS software. The results of the prediction show an increase in built-up areas in the west and northeast regions, a conversion of water bodies into built-up areas, and negligible changes in vegetation in the central part of the district. The impervious surface ratio in the study area does not change significantly. The LULC changes in the study area appear to vary linearly. QGIS played a crucial role in this study, helping to address the problems arising from socio-economic activities and environmental changes.

References

1. Majid MR, Jamaludin JA, Ibrahim WYW (2013) Estimation of residential impervious surface using GIS technique. Plan Malaysia 11:23–38. https://doi.org/10.21837/pmjournal.v11.i2.114
2. Ongsomwang S, Pimjai M (2014) Land use and land cover prediction and its impact on surface runoff
3. Zhang H, Gorelick SM, Zimba PV (2020) Extracting impervious surface from aerial imagery using semi-automatic sampling and spectral stability. Remote Sens 12(3). https://doi.org/10.3390/rs12030506

4. Kamaraj M, Rangarajan S (2022) Predicting the future land use and land cover changes for Bhavani basin, Tamil Nadu, India, using QGIS MOLUSCE plugin. Environ Sci Pollut Res 29(57):86337–86348. https://doi.org/10.1007/s11356-021-17904-6
5. Hu X, Weng Q (2009) Estimating impervious surfaces from medium spatial resolution imagery using the self-organizing map and multi-layer perceptron neural networks. Remote Sens Environ 113(10):2089–2102. https://doi.org/10.1016/j.rse.2009.05.014
6. Feng S, Fan F (2021) Impervious surface extraction based on different methods from multiple spatial resolution images: a comprehensive comparison. Int. J. Digit. Earth 14(9):1148–1174. https://doi.org/10.1080/17538947.2021.1936227
7. Kaspersen PS, Fensholt R, Drews M (2015) Using Landsat vegetation indices to estimate impervious surface fractions for European cities. Remote Sens 7(6):8224–8249. https://doi.org/10.3390/rs70608224
8. Zhang X et al (2020) Development of a global 30m impervious surface map using multisource and multitemporal remote sensing datasets with the Google Earth Engine platform. Earth Syst. Sci. Data 12(3):1625–1648. https://doi.org/10.5194/essd-12-1625-2020
9. Lu D, Moran E, Hetrick S (2011) Detection of impervious surface change with multitemporal Landsat images in an urban-rural frontier. ISPRS J Photogramm Remote Sens 66(3):298–306. https://doi.org/10.1016/j.isprsjprs.2010.10.010
10. Li W, Wu C, Choi W (2018) Predicting future urban impervious surface distribution using cellular automata and regression analysis. Earth Sci Inform 11(1):19–29. https://doi.org/10.1007/s12145-017-0312-8
11. Al Kafy A et al (2021) Predicting changes in land use/land cover and seasonal land surface temperature using multi-temporal landsat images in the northwest region of Bangladesh. Heliyon 7(7). https://doi.org/10.1016/j.heliyon.2021.e07623
12. Satya BA, Shashi M, Deva P (2020) Future land use land cover scenario simulation using open source GIS for the city of Warangal, Telangana, India. Appl Geomatics 12:281–290. https://doi.org/10.1007/s12518-020-00298-4/Published
13. Nair HM (2013) Estimation of effective impervious surface area of cochin using satellite images. [Online]. www.isca.in
14. Rahnama MR (2021) Forecasting land-use changes in Mashhad metropolitan area using cellular automata and Markov chain model for 2016–2030. Sustain Cities Soc 64. Elsevier Ltd. https://doi.org/10.1016/j.scs.2020.102548
15. Liping C, Yujun S, Saeed S (2018) Monitoring and predicting land use and land cover changes using remote sensing and GIS techniques—a case study of a hilly area, Jiangle, China. PLoS One 13(7). https://doi.org/10.1371/journal.pone.0200493
16. Muhammad R, Zhang W, Abbas Z, Guo F, Gwiazdzinski L (2022) Spatiotemporal change analysis and prediction of future land use and land cover changes using QGIS MOLUSCE plugin and remote sensing big data: a case study of Linyi, China. Land 11(3). https://doi.org/10.3390/land11030419
17. Alshari EA, Gawali BW (2022) Modeling land use change in Sana'a city of Yemen with MOLUSCE. J Sensors 2022. https://doi.org/10.1155/2022/7419031
18. Parekh JR, Poortinga A, Bhandari B, Mayer T, Saah D, Chishtie F (2021) Automatic detection of impervious surfaces from remotely sensed data using deep learning. Remote Sens 13(16). https://doi.org/10.3390/rs13163166
19. Kafy AA et al (2021) Cellular automata approach in dynamic modelling of land cover changes using RapidEye images in Dhaka, Bangladesh. Environ. Challenges 4. https://doi.org/10.1016/j.envc.2021.100084
20. Li F et al (2021) Estimating artificial impervious surface percentage in asia by fusing multi-temporal modis and viirs nighttime light data. Remote Sens 13(2):1–23. https://doi.org/10.3390/rs13020212
21. Das N, Mondal P, Sutradhar S, Ghosh R (2021) Assessment of variation of land use/land cover and its impact on land surface temperature of Asansol subdivision. Egypt J Remote Sens Sp Sci 24(1):131–149. https://doi.org/10.1016/j.ejrs.2020.05.001

22. Sajan B, Mishra VN, Kanga, Meraj G, Singh SK, Kumar P (2022) Cellular automata-based artificial neural network model for assessing past, present, and future land use/land cover dynamics. Agronomy 12(11). https://doi.org/10.3390/agronomy12112772

Influence on Water Characteristics Away from Various Sources of NIT Kurukshetra Using ArcGIS

Rahul Deopa and K. K. Singh

Abstract In this paper we assess the characteristics of various water sources of the NIT Kurukshetra campus using ArcMap 10.8 Software. Water Quality Index (WQI) is used to evaluate the drinking water quality from different sources on the NIT Kurukshetra Campus using the Weighted Arithmetic Method. A way of rating water quality is the Water Quality Index. It was done to determine the area's overall groundwater quality significance. It is a useful instrument for determining the quality of groundwater. Electroconductivity, color, odor, taste, pH, Chlorides (Cl), total dissolved solids (TDS), total alkalinity (TA), and total hardness (TH) are used in this study to calculate the quality of water at NIT Kurukshetra Campus. Water quality measurements were gathered at five distinct places on the NIT Kurukshetra Campus. The results revealed that all of the stations have Excellent water quality, according to the values of the WQI although the readings of most of the sources measured are above acceptable and occasionally permissible limits also but the water quality still comes out as excellent. The data from the evaluation was then processed using ArcGIS' spatial analysis function in which the Kriging technique of Interpolation is used from which various spatial distribution maps for various water source locations were constructed after selecting the appropriate interpolation strategy. Using Image Overlapping, the variation of physicochemical characteristics of some parameters with length is also determined.

Keywords Water quality · WQI · Kriging · Interpolation · ArcMap 10.8

1 Introduction

Water is a transparent, odorless, colorless liquid that forms the lakes, seas, ponds, rivers, and rain and is the basic requirement of living organisms and also a vital source of water supply [1]. Water covers 71% of the earth's surface. On Earth, 96.5% of the planet's crust water is found in seas and oceans which are saline, 1.7% in groundwater,

R. Deopa (✉) · K. K. Singh
Department of Civil Engineering, NIT Kurukshetra, Haryana, India
e-mail: rahul_32012513@nitkkr.ac.in

S. Mesapam et al. (eds.), *Developments and Applications of Geomatics*,
Lecture Notes in Civil Engineering 450, https://doi.org/10.1007/978-981-99-8568-5_24

1.7% in glaciers and ice caps of Antarctica and Greenland, and less than 0.3% of all freshwaters is in rivers, lakes, and ponds. All this water can be optimally sustained only when the quantity and quality are properly assessed.

For healthy living, drinkable safe water is very necessary. An adequate quantity of clean and fresh drinking water is a basic human necessity. The water quality index method is one of the most effective methods to determine the water quality in which water quality is accessed with the help of water quality indices [2]. The water quality index method is one of the most effective methods for determining water quality. Water quality indices are used to access water quality. Water quality is determined by a variety of physical, chemical, and biological factors [3]. Horton in 1965 was the first to attempt to combine the several characteristics that characterize the status of water into a single value known as the Water Quality Index [4]. To create his index, Horton chose ten of the most often measured variables. He chose rating scales for each variable that range from 0 to 100, with 100 representing the highest quality rate. The weighing parameters vary between 1 and 4. The weighted sum of sub-indices is divided by the sum of weights, and two coefficients based on water temperature and pollution level are multiplied by the final index score.

2 Study Area

Kurukshetra, considered one of the most religious and holy cities of India with immense reference to mythological and ancient India, is located in the state of Haryana. NIT Kurukshetra is one of the Prestigious Premier Technical Institutes situated in Kurukshetra, Haryana. Geographically, it has its locations defined at a latitude of 29°56′56.616″ and longitude of 79°49′2.114″. The campus extends over the area of 300 acres laid down on an attractive landscape. It is impressive in architecture and natural beauty (Fig. 1).

3 Materials and Methods

In the presented research work, the NIT Kurukshetra Campus is selected to assess the drinking water quality at various points of the source. There are 5 points of water source which are taken for the sampling method which are located in the different regions of the campus so that they can cover the whole campus. These sampling stations are: Tubewell 6, Tubewell 8, Tubewell 7, Tubewell 9 and Academic Section. The samples are collected from each location in December 2021 and tests are performed and analyzed in the laboratory. Color, odor, taste, electroconductivity, total hardness, total alkalinity, pH, salinity, chlorides, and total dissolved solids are among the criteria examined (TDS). The geographical distribution map of various

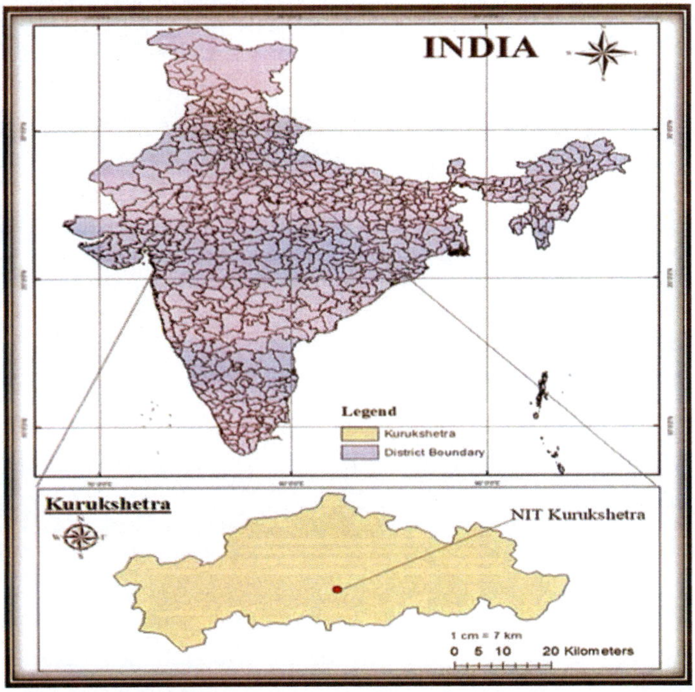

Fig. 1 Study area (NIT Kurukshetra)

physicochemical parameters is likewise created in ArcGIS utilizing GPS coordinates and interpolation techniques in the software's Spatial Analysis tool for various specified places (Table 1 and Fig. 2).

Water Quality Index (WQI) Method

This method is used to connect a group of water quality parameters and integrate them into a single number in accordance with a selected method of computation [5]. It represents the complex data in a simple and understandable way and provides a common framework for comparing a wide range of measured data with prescribed standard limits. The steps to find out WQI are as follows:

Table 1 Sampling stations of NIT Kurukshetra

Location L1	Tubewell No.6
Location L2	Tubewell No.9
Location L3	Tubewell No.8
Location L4	Tubewell No.7
Location L5	Academic section

Fig. 2 Sampling stations of NIT Kurukshetra

- Selection of parameter
- Determination of sub-index of the quality parameter
- Determination of Weightage factor
- Integration of sub-Indices in a mathematical expression.

Sub Index of Quality Parameter (q_n):

$$q_n = ((V_n - V_i) / (S_n - V_i))x100$$

where,

V_n = Observed Actual Value

S_n = Standard Value

V_i = Ideal Value

Note: Except for DO, pH, and temperature, all Ideal Values for water are set to zero.

Weightage Factor (W):

$$W_n = (K / S_n)$$

where,

$Sn = (1/\Sigma(1/S_n))$.
K = Proportionality Constant.
Note: $\Sigma W = 1$.

Water Quality Index (WQI):

$$\mathrm{WQI} = ((\Sigma w_i q_i)/(\Sigma w_i))$$

4 Results and Discussion

The values of various physicochemical parameters at various sampling stations (L_1, L_2, L_3, L_4, L_5) are given in Table 2. Various drinking water standards are given by the IS 10500:2012 [6].

Color: Color in water is generally due to dissolved organic matter from decaying vegetation or some inorganic materials lsuch as colored soil etc. 1 TCU is equal to the color produced by a 1 mg/L of platinum in the form of chloroplatinate ion.

Taste and Odor: Odor is measured by an instrument called Osmoscope. Taste and Odor are caused by dissolved gasses. Algae secrete oily substances which may result in bad taste and odor.

Electroconductivity: Electrical conductivity (EC) is related to total dissolved solids (TDS). The significance of EC and TDS resides in their impact on a water sample's corrosivity as well as the solubility of barely soluble molecules like CaCO3.The conductivity increases as the ion concentration grows. It's one of the most essential considerations when deciding whether or not water is suitable for irrigation. According to WHO recommendations, the EC value should not exceed 400 S/cm. According to the current investigation, the minimum EC value was 1373 S/cm, with a maximum of 3548 S/cm.

Total Dissolved Solids: TDS is made up of dissolved inorganic salts (mostly sodium, magnesium, bicarbonates, potassium, calcium, chlorides, and sulfates) as well as minor amounts of organic stuff. TDS concentrations in water vary significantly across geological locations due to changes in mineral solubilities (WHO (World Health Organization)) [7]. A high TDS value in water indicates that it has been severely mineralized. TDS has a Desired limit of 500 mg/l and a Permissible limit

Table 2 Indian ratings of water quality index

WQI value	Rating of quality	Grade
1–25	Excellent	A
26–50	Good	B
51–75	Poor/bad	C
76–100	Very poor/very bad	D
>100	Unfit for drinking	E

of 2000 mg/l, both of which are prescribed for drinking. TDS concentrations value from 747 mg/L to 1766 mg/L in this investigation.

pH: There has been no health-based pH recommended value proposed. Although pH has little direct impact on consumers, it is one of the most important operational water quality factors, with the optimal pH requiring a range of 6.5 to 9.5 according to WHO. According to IS:10,500 (2012) [6], the acceptable pH range for drinking water is 6.5–8.5 for household use and living organisms. It is defined as the negative logarithm of the hydrogen ion concentration and is one of the most essential water quality measures. At 25 °C, pure water has a pH of approximately 7.0. The minimum and maximum values of pH were measured as 6.83 and 7.41, respectively. There was no well in which the pH exceeds 8.5 as given by IS 10500(2012).

Alkalinity: The total of all titratable bases is the alkalinity of water, which is its acid neutralizing capacity. The presence of hydroxide ions, bicarbonate ions, and carbonate ions are the main causes. The permitted limit for hardness is 200 mg/L, and the reason for rejection is 600 mg/L, according to IS:10,500 (2012). The hardness value in our study region ranges from 65 to 110 mg/L.

Hardness: It's a term that describes the characteristics of highly mineralized waters. It is the concentration of multivalent metallic cations in solution. It is divided into two parts: carbonate hardness and non-carbonate hardness. The permitted limit for hardness is 200 mg/L, and the reason for rejection is 600 mg/L, according to IS:10,500 (2012). The value of hardness in our study region ranges from 125 mg/L to 411 mg/L.

Chlorides: Chlorides can be found in different quantities in all natural waters. As the mineral concentration rises, the chloride content rises as well. Unless the water is brackish or saline, the chloride ion is present in natural waterways in relatively low amounts, usually less than 100 mg/L. Water and beverages with high chloride concentrations have a salty taste (World Health Organization). Chloride concentrations greater than 250 mg/L, on the other hand, can cause a perceptible taste in water. The chloride content of water samples in our study area is quite low, ranging from 7.5 to 35 mg/L (Fig. 3, Tables 3, 4, 5, 6, 7, 8 and 9).

Among all physicochemical parameters Electroconductivity and Total Dissolved Solids have their values beyond acceptable limits for a particular time at that particular day. Hence, the relation between physicochemical parameters and the length is determined using image overlapping. The variation with distance is taken diagonally towards the centre. The results obtained are shown below (Figs. 4, 5 and 6 and Tables 10 and 11).

5 Conclusion

On the basis of the finding, it was concluded that the water quality index of NIT Kurukshetra Campus is found to be Excellent. Although at every sampling location the Electroconductivity, Total Dissolved Solid (TDS) value is beyond the acceptable limit on the particular collected sample for that particular day although more samples

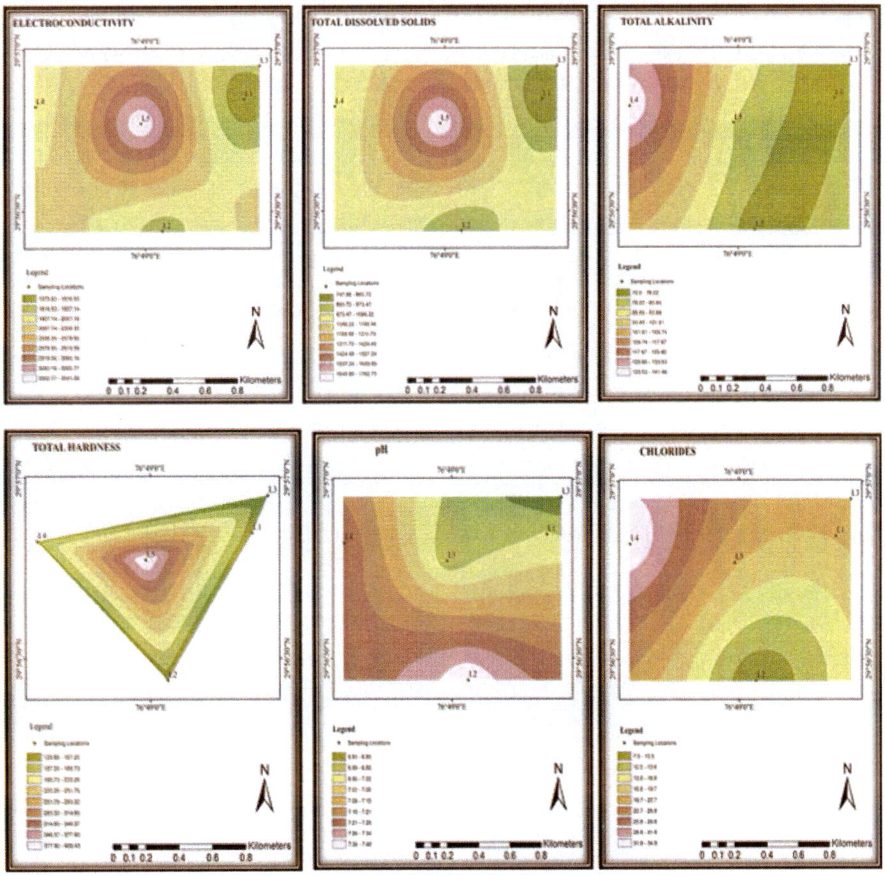

Fig. 3 Spatial distribution maps for various physicochemical parameters

need to be collected for more concrete information but still the water quality index is found as excellent.

From the various spatial distribution maps which we got through Kriging Technique of Interpolation by ArcGIS Software, we found that in case of Electroconductivity, Total Hardness and Total Dissolved solids, it is increasing beyond acceptable limits as we are moving away from the sources of water and getting maximum at the center of the NIT Kurukshetra campus. While Alkalinity, Chloride and pH values are well within permissible limits as suggested by the Indian Standard Code. It is found that Alkalinity is low at the Eastward side of the campus near Tubewell No.8, Tubewell No.6 and Tubewell No. 9 but as we move toward the westward side alkalinity is increasing at Tubewell No.7. The pH was found lowest at Tubewell No. 8 which is located at North-Eastward end of the campus and highest at southern source which is Tubewell No. 9 (at Harihar Bhawan). Chlorides values are found as lowest

Table 3 Physico-chemical parameters of waters at location L_1, L_2, L_3

S.No	Parameters	Acceptable limits (As per IS 10500 (2012)	Loc L_1	Loc L_2	Loc L_3
1	Color	–	Clear	Clear	Clear
2	Odor	–	Agreeable	Agreeable	Agreeable
3	Taste	–	Agreeable	Agreeable	Agreeable
4	Salinity	–	1.85	1	0.86
5	Electroconductivity	1500 micromho/cm (WHO)	3548	1975	1727
6	TDS	500 mg/l	1766	983	859
7	Total Alkalinity	200 mg/l	83	110	71
8	Total Hardness	200 mg/l	411	150	139
9	pH	6.5–8.5	6.92	7.27	7.41
10	Chlorides	250 mg/l	20.2	35	7.5

Table 4 Physico-chemical parameters of waters at site L_3 and L_5

S.No	Parameters	Acceptable limits (As per IS 10500 (2012)	Loc L4	Loc L5
1	Color	–	Clear	Clear
2	Odor	–	Agreeable	Agreeable
3	Taste	–	Agreeable	Agreeable
4	Salinity	–	0.65	0.86
5	Electroconductivity	1500 micromho/cm (WHO)	1373	1692
6	TDS	500 mg/l	747	847
7	Total Alkalinity	200 mg/l	80	65
8	Total Hardness	200 mg/l	135	125
9	pH	6.5–8.5	6.97	6.83
10	Chlorides	250 mg/l	20	22

Table 5 Water quality index at location L_1 (Tubewell No.6)

S.No	Parameter	Standard value (Sn)	Observed value (Vn)	Sub-index of quality parameter (qn)	Weightage factor (W)	Water quality index
1	Electroconductivity	1500	3548	236.53	0.00496350365	20.34
2	Total dissolved solids	500	1766	353.20	0.01489051095	
3	Total hardness	200	411	205.50	0.03722627737	
4	Total alkalinity	200	83	41.50	0.03722627737	
5	pH	8.5	6.92	5.33	0.8759124088	
6	Chlorides	250	20.2	8.08	0.0297810219	

Table 6 Water quality index at location L_2 (Tubewell No.6)

S.No	Parameter	Standard value (Sn)	Observed value (Vn)	Sub-index of quality parameter (qn)	Weightage factor (W)	Water quality index
1	Electroconductivity	1500	1975	131.67	0.00496350365	18.69
2	Total dissolved solids	500	983	196.60	0.01489051095	
3	Total hardness	200	150	75.00	0.03722627737	
4	Total alkalinity	200	150	75.00	0.03722627737	
5	pH	8.5	7.27	18.00	0.8759124088	
6	Chlorides	250	35	14.00	0.0297810219	

Table 7 Water quality index at location L_3 (Tubewell No.8)

S.No	Parameter	Standard value (Sn)	Observed value (Vn)	Sub-index of quality parameter (qn)	Weightage factor (W)	Water quality index
1	Electroconductivity	1500	1727	115.13	0.00496350365	31.06
2	Total dissolved solids	500	859	171.73	0.01489051095	
3	Total hardness	200	139	69.33	0.03722627737	
4	Total alkalinity	200	71	35.33	0.03722627737	
5	pH	8.5	7.41	27.33	0.8759124088	
6	Chlorides	250	7.5	3.00	0.0297810219	

Table 8 Water quality index at location L_4 (Tubewell No.9)

S.No	Parameter	Standard value (Sn)	Observed value (Vn)	Sub-index of quality parameter (qn)	Weightage factor (W)	Water quality index
1	Electroconductivity	1500	1373	91.53	0.00496350365	8.67
2	Total dissolved solids	500	747	149.40	0.01489051095	
3	Total hardness	200	135	67.50	0.03722627737	
4	Total Alkalinity	200	80	40.00	0.03722627737	
5	pH	8.5	6.97	2.00	0.8759124088	
6	Chlorides	250	20	8.00	0.0297810219	

Table 9 Water quality index at location L_5 (academic section)

S.No	Parameter	Standard value (Sn)	Observed value (Vn)	Sub-index of quality parameter (qn)	Weightage factor (W)	Water quality index
1	Electroconductivity	1500	1692	112.80	0.00496350365	77.26
2	Total dissolved solids	500	847	169.40	0.01489051095	
3	Total hardness	200	125	62.50	0.03722627737	
4	Total alkalinity	200	65	32.50	0.03722627737	
5	pH	8.5	6.83	80.35	0.8759124088	
6	Chlorides	250	22	8.80	0.0297810219	

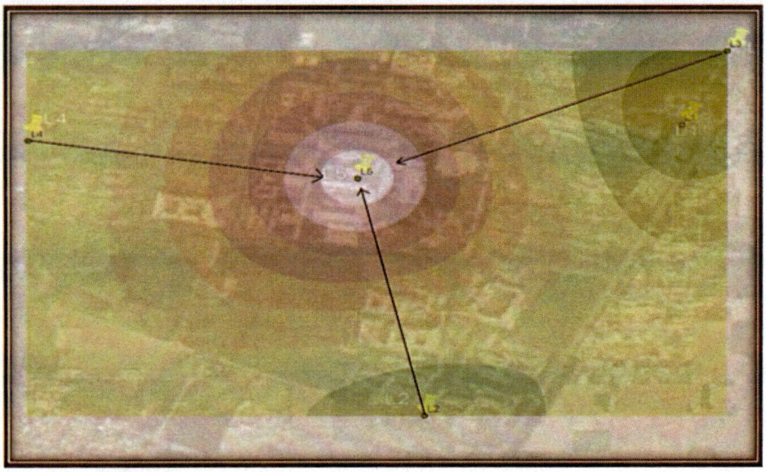

Fig. 4 Image overlapping

at Tubewell No.9 and its value increasing as we move away from this source and found highest at Tubewell No.7 at the Western side of the campus.

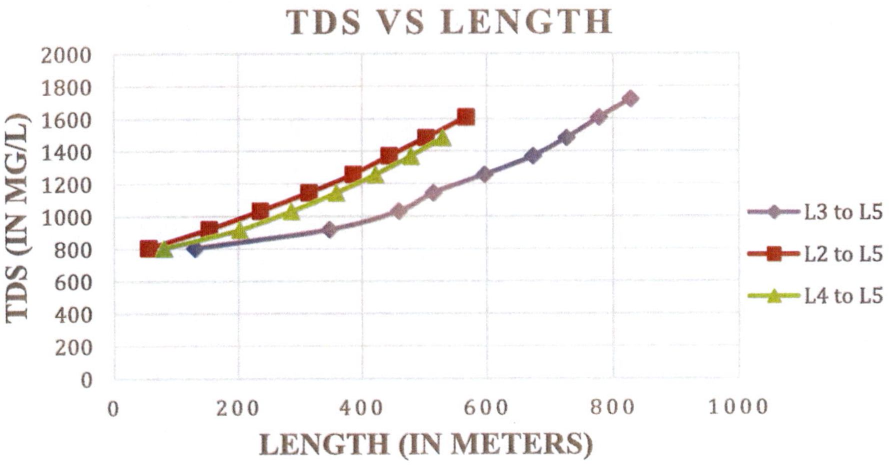

Fig. 5 Plot showing variation between TDS and length from various sources towards center location point L5

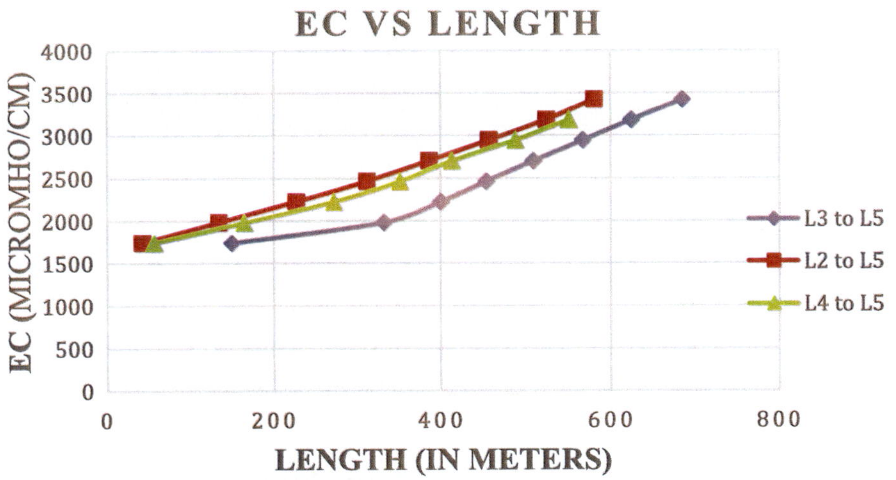

Fig. 6 Plot showing variation between EC and Length from various sources towards center location point L5

Table 10 Variation of TDS with length from various stations towards the centre L5 (in meters)

TDS value (in mg/L)	Distance (L3–L5)	Distance (L2–L5)	Distance (L4–L5)
803.5	130.5	0	0
916.5	348	56.5	0
1029.5	460	153	80.5
1142	516	237	203
1254.5	597.5	315.5	286.5
1367.5	674	387	359.5
1480.5	726.5	445	422.5
1608	778	504.5	480
1720.5	828	568	531

Table 11 Variation of EC with length from various stations towards the centre L5 (in meters)

EC value (in mg/L)	Distance (L3–L5)	Distance (L2–L5)	Distance (L4–L5)
1736.5	149.5	42.5	0
1977	332.5	134	56.5
2217.5	400.5	227.5	165
2458	454	312.5	272.5
2698.5	509.5	387	352
2939.5	567.5	457.5	413.5
3180	624.5	525	488.5
3420.5	685	581	551

References

1. Munavalli GR, Mohan Kumar MS (2005) Water quality parameter estimation in a distribution system under dynamic state. Water Res 39(18):4287–4298. https://doi.org/10.1016/j.watres.2005.07.043
2. Kirubakaran M, Ashokraj C, Colins Johnny J, Anjali R (2016) Groundwater quality analysis using WQI and GIS techniques : a case study of Manavalakurichi in Kanyakumari District, Tamilnadu, India groundwater quality analysis using WQI and GIS techniques : a case study of Manavalakurichi in Kanyakumari district, T. Int J I 2(February):341–347
3. Chabuk A, Al-Madhlom Q, Al-Maliki A, Al-Ansari N, Hussain HM, Laue J (2020) Water quality assessment along Tigris River (Iraq) using water quality index (WQI) and GIS software. Arab J Geosci 13(14). https://doi.org/10.1007/s12517-020-05575-5
4. Dhanraj MR, Rajesha R (2017) Assessment of ground water quality in Deralakatte Belma Panchayath Mangalore Dakshina Kannada. Int J Eng Trends Technol 44(1):12–16. https://doi.org/10.14445/22315381/ijett-v44p203
5. Naik V, Naik D, Naik G, Moger K, Yallurkar S (2019) Characterization of ground water based on water quality index in Bhatkal, Uttara Kannada District, Karnataka. pp 2199–2206
6. BIS (2012) Indian standard drinking water specification (Second Revision). Bur Indian Stand IS 10500(May):1–11 [Online]. http://cgwb.gov.in/Documents/WQ-standards.pdf
7. Anesthesia N et al (2010) Guidelines for 2010

Landslide Hazard Zonation Mapping Using Remote Sensing and GIS in Mountainous Terrain

Dolonchapa Prabhakar, Anoop Kumar Shukla, Babar Javed, and Satyavati Shukla

Abstract Landslides are a devastating natural phenomenon that can cause significant damage to people, property, and infrastructure. Various physical, geological, climatic, and tectonic factors can contribute to landslides in different parts of the world. In addition to natural causes, improper construction and chaotic development can also lead to landslides and the destruction of property and lives. Landslide hazard zonation (LHZ) is a method of dividing a landmass into homogeneous zones and ranking. Landslides are a pervasive and destructive hazard in the Himalayan regions. While they cannot be completely eliminated, their effects can be minimized. This study focuses on the Garhwal Himalaya region to identify areas prone to landslides. Geographical Information System (GIS) was used to create a database, analyze, and generate output. LISS III images were used to create the land use land cover (LULC) map and Landsat satellite data was used to identify. This research paper aims to integrate various techniques and tools to generate a geospatial database for landslide hazard zone delineation in the study area. LISS III images were used for land use land cover (LULC) mapping and Landsat satellite data for lineaments identification. The Analytical Hierarchy Process (AHP) approach was used to determine the weights of the landslide influence parameters. ERDAS Imagine and ArcGIS software were used to integrate the input layers after assigning them suitable weights. The primary objective of this research paper is to combine the various existing methods and tools to create a Geo-spatial database for delineating landslide hazard zones in the study area. The resulting map's susceptibility to landslides has been divided into five categories, namely very low, low, moderate, high, and very high. This LHZ map can be of great benefit to planners and designers in selecting the most appropriate route paths.

D. Prabhakar
Department of Civil Engineering, Indian Institute of Technology, Roorkee, Uttarakhand, India

A. K. Shukla (✉) · B. Javed
Manipal School of Architecture and Planning, Manipal Academy of Higher Education, Manipal, Karnataka, India
e-mail: anoop.shukla@manipal.edu

S. Shukla
Key Laboratory of Geospatial Informatics, College of Earth Sciences, Guilin University of Technology, Guilin, PR China

S. Mesapam et al. (eds.), *Developments and Applications of Geomatics*,
Lecture Notes in Civil Engineering 450, https://doi.org/10.1007/978-981-99-8568-5_25

Keywords Analytical Hierarchy Process (AHP) · GIS · Satellite data

1 Introduction

Himalayan ranges extend up to a length of 2500 km from the East to West with mean width along full longitudinal extension ranging from 110 to 400 km. The massive area covers the mountain ranges of Geological younger ages. These ranges contain some fantastic altitude gradients. The entire Himalayan region is inherently unstable. Every year, landslides occur in the area stretching from Jammu & Kashmir to the North-eastern states, including Himachal Pradesh, Uttarakhand, and Sikkim Himalayas, resulting in significant loss of life and property. Various regions of India get affected differently by landslides due to the characteristic, physiographic, geological, climatological, and tectonic conditions. In count to natural occurrence to landslides, improper arrangement and chaotic construction are also responsible for increased incidences of landslides and consequent damages of properties and human lives. However recurrent landslides occur due to different causative factors like topographic, lithological, structural features, natural calamities like earthquake, heavy rainfall, cloud bursts, human activities, and improper development activities without considering the depth of natural factors, unscientific formation cutting for road construction, and use of heavy high explosives in blasting that produces instability. This increases the failure of slope stability and triggers mass movement resulting in landslides. The brunt of landslides is mainly felt on national highways which get blocked during rainy season when more and more landslide areas get triggered. The landslide prompts demolition of roadways and progressing territories where escape for individuals and vehicles and supply of help material in routine is relatively incomprehensible. Besides a huge sum is also required for removal of debris, prevention of further landslide and planning activities increases work pressure over the Border Road Organizations (BRO), which is involved in construction, development and maintaining the roads in hilly terrain of the Himalayan region. Landslide chance alludes to the idea of harm liable to be caused if failure happens. The harm caused might be in the form of death toll and wounds or potential loss of land and property. The degree of harm is subject to existing areas that utilize examples of the region prone to be influenced and its populace. For e.g., a littler landslide in thickly populated regions may cause broad harm as thought about a noteworthy avalanche in remote zones. Therefore, the risk due to landslide is a function of hazard probability and the damage potential [1]. With the application of satellite imagery and GIS data, combined with ground information, several qualitative and quantitative approaches have been used in different studies to assess landslide susceptibility. These include frequency ratio (FR) [2, 3], index of entropy (IOE) [4], fuzzy logic [5], weight of evidence (WOE) [6], logistic regression (LR) [7], analytical hierarchy process (AHP) [8], and a range of machine learning approaches. Even though there are various approaches for assessing landslide susceptibility, there is still no conclusion over the best approach in different regions. In the qualitative approaches AHP has been regarded as the best approach to analyze landslide susceptibility, which is based on the knowledge and experts' experiences. Many applications have been reported from several parts of the world [9–11]. Some works have also reported on the comparisons of the FR and AHP methods for landslide susceptibility assessment worldwide [12–18]. However, no study has been reported from the Parts of the Tehri and Pauri Garhwal districts in India's Uttarakhand state in this aspect.

Landslide hazards refer to the probability of occurrence of landslide danger. The probability of occurrence of landslide in a specific terrain can be determined, predicted, and mapped with density of distribution and the terrain can be divided in terms of hazard zones. The causative factors are natural and anthropogenic and their intensity triaging landslide can be categorized, and weight assigned to them can be considered as important factors. The zones of landslide density can be classified from Very Low Hazard (VLZ) to Very High Hazard (VHH). The identified Hazard zones can be considered for preliminary planning activity of development schemes and help to avoid unstable areas during the planning stage. Even though it may be unavoidable, recognizing the potential for landslides in the early stages of planning will help to develop better preventive measures as a priority. High hazard zones should be given the highest priority.

In this regard, landslide hazard zonation mapping is an essential component of such a study. The landslide hazard zonation mapping in the vicinity of Rishikesh and along National Highway-58 can predict the zones of hazard in the region. This study will likewise be useful for doing the point-by-point mapping of the landslide risk appraisal mapping along the street which will feature danger of harm to the street and encompassing property, line of correspondences, schools, buildings, and neighborhoods independently. The research basically has focused on the landslide hazard zonation mapping around Rishikesh and the surrounding area which is covered within the Toposheet-53 J/8 of 1:50,000 scale of Survey of India. Validation of the result was carried out by landslide inventory mapping using GPS, Maps & other instruments along part of NH-58 from Rishikesh to Kaudiyala village that falls on the road 31 km from Rishikesh. The objectives of the study are to generate a Landslide Hazard Zonation map of the study area for effective and efficient disaster management in the future, and to determine the various parameters and importance of remote sensing and GIS in LHZ in the present study.

2 Objective of the Study Area

The main aim of the study is to quantify the landslide Hazards around the surrounding area of Rishikesh and National Highway-58, by preparing the Landslide Hazard Zonation Map (LHZ) with the aid of remote sensing and GIS and validating the result by conducting a GPS field survey along a portion of the national highway that runs through the research area to map the inventory of landslides. The main objective of the research is as below.

- To create a detailed GIS database of different thematic parameters depending upon the degrees of influence on the landslide process.
- To use GIS analysis techniques to evaluate the importance of the factors.
- To prepare landslide hazard zonation map for the Rishikesh area.
- Landslide hazard zonation map is to be prepared for part of National Highway-58 falling in the study area with a buffer zone of 2 km.
- Preparation of landslide inventory map along NH-58 and validation of result carried out by using GPS field survey.

3 Description of Study Area and Datasets Used

Parts of the Tehri and Pauri Garhwal districts in India's Uttarakhand state make up the research area. The study area has a perimeter of approximately 100 km and a surface area of 625 km^2. The following Toposheet No: - 53 J/8 of Survey of India is covered. The area extent is bounded between longitude 78° 15′ 00″ E and 78° 30′ 00″ E and latitude 30° 00′ 00″ N and 30°15 ′00″ N (Fig. 1).

Landslide Hazard Zonation Mapping utilizing Remote Sensing and GIS in the mountainous terrain involves the integration of various datasets to analyse factors contributing to landslide susceptibility. Some essential datasets used for this study are:

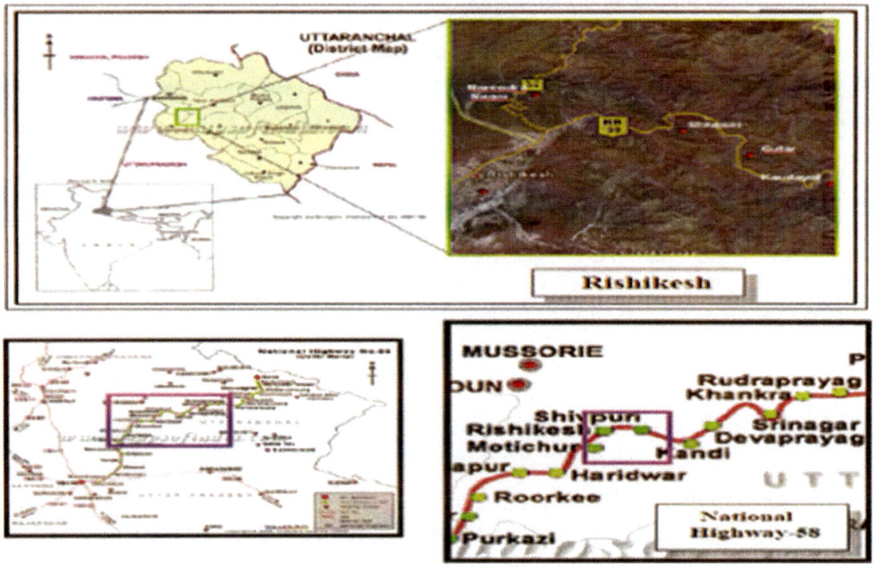

Fig. 1 Map showing location of the study area

- Satellite Imagery: Satellite images of LISS-III and Landsat series, provide valuable information about land cover, land use, and vegetation characteristics.
- Digital Elevation Model (DEM): DEM data captures the terrain's elevation and helps create slope, aspect, and curvature maps, which are essential for understanding the topographic influence on landslides.
- Land Cover Land Use Maps (LULC): These maps categorize land cover types such as forests, urban areas, agricultural lands, and bare soil. They contribute to identifying areas prone to landslides due to land use changes.
- Drainage Maps: Maps indicating drainage patterns, watershed boundaries, and proximity to water bodies help assess areas vulnerable to landslides due to erosion and saturation.
- Road and Infrastructure Maps: Information on roads, highways, and other infrastructure aids in identifying areas where construction and human activities might influence landslide susceptibility.
- Landslide Inventory Maps: Historical records of landslides, including their locations, types, and characteristics, contribute to understanding past events and validating hazard zonation results.

These datasets, when integrated and analysed using remote sensing and GIS techniques, contribute to a comprehensive understanding of landslide susceptibility and facilitate the creation of accurate Landslide Hazard Zonation maps in the mountainous terrain.

4 Methodology

The methodology adopted in this study are as follows:

- Study Area Selection: Selection of area for study where landslide problems are in existence.
- Data Collection and Compilation: Data availability /data procurement and collection.
- Data Preprocessing: Selection of satellite images suitable for regional and local level landslide study for land use identification and image processing.
- Collection of topographic maps relevant to satellite images and scale of study.
- Collection of relevant secondary data from all sources and access previous studies conducted on the area.
- Data organization, data pre-processing, import satellite data and Raster scan Topomap in GIS environment and provide projection system.
- Generation of relevant thematic data like Contour map, DEM, Drainage map, Slope map, Aspect map, Drainage frequency map.
- Lineament occurrence map, Lineament intersection map, Road map, Landslide inventory map and Location map.
- Conversion of vector thematic to raster format using raster conversion tool.

- Weighting and Factor Ranking: Definition of landslide hazard classification and rating/weight for each thematic class as per their role in landslide triggering intensity in attributed information.
- GIS analysis: Select GIS analysis tools for overlay operation by providing weight calculated for each thematic and prepare landslide zonation map.
- Verification of the errors and consequent modification to appropriate levels.
- Estimate geometrically, geographically, and statically landslide prone area with their hazard intensity class.
- Verification of the land use land error map, geography of different hazard zones obtained by the GIS analysis methodology.
- Map Generation and Visualization: Preparation of final thematic maps and landslide hazard zonation maps by reanalysis in GIS environment with symbolization.
- Preparation of final print out maps at 1:50,000 or as needed scale.

4.1 GIS Analysis for Landslide Hazard Zonation Mapping

Landslide Hazard Zonation Map is prepared in this work by using the weighted overlay method of raster analysis available in ArcGIS software. The Analytic Hierarchy Process (AHP) is used for the calculation of weight and rank assignment to different thematic parameters of landslide importance.

4.2 Analytic Hierarchy Process (AHP)

Geographers and spatial organisers are typically intrigued by choice problems that involve a large number of plausible options as well as several evaluation criteria that are incompatible and conflicting. The options are typically examined by a variety of individuals (decision-makers, managers, subject matter experts, and interest groups), each of whom has their own unique preferences regarding the relative importance of the criteria used to evaluate the options. On the other hand, Spatial Multi-criteria Decision Analysis (SMCDA) or Multi-criteria Decision Making (MCDM) can be used to identify the best solution to a given problem by taking into account multiple criteria. By combining GIS and MCDM, it is possible to develop a more comprehensive approach to spatial planning and management. This approach can be used to identify the most suitable solution to a given problem, taking into account a range of geographical data and multiple criteria. Contrarily, MCDM and a wide range of related methodologies, including multi-attribute utility theory (MAUT), public choice theory, and collaborative decision-making, offer a collection of techniques and procedures to reveal decision makers' preferences and to incorporate them into GIS-based decision-making. To this purpose, a collection for studying geographic events can be referred to as GIS-based (or spatial) multi criteria decision analysis, where the outcomes of the analysis (decisions) depend on the spatial arrangement of the events. Spatial analysis, as it is generally characterised, includes spatial multi-criteria analysis.

Multi-criteria decision-making is not a well-established process or effectively integrated into the field of spatial analysis and GIS, despite the fact that the majority of spatial decision problems are multi-criteria in nature and involve economic, social, environmental, and political dimensions as well as conflicting values. The Spatial Multi-criteria Decision Analysis (SMCDA) is a kind of multi attribute decision, making the (MADM) process. The main elements of SMCDA are evaluation criteria, alternatives, and decision maker's preferences. SMCDA issues ordinarily include criteria of changing significance to decision makers. Thus, data about the relative significance of the criteria is required. This is generally accomplished by doling out a weight to basis. The deduction of weights is a focal advance in inspiring the chief's inclinations. A weight can be characterized as esteem allocated to an assessment foundation that shows its significance with respect to other rule in the general utility. In multi-criteria decision-making, a variety of criteria-weighing techniques have been developed. Some of the most common techniques are rating, ranking, trade-off analysis, and pairwise comparison. Authors in [19] created the pairwise comparison technique within the framework of the Analytic Hierarchy Process (AHP) as shown in Table 1. Pairwise correlations are used in this method to create a percentage network. It generates the relative weights as yield using pairwise examinations as input. The weights are specifically determined by normalising the eigen vector related to the extreme Eigen value of the (reciprocal) ratio matrix. The present work involved six parameters: slope, aspect, drainage frequency, lineament occurrences, Lineament intersections and land use. It necessitated determining their relative importance. This was accomplished by comparing each pair of parameters pairwise. There are three main steps in the process [20].

1. Pairwise comparison-matrix development
2. Computation of the criterion weights
3. Estimation of the consistency ratio.

Table 1 Pairwise-comparison scale (Satty, 1980)

Significance of Issue	Description
1	Equal significance
2	Equal to moderate significance
3	Moderate significance
4	Moderate to strong significance
5	Strong significance
6	Strong to very strong significance
7	Very strong significance
8	Very to extremely strong significance
9	Extreme significance

A weighted evaluation system was employed in this investigation. The ArcGIS software's built-in arithmetic weighted overlay method, which accepts both continuous and discrete grid layers and produces continuous grid data layers, was used to integrate all of the layers. The complete methodology is shown in the methodology section Fig. 2. Using the ArcGIS spatial analyst module, all the components are categorised according to the significance of each element impacting landslide hazard. Each subclass is assigned a rating between 0 and 9 in order of increasing risk; zero denotes little risk and 9 denotes extreme risk. All the rating classes are multiplied by the corresponding weight to obtain the landslide potential index map. Table 2 displays the weights and their ranking of the landslide influence elements. The Analytical Hierarchy process (AHP) is used to calculate the weights. High weights imply high landslides, whereas the lowest weights suggest low landslides. The resulting values are obtained the range from 32 to 331. It should be classified into five classes very low, low, moderate, high and very high, moving average with window size of 3, 7 and 9 were considered.

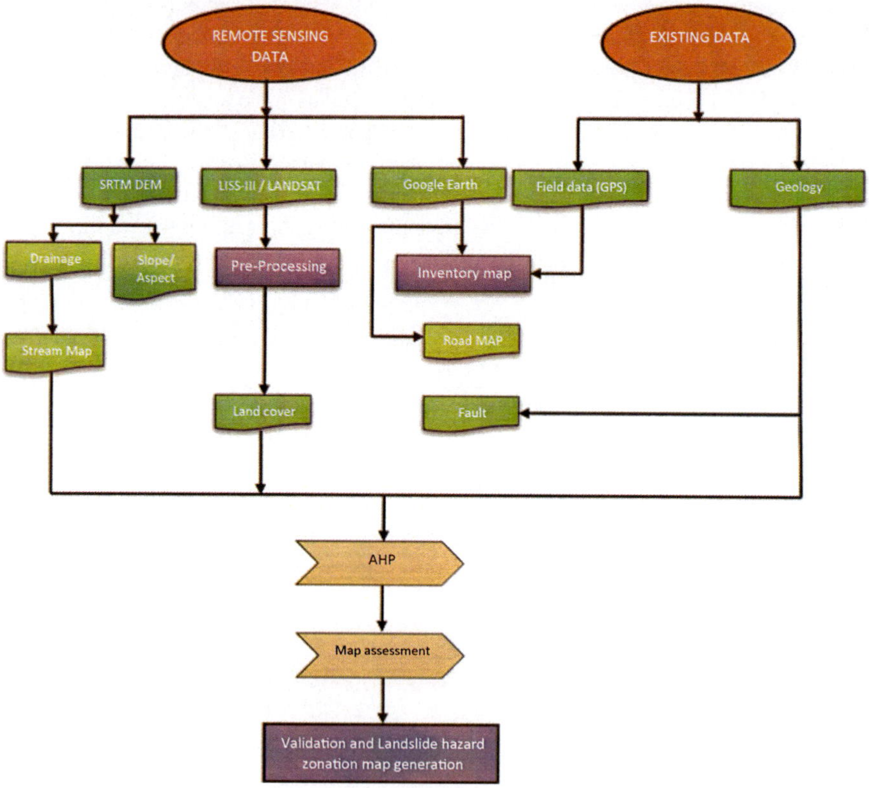

Fig. 2 Methodology flowchart for landslide hazard zonation mapping

Table 2 Computation of criterion weights

Factor	Slope	Drainage frequency	Land use	Lineament intersection	Lineament occurrence	Aspect
Slope	1	2	2	2	8	9
Drainage frequency	1/2	1	2	3	6	9
Land use	1/2	1/2	1	2	9	6
Lineament Intersection	1/2	1/3	1/2	1	8	7
Lineament Occurrence	1/8	1/6	1/9	1/3	1	4
Aspect	1/9	1/9	1/7	1/6	1/4	1

The Analytical Hierarchy Process (AHP) offers a structured and systematic approach to decision-making, allowing for the effective prioritization of factors in complex problems. It enables the integration of both objective data and subjective inputs, enhancing the comprehensiveness and accuracy of analyses. However, AHP relies on expert judgments and pairwise comparisons, which can introduce subjectivity and bias. Its complexity in constructing hierarchies and assigning weights can be time-consuming, and it may struggle to handle uncertainties and sensitivity to weight allocation. Careful consideration of these pros and cons is essential when applying AHP in decision-making contexts.

5 Results and Discussions

5.1 Thematic Parameters and Their Relationship with Landslide Causes

Slope Map

Slope is a crucial factor in landslide research since its stability determines how frequently and how intensely landslides occur. According to [21], landslides occur most frequently when the slope angle is between 35° and 40°, and less frequently when the slope is more than 40°. According to reports, landslides occur more frequently in regions with slopes between 6° and 45° [22]. The Song River portion of Dehradun experienced numerous landslides that occurred at angles ranging from 18 to 45 degrees [23]. The Marapalam landslide in 1993, which occurred close to Coonoor, occurred on a slope of roughly 30° in the slope-forming material [24]. Slope angle was calculated and used in this study as a one theme layer for landslide analysis. The digitised contour map has been used to create the TIN (Triangulated irregular network) model for the research region. The Delaunay triangulation of irregularly distributed nodes or observation points serves as the foundation for the TIN terrain model, which employs a sheet of continuous, connected triangular facets [25]. The TIN model was used to create the slope map. The majority of the research region, it was concluded, is sloped between 25° and 90°. There are four different types of slopes. The slope of 25° to 45° is more prone to landslides, according to observations made in the field using a handheld GPS and compass and from information accessible in a Geological Survey of India (GSI) article [26]. Therefore, this category is given a higher rank value. The largest frequency of landslides occurs on bare soils, crop and grass fallow areas, degraded forests, and barren land types as slopes rise. In locations with steep slopes, forests reduce the risk of landslides, but cultivation raised the risk. As the slope rises, vegetation clearing increased the likelihood of landslides. The requirement for a sufficiently thick mantle of residual soil prevents steep slopes from failing in reaction to rainfall. Rain is needed more to wet the bottom half of a slope where the weathered profile is dryer. Under the combined influences of comparable transpiration losses from a smaller moisture store and the increased gravitational pressure on the unsaturated down slope drainage, the steeper slope loses water more quickly. The roots close to the surface of the ground offer a component of strength adequate to withstand movement brought on by collapse of the deep residual soil.

Aspect map

The aspect map is yet another crucial element that affects the stability of the slope and ultimately landslides. It designates the slope's orientation and relates to the potential exposure to light and shade that a mountain or hill slope may have. The terrain's exposure to storm fronts is affected by the slope aspect. One of the main contributing reasons to a landslide is the slope's aspect. Additionally, it has an impact on pore water pressure and alternation variations.

Land use Land Cover (LULC) map

The presence and movement of rainfall-induced landslides are significantly influenced by the vegetation cover. The behaviour of landslides is frequently altered by changes in plant cover [27]. It has been determined that via numerous examinations that land use and land cover, particularly of the woody variety with robust and substantial root systems, contribute to the improvement of slope stability [28]. As a result of changes in land use over time, it is possible that forest land has been converted to another land use, such as farmland. In southern Honduras, especially when the slope increased, deep-rooted vegetation that stabilises the topsoil, such as shrubbery or forest, was a very significant role in lessening landslide hazard [22]. The shear strength of the soils is increased, according to [29], by vegetation roots that saturate the entire surface. In comparison to regions with less or no vegetation, places with denser vegetation were thought to be less prone to sliding [30]. However, at Dehradun and Mussoorie, 91% of the landslides happened in regions that weren't covered in trees, indicating the role that vegetation plays in the beginning of slope instability [31]. A land use land cover map on a scale of 1:50,000 was created using

LISS III images and supervised categorization. In this study area, land use is categorized into eight types. (1) Dense Vegetation (More than 40% Canopy), (2) Medium Vegetation (10 to 20% Canopy), (3) Degraded Vegetation (More than 40% Canopy), (4) Barren Land (Non-Rocky, wasteland), (5) Agriculture land, (6) Build up area, (7) River Sand and (8) Water Body. For validation purpose, land use and land cover classification accuracy assessment were performed which was found to be the Overall Accuracy which is 91.50 and Kappa Coefficient being 0.90.

Map of Drainage Frequency

In hilly places, drainage is a significant cause of landslides. Landslides that tend to occur less frequently are from the drainage network because shear strength is low and close to the drainage network due to the greater amount of water seepage there. Landslides occur more frequently close to the drainage network than farther from it in the study area. The initiation of a landslide may also be influenced by the terrain modification brought on by gully erosion [32]. A drainage map on a scale of 1: 50,000 was created from a topographic map in order to understand the function of drainage. The drainage map was superimposed over a 0.50 × 0.50 km grid cover. Each grid's drainage lines are tallied in order to determine the drainage frequency value. Based on the quantity of drainage incidents, the study region has been divided into five types.

Lineament incidence map

Fracture, joints, faults, bedding, and other features that are represented by lineament. These structures' influence encourages infiltration and the growth of hydrostatic pressure on the material that forms slopes [33]. Landslides and faults are closely related; 88% of all landslides were found within 250 m or less of major faults [34]. All the landslides in the Song River region happened on fault scarps on the escarpment side, with the formations descending into the hill [35]. By using visual interpretation, a lineament map has been created from IRS IC- LISS III satellite images (see Fig. 3a and b). The map was superimposed on a grid with a size of 0.50 × 0.50 km, and the number of lineaments found within each grid was counted to determine the lineament frequency. The final map was digitally transformed and digitised. Based on the density of lineaments, an order of importance has been determined, and ranks have been allocated in accordance.

Lineament intersection map

Each grid's lineament intersection count was calculated to produce the lineament intersection map. This map has been transformed into a digital file. The likelihood of a landslide occurring is increased with more intersections. Thematic parameters and their relationship with landslide cause the following criterion generated in Table 3 was generated.

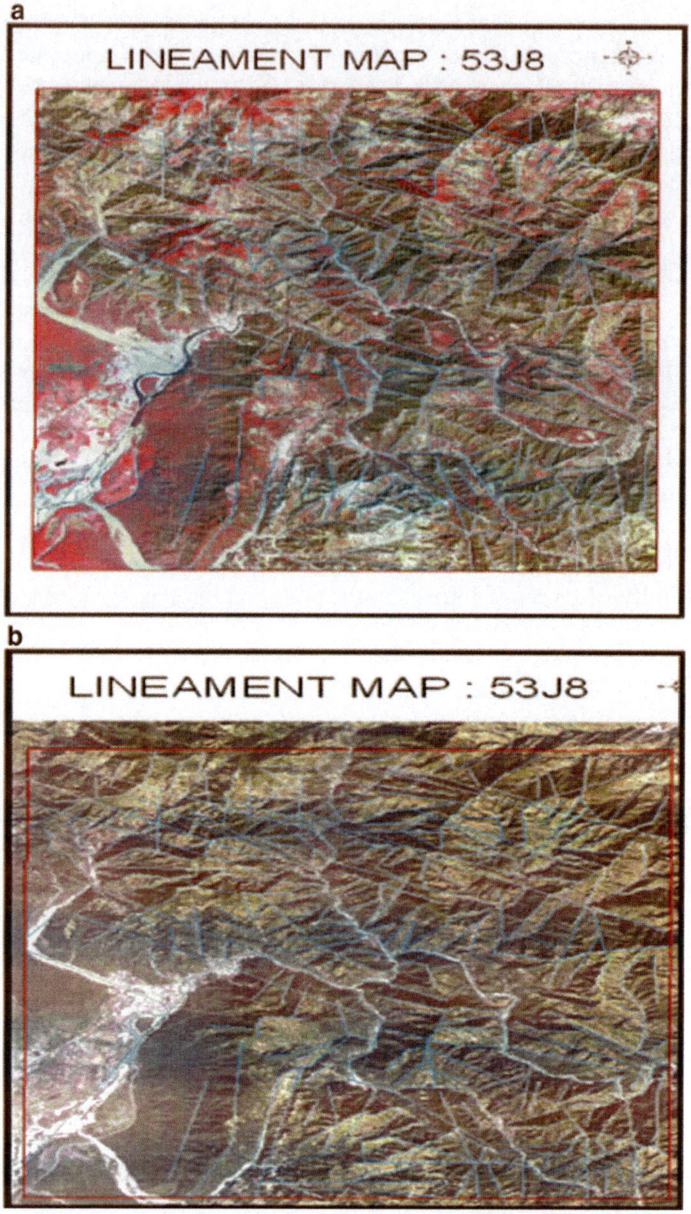

Fig. 3 **a** Showing lineament map with false colour composite (FCC) image along NH-58. **b** Showing lineament map natural colour composite image along NH-58

Table 3 Thematic/parameters and rank values

S.No	Thematic/Parameters	Categories	Rank
1	Slope map	0°–15°	2
		16°–25°	7
		26°–45°	9
		>45°	3
2	Aspect	Flat	1
		North, NE facing	2
		East, SE, NW facing	3
		West facing	4
		SW facing	5
		South facing	6
3	Land use	Dense vegetation (More than 40% canopy)	0
		Medium vegetation (10 to 20% canopy)	3
		Degraded vegetation (More than 40% canopy)	6
		Barren land (Non-Rocky, wasteland)	5
		Agriculture land	6
		Built up area	5
		River sand	2
		Waterbodies	0
4	Lineament occurrences	0 (Number of lineament occurrence within 1 km^2)	0
		1–2	2
		3–4	4
		5 & >	8
5	Lineament intersection map	0 (Number of lineament intersection)	0
		1–2	3
		3–4	6
		<5	9
6	Drainage frequency map	0 (No drainage/streams within 1 Km^2)	1
		1–2	3
		3–4	5
		5–6	7
		<7	9

Table 4 Factors and weights

S. no	Factors	Weighted by pairwise comparison method (AHP)
1	Slope	26.45
2	Aspect	3.43
3	Land use	6.55
4	Drainage frequency	21.72
5	Lineament occurrence	16.33
6	Lineament intersection	25.52

5.2 Pairwise Comparison-Matrix Development

To rate the relative preferences for two criteria, the approach uses an underlying scale with values ranging from 1 to 9. The scale is displayed in Table 1. As a result, if paradigm A acquires a score of 2 in comparison to paradigm B, then paradigm B should earn a score of 1/2 in comparison to paradigm A. The lower left half of the pairwise comparisons matrix can also be completed using the same reasoning. The evaluation scale for anything being compared to it must be 1, signifying equally desirable criteria. The software for multi-criteria weighting received this pairwise comparison matrix of the criteria parameters. Using these score values, criteria weights were now calculated and are shown in Table 2.

5.3 Computation of Criterion Weights

The criteria weights were computed using multi-criteria weighted software. The normalized weights are estimated by computing the Eigen values and Eigen vectors of opinion matrices created by the domain experts while deciding the order of preference of parameters categories for a particular decision problem on a nine-point scale Table 4.

5.4 Consistency-Ratio Estimation

To determine the consistency of the comparison criterion, the consistency ratio was estimated. The calculation is established on consistency-vector, consistency index, and lambda. In pairwise comparisons, a consistency ratio of 0.10 or less suggests a rational level of consistency; if the consistency ratio exceeds 0.10, the values of the ratio are indicative of inconsistent judgement. Using the multi-criteria weighted software, Criterion Ranking Module, the consistency ratio for the weights was determined; the resultant value was 0.0985, which indicated good consistency in pairwise comparisons.

6 Landslide Hazard Zonation (LHZ) Map: 53 J/8

The following graph is showing the category wise distribution of the percentage area (%) of hazard in the study area carried out from the GIS Analysis. Most of the study area is falling in the low category (46.75%), Moderate (44%), Very low (8.25%), High (5.5) and Very High (3%) Fig. 4. Even though the very high-risk hazard area distribution is comparatively lower, most of the landslides are distributed along the road sites, and this is the most important area where the human and other properties' distribution is less.

The result obtained for the output (LHZ) map of the buffer area of 2 km along National Highway-58 is that most of the landslides are falling in the high hazard zone Fig. 5. This verification is carried out by overlying the GPS data of landslide inventory mapping carried out along the road site during field survey. The resultant LHZ map is shown in Fig. 6.

6.1 Landslide Inventory Mapping

To prepare landslide inventory map along National Highway-58, a field tour is arranged, and the landslide locations are recorded with the help of handheld GPS. This attribute information is entered in ArcGIS with the road network map by creating a point data layer. Inventory map is created by using the symbolizing of the layer.

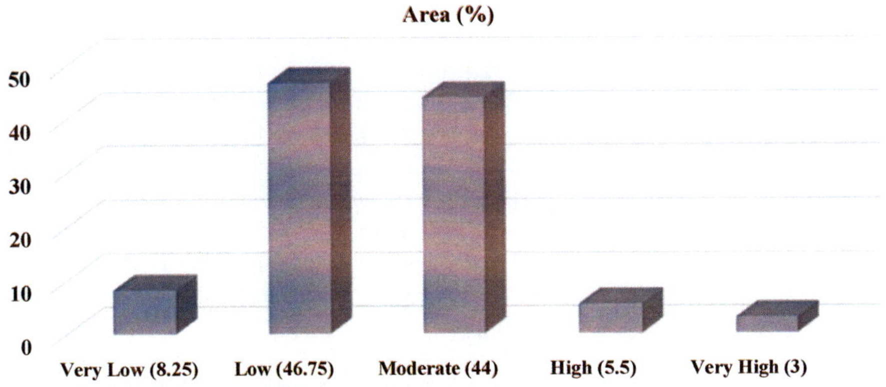

Fig. 4 Categorization of the study area in to varying zones of hazards

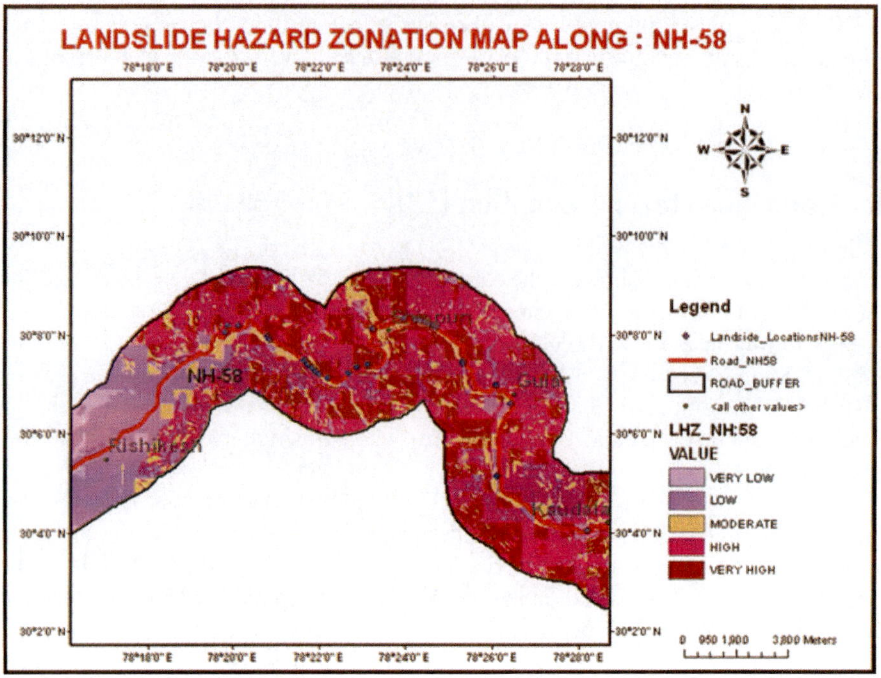

Fig. 5 Showing landslide hazard zonation mapping along NH-58

Most of the landslides are observed at the longitudinal direction between 78° 21" and 78° 25" before and after Shivpuri Village. For validation and visualisation purposes, online Google earth software and topographic map are used.

This inventory map is useful for indicative information of the landslides and distribution Fig. 7. This will not give any hazard information. This is also called Danger map i.e., this is a theme-based landslide mapping method.

Site-specific landslide risk identification using Google earth and data is carried out for two sites near Rishikesh location (Lat. 30° 08′ 8.72", Long. 78° 19′ 57.19"), where one Rafting camp is constructed around 100 m deep in the valley portion at north site of road. The upside portion of the road mountain portion is having very poor vegetation cover and very thin soil cover. Lithology is not too stable; it is highly fractured and weathered and well exposed due to road cutting and widening. Already along one site, a resent landslide is observed which is just above the campsite. The field photograph shows the recent landslide, and it is well observable in the satellite data Fig. 8a and b. This location is very prone to landslides due to the nature of ground conditions and deforestation over the ground. This may damage the road and affect the campsite and people coming to the place.

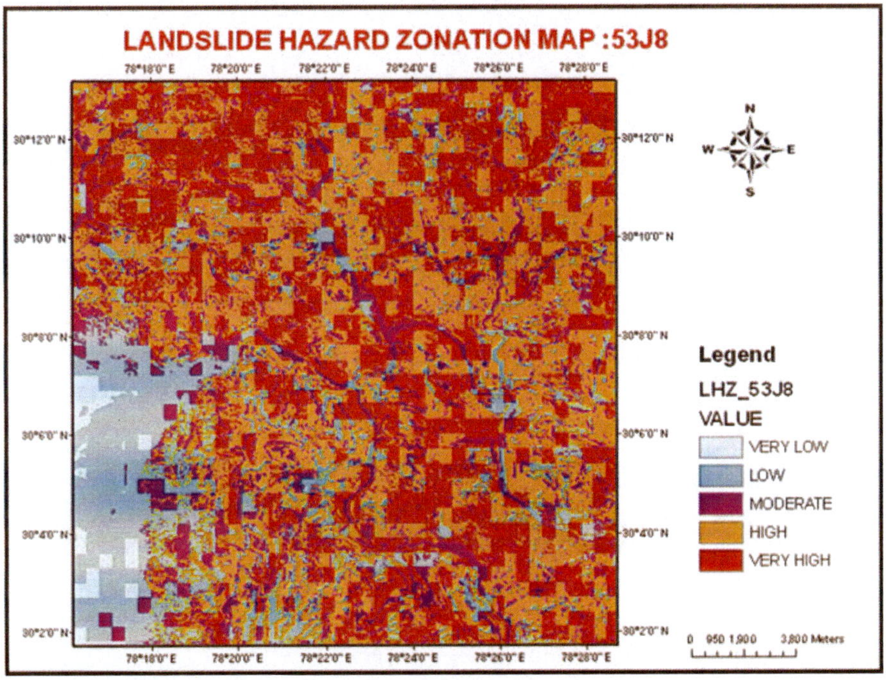

Fig. 6 Showing landslide hazard zonation mapping

7 Summary and Conclusions

The key findings and implications of this study are:

1. During this research work focus has been given on the preparation of Landslide Hazard Zonation map (LHZ) around the Rishikesh surrounding area and within the buffer zone of 2 km around part of National Highway-58, which falls within the Topomap-53 J/8 of 1:50,000 scale of Survey of India.
2. The LHZ is prepared by weighted overlay analysis and the thematic maps are prepared by the weight and rank method using decision rules such as the Analytical Hierarchy Process (AHP) method of decision-making analysis in GIS environment.
3. The different input thematic database are generated by manual digitization method using LISS-III, Landsat and high-resolution images available on Google earth.
4. The primary objective of this research paper is to combine the various existing methods and technologies for creating a Geo-spatial database for delineating landslide hazard zones in the study area.

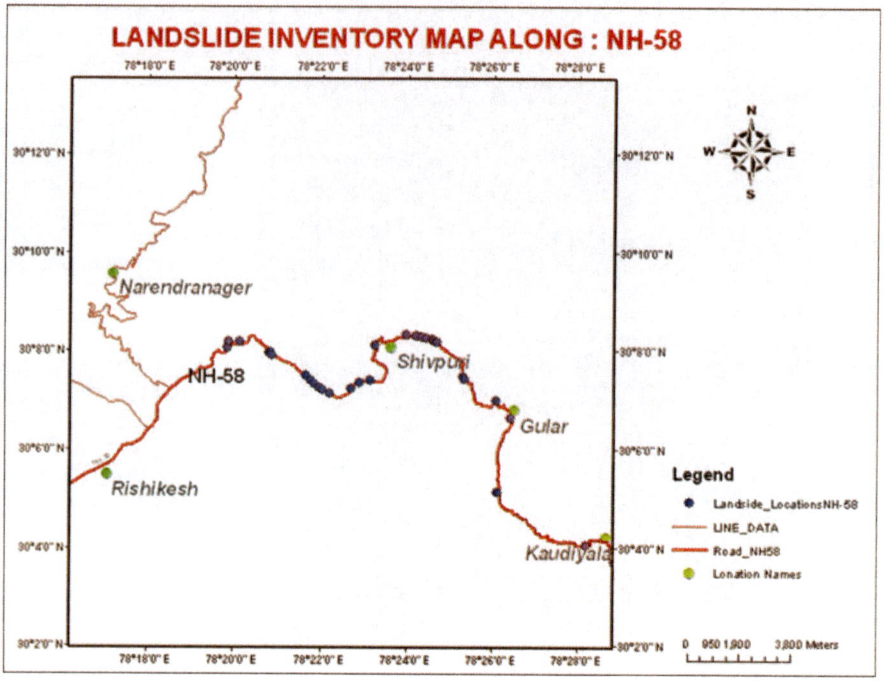

Fig. 7 Showing landslide inventory map along NH-58

5. The online Google database is found to be very useful for visualization and interpretation of the site-specific mapping. Using this some site-specific landslide risk identification is carried out; this kind of work may be very useful for site specific risk database generation, visualization and utilization in local level planning activities.
6. For validation of result, landslide inventory mapping is carried out using GPS, maps, and other instruments by field survey. Fieldwork is carried out only along the road sites from Rishikesh to Kaudiyala village, which is 31 km from Rishikesh.
7. The identified Hazard zones can be considered for preliminary planning activity of development schemes and help to avoid unstable areas during the planning stage. Even if unavoidable, recognizing them in the early stages of planning will help to develop better preventive measures as a priority.
8. The high hazards can be considered on high priority basis. In this regard, landslide hazard zonation mapping is the foremost component of such a study. The landslide hazard zonation mapping in the vicinity of Rishikesh identifies the areas of risk in the region and along National Highway-58.
9. This study will also be helpful for carrying out further detailed mapping on priority for preparation of the landslide risk assessment mapping along the road, which will highlight the risk of damage to the roads and the surrounding

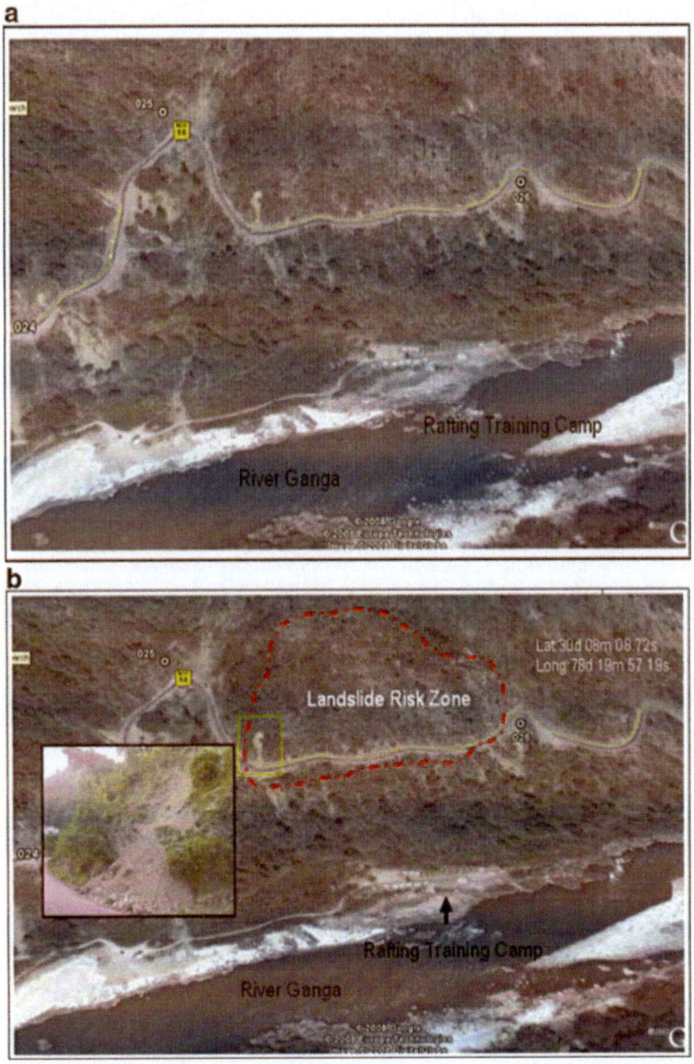

Fig. 8 **a** High-resolution satellite data **b** Site-specific Landslide risk identification, risk zone, field photograph, landslide location, road site and Rafting camp site at Rishikesh

property, line of communications, schools, buildings, restaurants, resorts, tourist and rafting camps along river, and residential areas individually. The LHZ map can be a valuable tool for planners and designers when deciding on the most appropriate route paths and in addition proper modelling and mitigation policies are required in this study region.

10. The study aligns with the United Nations' Sustainable Development Goals, particularly those related to disaster risk reduction, environmental sustainability, and resilient infrastructure. The outcomes of this research contribute to the creating safer and more sustainable habitats in mountainous areas worldwide.
11. In summary, the future scope of this work extends to incorporating data from various remote sensing platforms, such as satellite imagery, LiDAR, unmanned aerial vehicles (UAVs), and leveraging machine learning algorithms, such as artificial neural networks, support vector machines, and random forests which can provide a more comprehensive view of the terrain's characteristics. This approach could improve the accuracy and reliability of landslide hazard assessment.

References

1. Anbalagan R, Singh B (1996) Landslide hazard and risk assessment mapping of mountainous terrains—a case study from Kumaun Himalaya, India. Eng Geol 43(4):237–246
2. Khan H, Shafique M, Khan MA, Bacha MA, Shah SU, Calligaris C (2019) Landslide susceptibility assessment using frequency ratio, a case study of Northern Pakistan. Egypt J Remote Sens Space Sci 22(1):11–24
3. Li L, Lan H, Guo C, Zhang Y, Li Q, Wu Y (2017) A modified frequency ratio method for landslide susceptibility assessment. Landslides 14(2):727–741
4. Pourghasemi HR, Pradhan B, Gokceoglu C (2012) Application of fuzzy logic and analytical hierarchy process (AHP) to landslide susceptibility mapping at Haraz watershed, Iran. Nat Hazards 63:965–996
5. Ercanoglu M, Gokceoglu C (2002) Assessment of landslide susceptibility for a landslide-prone area (north of Yenice, NW Turkey) by fuzzy approach. Environ Geol 41:720–730
6. Basharat M, Rohn J, Baig MS, Khan MR (2014) Spatial distribution analysis of mass movements triggered by the 2005 Kashmir earthquake in the Northeast Himalayas of Pakistan. Geomorphology 206:203–214
7. Lombardo L, Mai PM (2018) Presenting logistic regression-based landslide susceptibility results. Eng Geol 244:14–24
8. Ali S, Biermanns P, Haider R, Reicherter K (2019) Landslide susceptibility mapping by using a geographic information system (GIS) along the China-Pakistan Economic Corridor (Karakoram Highway), Pakistan. Nat Hazard 19(5):999–1022
9. Vojteková J, Vojtek M (2020) Assessment of landslide susceptibility at a local spatial scale applying the multi-criteria analysis and GIS: a case study from Slovakia. Geomat Nat Haz Risk 11(1):131–148
10. Moragues S, Lenzano MG, Lanfri M, Moreiras S, Lannutti E, Lenzano L (2021) Analytic hierarchy process applied to landslide susceptibility mapping of the North Branch of Argentino Lake, Argentina. Nat Hazards 105(1):915–941
11. Strokova L (2022) Landslide susceptibility zoning in surface coal mining areas: a case study Elga field in Russia. Arab J Geosci 15(2):146
12. Thomas AV, Saha S, Danumah JH, Raveendran S, Prasad MK, Ajin RS, Kuriakose SL (2021) Landslide susceptibility zonation of Idukki district using GIS in the aftermath of 2018 Kerala floods and landslides: a comparison of AHP and frequency ratio methods. J Geovisualization Spat Anal 5:1–27
13. Babitha BG, Danumah JH, Pradeep GS, Costache R, Patel N, Prasad MK, Kuriakose SL et al (2022) A framework employing the AHP and FR methods to assess the landslide susceptibility of the Western Ghats region in Kollam district. Saf Extreme Environ 4(2):171–191

14. Asmare D (2023) Application and validation of AHP and FR methods for landslide susceptibility mapping around choke mountain, northwestern ethiopia. Sci Afr 19:e01470
15. Tesfa C (2022) GIS-Based AHP and FR methods for landslide susceptibility mapping in the Abay Gorge, Dejen-Renaissance Bridge, Central, Ethiopia. Geotechn Geol Eng 40(10):5029–5043
16. Mao Z, Shi S, Li H, Zhong J, Sun J (2022) Landslide susceptibility assessment using triangular fuzzy number-analytic hierarchy processing (TFN-AHP), contributing weight (CW) and random forest weighted frequency ratio (RF weighted FR) at the Pengyang county, Northwest China. Environ Earth Sci 81(3):86
17. Gupta N, Pal SK, Das J (2022) GIS-based evolution and comparisons of landslide susceptibility mapping of the East Sikkim Himalaya. Ann GIS 28(3):359–384
18. Rehman A, Song J, Haq F, Mahmood S, Ahamad MI, Basharat M, Mehmood MS et al (2022) Multi-hazard susceptibility assessment using the analytical hierarchy process and frequency ratio techniques in the Northwest Himalayas, Pakistan. Remote Sens 14(3):554
19. Satty TL (1980) The analytical hierarchy process: planning, priority setting, resource allocation. RWS Publication, Pittsburg
20. Nagarajan R, Khire MV (1998) Debris slides of Varandh Ghat, west coast of India. Bull Eng Geol Env 57(1):59–63
21. Glade T (2003) Landslide occurrences as a response to land use change: a review of evidence from New Zealand. CATENA 51:297–314
22. Rajakumar P, Sanjeevi S, Jayaseelan S, Isakkipandian G, Edwin M, Balaji P, Ehanthalingam G (2007) Landslide susceptibility mapping in a hilly terrain using remote sensing and GIS. J Indian Soc Remote Sens 35(1):31–42
23. Paul SK, Bartarya SK, Rautela P, Mahajan AK (2000) Catastrophic mass movement of 1998 monsoons at Malpa in Kali Valley, Kumaun Himalaya (India). Geomorphology 35(3):169–180
24. Burrough PA, McDonnell RA (1998) Principles of geographic information systems, pp 122–127. Oxford University Press
25. Dai FC, Lee CF (2002) Landslide characteristics and slope instability modeling using GIS, Lantau Island, Hong Kong. Geomorphology 42(3):213–228
26. Seshagiri DN, Badrinarayanan S, Upendran R, Lakshmikantham CB, Srinivasan V (1982) The Nilgiris landslide. Miscellaneous publication no. 57. Geological Survey of India
27. Gokceoglu C, Aksoy H (1996) Landslide susceptibility mapping of the slopes in the residual soils of the Mengen region (Turkey) by deterministic stability analyses and image processing techniques. Eng Geol 44:147–161
28. Greenways DR (1987) Vegetation and slope stability. In Anderson MF, Richards KS (eds) Slope stability. Wiley and Sons, New York
29. Malczewski J (1999) GIS and multicriteria decision analysis. John Wiley & Sons
30. Nagarajan R, Mukherjee A, Roy A (1996) Landslide hazard assessment using remote sensing and GIS—a case study. In Proceedings of the International conference on disasters and mitigation, vol I, pp A4 78–84. Madras, India, 19–22 Jan 1996
31. Gray DH, Leiser AT (1982) Biotechnical slope protection and erosion control. Van Nostrand Reinhold Company, New York
32. Saha AK, Gupta RP, Arora MK (2002) GIS-based landslide hazard zonation in the Bhagirathi (Ganga) valley, Himalayas. Int J Remote Sens 23(2):357–369
33. Sarkar S, Kanungo DP (2004) An integrated approach for landslide susceptibility mapping using remote sensing and GIS. Photogramm Eng Remote Sens 70(5):617–625
34. Perotto-Baldiviezo HL, Thurow TL, Smith CT, Fisher RF, Wu XB (2004) GIS-based spatial analysis and modeling for landslide hazard assessment in Steeplands, Southern Honduras. Agr Ecosyst Environ 103(1):165–176
35. Rautela P, Lakhera RC (2000) Landslide risk analysis between Giri and Tons rivers in Himachal Himalaya (India). Int J Appl Earth Obs Geoinf 2(3–4):153–160

Modeling Daily Streamflow from Idamalayar Catchment Using SWAT

C. Reshma and R. Arunkumar

Abstract SWAT, a semi-distributed, continuous, and process-based hydrologic model, is used for many applications such as streamflow simulations, water quality modeling, sediment yield modeling, etc. It requires spatial as well as temporal data as input. The present study examines the application of SWAT in simulating the daily streamflow from the Idamalayar catchment, Kerala. The model was simulated from 1987 through 2017 using daily meteorological data keeping three years as a warm-up. To calibrate and validate the model, a stand-alone tool, SWAT-CUP, is used. The model's capability to reproduce daily streamflow was assessed by means of the four performance- measures, which are R^2, NSE, PBIAS, and RSR. The values of these indicators turned out to be 0.64, 0.61, 22.5, and 0.63, respectively, in calibration, and while validating, these values were found to be 0.75, 0.67, 36.9, and 0.58, respectively. These values show that the model reasonably simulated the streamflow, especially at a daily time scale. Global sensitivity analyses showed that the effective hydraulic conductivity in main channel alluvium is the most sensitive parameter, followed by the SCS runoff curve number.

Keywords Streamflow simulation · SWAT · Sensitivity analysis · Idamalayar

1 Introduction

Water resource assessment and management are essential for initiatives concerning sustainable development [7]. Knowledge of runoff from a catchment is crucial for proper planning, designing, and operating different hydro-infrastructure systems. There are various hydrologic models for simulating the streamflow from a catchment. The physical or process-based hydrological models simplify the hydrological cycle at the watershed level and aim to represent the main hydrological processes [5].

C. Reshma (✉) · R. Arunkumar
Department of Civil Engineering, National Institute of Technology Calicut, Kozhikode, Kerala 673601, India
e-mail: reshmanarayananc@gmail.com

S. Mesapam et al. (eds.), *Developments and Applications of Geomatics*,
Lecture Notes in Civil Engineering 450, https://doi.org/10.1007/978-981-99-8568-5_26

Hydrologic models have been categorized into three distinct groups. They are classified into lumped, distributed, and semi-distributed models. The entire watershed is considered a homogeneous unit in lumped models, and a single outflow hydrograph is generated. In the semi-distributed models, the catchment's topography is made use of to partition the entire basin into homogeneous sub-basins. In distributed models, the whole catchment is divided into small units so that inputs, outputs, and parameters can vary spatially. These hydrologic models require both spatial and temporal data as input.

Soil and Water Assessment Tool (SWAT) is a physical, semi-distributed, and continuous hydrologic model developed by the United States Department of Agriculture (USDA). It utilizes process-specific equations for simulating different hydrological responses [14]. SWAT is effectively used for simulating the streamflow, modeling water quality, sediment yield, etc. [12]. The Digital Elevation Model (DEM), land use categories, and soil types are among the spatial data inputs needed for this model. It also requires temporal meteorological data such as minimum temperature, maximum temperature, precipitation, wind speed, solar radiation, and relative humidity. SWAT uses its weather generator to infill the missing data or supply mean data for different climatic variables.

SWAT has been widely used for the simulation of discharges from different catchments across the world [8]. Due to its effectiveness in simulating catchment hydrology, it is used widely for environmental and climate change studies [6]. SWAT has been effectively used in modelling streamflow from Indian basins. Anand et al. [1] and Yaduvanshi et al. [14] used SWAT for streamflow simulations. The present study intends to use the SWAT model to simulate the daily streamflow from the Idamalayar catchment, Kerala.

2 Study Area and Data Used

2.1 Study Area

The Idamalayar river is a main tributary of the Periyar river. It originates from the Anamala hills, which lie at an elevation of approximately 2520 m above the mean sea level. Anamalayar, Manaliyar, and several other small streams join to form the river Idamalayar. The Idamalayar river merges with the Periyar near the Bhoothathankettu barrage of the Periyar Valley Irrigation Project (PVIP). The point of confluence is approximately 1.65 km upstream of the barrage. The river Idamalayar is 74 km long up to its confluence with the Periyar.

The Idamalayar watershed is located between the latitudes 10° 10' N and 10° 20' N and longitudes 76° 20' E and 77° 40' E. The total area of the Idamalayar catchment is 380.79 km^2, excluding 101 km^2 of the catchment of Nirar that is diverted to Tamil Nadu. The average annual rainfall received by the watershed is 475 cm. Figure 1 shows the location and DEM of the Idamalayar catchment.

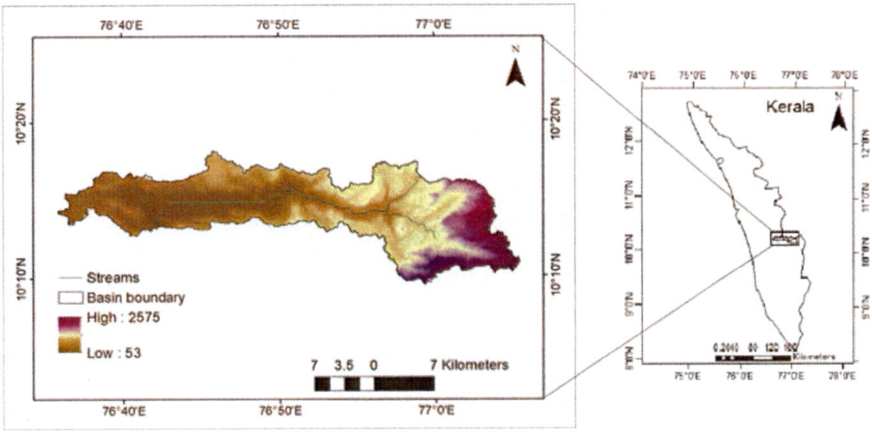

Fig. 1 Map showing the location, and DEM of Idamalayar basin

2.2 Data Used

The daily meteorological data from 1987–2017 obtained from the Indian Meteorological Department (IMD) is the primary data source. The daily rainfall was obtained as 0.25° × 0.25° gridded data from IMD for the two grid points in the basin. The daily temperature was received at 1° × 1° resolution, which was regridded into 0.25° × 0.25° grid size by employing the bilinear interpolation method. The streamflow from the Idamalayar catchment for the period of 1987–2017 was collected from KSEB, Idamalayar division.

Soil Maps, DEM, and Land Use maps are among the geographic inputs used in the study. Cartosat DEM was obtained from BHUVAN. The resolution of the DEM was 30 m. The land use map was derived from Landsat 8 OLI images. The soil map was obtained from the United Nations' Food and Agriculture Organization (FAO).

For delineating the watershed and defining the slopes, the DEM has been utilized. The altitude of the catchment ranged from 53 m at its lowermost elevation to 2575 m at its highest elevation. The land use categories in the Idamalayar catchment are shown in Fig. 2.

Three land use classes observed in the basin are forest, water, and rangeland. The predominant land use class in the Idamalayar catchment was forest, which constitutes up to 87.25% of the total area, followed by water, about 6.92%, and rangeland, the remaining 5.83%.

The soil categories of the Idamalayar basin are shown in Fig. 3. There are two soil classes in the study area. The two classes were clay loam (63.58%) and sandy clay loam (36.42%).

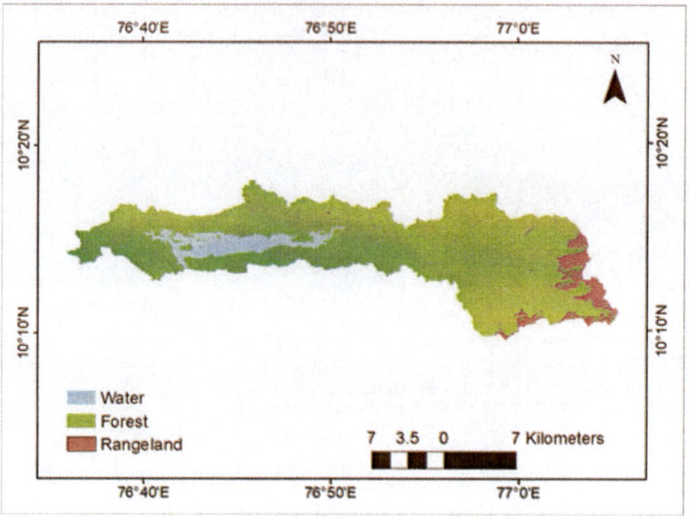

Fig. 2 Land use categories of Idamalayar catchment

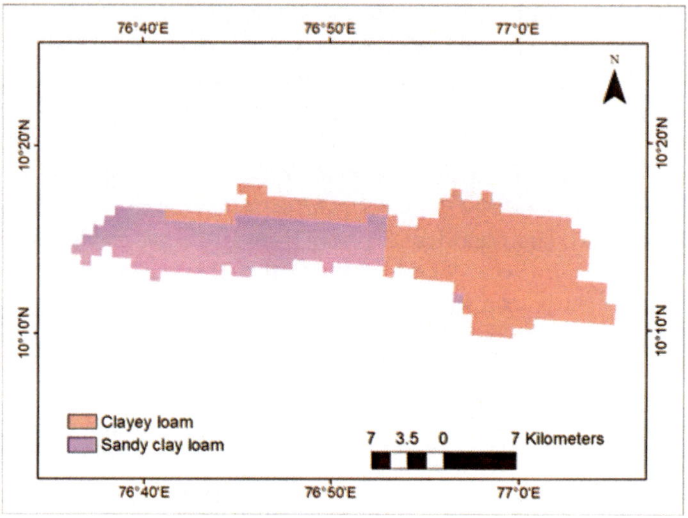

Fig. 3 Soil categories in the study area

3 Methodology

3.1 SWAT Model Description

SWAT is a hydrological model that uses mathematical equations relevant to each process to model the diverse hydrological responses of a watershed. The entire basin area is split up into multiple sub-basins in SWAT. The land use, slope, and soil classes are used to further divide the sub-basins into homogenous regions known as Hydrologic Response Units (HRUs). The fundamental equation supporting the SWAT model is the water balance equation. It is given by [2],

$$SW_t = SW_0 + \sum_{i=1}^{t}\left(R_{day,i} - Q_{Surf,i} - E_{a,i} - W_{seep,i} - Q_{gw,i}\right) \quad (3.1)$$

where SW_0 is the initial moisture content in the soil, and SW_t is the moisture content in soil on day t. In this equation, the net quantity of water lost from precipitation on the day i is subtracted from that day's total amount of precipitation. It includes surface runoff ($Q_{Surf,i}$), evapotranspiration ($E_{a,i}$), return flow ($Q_{gw,i}$) and water lost to the Vadose zone ($W_{seep,i}$) [10, 14]. All units are in mm of water. This equation is used for the prediction of changes in variables like runoff. The runoff using the runoff curve number (CN) approach is given by Eq. 3.2,

$$Q_{surf} = \frac{\left(R_{day} - I_a\right)^2}{\left(R_{day} - I_a + S\right)} \quad (3.2)$$

where I_a is the initial abstraction that depends on surface storage, infiltration, and interception. Q_{surf} is the excess rainfall that has accumulated (also known as runoff). The retention parameter, S, is calculated using curve number (CN).

3.2 Calibration and Validation of the Model

The procedure of changing the parameters of a model to ensure that the result of a simulation matches the data that was observed is known as model calibration. Uncertainty analysis refers to quantifying and propagating errors in the model parameters through calibration [11]. In this study, the calibration and validation of the model were carried out using a stand-alone tool named SWAT—Calibration and Uncertainty Program (SWAT-CUP). The Sequential Uncertainty Fitting Algorithm (SUFI-2) is employed from the different algorithms present. The model was executed from 1987 to 2017, with a three-year period for warm-up taken into account. For calibrating and validating, the observed streamflow from the Idamalayar catchment for the same period is used. The calibration period is taken as 1990–2007, and the validation

Table 1 Descriptions of selected parameters

Sl. No	Parameter name	Description
1	CH_K2	Effective hydraulic conductivity in main channel alluvium (mm/hr)
2	CN2	SCS runoff curve number
3	SOL_BD	Moist bulk density mm layer (g/cm^3)
4	GW_ DELAY	Groundwater delay (days)
5	ALPHA_ BNK	Base flow alpha factor for bank storage (days)
6	SURLAG	Surface runoff lag coefficient
7	ALPHA_ BF	Base flow alpha factor (days)
8	EPCO	Plant uptake compensation factor
9	ESCO	Soil evaporation compensation factor
10	CH_N2	Manning's n value for the main channel
11	GW_ REVAP	Groundwater revap coefficient
12	GWQMN	Threshold depth of water in the shallow aquifer required for return flow to occur (mm)
13	SOL_AWC	Available water capacity of the soil layer (mm H2O/ mm soil)
14	SOL_K	Saturated hydraulic conductivity (mm/hr)

period is taken as 2008–2017. Establishing a number of sensitive parameters is the first step in calibrating the model. Sharma et al. [11] identified fourteen parameters as significant for Indian basins. The same parameters are chosen in the present study. Table 1 shows these parameters and their description.

3.3 Model Evaluation Using Statistical Performance Measures

The effectiveness of the model is assessed using the coefficient of determination (R^2), Nash–Sutcliffe Efficiency (NSE), percent bias (PBIAS), and root mean square error-standard deviation ratio (RSR).

R^2 can be obtained as the square of the correlation coefficient. R^2 (Eq. 3.3) varies from zero to one. The variance of the actual observed values explained by the simulated values is indicated by this measure. A 100% explanation is represented by a value of 1 [3].

NSE (Eq. 3.4) expresses the extent of the residual variance with respect to the variance of the actual data. The value of NSE lies between $-\infty$ and 1.0. The ideal value of NSE is 1, which represents the model's effectiveness in replicating the observed conditions well. Moriasi et al. [9] state that an NSE value that exceeds 0.50 is satisfactory for a physical based model at a daily time scale.

PBIAS (Eq. 3.5) represents the mean trend of the model-predicted data to be greater or lower than the observed data [4]. Zero is the best value for PBIAS. Model underestimation bias is indicated by positive values, and overestimation bias is indicated by negative values.

Another model evaluation statistic, i.e., RSR, has been recommended by [13]. Lower values of RSR indicate better performance of the model.

$$R^2 = \frac{\left[\sum_{i=1}^{n}\left(Q_{m,i} - \overline{Q_m}\right)\left(Q_{s,i} - \overline{Q_s}\right)\right]^2}{\sum_{i=1}^{n}\left(Q_{m,i} - \overline{Q_m}\right)^2 \times \sum_{i=1}^{n}\left(Q_{s,i} - \overline{Q_s}\right)^2} \quad (3.3)$$

$$NSE = 1 - \frac{\sum_{i=1}^{n}\left(Q_{m,i} - Q_{s,i}\right)^2}{\sum_{i=1}^{n}\left(Q_{m,i} - \overline{Q_m}\right)^2} \quad (3.4)$$

$$PBIAS = \frac{\sum_{i=1}^{n}\left(Q_{m,i} - Q_{s,i}\right) \times 100}{\sum_{i=1}^{n}\left(Q_{m,i}\right)} \quad (3.5)$$

$$RSR = \frac{RMSE}{STDEV_m} = \frac{\sqrt{\sum_{i=1}^{n}\left(Q_{m,i} - Q_{s,i}\right)^2}}{\sqrt{\sum_{i=1}^{n}\left(Q_{m,i} - \overline{Q_m}\right)^2}} \quad (3.6)$$

where Q_m and Q_s stand for observed and predicted streamflow.

4 Results and Discussions

For delineating the basin and defining slopes, Cartosat DEM was used. The basin was divided into five sub-basins. It is then further separated into HRUs depending on land-use categories, soil, and slope classes. The Idamalayar basin contains twenty-one HRUs. SWAT was run at daily time steps from 1987–2017. The years 1987–1989 were taken as a warm-up period. For calibrating and validating the model, the SWAT-CUP SUFI-2 algorithm is adopted. The model calibration was done for 1990–2007, and the model was validated for 2008–2017. In Fig. 4, the observed and predicted streamflows calibration are displayed. It can be perceived that the model predicts the pattern of the daily streamflow satisfactorily. It is also observed that SWAT overpredicts some of the peaks during calibration.

The observed streamflow and the predicted streamflow for validation are displayed in Fig. 5. As observed, the streamflow's pattern and peaks are satisfactorily predicted by the model.

R^2, NSE, PBIAS, and RSR were the four metrics employed to assess the SWAT model's capacity to replicate observed streamflow. Table 2 displays the values of these performance indices for the model's calibration and validation.

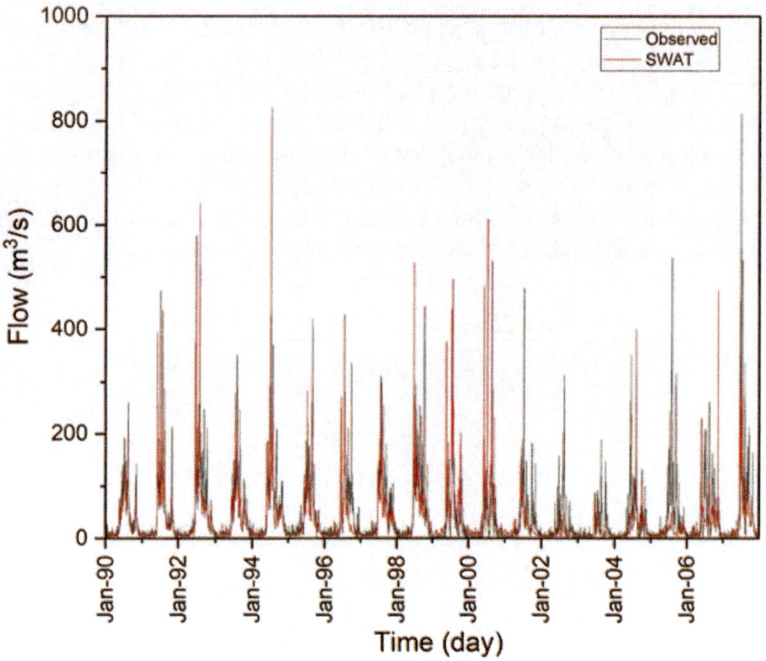

Fig. 4 Comparison of observed daily streamflow and simulated daily streamflow for calibration

Global sensitivity analysis was made use of to determine which parameters were sensitive for this specific catchment. The results are included in Table 3. It was observed that the Effective Hydraulic conductivity in main channel alluvium (CH_K2) is the highest sensitive parameter, followed by SCS runoff Curve Number (CN). The least sensitive parameter was the saturated hydraulic conductivity.

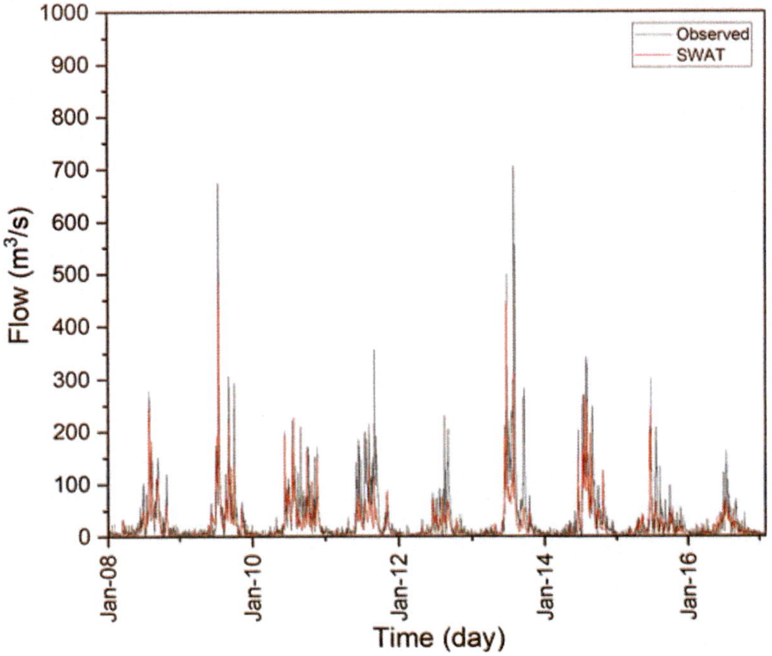

Fig. 5 Comparison of observed daily streamflow and simulated daily streamflow for validation

Table 2 Values of performance indices for calibration and validation

Performance index	Calibration	Validation
R^2	0.64	0.75
NSE	0.61	0.67
PBIAS	22.5	36.9
RSR	0.63	0.58

Table 3 Details of parameters and fitted values

Parameter name	Bound	Fitted value	Rank
CH_K2	(5, 130)	81.25	1
CN2	(-0.3, 0.3)	-0.258	2
SOL_BD	(-0.5, 0.6)	0.281	3
GW_DELAY	(30, 450)	126.6	4
ALPHA_BNK	(0, 1)	0.45	5
SURLAG	(0.05, 24)	18.4915	6
ALPHA_BF	(0, 1)	0.23	7
EPCO	(0.05, 1)	0.1735	8
ESCO	(0.8, 1)	0.866	9
CH_N2	(0, 0.3)	0.207	10
GW_REVAP	(0, 0.2)	0.018	11
GWQMN	(0, 2)	1.94	12
SOL_AWC	(-0.2, 0.4)	0.334	13
SOL_K	(-0.8, 0.8)	0.72	14

5 Conclusion

The present study investigated the effectiveness of SWAT in simulating the daily streamflow from the Idamalayar catchment, Kerala. The model was simulated from 1998–2017, with the first three years as a warm-up. Calibration was done from 1990–2007, and validation was done from 2008–2017. The efficiency of the model was studied using R^2, NSE, PBIAS, and RSR. During calibration, these values were found to be 0.64, 0.61, 22.5, and 0.63, and during validation, these values were found to be 0.75 and 0.67, 36.9, and 0.58, respectively. The results show that the performance of SWAT in modelling daily streamflow from the Idamalayar catchment was good. From the results of the global sensitivity analysis of the parameters, it is clear that CH_K2 is the most sensitive parameter, followed by CN.

References

1. Anand J, Gosain AK, Khosa R (2018) Optimisation of multipurpose reservoir operation by coupling soil and water assessment tool (SWAT) and genetic algorithm for optimal operating policy (case study: Ganga River Basin). Sustainability 10(5):1660
2. Arnold JG, Srinivasan R, Muttia RS, Williams R (1998) Large area hydrologic modeling and assessment part I: model development. J Am Water Resour Assoc 34(1):73–89
3. Bhadoriya UPS, Mishra A, Singh R, Chatterjee C (2020) Implications of climate change on water storage and filling time of a multipurpose reservoir in India. J Hydrol 590:125542
4. Gupta HV, Sorooshian S, Yapo PO (1999) Status of automatic calibration for hydrologic models: comparison with multilevel expert calibration. J Hydrol Eng 4(2):135–143

5. Hakala K, Addor N, Teutschbein C, Vis M, Dakhlaoui H, Seibert J (2019) Hydrological modeling of climate change impacts. Encycl Water:1–20
6. Haro-Monteagudo D, Palazón L, Beguería S (2020) Long-term sustainability of large water resource systems under climate change: a cascade modeling approach. J Hydrol 582:124546
7. Jaiswal RK, Lohani AK, Tiwari HL (2020) Development of framework for assessment of impact of climate change in a command of water resource project. J Earth Syst Sci 129:58
8. Luo M, Liu T, Meng F, Duan Y, Bao A, Xing W, Feng X, De Maeyer P, Frankl A (2019) Identifying climate change impacts on water resources in Xinjiang, China. Sci Total Environ 676:613–626
9. Moriasi DN, Arnold JG, Van Liew MW, Bingner RL, Harmel RD, Veith TL (2007) Model evaluation guidelines for systematic quantification of accuracy in watershed simulations. Trans ASABE 50(3):885–900
10. Neitsch S, Arnold J, Kiniry J, Williams J (2011) Soil & water assessment tool theoretical documentation version 2009. Texas Water Resources Institute, Texas
11. Sharma A, Patel PL, Sharma PJ (2022) Influence of climate and land-use changes on the sensitivity of SWAT model parameters and water availability in a semi-arid river basin. CATENA 215:106298
12. Shivhare N, Dikshit PKS, Dwivedi SB (2018) A comparison of SWAT model calibration techniques for hydrological modeling in the ganga river watershed. Engineering 4(5):643–652
13. Singh J, Knapp HV, Arnold JG, Demissie M (2005) Hydrological modeling of the Iroquois River watershed using HSPF and SWAT. J Am Water Resour Assoc 41(2):343–360
14. Yaduvanshi A, Sharma RK, Kar SC, Sinha AK (2018) Rainfall–runoff simulations of extreme monsoon rainfall events in a tropical river basin of India. Nat Hazards 90(2):843–861

Modelling the Low Impact Development Alternatives for Rainfall Runoff Reduction

B. Aneesha Satya, M. Shashi, and Allu Pavan Kumar Reddy

Abstract In order to control storm water overland runoff, which in turn regulates flood volume and flow rate, one of the surface runoff mitigation and treatment approaches is Stormwater Best Management Practice (BMP). Sustainable drainage systems (LID) are widely recommended and used in many parts of the world. In the United States and Canada, LID is called Low Impact Development (LID), which is an approach that encourages the interaction of natural processes with the urban environment to keep and rebuild ecosystems for water management. LID can be used to manage stormwater runoff at the property level. In this study four different elements in LID were studied for their efficiency in reducing runoff that include permeable pavement (PP), infiltration trenches (IT), bioretention cells (BC), and rainwater barrel (RB) storage. Hydrological performance of these LID practises in the storm water network were performed using PCSWMM which is considered one of the most promising hydrological models.

Keywords Low impact development · PCSWMM · Hydrological modeling

1 Introduction

In urbanizing watersheds, where growth-related increases in impervious areas lead to increases in flood flows, storm water management is a prevalent topic [1, 4]. The term Low Impact Development (LID) refers to a wide range of suitable technologies and practices that are meant to reduce the impact of watershed development on flow regimes while maintaining the current impermeable coverage. Infiltration trenches, porous pavement, grass swales, green roofs, rainwater barrel storage, and bio retention cells are a few examples of LID components that reduce peak and storm flow. BMPs are an engineered and built system that is intended to offer storm water quantity and quality control, according to the US Environmental Protection Agency [2]. LID aims to return the site's hydrologic conditions to their pre-development state.

B. A. Satya (✉) · M. Shashi · A. P. K. Reddy
Department of Civil Engineering, National Institute of Technology, Warangal, India
e-mail: abandi@student.nitw.ac.in

S. Mesapam et al. (eds.), *Developments and Applications of Geomatics*,
Lecture Notes in Civil Engineering 450, https://doi.org/10.1007/978-981-99-8568-5_27

Typically, the analysis of LID performance concentrates on factors of water quality such pollution loads and concentrations. Site hydrology management and stormwater pollutant load control have recently been centered on volume-based reduction [6]. The use of LID can offer a solution for the on-site management of stormwater runoff after construction [5]. One of the more straightforward methods for reducing urban flooding is storm water LID [3]. Urban stormwater management's most important tools are stormwater best practices that make use of detention and infiltration to lessen the negative effects of stormwater runoff from growing impermeable surfaces (Paula and Jerry 2009, Ryan et al. 2016).

The objective of the present investigation is to determine how well the planned LID can reduce onsite runoff that discharges from the study site. The objectives of this research is to propose simulation techniques to evaluate typical LID components' in volume reduction and attenuation performance utilising PCSWMM in a densely populated area.

2 Data

The datasets used is the half hourly precipitation data from the IMERG algorithm that combines information from the GPM satellite constellation to estimate precipitation over the majority of the Earth's surface that lacks precipitation-measuring instruments on the ground. The data is inter-calibrated, merged, and interpolated "all" satellite microwave precipitation estimates, together with microwave-calibrated infrared (IR) satellite estimates, precipitation gauge analyses, and potentially other precipitation estimators at fine time and space scales over the entire globe. The study area is confined to downstream to Waddepally lake, and undergo through inundation up to Gopalpuram cheruvu due to inadequacy of carriage capacity and overflow of water from Waddepally lake. The same problem is observed in storm drains of Gopalapuram cheruvu. Waddepally Lake is one of the three major water supply sources of Warangal Municipal Corporation. Its storage capacity is 132 million Cubic Feet (0.132) TMC. The water spread is over an area of about 600 acres and the catchment area of Waddepally Lake is 61.2 Sq. km. Water flows from Waddepally tank to Gopalapuram tank along North West direction. The length of drain from Waddepally to Gopalapuram Tank is 2040 m and the width of drain varies from 11 m to17 m. The catchment area of Gopalapuram Tank is 14.45 Sq. km (Fig. 1).

3 Methodology

The methodology followed for the current study is shown in Fig. 2. Firstly, the files in Net CDF format are collected from the IMERG GPM platform and the precipitation data is extracted. The IDF relations were calculated for the data extracted. PCSWMM, a GIS version of the EPA Storm Water Management Model (EPA SWMM) was

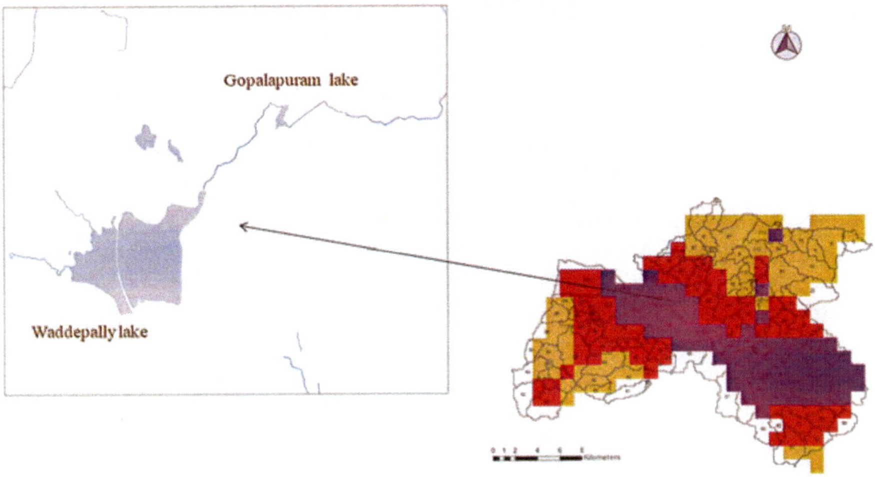

Fig. 1 Study area showing Waddepally and Gopalapuram lakes in the highly vulnerable zone

chosen to develop hydrological model of the study area and to evaluate the effects of LID on flood reduction in the study. PCSWMM consists of a dynamic rainfall-runoff model and a hydraulic model for piped systems. Different scenarios were applied for each element of LID in the storm water network and the simulations were carried to estimate the runoff quantity. Finally the reduction in runoff quantity is estimated. The study area is referred to as No-LID scenario prior to the introduction of elements of LID. Post LID development refers to the incorporation of the selected LID into both the impervious and pervious area.

Fig. 2 Methodology

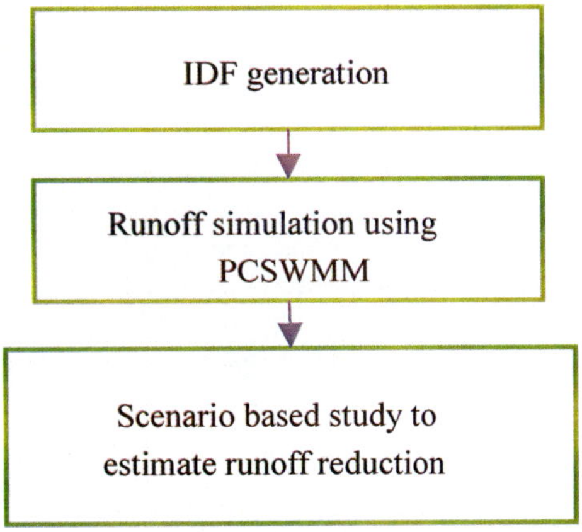

3.1 Implementation of LID Controls in PCSWMM

PCSWMM was used in this study to assess the impacts of urbanization on runoff and flood reduction effects of LID practices. PCSWMM allows modelling green infrastructure units like LID controls. It is used to model the amount of runoff that comes from the impervious area, mostly. Vertical layers are used to represent LID controls and their parameters (such as thickness and infiltration rate) are defined for each unit area of the model. Sub-catchments are selected based on the size that can be fitted with these LID controls. According to hydrologic features, the whole catchment is delineated to 23 sub-catchments for simulation purposes. The rainfall-runoff process was simulated using a nonlinear reservoir technique, which included infiltration, depression storage, evaporation, and surface runoff. The Curve Number approach was used to mimic infiltration in the model.

4 Runoff Simulation for Different Scenarios Analysis (W.R.T to 2 yr Design Storm and Combination of LIDs)

LID scenarios were simulated using PCSWMM to evaluate the projected runoff volume reduction. The model was re-run for multiple scenarios with the individual LID and combined LID scenarios in place. As per the Table 1, the individual placement of the Bio retention cell has reduced the volume of runoff by 19% in sub-catchment S21 and 33% in outfall. The mixed combination of four LIDs has also reduced the runoff by 11% in S23 and 22% in outfall. Table 1 shows the peaks in runoff volume for different combinations for a 2-year event. As per the Table 1, the individual placement of the Bio retention cell has reduced the volume of runoff by 19% in conduit c31 and 21% in outfall. The Permeable pavement placement also reduced the volume by 20% in S23. Table 2 shows the peaks in runoff volume for different combinations for a 2-year event. Permeable pavement scenarios typically have the biggest impact on lowering flood volume during storms. It has the least amount of storage because it takes up a fifth of the total amount of space that is available (150 mm). Using an infiltration trench with an effective storage depth of 750 mm, it is possible to store all stormwater flow. The infiltration trench is one of the most successful LID for lowering flood volume in the three storm events under study, despite having the smallest installed area (0.2 percent). Rain barrel storage, despite its tiny area of 0.5 percent, has the second-highest influence on flood volume reduction in all types of storm events due to the biggest effective storage depth of 1000 mm. Efficiency of the LID in reducing runoff volume is influenced by surface area, vertical storage, and drainage area. High catchment area values that are more than the drainage capacity are taken into account by the model. Stormwater can only be temporarily stored in LID up to a specific point before the storage layer infiltrates it.

Table 1 Runoff comparison within sub-catchments for LID controls placed in impervious area for 2-year storm event

Sub-catchment	NO LID	BC	IT	PP	Mixed
S1	1.077	1.062	1.017	1.059	1.017
S10	0.2358	0.2023	0.2089	0.2091	0.2089
S14	0.1889	0.1857	0.1858	0.1687	0.1859
S15	0.434	0.4211	0.4341	0.4341	0.3509
S16	0.141	0.1369	0.1413	0.1251	0.1414
S17	0.857	0.8358	0.7335	0.8569	0.8353
S19	0.348	0.3484	0.3603	0.3606	0.3118
S21	0.335	0.1763	0.1763	0.1698	0.176
S23	0.329	0.3194	0.329	0.329	0.2685
C31	3.175	2.416	2.901	3.016	2.983
OF	7.539	4.269	6.003	7.273	6.173

Table 2 Runoff comparison within sub-catchments for LID controls placed in pervious area for 2-year storm event

Sub-catchment	NO LID	BC	IT	PP	Mixed
S1	1.317	1.296	1.059	1.317	1.296
S10	0.2358	0.2089	0.2161	0.2161	0.2161
S14	0.1889	0.1763	0.1825	0.1825	0.1825
S15	0.3542	0.3542	0.3542	0.3204	0.3509
S16	0.147	0.1423	0.125	0.1471	0.1471
S17	0.923	0.8986	0.9234	0.734	0.9234
S19	0.3661	0.3118	0.3661	0.3118	0.3118
S21	0.335	0.2088	0.2136	0.2137	0.2137
S23	0.3453	0.3349	0.3453	0.2685	0.3349
C31	3.461	2.964	2.416	3.095	3.455
OF	8.89	5.328	8.75	5.519	4.269

5 Conclusions

Runoff from impermeable surfaces is the primary source of stormwater flow from the urban area [9]. Using LID procedures on impermeable lots can reduce many of these negative effects. In order to lessen runoff from impermeable surfaces, LID are being investigated as an alternative to traditional runoff management strategies. The LID material absorbs the collecting runoff, which subsequently seeps into the groundwater. Small local changes can significantly increase the catchment's ability to reduce runoff, lowering the risk of flooding. LID has a considerable impact on water runoff in this particular catchment region, with average reductions of 10%, 21%, and 32%, respectively, for the three techniques. When it came to reducing runoff, Bio retention cells outperformed the combinations. When selecting LID procedures, cost-effectiveness is critical. In addition to LID components, it's important to improve storm drainage systems. When linked to traditional drainage systems, LID can control stormwater runoff well in a future with more people living in cities. There are some disadvantages to this study that involve addressing the uncertainties caused by global climate models and a lack of data, as well as taking into account extreme rainfall events in urban flooding applications. In this study, the drainage area is studied on a site-scale instead of a catchment-scale, so the individual LID efficiency estimates are very different from those in the literature. The study's conclusion is that adopting many LID controls at once is more effective than adopting just one LID at a time. One way to keep all of the runoff from the catchment is to make the LID storage volume bigger.

References

1. Misra AK (2011) Impact of urbanization on the hydrology of Ganga Basin (India). Water Resour Manage 25(2):705–719. https://doi.org/10.1007/s11269-010-9722-9
2. USEPA (United States Environmental Protection Agency) (1999) Preliminary data summary of urban storm water best management practices, Office of Water (4303) Washington, EPA-821-R-99-012
3. USEPA (United States Environmental Protection Agency) (2000) Low Impact Development (LID): a literature review. USEPA, Washington, DC. http://water.epa.gov/polwaste/green/upload/lid.pdf
4. Poresky A, Bracken C, Strecker E, Clary J (2011) International stormwater best management practices (BMP) database technical summary: volume reduction. GeoSyntec Consultants & Wright Water Engineers, Inc.
5. BWM (Bureau of Watershed Management) (2006) Pennsylvania stormwater best management practices manual. Department of Environmental Protection, vol 34, document number: 363-0300-002
6. Dietz ME, Clausen JC (2005) A field evaluation of rain garden flow and pollutant treatment. Water Air Soil Pollut 167(2):201–208
7. Winston RJ, Dorsey JD, Hunt WF (2016) Quantifying volume reduction and peak flow mitigation for three bioretention cells in clay soils in northeast Ohio. Sci Total Environ 553:83–95. https://doi.org/10.1016/j.scitotenv.2016.02.081
8. Rees P, Schoen J (2009) PCSWMM evaluation. Water Resources Research Centre. University of Massachusetts, Project#08-08/319
9. NRCS (Natural Resources Conservation Service) (1986) Urban hydrology for small watershed, 2nd edn. United states Department of Agriculture. Technical Release 55, ISBN: 210-VI-TR-55

Performance Evaluation of Support Vector Machine and Random Forest Techniques for Land Use-Land Cover Classification—A Case Study on a Mili Scale Agricultural Watershed, Tadepalligudem, India

Chirasmayee Savitha and Talari Reshma

Abstract Land Use-Land Cover Mapping, obtained at a specific time, plays a crucial role in planning and monitoring both regional and global surfaces of the earth. For monitoring and analysis of small-scale agricultural watersheds with an area of less than 10,000 ha, a Land Use-Land Cover (LULC) classification that is nearly accurate is required. With the availability of Sentinel-2 datasets with spatial and temporal resolutions of 10 m and 5 days, LULC mapping and monitoring can be improved. However, conventional methods of classification are unable to demonstrate real features. In recent years, the introduction of machine learning techniques has made it possible to classify LULC effectively and efficiently, particularly for watersheds on the milli-scale. Therefore, in the present study, LULC maps are obtained for an agricultural watershed in Tadepalligudem, West Godavari district, Andhra Pradesh, using supervised Machine Learning methods. To highlight the importance of machine learning in LULC classification for a milli-scale watershed, Support Vector Machines (SVM) and Random Forests (RF) are compared with conventional Maximum Likelihood Classification methods. LULC was calculated using Sentinel-2 data acquired from the United States Geological Survey for three months in 2020. The field observations detected a total of four classes, and the training datasets were obtained using both visual inspection and field observations. Using Overall Accuracy, Kappa Coefficient, and R-Squared values, the accuracy of the LULC maps relative to the ground truth is determined. The study reveals that the overall accuracy of SVM and RF is 87.5 2.00 and 85.5 2.00, respectively, whereas it is 70.87 1.80 for Maximum Likelihood classification. The mean Kappa coefficient values for SVM and RF are 0.86 and 0.85, respectively, while for maximum likelihood classification, it is 0.67. The

C. Savitha (✉) · T. Reshma
Department of Civil Engineering, National Institute of Technology Andhra Pradesh, Tadepalligudem, India
e-mail: pj032101@student.nitandhra.ac.in

T. Reshma
e-mail: reshma@nitandhra.ac.in

S. Mesapam et al. (eds.), *Developments and Applications of Geomatics*, Lecture Notes in Civil Engineering 450, https://doi.org/10.1007/978-981-99-8568-5_28

observed average R-squared value for the support vector machine is 0.67, whereas it is 0.69 for the RF. In addition, it is observed that SVM performs admirably among the three classification methods.

Keywords Accuracy assessment · LULC · Machine learning · Random forests · Support vector machines

1 Introduction

Identifying and classifying the physical features covering the Earth's surface and the way they are utilized together can be termed as Land Use-Land Cover (LULC) classification. Classified LULC maps obtained at a specific time frame play a predominant role in planning and monitoring the Earth's surface at the regional and global scale [1, 2]. Acquiring LULC maps and detecting changes that occurred over some time has an indispensable role in several domains, such as urban and regional planning, climate studies, hydrological modeling, crop yield studies, and more [3–6]. The water resource is one of the integrated approaches that includes all hydrological components and their linkages. Wind patterns, topography, and land cover, along with land-use changes, play a significant role in rainfall-runoff patterns and the distribution of global water resources. Mapping LULC is therefore considered to be a prominent and integral segment in various hydrological models such as the Soil Water Assessment Tool (SWAT), Variable Infiltration Capacity models (VIC), etc. [7]. It also affects Land surface temperature (LST), one of the major indicators in climate change studies [8]. Anthropogenic changes induced, such as increased agricultural activity and infrastructural development, lead to changes in LULC over a greater spatial extent and have a significant impact on rainfall trends. Changes induced in LULC are shown to have a notable impact on climate across multiple paths that accentuate land surface temperature and rainfall, thus playing a prominent role in climate modeling. It is also one of the key drivers of global change and plays a prominent role in many international policy issues. Quantitative and qualitative assessment and foresight of changes occurring at spatial and temporal scales are therefore the most effective mechanisms for managing and understanding Earth's surface alteration. Therefore, assessment of LULC changes is unavoidably required for a wide range of applications.

Geospatial technologies are the most well-known approaches for mapping and analyzing changes in LULC due to their accurate digital image processing and the significant advantage of acquiring data at various temporal scales[9]. Landsat, ResourceSat, and Sentinel data [10–12] are among many others that are frequently used to map, classify, and monitor LULC change data due to their relatively high spatial resolution. The classification of remotely sensed data is an intricate process that requires the attention of many factors. The significant steps of LULC classification may include the determination of an appropriate classification method, training sample selection, classification processing, and assessing accuracy. A proper classification method requires an adequate number of training samples. A classification method is devised based on the need, the resolution of the chosen data, agreement with previous work, the algorithms available, and time constraints. Based on the literature above, it is noted that most conventionally used methods of LULC classification, such as Maximum likelihood classification (MLC), k means, and parallelopiped classification derived from coarser spectral and spatial resolution data, have limitations, thus influencing classification accuracy [13, 14].

Lately, the advent of machine-learning methods on remote sensing data for LULC mapping, monitoring, and classifying interest has drawn substantial attention [15–19]. The alignment of Machine learning and remote sensing is demonstrating great results due to high classification accuracies, the availability of satellite data, and the increase in computational power of the user group. Signature files, the topography of the study area, sensor parameters, and the level of classification are some of the factors influencing the accuracy of classification methods. The two most used methods in the classification of images are supervised classification and unsupervised classification. The supervised classification method utilizes trained samples, which play a crucial role in classifying images. The most popularly used supervised classification methods are support vector machine (SVM), decision tree (DT), and random forest (RF), namely [20–22]. In contrast, the unsupervised classification method classifies images based on similar spectral characteristics, where the classification of images is performed without training samples. A Few Unsupervised classifications include the K-means algorithm and ISODATA (Iterative Self-Organizing DATA) [23]. Most researchers have found that Artificial Neural networks (ANN), DT, SVM, and RF machine learning methods are frequently used [17, 22, 24].

For a small milli-scale agricultural watershed with an area of less than 10,000 Ha, the classification of fine-scale LULC is challenging, however, it is one of the important parameters of various applications. In recent years, LULC classification for small watersheds using machine learning has gained exceptional attention. Every method for machine learning has distinct accuracy levels. It has been discovered that ANN, SVM, and RF provide better precision than other traditional classifier techniques, whereas SVM and RF are the most effective techniques for LULC classification compared to all other machine-learning techniques. However, sensor characteristics and image data-related parameters, such as spatial and temporal resolution, processing software and hardware, etc., determine the precision of the classification of LULC. Due to the development of new algorithms, there is a need to identify the best algorithm for LULC mapping using Machine learning methods [25–27]. SVM is a nonparametric supervised classification method whose key concept is to find optimal hyperplanes from many hyperplanes between the two different classes. The training datasets or data samples that delineate the two distinguishable classes are termed support vectors. Kernels are one of the most essential parts of the SVM method. The most used kernels in remote sensing are linear, radial bias, and polynomial kernels. RF is a non-parametric supervised machine algorithm that can handle a variety of data, both remotely sensed images and numerical data. It mainly depends on regression and correlation between trees. Several trees (n-tree) and features in each split (m-try) are the two parameters termed the base of the method and are important parameters in the RF algorithm. It is one of the most widely used ML algorithms. SVM was formally presented in 1992, depending on a statistically learned method that determines the border between two different classes that are distinguishable from each other by either linearly separable or non-linearly separable data [19].

From the literature, it is notable that Machine learning studies are shown to perform satisfactorily when compared to conventional methods [21, 22]. The results of Machine-learning methods depend on the model setup, quality and quantity of training samples selected, and parameters that are used as input to achieve high-accuracy LULC maps. Although many studies in machine learning techniques are conducted to obtain LULC classification maps, there has been a narrow concentration on the development of LULC maps for milli-scale watersheds. Therefore, the present study employs two machine-learning techniques, SVM and RF, along with a traditional ML classifier to evaluate the performance of algorithms for a milli-scale agricultural watershed. Based on the kappa coefficient and R-squared values, the results obtained from SVM, RF, and ML algorithms are compared and evaluated.

2 Materials and Methods

2.1 Study Area

Classification of LULC in stable areas is more flexible than classifying in challenging landscapes. Classifying LULC units in a dynamic region emphasized different methods of classification approaches depending upon accuracy assessment. The present study is carried out in Tadepalligudem agricultural watershed, located in the West Godavari district, Andhra Pradesh, for the present research. It extends between the 16°48′0" N to 16°54′ 28.8 "N latitude and 81° 27′ 54" E to 81° 31′ 48" E longitudes covering an area of around 53.75 Km^2[Fig. 1] with water from Eluru canal and shallow aquifer a primary source. Paddy, followed by coconut, is the major crop contributing around 95% of the total cropped area together. This study area has seen a rapid change in LULC due to the state's bifurcation, leading to infrastructural activities from 2014. With an increasing population density, there has been an in LULC by reducing Agricultural land, expanding human settlement, and increasing Barren land.

2.2 Materials

In the present research, Sentinel-2 Multispectral Instrument (MSI) with a spatial resolution of 20 m having ten bands is used in mapping LULC classes of an agricultural watershed. Sentinel-2 data with tile number T44QND with cloud cover zero

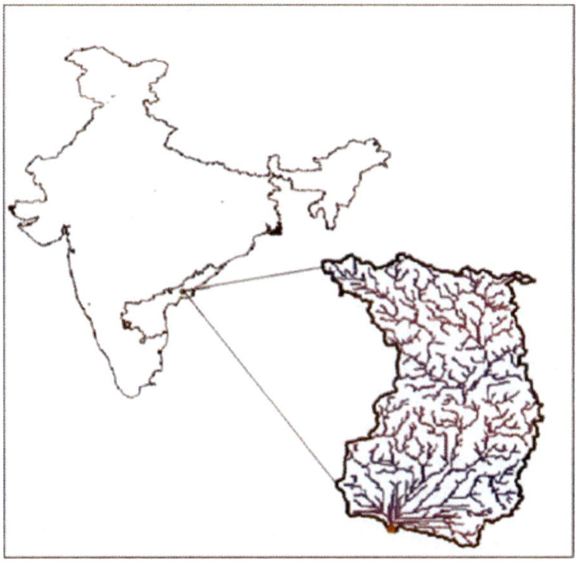

Fig. 1 Location map of study area

percent downloaded which is available freely from the United States Geological Survey (USGS) website (https://earthexplorer.usgs.gov) is downloaded for the three months of January, April, and December for the year of 2020.

2.3 Methodology

The LULC was classified into four different classes for the present study using two different non-parametric supervised algorithms (SVM, RF) and one parametric supervised classification (MLC). SVM, RF, and MLC methods have been selected for the classification of LULC maps. Four LULC classes were identified, and training samples are obtained from field surveys and visual interpretation using Google Earth Pro. Further, these helped to determine the accuracy of classified maps. Four LULC classes were used in the present study Water bodies, Built-up areas, Agricultural land, and Barren land. A description of the LULC classes classified and algorithms used is provided in Tables 1 and 2 and the methodology was followed in Fig. 2. LULC classification using machine learning algorithms (SVM, RF) is conducted to make use of the free statistical software tool R and Maximum Likelihood classification using Arc Map 10.7.1. The satellite data obtained clipped to the extent of the study area after pre-processing. For precise classification, many training samples are required as inputs for Machine-learning algorithms, which are essential elements in supervised learning. Training data were acquired and delineated from the satellite observations made from visual interpretation methods in Arc Map 10.7.1. Training and Testing samples are divided in the ratio Of 70:30 respectively. Digitized training polygon

Table 1 LULC type description

Class Name	Description
Water bodies	Areas covered by water Eg: Lakes, Canal
Built-up areas	Area consisting of settlements, roads
Agricultural land	Area covered with crops
Barren land	Area without vegetation

data is converted to point data in R 4.0.5. Later point data is used for classification using RF, SVM, and Maximum likelihood classification methods. The models were operated in RStudio and ArcMap, and the final classified LULC maps are obtained. RF and SVM were trained using train models by tags such as 'rf' and 'SVM Linear3'. 'Random forests 'rf' train models are based on classification, Regression techniques. The tuning parameter is 'mtry,' which selects randomly, and similarly, the tuning parameters for SVM 'SVM Linear3' are cost (Cost) and Loss (Loss Function). The number of trees (n-tree) used is 500 and the mtry is four.

Table 2 Description of algorithms

Algorithms	Description
Support vector machines	A non-parametric technique that makes assumptions about the statistical nature of the data
Maximum likelihood classification	It is a one parametric classifier taking statistical data within the class distributions and for data distributed normally into account
Random forest	It is a non-parametric technique that requires two parameters n-tree and m-try termed as the base of the method

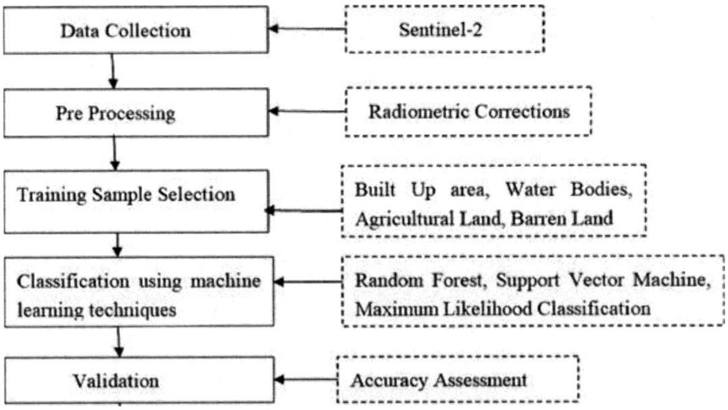

Fig. 2 Methodology flow chart

3 Results and Discussions

For the classification of the LULC of the agricultural watershed, machine learning techniques (SVM and RF) and one traditional classification method (MLC) are used. Maps are made for the months of April, January, and December 2020, as shown in Figs. 3, 4, and 5, respectively. Two machine learning methods that are incorporated into this study are RF and SVM. RF is one of the widely used non-parametric learning method that is based on a decision tree and Classification and Regression Techniques (CART). There are two crucial parameters in the RF method, the number of trees (n-tree) and several splits in each tree (m-try). The number of trees used in this algorithm is 500, while the split m-try is 4. Another machine learning technique used in this study is SVM, which is a non-parametric supervised classification method based on structural risk minimization. It can support linear and non-linear data from distinct classes. L2 regularized Support Vector Machine (dual) with a linear kernel based on regression and classification techniques is used in this model. Maximum likelihood classification is a traditional classification technique used in deriving LULC maps in the present study. Comparative analysis is performed between three techniques of classification based on accuracy assessment. The accuracy of each method used in classifying LULC has been assessed using the confusion matrix, such as Overall Accuracy (OA), Producer's Accuracy (PA), and User's Accuracy (UA). OA, PA, and UA were derived for the three techniques, as shown in Table 3. The Kappa coefficient measures the closeness of instances classified by the machine learning classifier to the ground truth data. R-square (R2) is one of the statistical methods demonstrating the proportion of the variance for a dependent variable by an independent variable or variables in a regression model. The average values of Kappa and R2 used to determine the accuracy of the model are presented in Table 5. From Table 3 UA of maps generated using SVM is displaying good results for all the classes when compared with MLC and RF. PA of built-up areas and water bodies is nearly the same for the three classification methods, while PA of agriculture and barren land is showing a better value for SVM classification when compared to MLC and RF as shown in Table 3. PA obtained from the perspective of the mapmaker (the producer) and UA obtained from the present study indicate SVM as a good classifier compared to MLC and RF. The overall accuracy values from Table 3 for SVM and RF are 87.5 ± 2.00 and 85.5 ± 2.00, respectively, while they are 70.87 ± 1.80 for MLC. OA of SVM and RF classification are nearby, indicating that both machine learning techniques are performing better when compared with MLC. The area obtained from each mapped class is shown in Table 4. From Table 5 the average Kappa coefficient values are 0.86, 0.67, and 0.85 for the classified maps using SVM, MLC, and RF methods, respectively. From the kappa coefficient, it can be inferred that LULC maps generated using machine learning techniques are outperforming MLC, a traditional classification method. From Table 5, the R2 value of 0.69 for the RF method is higher when compared to the SVM (0.67) and MLC (0.44). The results of the three methods employed show that the SVM classifier outperformed when compared to the RF and MLC classifiers.

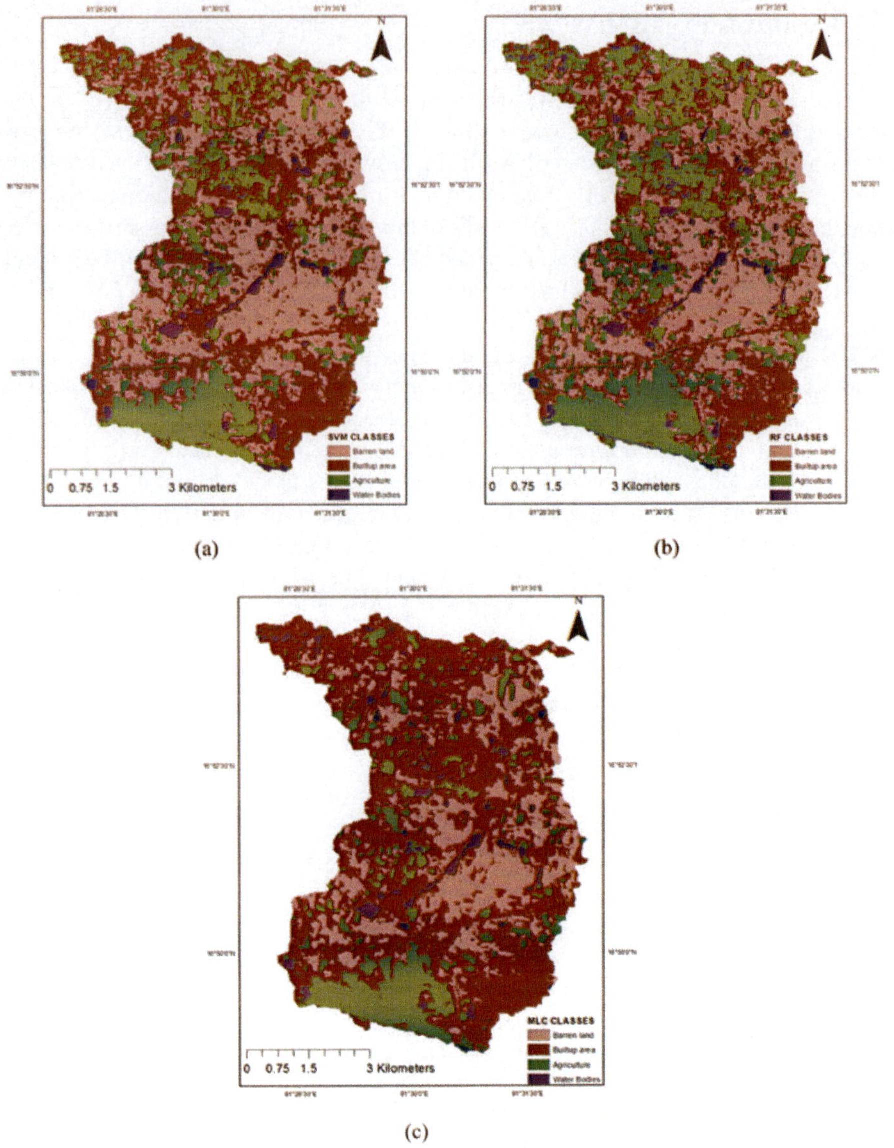

Fig. 3 Three LULC classified maps of the study area for **a** Support vector machines; **b** Random forests; **c** Maximum likelihood for April

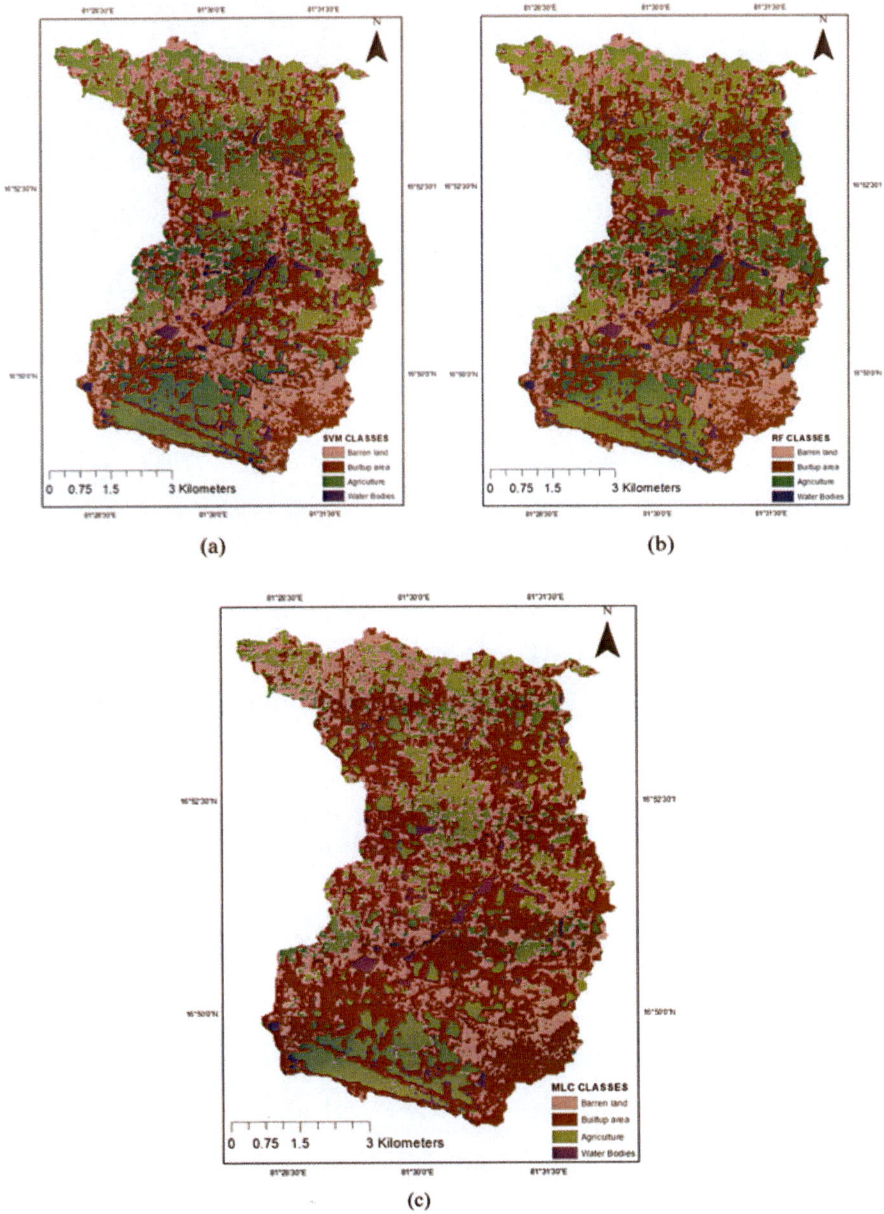

Fig. 4 Three LULC classified maps of the study area for **a** Support vector machines; **b** Random forests; **c** Maximum likelihood for January

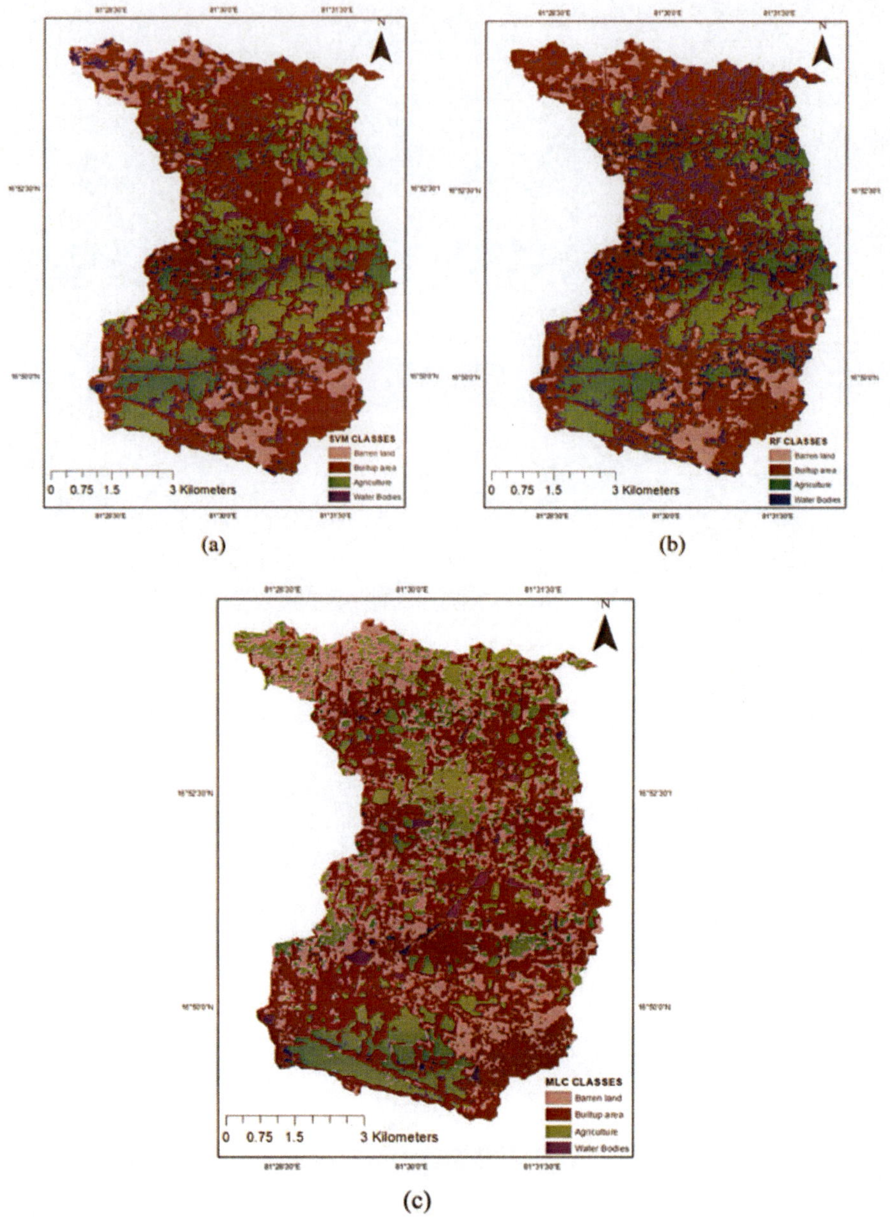

Fig. 5 Three LULC classified maps of the study area for **a** Support vector machines; **b** Random forests; **c** Maximum likelihood for December

Table 3 Average accuracy assessment for each class and method

Algorithm	LULC class	User's accuracy	Producer's accuracy	Overall accuracy (%)
Support vector machines	Built-up area	0.76 ± 0.02	0.96 ± 0.02	87.5 ± 2.00
	Water bodies	0.98 ± 0.01	0.88 ± 0.03	
	Agriculture	0.83 ± 0.04	0.89 ± 0.02	
	Barren land	0.85 ± 0.02	0.95 ± 0.02	
Maximum likelihood classification	Built-up area	0.55 ± 0.02	0.97 ± 0.02	70.87 ± 1.80
	Water bodies	0.96 ± 0.01	0.96 ± 0.03	
	Agriculture	0.78 ± 0.02	0.54 ± 0.04	
	Barren land	0.62 ± 0.03	0.62 ± 0.03	
Random forest	Built-up area	0.75 ± 0.01	0.96 ± 0.02	85.5 ± 2.00
	Water bodies	0.89 ± 0.02	0. 91 ± 0.03	
	Agriculture	0.77 ± 0.02	0.85 ± 0.02	
	Barren land	0.91 ± 0.01	0.72 ± 0.01	

Table 4 Individual and average areas of LULC classes

Algorithm	LULC class	Area in Sq. Km (January)	Area in Sq. Km (April)	Area in Sq. Km (December)	Average area in Sq. Km
Support vector machines	Built-up area	20.20	23.06	22.65	22.0
	Water bodies	2.34	1.50	4.7	2.8
	Agriculture	18.71	10.97	14.15	14.6
	Barren land	12.50	18.22	12.25	14.3
Maximum likelihood classification	Built-up area	20.20	20.04	27.06	22.43
	Water bodies	2.34	1.81	10.75	4.97
	Agriculture	18.71	15.56	10.07	14.78
	Barren land	12.51	16.34	5.88	11.58
Random forests	Built-up area	26.55	28.25	25.5	26.77
	Water bodies	1.4	1.17	2.45	1.67
	Agriculture	12.03	11.36	10.02	11.14
	Barren land	13.77	12.97	15.79	14.18

Table 5 Average Kappa and R square for different algorithms

Algorithm	Kappa statistics	R^2
Support vector machines	0.86	0.67
Maximum likelihood classification	0.67	0.44
Random forest	0.85	0.69

4 Conclusions

The present study improved the LULC monitoring for a small mill-scale agricultural watershed by developing a nearly accurate LULC map using machine learning algorithms in the R environment. In the present study, nearly accurate and high-resolution

LULC maps for a small agricultural watershed in Tadepalligudem, India was created using SVM and RF machine learning algorithms for three cropping seasons, kharif, rabi, and zaid for the year 2020. The objective of this research is to traverse two machine learning techniques and a traditional technique to classify LULC for a milli-scale agricultural watershed. From this study, it is observed that LULC maps obtained using SVM and RF exhibited an OA of 87.5 ± 2.00 and 85.5 ± 2.00 respectively while it is 70.87 ± 1.80 using the Maximum Likelihood classification method, indicating that SVM and RF classification were good classifiers considering OA compared to Maximum Likelihood Classification. Producer's Accuracy obtained in the view of the mapmaker (the producer) and Users Accuracy obtained from the present study indicate towards SVM and RF as good classifiers compared to Maximum Likelihood Classification. However, the average Kappa coefficient values of SVM and RF are 0.86 and 0.85 respectively. The R^2 values of SVM and RF are 0.65 and 0.69 respectively, indicating both RF and SVM are performing better than MLC. It is also observed that the area of each class varies under each classifying method. Built-up Area and Agriculture classes are shown to be exhibiting maximum variation in each classification method, while the water body class remained nearly the same. The results have shown that RF and SVM as qualitative classifiers, however, from this study based on accuracy assessment, SVM is found to be performing satisfactorily for the milli-scale agricultural watershed. From the study, machine learning algorithms have been shown to outperform MLC for the classification of LULC in small regions.

Acknowledgments This research is funded by the Department of Science & Technology-Science and Engineering Research Board (DST- SERB) sponsored project bearing the sanction number EEQ/2019/000463 undertaken in the Department of the Civil Engineering, NIT Andhra Pradesh titled Monitoring and Analysis of Non-Point Source (NPS) Pollutants in an Agricultural Watershed using Soil Water Assessment Tool (SWAT) under the Principal Investigation of Dr. Talari Reshma, Assistant Professor.

Funding This research is funded by the Department of Science & Technology-Science and Engineering Research Board (DST- SERB) sponsored project bearing the sanction number EEQ/2019/000463 undertaken in the Department of the Civil Engineering, NIT Andhra Pradesh titled Monitoring and Analysis of Non-Point Source (NPS) Pollutants in an Agricultural Watershed using Soil Water Assessment Tool (SWAT) under the Principal Investigation of Dr. Talari Reshma, Assistant Professor.

Conflicts of Interest "The authors declare no conflict of interest."

Data Availability Statement In the present research, Sentinel-2 Multispectral Instrument (MSI) with a spatial resolution of 20m was downloaded from the United States Geological Survey (USGS) website (https://earthexplorer.usgs.gov).

Code availability R code is available. Arc Map 10.7.1 Licensed version is used.

References

1. Kafy AA, Rahman ANMF et al (2021) Assessment and prediction of seasonal land surface temperature change using multi-temporal Landsat images and their impacts on agricultural yields in Rajshahi, Bangladesh. Environ Chall 4:100147. https://doi.org/10.1016/j.envc.2021.100147
2. Hashem N, Balakrishnan P (2015) Change analysis of land use/land cover and modelling urban growth in Greater Doha, Qatar. Ann GIS 21:233–247. https://doi.org/10.1080/19475683.2014.992369
3. Liou Y-A, Nguyen AK, Li M-H (2017) Assessing spatiotemporal eco-environmental vulnerability by landsat data. Ecol Ind 80:52–65. https://doi.org/10.1016/j.ecolind.2017.04.055
4. Nguyen K-A, Liou Y-A (2019) Mapping global eco-environment vulnerability due to human and nature disturbances. Methods X 6:862–875. https://doi.org/10.1016/j.mex.2019.03.023
5. Nguyen K-A, Liou Y-A (2019) Global mapping of eco-environmental vulnerability from human and nature disturbances. Sci Total Environ 664:995–1004. https://doi.org/10.1016/j.scitotenv.2019.01.407
6. Talukdar S, Pal S (2020) Wetland habitat vulnerability of lower Punarbhaba river basin of the uplifted Barind region of Indo-Bangladesh. Geocarto Int 35:857–886. https://doi.org/10.1080/10106049.2018.1533594
7. Koneti S, Sunkara SL, Roy PS (2018) Hydrological modeling with respect to impact of land-use and land-cover change on the runoff dynamics in Godavari river basin using the HEC-HMS model. ISPRS Int J Geo Inf 7:206. https://doi.org/10.3390/ijgi7060206
8. Majumder A, Kingra PK, Setia R et al (2020) Influence of land use/land cover changes on surface temperature and its effect on crop yield in different agro-climatic regions of Indian Punjab. Geocarto Int 35:663–686. https://doi.org/10.1080/10106049.2018.1520927
9. Chen X, Vierling L, Deering D (2005) A simple and effective radiometric correction method to improve landscape change detection across sensors and across time. Remote Sens Environ 98:63–79. https://doi.org/10.1016/j.rse.2005.05.021
10. Stefanov WL, Netzband M (2005) Assessment of ASTER land cover and MODIS NDVI data at multiple scales for ecological characterization of an arid urban center. Remote Sens Environ 99:31–43. https://doi.org/10.1016/j.rse.2005.04.024
11. Toure SI, Stow DA, Shih H et al (2018) Land cover and land use change analysis using multi-spatial resolution data and object-based image analysis. Remote Sens Environ 210:259–268. https://doi.org/10.1016/j.rse.2018.03.023
12. Usman M, Liedl R, Shahid MA, Abbas A (2015) Land use/land cover classification and its change detection using multi-temporal MODIS NDVI data. J Geogr Sci 25:1479–1506. https://doi.org/10.1007/s11442-015-1247-y
13. Halder A, Ghosh A, Ghosh S (2011) Supervised and unsupervised landuse map generation from remotely sensed images using ant based systems. Appl Soft Comput 11:5770–5781. https://doi.org/10.1016/j.asoc.2011.02.030
14. Wu L, Zhu X, Lawes R et al (2019) Comparison of machine learning algorithms for classification of LiDAR points for characterization of canola canopy structure. Int J Remote Sens 40:5973–5991. https://doi.org/10.1080/01431161.2019.1584929
15. Adam E, Mutanga O, Odindi J, Abdel-Rahman EM (2014) Land-use/cover classification in a heterogeneous coastal landscape using RapidEye imagery: evaluating the performance of random forest and support vector machines classifiers. Int J Remote Sens 35:3440–3458. https://doi.org/10.1080/01431161.2014.903435
16. Civco DL (1993) Artificial neural networks for land-cover classification and mapping. Int J Geogr Inf Syst 7:173–186. https://doi.org/10.1080/02693799308901949
17. Maxwell AE, Warner TA, Fang F (2018) Implementation of machine-learning classification in remote sensing: an applied review. Int J Remote Sens 39:2784–2817. https://doi.org/10.1080/01431161.2018.1433343

18. Teluguntla P, Thenkabail PS, Xiong J et al (2015) Global food security support analysis data (GFSAD) at nominal 1 km (GCAD) derived from remote sensing in support of food security in the twenty-first century: current achievements and future possibilities. In: Thenkabail PS (ed). CRC Press, pp 131–160
19. Zhang D, Pan Y, Zhang J et al (2020) A generalized approach based on convolutional neural networks for large area cropland mapping at very high resolution. Remote Sens Environ 247:111912. https://doi.org/10.1016/j.rse.2020.111912
20. Pal M (2005) Random forest classifier for remote sensing classification. Int J Remote Sens 26:217–222. https://doi.org/10.1080/01431160412331269698
21. Pal M, Mather PM (2005) Support vector machines for classification in remote sensing. Int J Remote Sens 26:1007–1011. https://doi.org/10.1080/01431160512331314083
22. Shih H, Stow DA, Tsai YH (2019) Guidance on and comparison of machine learning classifiers for Landsat-based land cover and land use mapping. Int J Remote Sens 40:1248–1274. https://doi.org/10.1080/01431161.2018.1524179
23. Camps-Valls G, Benediktsson JA, Bruzzone L, Chanussot J (2011) Introduction to the issue on advances in remote sensing image processing. IEEE J Sel Top Sign Process 5:365–369. https://doi.org/10.1109/JSTSP.2011.2142490
24. Dixit A, Agarwal S (2020) Super-resolution mapping of hyperspectral data using artificial neural network and wavelet. Remote Sens Appl Soc Environ 20:100374. https://doi.org/10.1016/j.rsase.2020.100374
25. Camargo FF, Sano EE, Almeida CM et al (2019) A comparative assessment of machine-learning techniques for land use and land cover classification of the brazilian tropical savanna using ALOS-2/PALSAR-2 polarimetric images. Remote Sens 11:1600. https://doi.org/10.3390/rs11131600
26. Jamali A (2019) Evaluation and comparison of eight machine learning models in land use/land cover mapping using Landsat 8 OLI: a case study of the northern region of Iran. SN Appl Sci 1:1448. https://doi.org/10.1007/s42452-019-1527-8
27. Rogan J, Franklin J, Stow D et al (2008) Mapping land-cover modifications over large areas: a comparison of machine learning algorithms. Remote Sens Environ 112:2272–2283. https://doi.org/10.1016/j.rse.2007.10.004

Photogrammetric Survey of an Intertidal Area: A Case Study in NW Spain

M. Gil-Docampo, S. Peña-Villasenín, S. Peraleda-Vázquez, R. Carballo, and N. Gómez-Conde

Abstract For years, bathymetry technique has been employed to obtain georeferenced data of the depths of the water using an echo sounder on board. This led to some problems in shallow water because it requires a shallow draft boat, but thanks to technological advances, in recent years, Unmanned Aerial Vehicles (UAVs) are used for a wide variety of applications, including bathymetry in these context. There are recent works that propose its use through multispectral data or photographic images. These methods allow much more complete and precise work to be carried out, since data collection could not be possible through traditional methodology. In the present work, a combination of the data obtained by photogrammetric survey with UAV is made for the zones that are characterized by a shallower depth and bathymetric survey with a low-cost sounder (fishing sounder) for the deeper zones. From the combination of both methods, the bathymetry of the intertidal zone of the Pontedeume estuary (Spain) is obtained. In this way, Digital Terrain Model (DTM) and bathymetry are performed, which provides complete information of the morphological characteristics created by sea currents.

Keywords Bathymetry · Digital Terrain Models (MDT) · Unmanned Aerial Vehicle (UAV) · Low-cost sounder

1 Introduction

At present, aerial photogrammetry using UAV (Unmanned Aerial Vehicle) has led to a reduction in costs mainly in areas of reduced extension, since the equipment used is cheaper than traditional methods. It means that it is available to a greater number of users and, besides obtains images with higher resolution than those taken traditionally using aerial photogrammetry. It is due to the possibility of performing low-altitude

M. Gil-Docampo (✉) · S. Peña-Villasenín · S. Peraleda-Vázquez · R. Carballo · N. Gómez-Conde
Agroforestry Engineering Department, University of Santiago de Compostela, Higher Polytechnic School of Engineering, 27002 Lugo, Spain
e-mail: ml.gil@usc.es

S. Mesapam et al. (eds.), *Developments and Applications of Geomatics*,
Lecture Notes in Civil Engineering 450, https://doi.org/10.1007/978-981-99-8568-5_29

flights and the possibility of correcting camera distortions. High precision is achieved, and consequently better results [3].

For decades, multibeam sonar systems, have been used to obtain the bathymetry of an area, sweeping the seabed transversely to the navigation path, which allows georeferenced depth data to be obtained from underwater topography of areas at low cost and high resolution [12].

With the appearance of UAVs, depth measurement methods have suffered a growing impact, providing certain alternative [1] by means of photogrammetry and remote sensing techniques that allow obtaining depth data for the generation of 3D models of the seabed [15]. Several works propose UAV bathymetry using hyperspectral data, which are increasingly capable of taking images of water and underwater areas [10], as well as multispectral data [13], or photographic images [5, 8]. Through Structure from Motion (SfM) photogrammetry, the geometric properties of a surface can be known and Digital Terrain Models (DTMs) can be obtained accurately [7]. In addition, ground control points (GCPs) are used to optimize the UAVs trajectory and the point cloud [4]. To obtain accurate results, the distribution, quantity and distance between them must be established [18]. It is known that a proper number of control points, provides more accurate the results [14].

Traditional bathymetry requires the use of specific vessels and equipment that entail a high cost and take a long time [20]. However, thanks to the use of UAVs, costs are reduced [13] since the use of vessels in shallow waters is not necessary. In addition, it allows obtaining hydrographic information directly and in real time. The use of UAVs facilitates surveying information on the depth water in a simple way to obtain data in shallow areas [9]. The multibeam echosounder provides high-precision, high-density, and high-efficiency deep water point clouds [11, 19]. However, they also have some drawbacks, such as the errors that are generated due to the refraction phenomenon of water waves [15].

The continuous search for renewable energies that are more competitive has given rise to the study of, among others, tidal energy. One of the ways to obtain this type of energy is through the morphological characteristics created by sea currents, which is why various studies are being carried out on the northwest coast of the Iberian Peninsula to obtain energy through these currents [2]. To carry out a study of the currents and determine the feasibility of taking advantage of tidal energy from them, it is necessary to know the morphology of the terrain. That is why this work is carried out in order to obtain the relief of the intertidal zone of an estuary. The objective is to determine the bathymetry of the area of the intertidal zone of the estuary by means of a photogrammetric survey with UAV in combination with the data obtained by means of a low-cost bathymetric system.

The work is carried out in the municipality of Pontedeume (Fig. 1), in the province of A Coruña, in northwestern Spain. The extension where the flight takes place (Fig. 2) has an area of approximately 114 hectares. The study area does not have large elevation variations, so it is characterized by being a flat area.

Fig. 1 Location of Pontedeume, Spain

Fig. 2 Study área

2 Material and Methods

2.1 *Equipment and Software Used*

Aircraft. The UAV used to carry out the flight in this work is the DJI Phantom 4 RTK (Fig. 3). It is characterized by a weight of 1,391 g, an RTK horizontal positioning accuracy of 1 cm + 1 ppm, RTK vertical positioning accuracy of 1.5 + 1 ppm and absolute horizontal accuracy in photogrammetric models. Achieves a maximum

Fig. 3 Phantom 4 RTK aircraft and control used to carry out the flight

speed of 6 m/s (automatic), 5 m/s (manual), maximum flight time of 30 min, high-sensitivity single-frequency GNSS and hover accuracy range with RTK enabled of ± 0.1 m vertical and horizontal.

Another feature is TimeSync, so that the photos use more precise meta-information and fixes the positioning data to the optical center of the lens, which optimizes the results of photogrammetric methods and the image achieves a centimetric level of precision. Such precision is also guaranteed through the D-RTK GNSS Mobile Station, which allows data to be obtained from the relative position of the drone in real time.

Since it has an OcuSync transmission system, it allows stable and reliable HD image transmission. It has a range of 7 km, which makes it very suitable for mapping large areas.

Aerial camera. The UAS incorporates an RGB camera (Fig. 4) with a 1" 20-megapixel CMOS sensor from the DJI brand. It has a high resolution, so ground pixel distance (GSD) of 2.74 cm at 100 m altitude is achieved. The camera features a field of view (FOV) 84 ° 8.8 mm/24 mm (35mm equivalent format), f/2.8–f/11, focus at 1m–∞; 4864 × 3648 (4:3) and 5472 × 3648 image size; GSD H/36.5 cm/pixel (where H is the height of the aircraft).

The taking of coordinates on the ground, which will be used as support points for the georeferencing of the images during their processing, will be carried out using a GPS of the Stonex brand and model S900T (Fig. 5). This equipment allows obtaining

Fig. 4 Camera with 1" Phantom RTK CMOS sensor

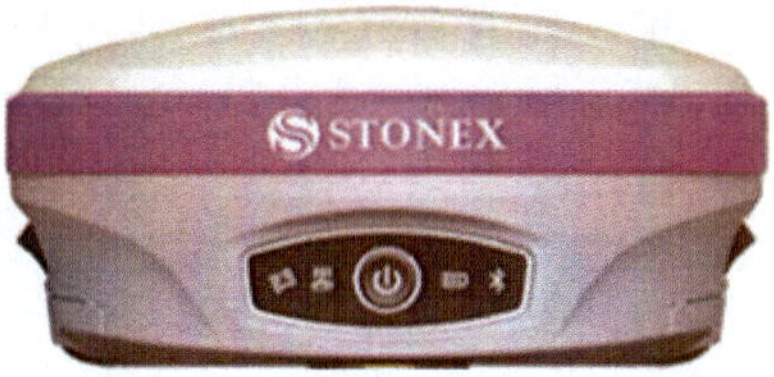

Fig. 5 GPS Stonex S900T used to take points

the positioning and elevation of a specific point precisely by means of the satellite signal.

Sonar. To carry out the bathymetry, a Deeper Smart Sonar CHIRP+2 sonar is used (Fig. 6). It is a low-cost probe used for fishing. It allows to design bathymetric maps, through its application (that receives the same name), thus obtaining the contour lines of the terrain. This equipment sends pulses of sound waves through the water. When the pulses reach the bottom, they reflect back to the surface. The sonar is responsible for determining the time that elapses since the wave goes down, reaches the bottom and bounces. In this way, the device determines the depth. As it receives the pulse, another one is sent. In the case of the Deeper CHIRP+2 sonar, it sends 15 pulses per second.

It is a 3-frequency CHIRP sonar—Narrow CHIRP 675 kHz (7°cone angle), Medium CHIRP 240 kHz (20° cone angle), Wide CHIRP 100 kHz (47° cone angle). Target separation: 1 cm Narrow Chirp; 2.4 cm Medium Chirp, 2.4 cm Wide Chirp. The depth range is 15 cm–100 m and a stable connection range of up to 120 m.

Flight planning software. The software used for flight planning allows planning of the mission to obtain the photogrammetric survey by UAV. In the case of this work, the application used is Groun Station (GS) PRO from DJI. In this application, the parameters of interest are introduced depending on the results that are intended to be obtained (longitudinal and transverse overlap, flight height…). The quality

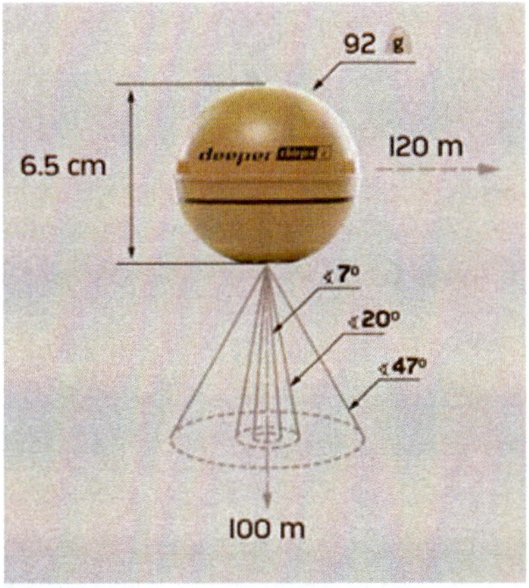

Fig. 6 Fishdeeper brand Sonar CHIRP+2

of the results obtained will depend on this. DJI GS RTK application has several flight planning modes: 2D and 3D photogrammetry, trajectory, constant height, area segmentation and such.

Image processing software using UAV. For the processing of all the images obtained during the flight, the Agisoft Metashape Professional photogrammetric software is used. This program is in charge of treating the images by means of image correlation algorithms, in such a way, it detects common points between them and calibration parameters are obtained. It allows obtaining high-resolution products. From it, the point cloud, the DTM, the high resolution orthomosaic, the contour lines and the 3D models are created.

Bathymetric data processing software using sonar. For the bathymetric survey by sonar, the Fish Deeper software provided by the sonar itself, has been used. This application is responsible for the synchronization of the data obtained and generates a contour map. It allows the export of the contour map to PDF format and the download of the coordinates in CSV format.

UAV data integration software—sonar. For the integration of the data obtained by bathymetric survey with sonar and the images obtained by photogrammetric survey with UAV, it is necessary to use a Geographic Information System (GIS). In the present work, it has been used Global Mapper.

2.2 *Methodology*

To carry out the work, the SfM methodology is applied (Fig. 7). This technique is based on the superposition of images to obtain a 3D model, as occurs with stereoscopic photogrammetry [17]. Unlike traditional photogrammetry, the SfM methodology automatically resolves the camera positions and orientation without the need to previously set control points of known 3D coordinates, being carried out through automatic correlation of a group of images with superimposition obtained in an unstructured way [6].

Flight planning. The phase prior to carrying out the flight is its planning. It is in the part of the work where the criteria that must be considered are established depending on the area to be flown, on which the quality of the results obtained depends. Mission planning should be based on the GSD needed and camera parameters such as focal length, sensor or pixel size.

Subsequently, the height at which the flight is determined, for which (1) formula is used:

$$\text{Flight Height (m)} = \frac{\text{GSD} \times \text{Focal Length}}{\text{Pixel Size}} \tag{1}$$

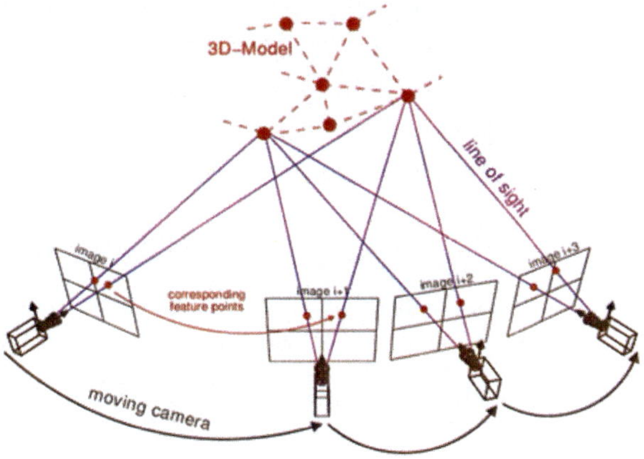

Fig. 7 Structure from motion photogrammetric principle [16]

On the other hand, the overlap of the images depends on the type of result that is intended to be obtained, and it should be a greater longitudinal and transversal overlap (>70%) in works where more precise results are needed. The application used performs the planning automatically depending on the parameters established in it.

Based on the above, a flight was made at a height of approximately 120 m above the study area, with a transverse overlap of approximately 70% and a longitudinal overlap of 84% (Figs. 8 and 9).

Regarding the flight height, the orography of the terrain does not present great unevenness, since it is a fairly flat terrain, so it has not been necessary to adapt the flight height according to the area to be flown, so it is expected a constant flight height of 120 m throughout its execution.

Fig. 8 Screenshot of flight planning in the DJI app

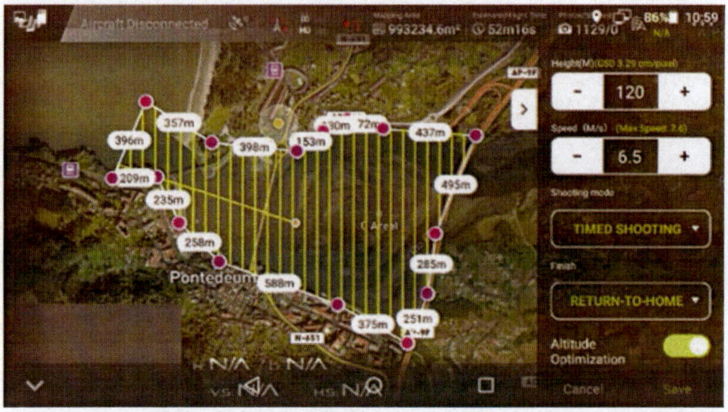

Fig. 9 The parameters selected in the app

GCPs allows to establish geographic coordinates with precision and checkpoints (CP) verify the final precision of results. The number of GCPs are established based on the surface to fly over.

Image processing. Images and GCPs are imported into the Agisoft Metashape software. From this program, an alignment of the photos is performed following the steps: *Dense point cloud,* that it is the first product obtained from digital photogrammetry. *Mesh*: which is made up of millions of points that are spatially arranged in a three-dimensional way; *Solid model of the study area,* for representation of the surface that is being studied. Each of the points that make up this grid includes the elevation of the terrain at that point Finally it is possible to analyze the *Confidence model* which allows knowing the reliability of the photogrammetry performed. The confidence of the model can be studied according to the scale that accompanies it, where the bluest colors reach the highest confidence percentages and the red areas, the least.

Bathymetric survey by sonar. To carry out sonar bathymetry, a kayak is used sweeping the seabed transversely to the navigation path. The sonar is placed in the kayak so the entire surface from the bottom data to be obtained is covered. In addition, the GPS has been used to be able to determine the coordinates of the surface of the sheet of water along the route, at the same time that the data from the probe was taken. It will allow to obtain the absolute coordinates of the terrain, by means of the subtraction of the model from the surface of the water sheet respect to the model of the bottom of the land generated from bathymetry. The result is the DTM of the bottom.

The Fish Deeper application generates PDFs and extracts the coordinates and depth of all points. The generated PDFs and coordinate files are loaded into the GIS software. In this way, the georeferencing of the PDFs takes place. The generation of PDFs was carried out by zones, to avoid loss of quality.

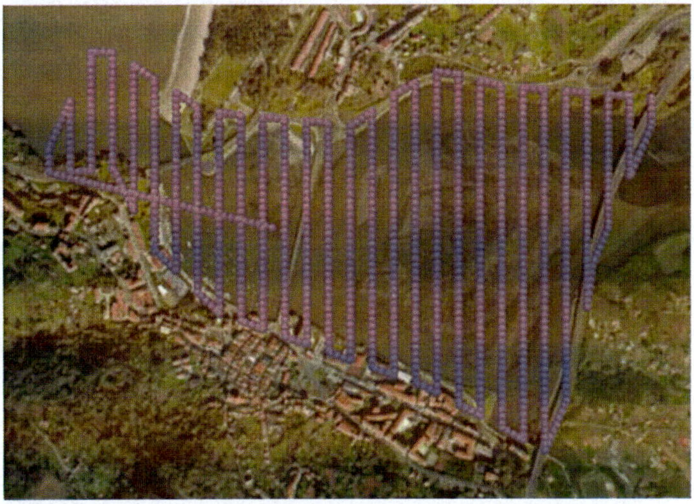

Fig. 10 Images taken over the area by flight planning

3 Results and Discussion

3.1 Flight Execution

Through the flight planning programmed in the application, the flight is executed. It is carried out during spring tides in the month of March, at the time of the lowest level to record the greatest amount of land through this method. In Fig. 10 can be seen that the arrangement of images, where each point represents an image. To cover the entire area with the aircraft, it was necessary to carry out two flights from different locations. The total number of photographs taken was 1095.

Likewise, after the execution of the flight, the GCPs and CPs on the ground were taken for greater accuracy. To do this, using GPS (Fig. 11), several points around the study area were taken, trying to distribute the points in the most balanced way possible, taking points on the margins and, to obtain points in the area central, some points were taken on the bridge located in the center.

3.2 Image Processing

The results of the point cloud (Fig. 12), solid model (Fig. 13), mesh (Fig. 14), DTM (Fig. 16), 3D model, Orthophoto (Fig. 17) and the Level Curves (Fig. 18) are obtained.

The visualization of the confidence model obtained by processing the images in the Agisoft Metashape can be also analyzed.

As expected, examining the confidence model in detail, in certain areas the UAV does not detect the bottom well, since it presents apparently more blurred zones in

Fig. 11 Taking control points

Fig. 12 Dense point cloud visualization

Fig. 13 Visualization of the solid model of the terrain in Agisoft Metashape

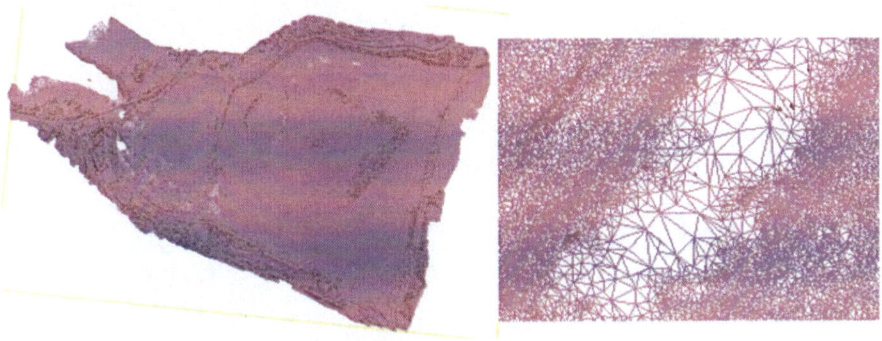

Fig. 14 On the left, visualization of the mesh in Metashape; on the right, mesh zoom

Fig. 15 Visualization of the confidence model in Agisoft Metashape

the model, whose information was complemented with the information obtained with sonar (Fig. 15).

It should be noted that in some of the border areas they have less reliability as we have been able to verify in the confidence model. As is known, photogrammetry is based on determining the coordinates of an element in a space by means of two photographs that are taken from different points of view, giving a 3D view to the model in overlap of these. The low confidence that occurs in the edges, is due to the fact that there is a low overlap, as the terrain of the borders are out of this study. So at the edges of the model, the image has greater distortion, due to the lack of overlap.

Subsequently, the DTM is obtained (Fig. 16). The model visually represents the different heights of a surface with respect to the mean sea level and allows characterizing the forms that its relief has, as well as the elements that make it up. The blue tones represent a lower altitude, reaching and, on the contrary, the areas of higher elevation, characterized mainly by greenish tones, reaching higher values in some areas. Where there are buildings, in the area on the left and, in the upper right area, coinciding with high-altitude trees (in red).

Fig. 16 Digital Terrain Model generated from the study area using Agisoft Metashape

When the conditions are optimal, that is, in calm waters, when there is no reflection of the sun on the water causing glare in the photographs and the wind does not cause waves that prevent the bottom from being seen, the photogrammetric survey with UAV works adequately up to a depth around 1 m.

An inconvenience that arises for taking data by means of UAV, is the roughness of the water caused by the wind, as well as the reflections caused by the incidence of the sun on the water. In such circumstances, it is not possible to obtain information, since the display of the bottom is not reached. It can be partially solved, by marking manually in the software, points in some areas and in the case of reflections caused by the incidence of the sun, by making slightly oblique flights to the north.

On the other hand, in the case of waves created by the wind in the water, a low-pass filter could be used, which reduces sharpness and minimizes the effect of the roughness of the water. This would allow, in some cases, to obtain more information on the bottom, because when they present stains, they would be more identifiable for the software used.

The orthophoto obtained (Fig. 17) provides quite complete information on the study area, since hardly any areas with missing data are observed, so it could be concluded that a result of good visual quality is obtained.

In Fig. 20 are the contour lines generated on the orthophoto determined from the photogrammetric survey by UAV (Fig. 18).

Fig. 17 Orthophoto generated by Agisoft Metashape software

Fig. 18 Level curves obtained from the Agisoft Metashape software

Fig. 19 Visualization of contour lines in Fish Deeper

3.3 Bathymetric Survey by Sonar

These data have been complemented with the information obtained by means of low-cost sonar bathymetry, since there are some areas in which a clear data cannot be obtained, fundamentally at the depth water.

From the data obtained through the bathymetric survey with sonar in the Fish Deeper application, the contour lines and the coordinates of the points measured with the sonar are obtained (Fig. 19).

From the data of both methods (photogrammetric and bathymetric survey), a joint and detailed model of the contour lines of the study area is obtained (Fig. 20). It presents a greater precision, since the areas (in which SFM photogrammetry was not able to obtain data from the bottom of the water by means of UAV), are complemented by those obtained by the sonar measurement.

Below (Fig. 20) are the contour lines generated on the orthophoto, where the topography of the study area can be seen, were determined from both the photogrammetric survey by UAV (Fig. 18) and the bathymetric survey by the sounder (Fig. 19). Making a comparison of the bathymetric curves result, it is observed that they visually coincide quite well with the orthophoto information in the overlap zone.

Fig. 20 Level curves on orthophoto

4 Conclusion

We can conclude that the data obtained are consistent with the shape of the river. In areas where the water is deep, it has not been possible to obtain data using only UAV, since it does not allow the texture of the bottom to be distinguished. This can be seen in the DTM, since the areas that do not present sharpness coincide with the images in which the texture of the terrain is not well distinguished. In these cases, sonar bathymetry is required to complete data in deeper areas.

The work performed joining UAV data and low-cost bathymetry surveying, allows obtaining a complete measurement of the area surveyed, which offers accurate results in areas of greater depth of water at low-cost.

Acknowledgements We thankfully acknowledge the financial support from Xunta de Galicia through the grant for the Consolidation and Structuring of Competitive Research Units within the Galician University System 2020-22 (ED431B 2020/25).

References

1. Aarnink JL (2017) Bathymetry mapping using drone imagery. Coastal Engineering MSc Thesis
2. Álvarez-Díaz J (2019) Tidal stream energy exploitation through marine spatial planning. Doctoral Thesis. Escola de Doutoramento Internacional. Programa de Doutoramento en

Enxeñería para o Desenvolvemento Rural e Civil (Lugo) 0169-2892. J Cult Heritage 38:195–203. https://doi.org/10.1016/j.culher.2019.01.001
3. Arza-García M, Gil-Docampo M, Ortiz-Sanz J (2019) A hybrid photogrammetry approach for archaeological sites: block alignment issues in a case study (the Roman camp of A Cidadela)
4. Brodie K, Bruder BL, Slocum RK, Spore NJ (2019) Simultaneous mapping of coastal topography and bathymetry from a lightweight multicamera UAS. IEEE Trans Geosci Remote Sens US Army
5. Chirayath V, Earle S (2016) Drones that see through waves—preliminary results from airborne fluid lensing for centimetre-scale aquatic conservation. Aquatic Conserv Mar Freshw Ecosyst 26(Suppl. 2):237–250. https://doi.org/10.1002/aqc.2654
6. Fonstad MA, Dietrich JT, Courville BC, Jensen JL, Carbonneau PE (2013) Topographic structure from motion: a new development in photogrammetric measurement. Earth Surf Proc Land 38(421–430):2013. https://doi.org/10.1002/esp.3366
7. Gil-Docampo M, Arza-García M, Ortiz-Sanz J, Martínez-Rodríguez S, Marcos-Robles J, Sánchez-Sastre L (2020) Above-ground biomass estimation of arable crops using UAV-based SfM photogrammetry. Geocarto Int 35(7):687–699. https://doi.org/10.1080/10106049.2018.1552322
8. Gonçalves H, Henriques R (2015) UAV photogrammetry for topographic monitoring of coastal areas. J Photogram Remote Sens 104:101–111
9. Hatanaka K, Toda M, Wada M (2007) Data analysis of a low-cost bathymetry system using fishing echo sounders. IEEE. ISBN: 0-92284-35-1. https://doi.org/10.1109/OCEANS.2007.4449181
10. Liu B, Liu Z, Men S, Li Y, Ding Z, He J, Zhao Z (2020) Underwater hyperspectral imaging technology and its applications for detecting and mapping the seafloor: a review. Sensors 20:4962. https://doi.org/10.3390/s20174962
11. Micallef A, Le Bas TP, Hubenne VA, Blondel P, Hühnerbach V, Deidun A (2012) A multimethod approach for benthic habitat mapping of shallow coastal áreas with high-resolution multibeam data. Cont Shelf Res 39(2012):14–26. https://doi.org/10.1016/j.csr.2012.03.008
12. Rodrigo C (2006) Seabed characterization and classification in Puerto Montt bay using multibeam bathymetry and backscatter data. Invest Mar Valparaíso 34(1):83–94. https://doi.org/10.4067/S0717-71782006000100007
13. Rossi L, Mammi I, Pelliccia F (2020) UAV-derived multispectral bathymetry. Remote Sens 12. https://doi.org/10.3390/rs12233897
14. Sanz-Ablanedo E, Chandler JH, Rodríguez-Pérez JR, Ordoñez C (2018) Accuracy of unmanned aerial vehicle (UAV) and SfM photogrammetry survey as a function of the number and location of ground control points used. Remote Sens 10(10):1606. https://doi.org/10.3390/rs10101606
15. Specht M, Wiśniewska M, Stateczny A, Specht C, Szostak B, Lewicka O, Stateczny M, Widźgowski S, Halicki A (2022) Analysis of methods for determining shallow waterbody dephts base don images taken by unmanned aerial vehicles. Sensors 22:1844. https://doi.org/10.3390/s22051844
16. Theia-sfm.org (2016) (Web Page) http://theia-sfm.org/sfm.html
17. Tomás R, Riquelme A, Cano M, Abellán A, Jordá L (2016) Structure from Motion (SfM): una técnica fotogramétrica de bajo coste para la caracterización y monitoreo de macizos rocosos. In: Conference: 10 Simposio Nacional de Ingeniería Geotécnica, vol 1 at: A Coruña, Spain
18. Villanueva JKS, Blanco AC (2019) Optimization of ground control point (GCP) configuration for unmanned aerial vehicle (UAV) survey using Structure-from-Motion (SfM). In: The international archives of the photogrammetry, remote sensing and spatial information sciences, vol XLII-4/W12, 2019 5th international conference on geoinformation science–geoadvances, Casablanca, Morocco, 10–11 Oct 2018
19. Wilson MF, O'Connell B, Brown C, Guinan JC, Grehan AJ (2007) Multiscale terrain análisis of multibeam bathymetry data for hábitat mapping on the continental slope. Mar Geodesy 30(1–2):3–35. https://doi.org/10.1080/01490410701295962
20. Zhang X, Ma Y, Zhang J (2020) Shallow water bathymetry base Don inherent optical properties using high spatial resolution multispectral imagery. Int Arch Photogramm Remote Sens Spatial Inf Sci XLII-2/W10:916. https://doi.org/10.5194/isprs-archives-XLII-2-W10-9-2019

Potential Zones Identification to Effectively Exploit Solar and Wind Energy in the State of Assam—A MCDA Approach Using GIS and Remote Sensing

P Taniya Raj and N. S. R. Prasad

Abstract With the frequent coal shortages, increased demand and rising global warming comes the need to switch to renewable sources of energy. Assam being rich in terms of Solar and Wind potential, the untapped energy needs to be harnessed in a sustainable manner. In this study the requirements for solar to produce maximum possible output will be understood and using the tools of GIS the best locations for solar and wind system installation will be found using the criterions for efficient harnessing of renewable energy. In order to accomplish this, 10 different criterions for Solar are considered that include Horizontal Irradiance, Average Annual Temperature, Distance from Power Transmission Lines, Distance from Residential Area, Elevation, Slope, Average Annual Cloud Cover, Humidity, Aspect. Similarly, for Wind 7 different criterions are considered which includes Wind Speed, distance from power transmission lines, Distance from Roads, Distance from Residential Area, Slope, Distance from Airports, Bird Sites. By performing weighted overlay analysis using AHP model, the suitable areas are defined and undesirable regions are excluded from suitability map for both Wind and Solar results.

Keywords MCDA-Multiple-Criteria Decision Analysis · AHP-Analytic Hierarchy Process

1 Introduction

The north-eastern India is facing cumulative peak time shortage of 4.1 with Assam and Manipur topping the list with deficits of 3.5 and 5.2 percentage respectively, the CEA (Central Electricity Authority) has found out in Load Generation Balance Report. This void can be filled with utilization of renewable energy. Therefore, it is

P. T. Raj (✉)
Central University of Karnataka, Kalaburagi, India
e-mail: 2017dgh06@cuk.ac.in

N. S. R. Prasad
NIRDPR, Hyderabad, India

S. Mesapam et al. (eds.), *Developments and Applications of Geomatics*,
Lecture Notes in Civil Engineering 450, https://doi.org/10.1007/978-981-99-8568-5_30

inevitable that the switch to renewable sources of energy is the need of the hour. The potential capacity is highest for Wind and Solar Energy in India, installing these green sources of energy generation will reduce the dependency on coal and keep power cuts at bay in Assam and other North Eastern States where there is shortage of electricity every year. Here comes the need of scientific assessment of the requirements for effective exploitation of the Solar and Wind Energy and overlaying the parameters to identify potential areas for installation of Solar and Wind energy units. GIS is an effective tool to identify such locations based on the importance that the user gives by classification, weightage analysis, area calculation, integrating the criterions and mapping.

India's renewable energy sources	Installed capacity (GW)	Estimated potential capacity (GW)
Wind	37.5	102
Solar	33.7	750
Biomass	9.9	25
Small hydropower	4.7	20

Source Central Electricity Authority (India), "Renewable Energy Sources," 2020; India Brand Equity Foundation, "Renewable Energy," 2017; and Ministry of New and Renewable Energy (India), "National Solar Mission Grid Connected Solar Rooftop Programme in India," 2019

2 Study Area

Assam is a state in the North east part of India lying south to the eastern Himalayas, and the state is drained by Brahmaputra and Barak rivers.

The state is experiencing a peak increase in demand of electricity, where the annual consumption for the year has been 1400 MW. According to official records, of the total 25,372 inhabited villages (as per the 2011 censes) in the state, 24,179 villages (95.2%) have been electrified so far. The renewable energy potential of the state is 14487, the largest share is of Solar followed by Biomass and Wind energy. Assam is rich in Solar Energy and Wind Energy, and this energy can be tapped to bring down electricity shortages and bring in electricity to the remote parts of Assam. This will bring down the dependence on thermal and import of coal (Fig. 1).

3 Literature Review

Similar study has been performed by Yassine Charabi a, Adel Gastli (1). The study identified suitability of solar farms using FLOWA Model for Oman. The FLOWA stands for Fuzzy Logic Ordered Weight Averaging module developed by Boroushaki and Malczewski. Of the total area 0.5% has been found suitable. The study considered

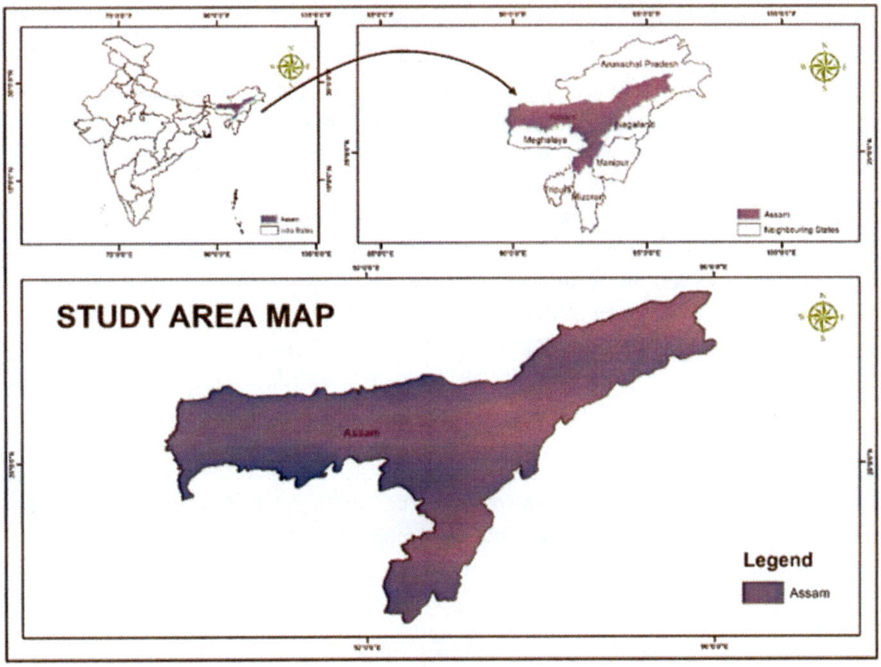

Fig. 1 Study area map

Sand and Sensitive areas too as criterions since installations can't be constructed in desert areas which is apt for countries like Oman which are dry. But the study neglects the temperature factor and the fact that with every 1 Degree the performance of PV cells declines by 0.5%.

Another study identified the suitability of solar mapping in South India. This study consisting of GHI, DEM, SLOPE, Land Use Land Cover, Protected Areas and Road and Railway Network which were taken as the required criterions. A non-suitability map was also created to restrict in the final suitability map. The credibility lies with the fact that the study has also taken into consideration the LULC features like waterbodies and forests and created a buffer to restrict. It was observed that a total of 46,500 sq.km of area is available for solar potential. Andhra Pradesh has the highest land availability for installation which is around 22,676 sq.km and Kerala has the least land available for solar potential zonation compared to the other states in the south India which is the study area of the work. The study appears as a much-generalised zonation where the potential of states is known in terms of solar output but the regional and specific potential in terms of availability of transmission lines, aspect, cloud cover, and other such important parameters are not considered.

4 Materials and Methods

Firstly, all the factors that are favourable and unfavourable have been identified. Data has been collected from reliable sources for the parameters that influence the Solar and Wind energy output. The collected data thereafter has been pre-processed based on requirement (Figs. 2 and 3).

Undesirable ranges have been restricted by creating a buffer zone. For both wind and solar studies, roads and settlement are restricted areas therefore, the buffer has been created up to 500m. Similarly, the waterbodies are restricted since construction is not feasible in such conditions. For the Solar Systems to work efficaciously the horizontal irradiance should not be less than 1300 kWh.m^{2}year^{-1} hence, areas having less than the required irradiation has also been restricted.

On the other hand, the Wind systems pose threat to bird life due to which it is desired that turbines are not located in the vicinity of the bird sites. Also, Wind turbines have impact on Radar Systems which affects the ATC services of Airports. Buffer has been created for both the parameters. All the restricted layers are thereafter overlaid and a common restriction map is prepared.

Thematic layers have been created for better understanding of range of individual factors along with a restriction Map. Classification has been performed along with labelling by assigning suitability based on the range. Ranges are decided based on literature.

The resolution for all layers has been kept at 30m to maintain accuracy. Maps have been thereafter reclassified using a Spatial Analysis tool based on the rated

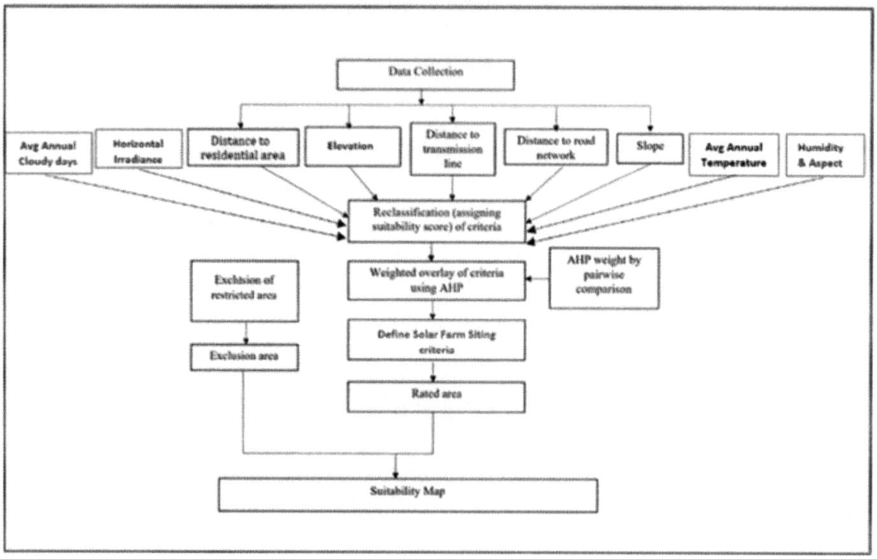

Fig. 2 Flow chart for site suitability for solar energy

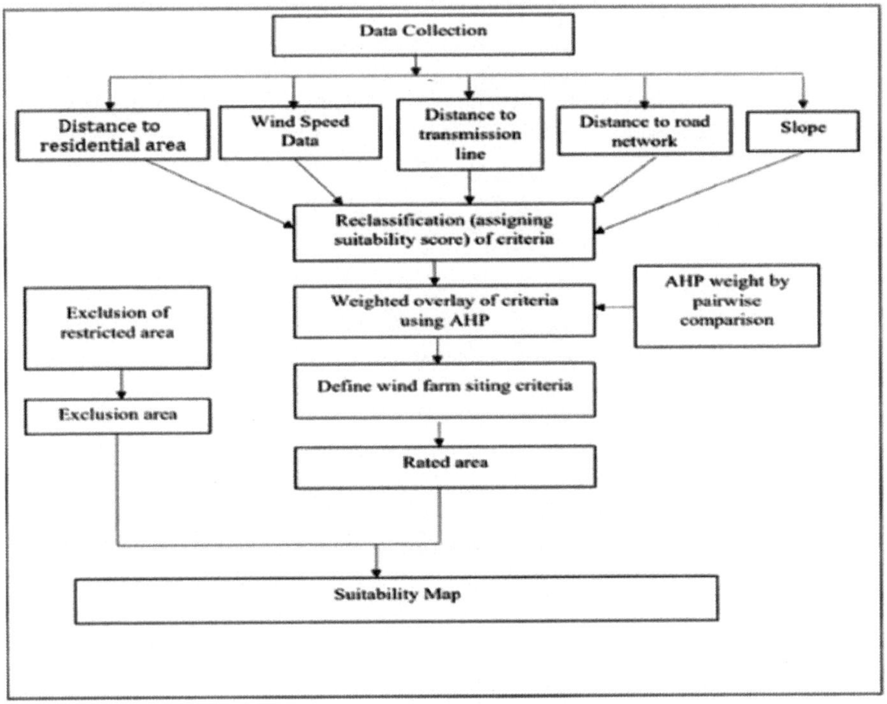

Fig. 3 Flow chart for site suitability for wind energy

value or importance for all maps. Before proceeding to Weighted Overlay, AHP has been used to assign weights to all the criterions. Verification of AHP has been done using CR. The CR being less than 10 percent is acceptable. Therefore, the weights obtained from AHP are used in Weighted Overlay.

4.1 Criteria for Solar Systems Installation

Horizontal Irradiance

The minimum required Horizontal Irradiance for the Solar Systems to produce optimum level of energy is 1300 $kW.m^{-2}.year^{-1}$ (3.5 $kWh.m^{-2}day^{-1}$). Assam receives an efficacious amount of radiation which is 1505 $kW.m^{-2}.year^{-1}$. Therefore, regions having radiation less than this are given least suitability. The Average Radiation received by the state is 4.29 $kWh.m^{-2}day^{-1}$.The highest horizontal irradiance observed in the data was 1723.98 and the lowest horizontal irradiance observed is 1028.91Therefore, areas having radiation less than 1300 $kWh.m^{-2}.year^{-1}$ are restricted.

Average Annual Temperature

Solar Panels need a temperature between 15 and 25 °C. The output power generation is highest at 25 °C. The performance of PV cells decline with increase in temperature above 25 °C. The amount generated decreases by 0.5% with increase in every 1 °C. Therefore, a temperature around 15 °C is given the least suitability and temperature ranging from 24.4 to 25.6 that is given the highest suitability.

Distance from Power Transmission Lines

Distance from Power Transmission Lines is an important criterion considering the costs associated with accessibility for equipment, network security, safety and repairs. Studies have shown that the distance of 0–5 Kms from PTL is best suited for installing Solar Systems and areas having distance greater than 20 Kms that have been given the least suitability for economic reasons and accessibility.

Distance from Major Roads

Transportation of equipment, construction and installation materials are fundamental to Solar Systems therefore, road networks are an important aspect. Construction of new roads, only for setting solar systems is very expensive and uneconomical. Therefore, accessibility to road network forms an important criterion to solar installation. Distance from roads less than 5 Kms is classified as most suitable and distances greater than 20 Kms as least suitable.

Distance from Residential Area

As distance from the residential Area increases, the power has to be transmitted to a larger distance, which in turn leads to loss of energy. Higher the distance, higher is the energy loss due to transmission. Therefore, distances less than 5km from the residential area is given the highest suitability. Moreover, distances greater than 20 km are given the least suitability. When transmitted to distant locations, there occur huge expenses and losses in electricity due to resistance.

Elevation

The Solar Radiation received is determined by the thickness of the atmosphere the radiation has to pass through. Radiation received at low elevated areas is lesser due to the fact that it has to pass through more thickness of the atmosphere. On the other hand, high elevated areas receive greater number of short waves as they need to pass through atmosphere of less thickness. In Assam, elevations greater than 400m are therefore classified as most suitable and areas less than 100m elevation are classified as the least suitable.

Slope

The performance of a solar collector is determined by the slope angle. The shadow cast of one panel of Solar would tend to compromise the energy produced, thereby affecting the output. Slopes less than 3% are given the highest priority, whereas the slopes greater than 15% are given the least suitability.

Average Annual Cloudy Days

Number of cloudy days affects the amount of insolation received and thereby the temperature of the area. The PV cells depend on radiation; therefore, the cloudy weathers decrease the probability. The output depends on the density of the cloud cover too. It should also be noted that the solar panels still function on cloudy days.

Humidity

Humidity is the concentration of water vapour present in the air. Humidity decreases the efficiency by 10–20% of the output (http://aip.scitation.org). Researchers have shown that the PV Module temperature and humidity are inversely proportional to the power generated by the solar panel. Tiny water droplets or water vapours can collect on the solar panels and they tend to reflect or refract the insolation away from the solar cells.

Areas having Relative Humidity less than 75% are classified as most suitable and areas having humidity greater than 78% are classified as highly suitable or most suitable.

Aspect

Aspect is another criterion that has been considered in the parameters. The output energy produced is maximum when the radiation reaches the panels perpendicularly.

Researchers have found that panels directly facing East or West produce 20 percent less electricity than those facing directly south. North is the most unproductive direction in terms of the solar output that can be received. South East and South West directions have fair productivity.

4.2 Criteria for Wind Farm Siting

Wind Speed (M/s)

Wind turbines should be located preferably in areas where the winds are strong for a longer duration. Even a small increase in the amount of velocity received highly affects the power generated. Wind Speeds greater than 4.6 m/s are given Extreme suitability, whereas areas having wind speed less than 1.6 m/s are given the least suitability. This parameter should be given the highest weightage as it is fundamental to site location.

Distance from Power Transmission Lines

Longer the distance between Power Transmission Lines and the turbines, greater are the energy losses and economic costs for cabling. Locating the farms closer to electricity grid makes the process of installation smoother and efficient, reducing the cabling costs at the same time reducing the power loses due to resistance.

Areas having a distance of less than 5 kms from the power transmission lines have been classified as extremely suitable and those having a distance greater than 20 km from the transmission lines is given the least suitability.

Distance from Roads

Setting up of Wind farms, their maintenances, transport of labour and resources, safety, all require a good road network. However, the road itself and the area in the vicinity i.e. around 500 m cannot be used for the construction of wind farms therefore, a buffer is created and this area would be restricted for suitability assessment. For the remaining areas, Euclidian distance has been performed where they are classified into 5 classes. Areas having a distance greater than 5 km is extremely suitable and those having distance greater than 20 km as least suitable.

Distance from Residential Area

Wind Farms are a source of noise and can cause disturbance to population, thus locating farms in the vicinity of the settlement is not desirable. A buffer has been created around the residential area and restricted in the map, so that suitability does not occur in the settlement areas. At the same time, the distance of wind farms should be at a reasonable distance so that the transmission losses can be minimized.

Slope

Slope is another important factor in turbine installation and working. Slopes have a hindering effect thereby affecting the output considerably. Moreover, construction on the slope increases the cost of installation. Areas with minimum slope is therefore desirable. Slopes having less than 5° are given highest suitability. On the other hand, areas having slope greater than 40° are given the least suitability.

Distance from Airports

Wind turbines pose a threat to low flying aircrafts. Also, the rotating blades of the wind turbines create turbulences downwards, such turbulences are hazardous to small and light aircrafts.

Plus, wind turbines can have an impact on communication, navigation and surveillance services like the radar functioning. The radar identifies the wind turbines as airplanes when the signals are tracked (Doppler effect). They impact the ability of ATC services.

Similar studies [1] have taken 2500 m from airports as restriction but based on some research papers and international standards [2] 40% of all US wind turbines exiss within 10 km of a small airports and about 5% exist within 5 km of a small airport (based on data from OurAirports, 2016, USGS, 2014). Therefore in this study an area of 5 km is used as restriction by creating the buffer.

Bird Sites

Studies on the impact of Wind turbines on birds have shown that an average number of collision fatalities in different European wind farms varies between a few birds per turbine per year to upto 64 birds per turbine per year [3].

In another study 139 bird species include 6 Near Threatened Species as per IUCN Red List in the Kutch region [4]. While some studies have ignored this parameter [1]. This study has restricted an area of 5 kms around the Bird sanctuaries by creating a buffer.

4.3 Analytic Hierarchy Process (AHP)

In our study Insolation is a parameter that has greater impact on solar output therefore, this parameter would be given high importance. But in comparison with other parameters what is the weight that it should be given and what is the weightage that must be given to other parameters? This is answered by the AHP model. It is obvious that all parameters do not hold equal importance. Certain criteria hold more importance than others. AHP helps assign weightage to individual criteria and organises the data in a structured way. To identify the suitability, it is necessary that we prioritise the parameters. AHP makes this process easier by asking the user the pair wise importance, and with this the relative importance is known (Tables 1 and 2).

Table 1 Classification and AHP weightages for solar site suitability

S.No	Criteria	Classification	Rated value	Remark	AHP weight (%)
1	Global horizontal irradiance (kWh/m^2)	<1505	1	Least suitable	18
		1505–1555	2	Less suitable	
		1555–1598	3	Moderately suitable	
		1598–1647	4	Suitable	
		>1647	5	Most suitable	
2	Average annual temperature (degree Celsius)	<19.6	1	Least suitable	12
		19.6–21.7	2	Less suitable	
		21.7–23.2	3	Moderately suitable	
		23.2–24.2	4	Suitable	
		>24.2	5	Most suitable	
3	Average annual relative humidity (%)	<75	5	Most suitable	8
		75–76	4	Suitable	
		76–77	3	Moderately suitable	
		77–78	2	Less suitable	
		>78	1	Least suitable	
4	Average annual cloud cover (%)	<33	5	Most suitable	8
		33–35	4	Suitable	
		35–38	3	Moderately suitable	
		38–40	2	Less suitable	
		>40	1	Least suitable	

(continued)

Table 1 (continued)

S.No	Criteria	Classification	Rated value	Remark	AHP weight (%)
5	Elevation (Meter)	>100	1	Least suitable	10
		100–200	2	Less suitable	
		200–300	3	Moderately suitable	
		300–400	4	Suitable	
		>400	5	Most suitable	
6	Slope (%)	<3	5	Most suitable	8
		3–5	4	Suitable	
		5–8	3	Moderately suitable	
		8–15	2	Less suitable	
		>15	1	Least suitable	
7	Aspect	S	5	Most suitable	8
		SE + SW	4	Suitable	
		E + W	3	Moderately suitable	
		NE + NW	2	Less suitable	
		N	1	Least suitable	
8	Distance from major roads (Meter)	<5000	5	Most suitable	12
		5000–10,000	4	Suitable	
		10,000–15,000	3	Moderately suitable	
		15,000–20,000	2	Less suitable	
		>20,000	1	Least suitable	
9	Distance from build-up area (Meter)	<5000	5	Most suitable	8
		5000–10,000	4	Suitable	
		10,000–15,000	3	Moderately suitable	
		15,000–20,000	2	Less suitable	
		>20,000	1	Least suitable	
10	Distance from PTL(Meter)	<5000	5	Most suitable	14
		5000–10,000	4	Suitable	
		10,000–15,000	3	Moderately suitable	
		15,000–20,000	2	Less suitable	
		>20,000	1	Least suitable	

Table 2 Classification and AHP weightages for wind site suitability

Criteria	Classification	Rated value	Remark	AHP weight (%)
Wind speed (m/s)	<1.6	1	Less suitable	40
	1.6–2.6	2	Moderately suitable	
	2.6–3.6	3	Suitable	
	3.6–4.6	4	Highly suitable	
	>4.6	5	Extremely suitable	
Slope (Degree)	<5	5	Extremely suitable	20
	5–10	4	Highly suitable	
	10–20	3	Suitable	
	20–40	2	Moderately suitable	
	>40	1	Less suitable	
Distance from roads (Meter)	<5000	5	Extremely suitable	15
	5000–10,000	4	Highly suitable	
	10,000–15,000	3	Suitable	
	15,000–20,000	2	Moderately suitable	
	>20,000	1	Less suitable	
Distance from residential (Meter)	<5000	2	Moderately suitable	10
	5000–10,000	5	Extremely suitable	
	10,000–15,000	4	Highly suitable	
	15,000–20,000	3	Suitable	
	>20,000	1	Less suitable	
Distance from PTL(Meter)	<5000	5	Extremely suitable	15
	5000–10,000	4	Highly suitable	
	10,000–15,000	3	Suitable	
	15,000–20,000	2	Moderately suitable	
	>20,000	1	Less suitable	

5 Results and Concluding Remarks

The study concludes that 1341.50 sq.km of the total area is the most suitable for Solar energy harnessing. 34,591.90 sq.km of the total is Suitable for Solar Potential Harnessing while 20,964.32 sq.km is Moderately Suitable. Karimganj, Silchar and Jorhat are the districts where Solar Harvesting suitability (Most Suitable) regions have been found.

For Wind Potential totally 666.73 sq.km of the area is extremely suitable and 35,888.08 sq.km of the area is highly suitable and. 21,290.54 sq.km is suitable. The suitability map resulted in suitability in fragments majorly over the North-western part of Assam. Extreme Suitability is found in the North western region of Assam and Northern part of Assam that include districts like Bongaigoan, Barpeta, Nalbari, Goalpara, Darrang, Northern part of Karbi-Anglong Region and other districts around them.

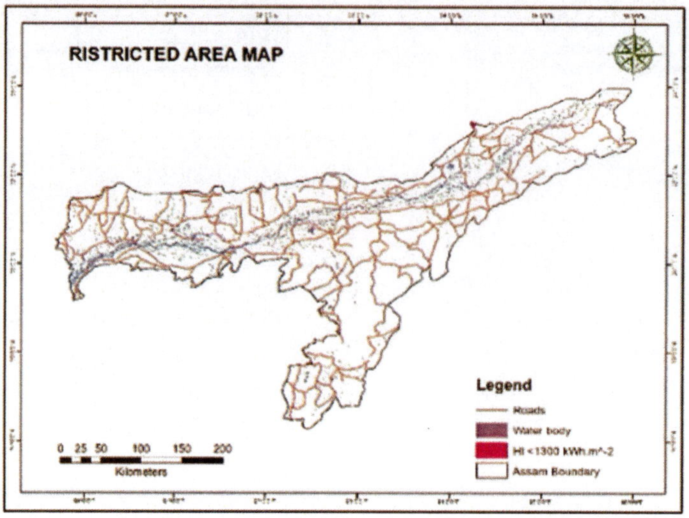

Fig. 4 Restricted area map for solar suitability

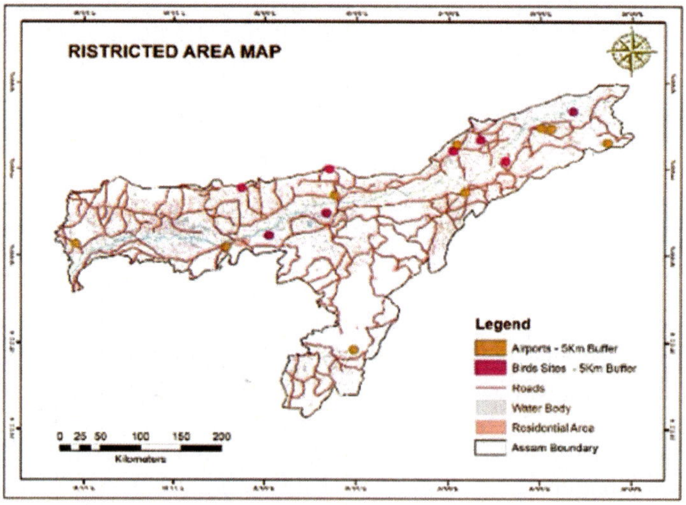

Fig. 5 Restricted area map for wind Suitability

North Eastern states especially Assam have hilly terrains making transmissions difficult, therefore medium scale installations in the remotest parts will avoid the need of long-distance transmission and energy losses. Bringing electricity to backward and tribal regions becomes an important aspect towards development.

The basic necessity to find the suitable sites is fulfilled by this study by considering all the factors that impact the productivity and output of the renewable systems. Giving highest weightages to the ones that have high priority is made possible using the GIS and AHP model.

With increased energy consumption and demand, finding an alternative source of energy is a necessity. Indiscriminate use of non-renewable sources of energy not only leads to exhaustion of the resource but also brings the world closer to global warming. While there are numerous villages and households in our country having no electricity there are also frequent fluctuations and electricity cuts.

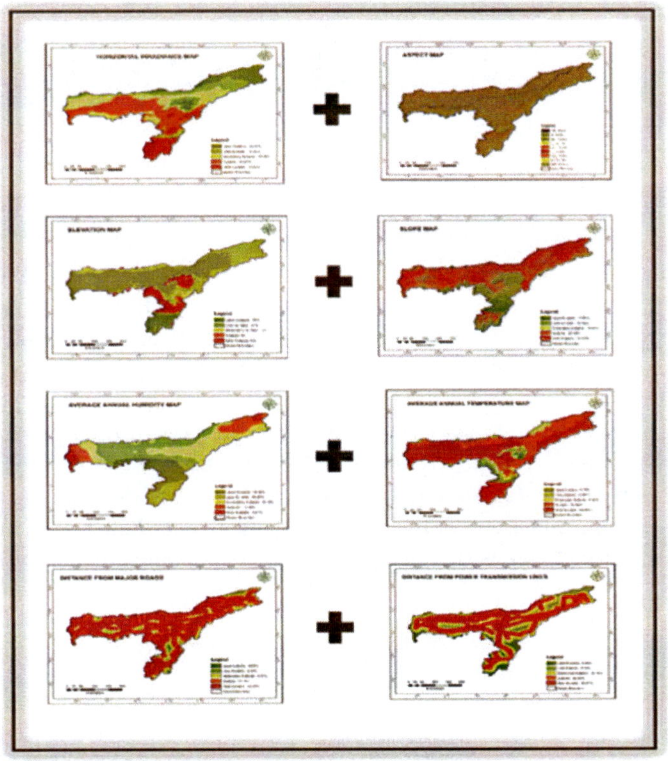

Fig. 6 Reclassified criteria maps for solar suitability

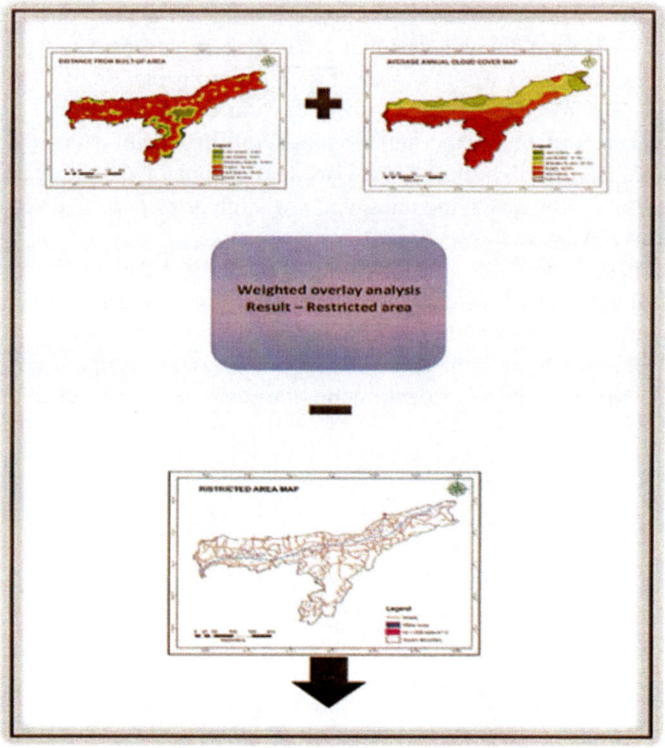

Fig. 7 Reclassified criteria maps for solar suitability

Large scale and medium scale solar plantation along with wind turbine installations would create a hybrid approach to mitigate the electricity crisis being faced by the country. This study gave a comprehensive understanding of the conditions that solar and wind energy sources demand for effective working at the same time ensuring the non-feasible areas are excluded from suitability.

The study is oriented at the MCDA integrated approach by taking into consideration the environmental, economic and technical limitations making it a multi-aspect study. The technical and economic limitations are also included making it desirable in terms of planning and execution. Therefore, this study can help the energy planners take decisions (Figs. 4, 5, 6, 7, 8, 9, 10, 11 and 12).

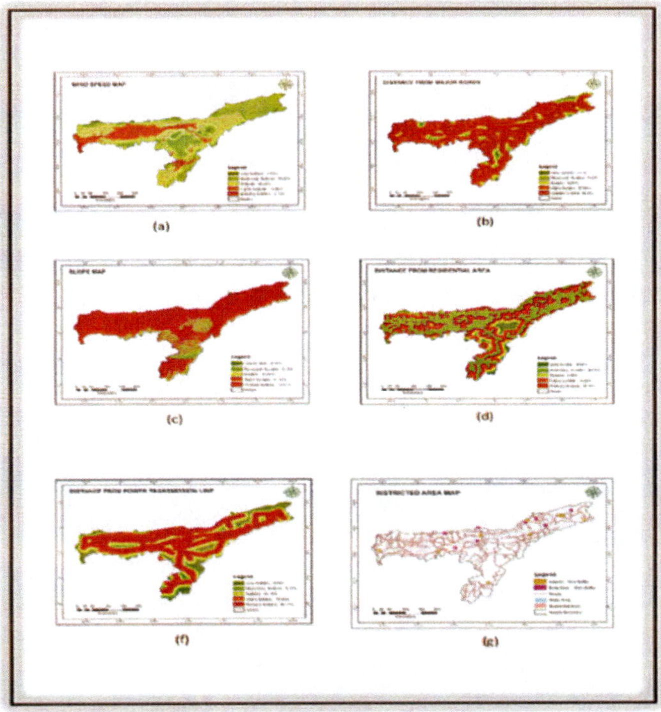

Fig. 8 Reclassified criteria maps for wind suitability

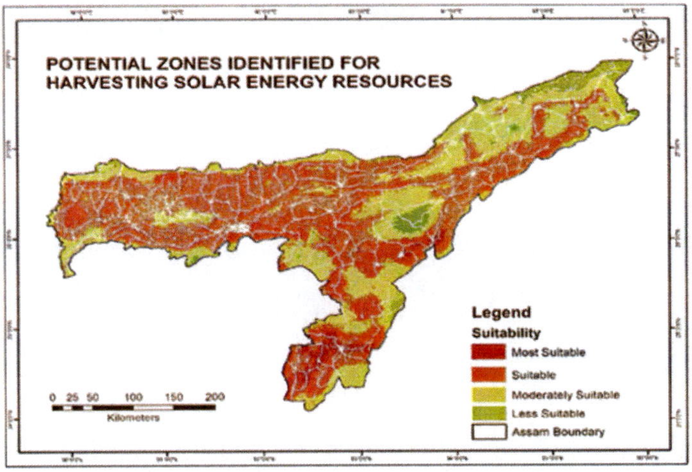

Fig. 9 Result: potential zones identified for solar site suitability in Assam

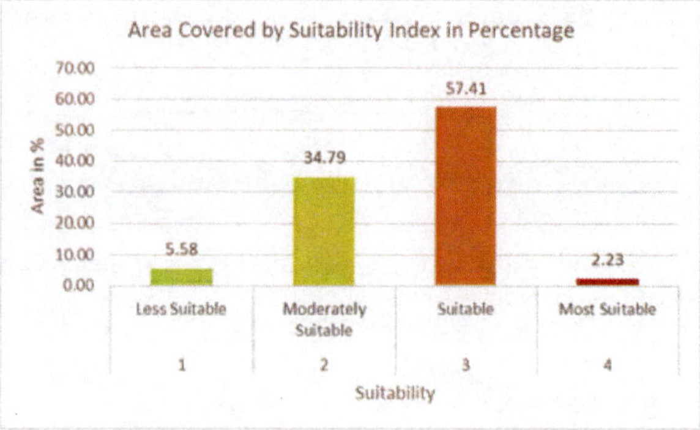

Fig. 10 Area covered by solar suitability classes in percentage

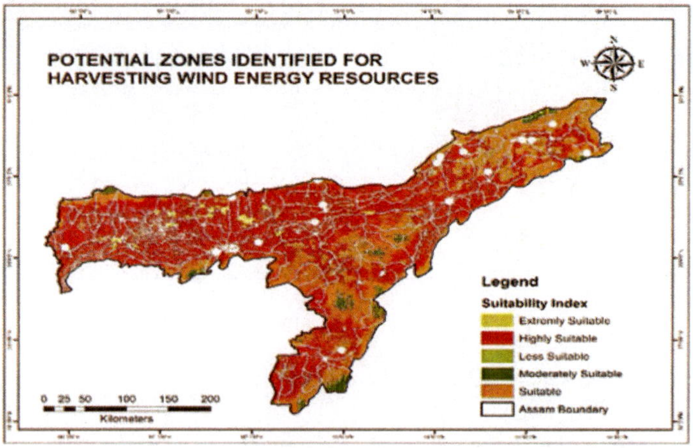

Fig. 11 Result: potential zones identified for wind site suitability in Assam

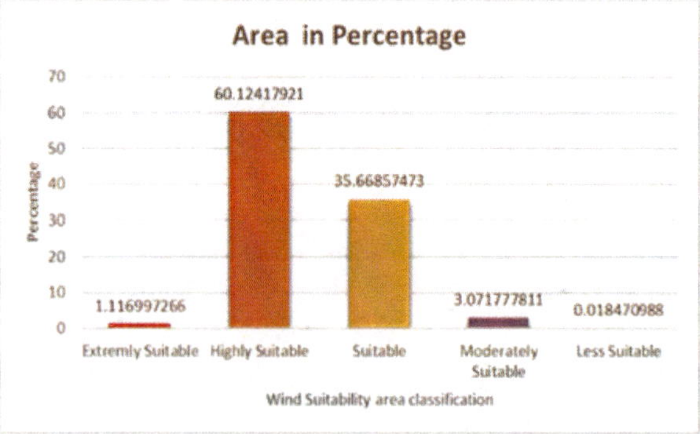

Fig. 12 Area covered by wind site suitability classes in percentage

References

1. Baseer MA, Rehman S, Meyer JP, Alam MdM. https://www.researchgate.net/publication/320328038_GIS-based_site_suitability_analysis_for_wind_farm_development_in_Saudi_Arabia
2. Tomaszewski JM, Moriarty JK. https://www.researchgate.net/publication/328707431_Do_wind_turbines_pose_roll_hazards_to_light_aircraft
3. Langston RHW, Pullan JD. https://tethys.pnnl.gov/sites/default/files/publications/Langston%20and%20Pullan%202003.pdf
4. Kumar S, Samsoor Ali MS, Hameed A, Arun PR. https://www.researchgate.net/publication/259390231_Impact_of_wind_turbines_on_birds_a_case_study_from_Gujarat_India

Prediction of Soil Organic Carbon in Unscientific Coal Mining Area Using Landsat Auxiliary Data

Naorem Janaki Singh, Lala I. P. Ray, Sanjay-Swami, and A. K. Singh

Abstract Unscientific coal mining is affecting soil attributes, deteriorating soil health and crop productivity, and is reflected by the soil organic carbon (SOC) content. The quantification of SOC is challenging with limited resource availability; however, satellite covariance is the alternate source of SOC determination with minimum labour requirement and limited laboratory facilities. An attempt is made to estimate the SOC using Landsat data and a model is developed by evaluating stepwise and enter/removal regression approaches. Fourteen predictor variables were used to build models and evaluate the prediction accuracy. Results showed that the SOC ranges 0.81–2.41% under unscientific coal mine affected sites; NDVI and BSI range 0.16–0.61 and −0.35–0.12 with mean 0.32 and −0.03, respectively. SOC is correlated with RI (r −0.33) and GRVI (r 0.34). The enter (all variables in a block enter in a single step) approach linear regression model (Model 3) using multiple variables can explain only 43% SOC (RMSE 0.66 and R^2 0.43). The stepwise linear regression model (Model 2) and Model 3 predicted the SOC as 6.94 and 3.78% higher than the actual SOC data. The model performance is increased by using multiple variables which may subside the less number of soil samples.

Keywords Unscientific coal mine · Soil organic carbon · Indices

1 Introduction

India is the fifth largest coal reserve and fourth largest coal producer (730.87 million tons in 2019–2020) in the world [1]. The coal mining in Meghalaya, NE India, was started on a small scale in the nineteenth century during the British period. The active mining was begun in the mid-1970s. Large-scale unscientific coal mining (such as rat-hole or open mining) was begun since 1980s in Jaintia hills without the permission of environmental clearance and safety measures of mining workers [2]. The Coal Mining Act, 1973, was not followed in Meghalaya due to the 6th schedule of

---

N. J. Singh (✉) · L. I. P. Ray · Sanjay-Swami · A. K. Singh
College of PG Studies in Agricultural Sciences, CAU (Imphal), Umiam, Meghalaya 793103, India
e-mail: naorem2005@mail.com

S. Mesapam et al. (eds.), *Developments and Applications of Geomatics*,
Lecture Notes in Civil Engineering 450, https://doi.org/10.1007/978-981-99-8568-5_31

Constitution of India (state where the landowners were the owners of the minerals below their land) and Article 371 (which allowed state government to formulate its own coal policy in order to recognize customary tribal laws). The increased mining activity from 3.26% in 1975 to 10.75% in 2001 caused a decease in the forest cover from 22.5% in 1975 to 12.34% in 2001 and also crop area from 2.65% in 1975 to 1.62% in 2001 [3]. The mine spoils such as gravels, rocks, sand, and soil were piled and dumped causing damage to the surrounding vegetation, every million ton of coal extraction could damage about 4 ha of surface area [4]. The open cast mining extracted about 250 million tons and about 100 million tons of mine spoil are overburden the nearby land as a dumping site. Subsequently, the acid mine drain from such dumping site affect the soil properties of agricultural field [5, 6]. Most of the agricultural lands were turned into the meadow (grass land). The quantification of key component of soil health—soil organic carbon (SOC) is urgently necessary for alternative suggestions for improving the livelihood of the society. Many authors attempted to study soil organic carbon (SOC) using equations derived from soil properties and related environmental variables derived from remote sensing and geographical database [7]. Mulder et al. [8] digitally mapped the SOC using the spectral signature of soil which is related to soil colour depending upon humic acid content and its composition [9]. Digital mapping covers larger areas and also overcome the limitations of laboratory facilities and intensive sampling with advantage of satellite data [10]. The spectral mainly used for SOC prediction are 450, 590 and 664 nm at a visible range, and 1600–1900 nm, 2100 nm and 2300 nm short-wave infrared (SWIR) [11, 12]. Various vegetation indices were derived to study the SOC [13, 14]; selection of the most suitable vegetation index in SOC prediction is a major challenge. In this study, it is aimed to assess the SOC accumulation pattern using various indices derived from Landsat data with objectives (i) to find out the model promising predictor variables and (ii) to construct the suitable regression model for mapping the SOC. The high-resolution data has limitation over the coverage of areas. The open cast and rat-hole coal mining is carried out in wide scattering throughout the Jaintia Hills district, Meghalaya. The multiple vegetation indices could be derived from the multi-spectral sensor data such as blue, green, red, near infrared (NIR), short-wave infrared (SWIR), and thermal band. These indices are the input variables (independent variable) for prediction of dependent variable soil organic carbon (SOC) in the statistical approach. A medium resolution, large area coverage with multiple spectral band satellite data could be the best option for achieving such study. The constraint in the ground sample collection may be partially overcome with the help of space data which is an alternate source of soil attribute analysis for the well-being of farmer community.

2 Material and Methodology

2.1 Study Area

Jaintia hills covers an area of 3,819 km^2 and lies between longitude 91°51'–92°45' E and latitude 25°5'–25°4' N with elevation 1627 m above sea level. The average annual rainfall is 1326 mm and temperature ranged from 7.8 to 24.5 °C. About 40 million tons of coal (thin layer (30–212 cm), sub-bituminous type with high sulphur content (2.7–5.0%), low ash content (1.3–24.7%), moisture content (0.4–9.2%), high volatile matter and high calorific value (5,694–8230 kilo calories/kg)) is reserved [2, 15]. The agricultural fields were turned into the meadow with the effect of mining activity, and crops in the lowland area are limited to paddy. NBBSLUP (1996) published soil map (1:500,000), physiographic map (1:1600, 000), agro-ecological region (1:1600, 000), and geology (1:1600, 000) are available for this area.

2.2 Soil Sampling and Process

The random ten (10) soil samples were collected at 0–20 cm from each soil mapping unit (SMU) occurring in a common area obtained after intersecting soil properties, climate, geology and physiography [16] in QGIS 2.18 on the basis of purposive sampling method [17] and the SCORPAN model [18]. The coordinate of each sample point was recorded using a Global Positioning System (GPS) with an accuracy of ±3 m. One composite sample was made from each ten random samples and 114 total composite samples are collected (Figs. 1 and 2). The average of ten coordinates represented the location coordinate of a composite sample. The soil samples were air-dried in the laboratory and sieved through 1 mm for soil organic carbon analysis by wet oxidized method (Walkley and Black, 1934). The SCORPAN model [18] is giving below

$$Sa = f(s, c, o, r, p, a, n)$$

where Sa is soil properties, f is the relationship function between soil conditions and the covariates, s is the information of soil, c is the climate at the site, o is an organism (vegetation, fauna or human activities), r is topography, p is parent materials, a is age and n is space.

The sampling sites occurred in two climatic zones (i.e. warm prehumid thermic and warm prehumid hyperthermic) and two physiographic units (i.e. intermediate plateau and lower plateau) and geology (Jaintia series and Disang series (ultrabasic in deep shades)). The SMU of sampling sites are described below

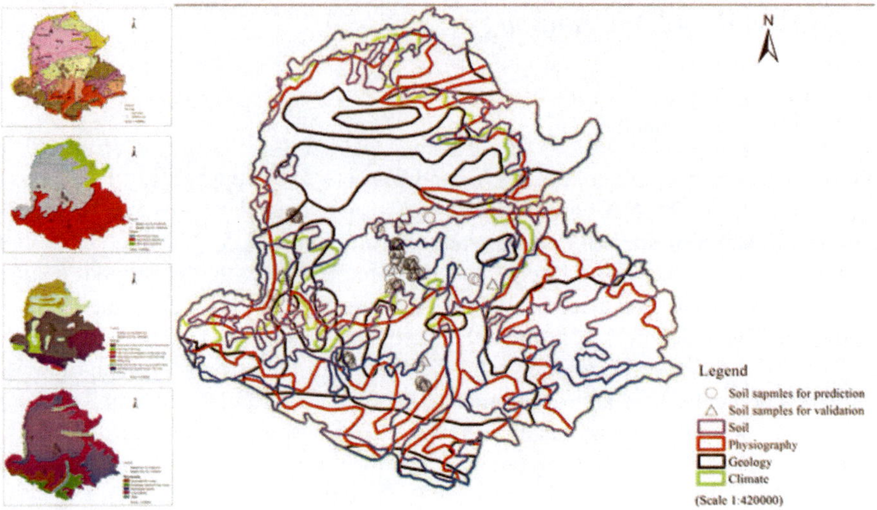

Fig. 1 The soil sampling sites from the common area obtained after intersection of soil map, climate, geology, and physiography. (*Source* https://megsoil.gov.in/images/meghalaya_soil_map.jpg)

2.2.1 Soils of Upper Plateau

SMU 01—Soils on moderately sloping are loamy texture, moderate erosion and deep and excessively drained, which is associated with coarse loamy, moderately deep and excessive drained, very severe erosion and strong stoniness on gently sloping. The taxonomy is *Typic Kandiudults* (covering 174900 ha).

SMU 02—Soils on gently sloping are loamy texture, moderate erosion, deep and excessively drained which is associated with loamy soil, very slight erosion, ground water level <1 m and deep and poorly drained on very gently sloping valleys, taxonomy is *Typic Haplohumults* (covering 49300 ha).

SMU 08—Soils on moderately steep are the loamy skeletal soils (sandy), moderately deep and excessively drained, very severe erosion and strong stoniness., The taxonomy is *Typic Dystrochrepts* (covering 22400 ha).

2.2.2 Soils of Hills of Middle and Lower Plateaus (Sub-Montane)

SMU 19—on steeply slope, loamy surface, very severe erosion, moderately deep and excessively drained, and strong stoniness; taxonomy is *Typic Dystrochrepts* (covering 65000 ha).

SMU 22—on moderately steep slope, loamy surface, moderate erosion, moderately deep, excessively drained, coarse loamy soils and strong stoniness; taxonomy is *Typic Dystrochrepts* (covering 25300 ha).

2.2.3 Satellite Image Acquisition

The satellite acquisition date (27th December 2015) was nearly synchronized with soil sampling date. Landsat 8-9 Operational Land Imager (OLI)—Thermal Infrared Sensor (TIRS) Collection 2 Level 2 (L2) was downloaded from the U.S. Geological Survey website (https://earthexplorer.usgs.gov). The coordinate system of data is WGS 1984 UTM Zone 46N, with spatial resolution of 30 m and 190 km swath. The spectral bands are B1 (435-451 nm) (Coastal Aerosol), B2 (452-512) (Blue), B3 (533-590) (green), B4 (636-673) (red), B5 (851-879) (NIR), B6 (1566-1651) (SWIR 1), B7 (2107-2294) (SWIR 2) and B10 (10600-11190) (Thermal Infrared Sensor (TIRS) 1). The DN values were converted to Top of Atmosphere Reflectance using the equation [19]:

$$\rho_\lambda = (\mathrm{M}\rho * \mathrm{Qcal} + \mathrm{A}\rho)/\sin(\theta_{\mathrm{SE}})$$

where ρ_λ is TOA spectral reflectance, $M\rho$ is reflectance multiplicative scaling factor for the band, $A\rho$ is reflectance additive scaling factor for the band, Qcal is pixel value in DN and θ_{SE} is solar elevation angle.

The fourteen indices were derived from the satellite data (Table 1).

2.2.4 Statistical Analysis

The statistical analysis was initiated to examine the nature, direction and strength of the association SOC and satellite covariance using the Microsoft Excel and SPSS. The Pearson correlation between the SOC and indices were tested for significance at 1 and 5% level of significance [32]. The SOC prediction models were built using the stepwise linear regression (SLR) and enter/removal regression approach. The prediction model is selected based on highest R^2 and lowest RMSE. The SLR model is expressed as

$$Z(x_0) = \beta_0 + \sum_{i=1}^{n} \beta_i * p_i + \varepsilon$$

where $Z(x_0)$ is the estimated SOC, β_0 is a constant term, β_i represents the regression coefficients, p_i denotes the independent variables and ε indicates the residuals.

The prediction is confirmed with a comparison between the actual SOC of validation points and corresponding predicted SOC.

Table 1 Indices used for soil organic carbon (SOC) prediction

Sl. no	Index	Definition	Details	Reference
1	NDVI	NIR−Red/NIR + Red	B5−B4/B5 + B4	Rouse et al. [20]
2	G ratio	NIR/Green	B5/B3	Buschmann and Nagel Tucker [21]
3	GRVI	Green−Red/Green + Red	B3−B4/B3 + B4	Tucker [22]
4	GDVI	NIR-G	B5−B3	Tucker [22]
5	GNDVI	(NIR−Green)/(NIR + Green)	B5−B3/B5 + B3	Gitelson et al. [23]
6	RI	$Red^2/Green^3$	$B4^2/B3^3$	[24]
7	SAVI	(1 + L)*(NIR−Red)/(NIR + Red + L) L = 0.5	(1 + L)*(B5−B4)/(B5 + B4 + L)	Huete [25]
8	MGSAVI	0.5*[(2*NIR + 1)−√(2*NIR + 1)^2−8*(NIR−G)]	0.5*[(2B5 + 1)−√(2*B5 + 1)^2–8*(B5−B3)]	Qi et al. [26]
9	GSAVI	1.5*[(NIR−G)/(NIR + G + 0.5)]	1.5*(B5−B3)/(B5 + B3 + 0.5)	Sripada et al. [27]
10	MSAVI	(1 + L)(NIR−Red)/(NIR−Red + L) L = 1−2*V* NDVI* WDVI V = NIR/Red	(1 + L)*(B5−B4)/(B5−B4 + L)	Qi et al. [26]
11	MSAVI2	[2*NIR + 1−√{(2*NIR + 1)^2−8 *(NIR-Red)}]/2	[2*B5 + 1−√{(2*B5 + 1)^2−8*(B4−B3)}]/2	Qi et al. [28]
12	BSI	(SWIR2 + R)−(NIR + B) (SWIR2 + R) + (NIR + B)	(B7 + B4)−(B5 + B2) (B7 + B4) + (B5 + B2)	Diek et al. [29]
13	WDVI	NIR + V*Red V = NIR/Red	B5 + V*B4	Clevers [30]
14	GWDRVI	(a*NIR−G)/(a*NIR + G) a = 0.12	(0.12*B5−B3)/(0.12*B5 + B3)	Gitelson [31]

3 Result and Discussion

3.1 General Statistics

The statistical descriptions of SOC and satellite covariance (predictors) were presented in Table 2. The SOC varied from 0.81-2.41% with mean 1.05±0.28% (mean ± standard deviation). The satellite covariance such as GWDRVI, MSAVI2, MGSAVI, MSAVI, WDVI, GSAVI, RI, GNDVI, GDVI, GRVI, G ratio, BSI, SAVI and NDVI varied from −0.7 to −0.39, 0.08 to 0.5, −0.49 to −0.05, −1.97 to 3.13, 0.35 to 0.97, 0.11 to 0.47, 0.76 to 1.52, 0.2 to 0.57, 0.06 to 0.35, −0.1 to 0.07, 1.5 to 3.63, −0.35 to 0.12, 0.13 to 0.64 and 0.16 to 0.61, respectively. The similarity between mean and median indicate the normal distribution (Table 2). The nature of the normal distribution curve was explained by the skewness and kurtosis. The SOC and covariance except RI and BSI were found to have positive skewness which were <3 acceptable range of normal distribution data in soil properties. The skewness values <3 indicates that the small SOC and covariance values were more concentrated. The SOC distribution is skewed due to the influence of some larger skewed SOC values. Hence, median is more representative than the mean of such skewed SOC data. Kurtosis was the characteristic of distribution peak corresponding to the average value in the probability density distribution curve. The kurtosis values were >0, and the peak value of probability density distribution curve was bigger than that of the normal distribution.

The NDVI and Bare Soil Index (BSI) estimated from DN values ranged −0.21–0.99 and −0.32–0.51 and represented a more reliable result for this soil (figure 2). The unscientific coal mining activity was concentrated on the low values of NDVI and BSI. The highest values of NDVI and BSI were found in the forest area. The figure clearly shows that the open cast and rat-hole coal mining was affected by the crop or vegetation (NDVI −0.21) and high soil bareness (BSI 0.51). This might be attributed to improper management of acid mine drains to low-lying areas and dumping of mine soils (Fig. 3).

Table 2 Statistical description (DN converted to reflectance)

	Range	Minimum	Maximum	Mean	Median	Std. deviation	Skewness	Kurtosis
SOC	1.60	0.81	2.41	1.05	1.00	0.28	2.77	8.06
GWDRVI	0.30	−0.70	−0.39	−0.58	−0.58	0.05	1.25	3.04
MSAVI2	0.42	0.08	0.50	0.22	0.21	0.07	1.51	3.54
MGSAVI	0.44	−0.49	−0.05	−0.32	−0.32	0.08	1.00	2.23
MSAVI	5.10	−1.97	3.13	0.48	0.40	0.61	0.21	8.71
WDVI	0.63	0.35	0.97	0.64	0.63	0.11	0.28	0.70
GSAVI	0.36	0.11	0.47	0.27	0.27	0.06	0.65	1.51
RI	0.76	0.76	1.52	1.25	1.29	0.16	−1.22	1.59
GNDVI	0.37	0.20	0.57	0.37	0.37	0.07	0.54	1.49
GDVI	0.29	0.06	0.35	0.18	0.17	0.05	0.84	1.73
GRVI	0.17	−0.10	0.07	−0.05	−0.06	0.04	1.66	2.99
G_RATIO	2.14	1.50	3.63	2.21	2.19	0.37	1.60	4.25
BSI	0.47	−0.35	0.12	−0.03	−0.01	0.09	−1.64	3.38
SAVI	0.50	0.13	0.64	0.35	0.35	0.09	0.68	1.38
NDVI	0.45	0.16	0.61	0.32	0.32	0.09	1.34	2.78

Fig. 2 The soil sample collection from the coal mine adjoining agricultural field converted to meadow (grassland)

Fig. 3 Effect of the unscientific coal mine on NDVI and Bare soil index (DN values not converted to reflectance)

3.2 Relationship Between Indices and SOC

Karl Pearson's correlation coefficient (r) was analysed between indices as predictor variables and SOC as a dependent or criterion variable (Table 3). The correlation coefficient (r) is categorized as 0–0.19 (very weak), 0.2–0.39 (weak), 0.40–0.59 (moderate), 0.6–0.79 (strong) and 0.8–1 (very strong). Among the covariance indices, only the redness index (RI) and green red vegetation index (GRVI) were significantly correlated with SOC content at a 1% significant level. There was a weak significant association of SOC with GRVI ($r = 0.337$) and negative with RI ($r = -0.333$). The green and red reflectance may be very much sensitive to the coal mine sulphur-rich soil.

Table 3 Correlation relationship between indices and SOC

	SOC	GWDRVI	MSAVI2	MGSAVI	MSAVI	WDVI	GSAVI	RI	GNDVI	GDVI	GRVI	G_ RATIO	BARESI	SAVI	NDVI
SOC	1.00														
GWDRVI	0.16	1.0													
MSAVI2	0.19	0.988**	1.00												
MGSAVI	0.14	0.998**	0.984**	1.00											
MSAVI	−0.05	0.05	0.05	0.06	1.00										
WDVI	0.08	0.848**	0.862**	0.852**	0.15	1.0									
GSAVI	0.11	0.976**	0.970**	0.980**	0.10	0.938**	1.0								
RI	−0.333**	−0.550**	−0.615**	−0.527**	0.13	−0.2	−0.409**	1.0							
GNDVI	0.11	0.991**	0.972**	0.995**	0.09	0.856**	0.981**	−0.490**	1.0						
GDVI	0.12	0.954**	0.956**	0.956**	0.11	0.967**	0.993**	−0.370**	0.953**	1.0					
GRVI	0.337**	0.596**	0.658**	0.571**	-0.13	0.245*	0.457**	−0.995**	0.532**	0.422**	1.0				
G_ RATIO	0.18	0.998**	0.989**	0.994**	0.03	0.840**	0.968**	−0.574**	0.982**	0.948**	0.622**	1.0			
BARESI	−0.22	0.890**	−0.900**	−0.878**	0.06	−0.601**	−0.801**	0.802**	−0.863**	−0.762**	−0.831**	−0.897**	1.0		
SAVI	0.14	0.974**	0.982**	0.977**	0.11	0.924**	0.993**	−0.484**	0.979**	0.984**	0.526**	0.966**	−0.833**	1.0	
NDVI	0.20	0.969**	0.976**	0.965**	0.0	0.741**	0.913**	−0.724**	0.956**	0.878**	0.756**	0.969**	−0.953**	0.937**	1.0

** significant at $p \leq 0.01$, * significant at $p \leq 0.05$

Table 4 Model summary

Sl. no	Regression method	Model	RMSE	R^2
1	Stepwise	SOC = 0.378 + 0.648 * Gratio (DN not converted to reflectance)	0.248	0.06
After DN converted to reflectance				
2	Stepwise	SOC = 1.198 + 2.709 * GRVI	0.34	0.12
3	Enter	SOC = −60.953 − 69.464 * MGSAVI − 0.037 * MSAVI − 39.560 * WDVI + 13.764 * RI − 52.639 * GDVI + 68.533 * GRVI + 17.762 * Gratio + 0.171 * BARESI + 132.377 * SAVI − 78.120 * NDVI	0.66	0.43

3.3 Spatial Modelling and Prediction of SOC Content

The outliers are removed, predictors suffer from multicollinearity causing unreasonably high the coefficient of estimators, and hence, it is unreliable inference and wider confidence interval [33]. Multicollinearity problem could be solved using stepwise multiple regressions [34]. The soils in the coal mining activitie site and also piling the mine spoils over the grassy land resulting similar spectral. Some of the indices behave similarly and have a strongly influence on other indices. If the DN values were not converted to the reflectance, the SOC (as dependent variable) prediction using fourteen remote sensing variables as independent in stepwise regress was found as SOC = 0.378+0.648*Gratio (Model 1) with R^2 0.06; this model could explain the SOC variability on 6% which is very poor; however, after DN is converted to the reflectance, model explanatory of SOC was improved as shown in Table 4. The stepwise linear regression model, SOC =1.198+2.709*GRVI (Model 2) with R^2 0.12 and RMSE 0.337, which was found weak (12% variation of SOC) association with SOC. Another model was attempted to be built using all fourteen indices in linear regression using the enter approach and found as SOC=-60.953−69.464* MGSAVI-.037* MSAVI-39.560* WDVI+13.764* RI-52.639* GDVI+68.533* GRVI+17.762* Gratio+0.171* BARESI+132.377* SAVI-78.120* NDVI (Model 3) with RMSE = 0.66 and R^2=0.43 (Table 4). This model could explain 43% of SOC variability. Similarly, [35] found that individual variables' effect on SOC prediction was not good, however, the model performance improved with variables summed together. Models 2 and 3 predicted the SOC as 6.94 and 3.78% higher than the actual SOC data (Fig. 4).

3.4 Validation

The accuracy of prediction is studied using root mean prediction error (RMSE) and coefficient of determination (R^2). A good model considers the lowest RMSE and highest R^2:

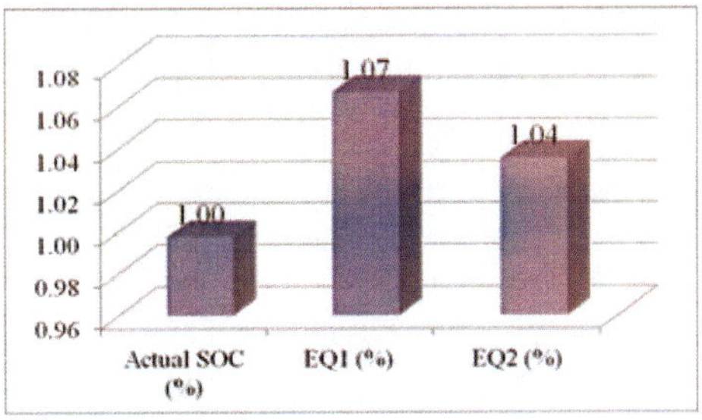

Fig. 4 Comparison of SOC accumulation using two obtained equations

Table 5 Model validation

Model	R^2	RMSE
1	−0.14	0.67
2	0.03	0.62
3	0.22	0.55

$$R^2 = 1 - \frac{\sum_{i=1}^{n} \left(\hat{y}_i - y_i\right)^2}{\sum_{i=1}^{n} \left(y_i - \overline{y}_i\right)^2}$$

where $\hat{y}$ represents the prediction, y_i represents the actual values and $\bar{y}$ represents the mean of all the values.

The R^2 for the Models 1, 2 and 3 were −0.14, 0.03 and 0.22, and RMSE were 0.67, 0.62 and 0.55, respectively (Table 5).

4 Conclusion

The medium resolution multi-spectral satellite data was a very good source for soil attribute analysis for a large area. The NDVI and BSI were the best vegetable indices for assessment of coal mining activity. The green and red reflectances were sensitive to the mining activity. The model performance improved using multiple variables for small sample size. The multicollinearity among the variables due to similar spectral over the small mining sites could be overcome by increasing the sample population soil and spectral. The spectral reflectance value at the same soil sampling site were extracted; hence the number of soil samples was equal to the number of spectral extracted from the satellite data. This dataset would be justified for the statistical analysis following stepwise linear regression. Further, it was suggested that the model developed through this research could be possible for refinement with increasing sample size.

Acknowledgements The authors thank the authority of CAU, Imphal, for financial assistance.

Declaration My team and I declare that there is no objection to use the photo of soil sample collection from the coal mine abundant agricultural field in the publication.

References

1. Ministry of Coal Homepage, Coal production, Government of India, Ministry of Coal, Shastri Bhawan, Dr. Rajendra Prasad Road, New Delhi-110001 (2020) http://www.coal.nic.in/major-statistics/production-and-supplies. Accessed 24 Mar 2021
2. NGT Homepage (2019) Fourth Interim Report. https://greentribunal.gov.in
3. Sarma K (2005) Impact of coal mining on vegetation: a case study in Jaintia Hills District of Meghalaya, India. Dissertation. International Institute for Geo-information Science and Earth Observation (ITC), Enschede, The Netherlands
4. Ghose AK (2000) Mining in 2000 AD challenges for India. Journal of the Institute of Engineers 39:1–11
5. Lyngdoh EAS, Sanjay-Swami (2020) Potential screening of phytoremediating crops and performance of maize in phytoremediated coal mined acid soil with phosphorus application. J Environ Biol 41(6):1788–1797. https://doi.org/10.22438/jeb/41/6/SI-283
6. Malang A (2010) Studies on fertility status and productivity potential of coal mine affected soils and their adjoining areas of Jaintia Hills district of Meghalaya. MSc. Thesis
7. Wang S, Adhikari K, Wang Q, Jin X, Li H (2018) Role of environmental variables in the spatial distribution of soil carbon (C), nitrogen (N), and C: N ratio from the northeastern coastal agroecosystems in China. Ecol Indic 84:263–272
8. Mulder VL, de Bruin S, Schaepman ME, Mayr TR (2011) The use of remote sensing in soil and terrain mapping–a review. Geoderma 162(1):1–19. https://doi.org/10.1016/j.geoderma.2010.12.018
9. Viscarra Rossel RA, Minasny B, Roudier P, McBratney AB (2006) Colour space models for soil science. Geoderma 133(3–4):320–337
10. Mandal UK (2016) Spectral color indices based geospatial modeling of soil organic matter in Chitwan district, Nepal. In: The International archives of the photogrammetry, remote sensing and spatial information sciences, 2016 XXIII ISPRS Congress, vol XLI-B2. Prague, Czech Republic, pp 43–48
11. Castaldi F, Chabrillat S, Chartin C, Genot V, Jones AR, van Wesemael B (2018) Estimation of soil organic carbon in arable soil in Belgium and Luxembourg with the LUCAS topsoil database. Eur J Soil Sci 69(4):592–603. https://doi.org/10.1111/ejss.12553
12. Castaldi F, Hueni A, Chabrillat S, Ward K, Buttafuoco G, Bomans B, Vreys K, Brell M, van Wesemael B (2019) Evaluating the capability of the Sentinel 2 data for soil organic carbon prediction in croplands. ISPRS J Photogramm Remote Sens 147:267–282. https://doi.org/10.1016/j.isprsjprs.2018.11.026
13. Kumar P, Sajjad H, Mahanta KK, Ahmed R, Mandal VP (2018a) Assessing suitability of allometric models for predicting stem volume of Anogeissus pendula Edgew in sariska Tiger Reserve, India. Remote Sens Appl Soc Environ 10:47–55
14. Kumar P, Sajjad H, Tripathy BR, Ahmed R, Mandal VP (2018b) Prediction of spatial soil organic carbon distribution using Sentinel-2A and field inventory data in Sariska Tiger Reserve. Nat Hazards 90(2):693–704
15. Nayak B (2013) Mineral matter and the nature of pyrite in some high-sulfur tertiary coals of Meghalaya, Northeast India. J Geologic Soc India 81:203–214
16. Soil and Water Conservation Department (2022) Basic information 08-Feb-2022 Map of soils of Meghalaya. Published by National Bureau of Soil Survey and Land use planning (NBSSLUP) in collaboration with Directorate of Agriculture, Meghalaya. https://megsoil.gov.in/images/meghalaya_soil_map.jpg

17. Zhu AX, Yang L, Li B, Qin C, English E, Burt JE, Zhou C (2008) Digital soil mapping: a state of the art. In: Hartemink AE, McBratney A, Mendonça-Santos M (eds) Digital soil mapping with limited data. Springer, Dordrecht, pp 233–245
18. McBratney AB, Mendonca Santos ML, Minasny B (2003) On digital soil mapping. Geoderma 117:3–52
19. Vaughn I (2019) Landsat 8 (L8) Data Users Handbook. Version 5.0, p. 54. U. S. Geological Survey
20. Rouse JWJr, Haas RH, Schell JA, Deering DW (1974) Monitoring vegetation systems in the Great Plains with ERTS. In: NASA Special Publication 351. 309 Greenbelt, MD, USA. NASA Goddard Space Flight Center
21. Buschmann C, Nagel E (1993) In vivo spectroscopy and internal optics of leaves as basis for remote sensing of vegetation. Int J Remote Sens 14:711–722. https://doi.org/10.1080/01431169308904370
22. Tucker CJ (1979) Red and photographic infrared linear combinations for monitoring vegetation. Remote Sens Environ 8(2):127–150. https://doi.org/10.1016/0034-4257(79)90013-0
23. Gitelson AA, Kaufman YJ, Merzlyak MN (1996) Use of a green channel in remote sensing of global vegetation from EOS-MODIS. Remote Sens Environ 58(3):289–298. https://doi.org/10.1016/S0034-4257(96)00072-7
24. Pouget M, Madeira J, Le Floch E, Kamal S (1990) Caracteristiques spectrales des surfaces sableuses de la region cotiere Nord-Ouest de l'Egypte: Application aux don andes satellitaires SPOT. In: 2eme JounCes de Tanddetection: Caracterisation et suivi des milieux terrestres en regions arides et tropicales. 4–6/12/1990. Collection Colloques et Seminaires, Paris, ORSTOM
25. Huete AR (1988) A soil-adjusted vegetation index (SAVI). Remote Sens Environ 25(3):295–309. https://doi.org/10.1016/0034-4257(88)90106-X
26. Qi J, Chehbouni Al, Huete AR, Kerr YH, Sorooshian S (1994a) A modified soil adjusted vegetation index (MSAVI). Remote Sensing Environ 48:119–126
27. Sripada RP, Heiniger RW, White JG, Meijer AD (2006) Aerial color infrared photography for determining early in-season nitrogen requirements in Corn. Agron J Remote Sensing 98:968–977. https://doi.org/10.2134/agronj2005.0200
28. Qi J, Kerr J, Chehbouni A (1994b) External factor consideration in vegetation index development. In: 6th International symposium on physical measurements and signatures in remote sensing, held in Val d' Isere, France, on 17–21 January 1994. CNES, Toulouse, France, pp. 723–730
29. Diek S, Fornallaz F, Schaepman ME, de Jong R (2017) Barest Pixel Composite for agricultural areas using landsat time series. Remote Sens 9:1245
30. Clevers JGPW (1988) The derivation of a simplified reflectance model for the estimation of leaf area index. Remote Sensing Environ 25:53–69
31. Gitelson AA (2004) Wide dynamic range vegetation index for remote quantification of biophysical characteristics of vegetation. J Plant Physiol 161:165–173
32. Gomez KA, Gomez AA (1984) Statistical procedure for agriculture research, 2nd edn. International Rice Research Institute, Los Banos, Philippines. John Wily and Sons, New York
33. Greene WH (2008) Econometric analysis, 4th edn. Dorling Kindersley, New Delhi, India, pp 56–61
34. Frisch R (1934) Statistical confluence analysis by means of complete regression systems. Oslo University, Institute of Economics, Publication No-5
35. Padilha MCC, Vicente LE, Demattê JAM, dos SantosWendriner Loebmann DG, Vicente AK, Salazar DFU, Guimarães CCB (2020) Using landsat and soil clay content to map soil organic carbon of oxisols and Ultisols near São Paulo, Brazil. Geoderma Reg 21:1–9

Rainfall Runoff Modeling Using HEC-HMS for Munneru River Basin, India

Eswar Sai Buri, Keesara Venkatareddy, and K. N. Loukika

Abstract The aim of this study was to assess the accuracy and suitability of the Hydrologic Engineering Center (HEC)—Hydrologic Modeling System (HMS) in predicting daily streamflows. In this study, we utilized the models that were integrated with the ArcGIS interface for hydrological research for the Munneru River basin, India comes under Lower Krishna River basin covering the drainage area of 10,390 square kilometers. The main focus of the streamflow analysis was to determine the effectiveness of the model when calibrated and optimized using observed flows in simulating gauged streamflows. To run the model, daily weather gauge station data for the 1993–2017 period were used as inputs, along with Land Use/Land Cover (LU/LC) classes generated from remote sensing satellite imagery, a soil map, and a Digital Elevation Model (DEM). To assess model performance and calibration, daily stream discharge data from the Keesara gauge station maintained by Central Water Commission (CWC) which is an outlet of the basin were utilized.

Keywords HEC-HMS · Land use · Land cover · Munneru · Streamflow

1 Introduction

Water management is crucial due to the increasing water demand with the growing population. To achieve a better understanding, watershed models are utilized, particularly for the estimation of the runoff depth, which is a significant and complex hydrological process in a watershed. Runoff simulations are essential for various purposes, such as producing hydropower, irrigation, and consumptive use. Thus, accurately estimating the rainfall converted into runoff is necessary, which can be achieved by executing water budget equations in hydrological models at a basin scale [1]. The streamflow pattern in hydrological modeling is influenced by land use/land cover (LU/LC), as it causes infiltration variations [2]. To capture the spatial diversity of the catchment, hydrological models require data inputs such as Digital

E. S. Buri (✉) · K. Venkatareddy · K. N. Loukika
Department of Civil Engineering, National Institute of Technology Warangal, Hanamkonda, India
e-mail: beswar@student.nitw.ac.in

S. Mesapam et al. (eds.), *Developments and Applications of Geomatics*,
Lecture Notes in Civil Engineering 450, https://doi.org/10.1007/978-981-99-8568-5_32

Elevation Model (DEM), soil map, weather gauge stations data, and LU/LC, and are then sub-divided into homogenous sub-watersheds. Hydrological models are used to analyze the changes in hydrologic parameters due to various anthropogenic and climate change impacts. However, it should be noted that no single model can perform well in all hydroclimatic regimes worldwide. Streamflow data is fundamental for hydrologic research, as it represents the runoff phase of the hydrologic cycle. However, continuous streamflow measurement is challenging and often limited or unavailable, making it difficult to assess water resources, particularly in developing countries [3, 4]. Therefore, the hydrologist's primary challenge is to establish the relationship between rainfall and streamflow conversions to manage water resources effectively.

Several hydrological models, including SWAT, HEC-HMS, PRMS, and WetSpa, utilize regularly available data to simulate the significant components of water balance. These models are freely accessible in the public domain. HEC-HMS, a hydrologic modeling system developed by the US Army Corps of Engineers Hydrologic Engineering Center [5], is a continuous and event-based model. It was initially designed to simulate precipitation-runoff processes in dendritic watershed systems but has since been improved to address a broader range of issues, such as flood hydrographs, water supply distribution in large river basins, and runoff in small urban or natural watersheds. It is physically based and requires portable data inputs to produce realistic results.

For estimating runoff, the hydrological model HECHMS was used, which uses DEM, LU/LC map, and soil map [6]. Land use and land cover constantly change due to increasing population taken as one of parameters; its influence on CN and their impact on surface runoff are analyzed [7]. Suitability of hydrological models to a particular study area needs to be studied for vulnerable basins which are prone to floods.

The primary objective of this study is to analyze the long-term daily streamflow which are simulated from the HEC-HMS model. The streamflow values were first calibrated and validated from the observed values taken at the gauge station at the outlet of the study area.

2 Study Area

Munneru River is a significant tributary of the Krishna River, located in south India. It originates in the Nallamala Hills in the state of Andhra Pradesh and flows through the districts of Khammam and Warangal in Telangana before joining the Krishna River near Vijayawada in Andhra Pradesh. The Munneru River basin is agriculturally dominant, with paddy fields, sugarcane plantations, and other crops. The river and its tributaries play a vital role in the socio-economic development of the region, providing irrigation water for agriculture, drinking water for villages, and serving as a source of hydropower generation. Its geographical coordinates are between 16.6°N and 18.1°N latitudes and 79.2°E and 80.8°E longitudes. Figure 1 depicts

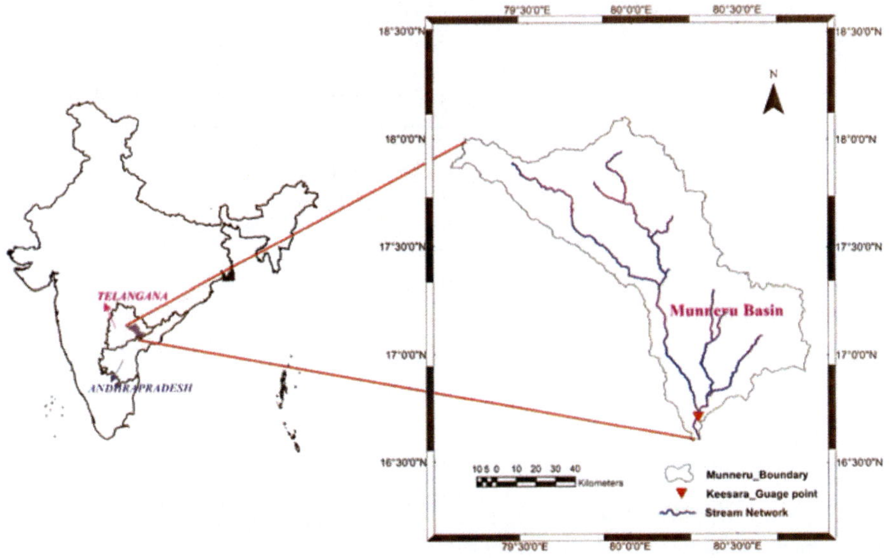

Fig. 1 Location map of study river basin in map of India

the location map of the study basin. The Munneru basin is selected as a part of the ongoing research project IWMM-BIS, titled "Integrated Water Management Model for Brazil, India, and South Africa under climate change scenarios [8, 9]".

3 Data Used

The observed daily gridded precipitation data for the period 1993–2017 was taken from Indian Meteorological Department (IMD), and it is averaged five subbasins considered in the present study. Digital Elevation Model (DEM) was prepared from CARTOSAT satellite data having a resolution of 30 m. The soil map was prepared from the map taken from International Soil Reference and Information Centre (ISRIC) having a resolution of 1 km. The LU/LC map is obtained from National Remote Sensing Center (NRSC) having scale of 1:250,000. Daily streamflow values are considered at the Keesara gauge station maintained by CWC. Figure 2 elucidates the DEM, Soil, and LU/LC maps of the study area.

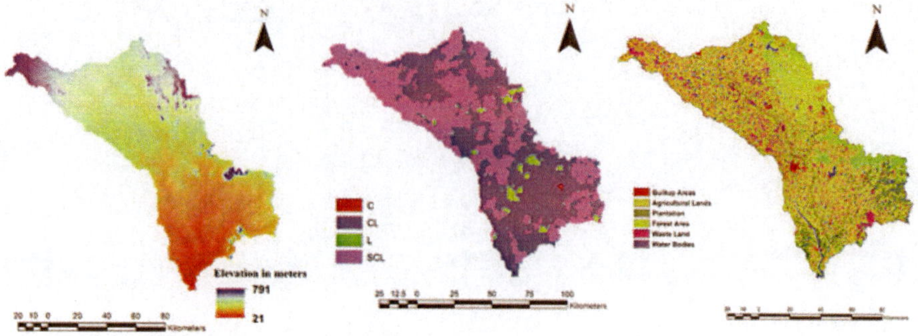

Fig. 2 DEM, soil, and LU/LC maps of the study area

4 Methodology

The methodology flowchart (refer to Fig. 3) outlines the steps involved in data collection, Curve Number map preparation, HEC-HMS model setup and calibration, validation, and analysis of daily streamflow. Figure 3 represents the methodology flowchart of the present study. After taking DEM, LU/LC, and soil maps, the curve number map was generated in ArcGIS software and is depicted in Fig. 4.

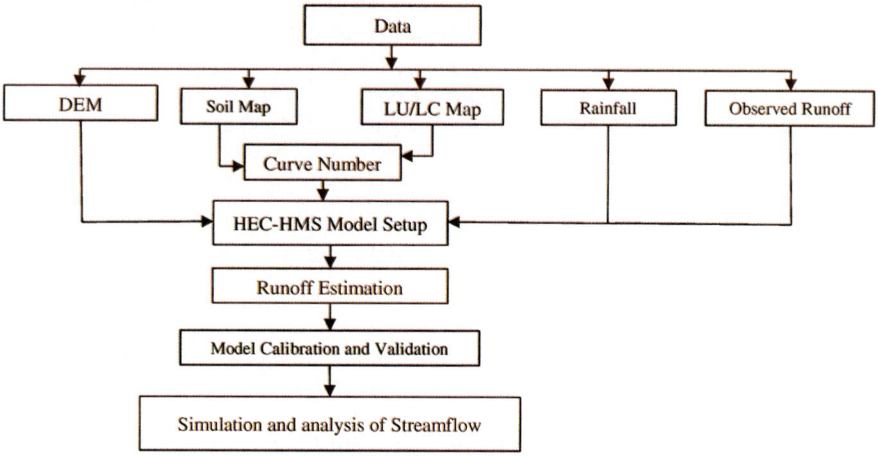

Fig. 3 Methodology flowchart of impervious surface map using landsat satellite images

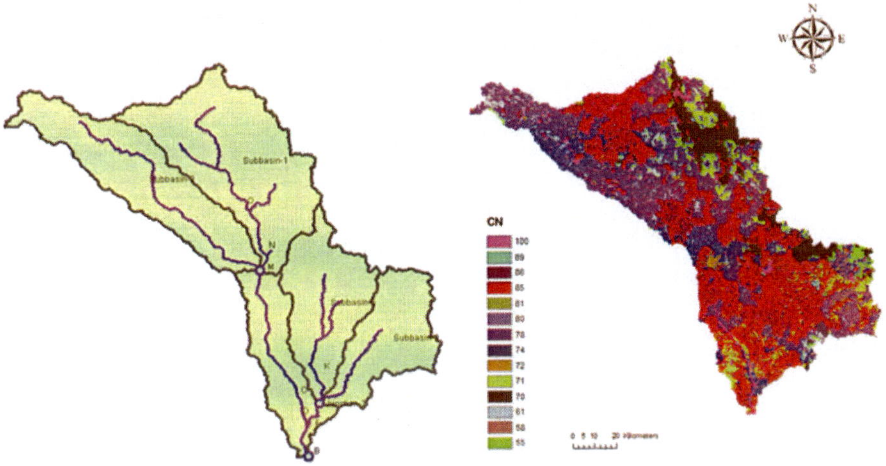

Fig. 4 Subbasin map and curve number map of the study area

4.1 Hec-Hms

HEC-HMS is public domain software most used for event and continuous simulation, which was developed by the US Army Corps of Engineers. Previous versions of HEC-HMS did not involve any GIS pre-processing but the latest version has GIS pre-processing capabilities. Various methods are available to find the infiltration loss like initial and constant, deficit and constant, exponential, Green and Ampt, SCS curve number, Smith-Parlange, and Soil Moisture Accounting (SMA). In the present study, SCS curve number method was used. To transform excess rainfall into direct runoff hydrograph, the SCS unit hydrograph method was used. For routing Muskingum routing was used. Daily data of discharge at the Keesara gauge point were considered for the calibration period of 1993–2007 and for the validation that of 2008–2017 were used.

5 Results and Discussion

The Munneru basin was sub-divided into five subbasins. Soil Conservation Service (SCS), Curve Number, SCS Unit Hydrograph—Lag time in minutes, Muskingum—x, and Muskingum—K are the parameters taken into consideration for calibration of the HEC-HMS model. The initial and optimal values of parameters which are considered for calibration purposes are shown in Table 1. The six reaches are named P, C, A, E, G, and J, respectively. The initial values of Muskingum -x and K values are assigned to each reach. The simulated flow and observed flow values for the calibration period (1993–2007) are depicted in Fig. 5, and similarly Fig. 6 elucidates the simulated flow and observed flow values for validation period (2008–2017). The coefficient of determination value (R^2) was calculated for both calibration and validation periods. The values come as 0.46 and 0.36, respectively. The highest average monthly streamflow (1013.44 m^3/s) was recorded for the entire observation period in August 2008. Over the years 1993 to 2017, there were 52 instances of average monthly observed streamflow values greater than 100 m^3/s, and there were 13 instances of streamflows larger than 300 m^3/s. Peak annual average monthly streamflow values were over- and under-simulated 8 and 16 times, respectively, during the entire period.

Table 1 Parameters considered for calibration

Subbasin/reach	Parameter	Initial value	Optimal value
Subbasin-1	SCS curve number	77.88	85.195
Subbasin-2		77.88	84.325
Subbasin-3		77.88	83.722
Subbasin-4		77.88	83.308
Subbasin-5		77.88	83.296
Subbasin-1	SCS unit hydrograph—lag time in minutes	1000	4777.4
Subbasin-2		1000	4648.5
Subbasin-3		1000	4186.4
Subbasin-4		9000	10,889
Subbasin-5		1000	2211.7
P	Muskingum–x	0.1	0.18353
C		0.1	0.12605
A		0.1	0.1149
E		0.1	0.32607
G		0.1	0.37322
J		0.1	0.41542
P	Muskingum–K	0.5	125.96
C		0.5	144.64
A		0.5	149.59
E		0.5	117.96
G		0.5	67.341
J		0.5	67.341

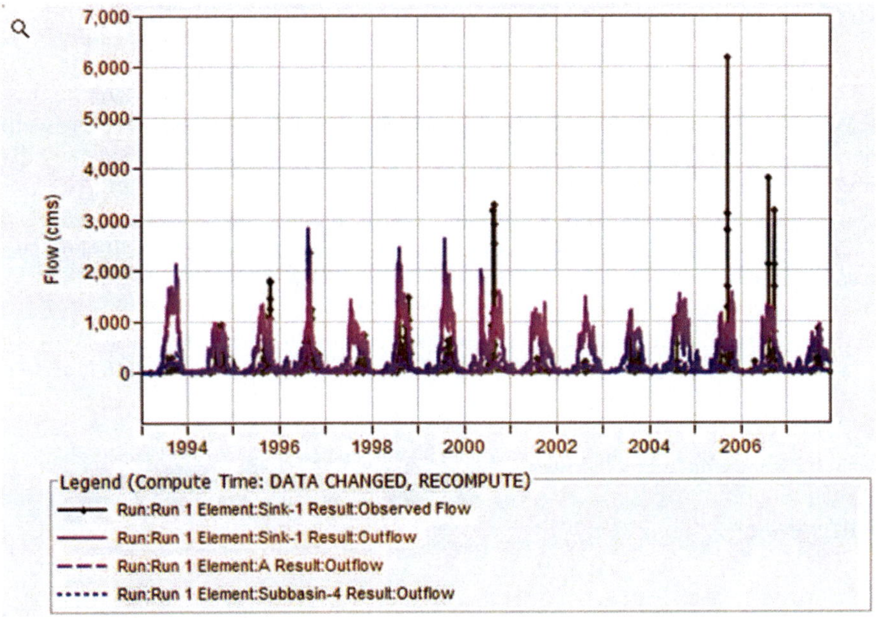

Fig. 5 Simulated flow versus observed flow for calibration period (1993–2007)

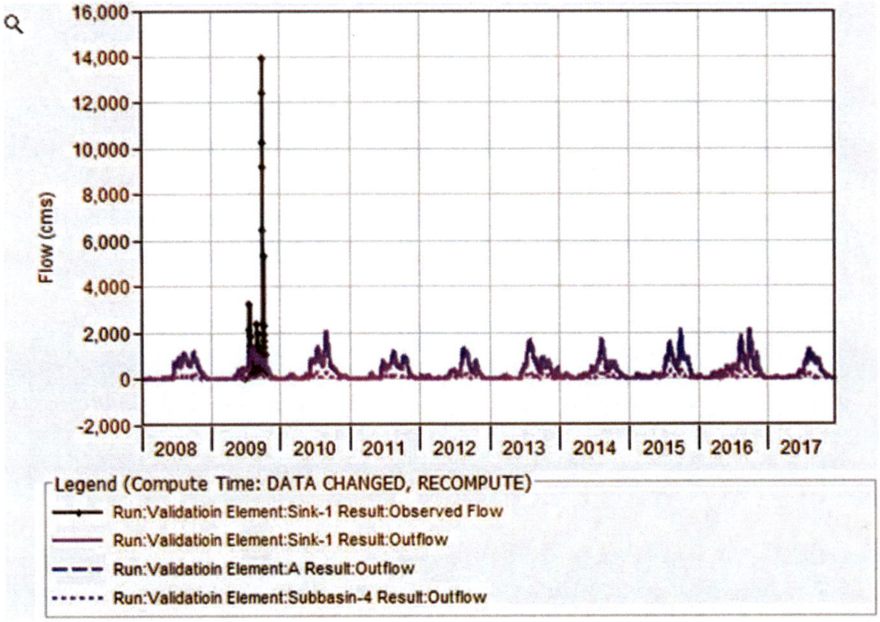

Fig. 6 Simulated flow versus observed flow Validation period (1993–2007)

6 Conclusion

The analyses of long-term daily streamflows which are simulated from the HEC-HMS model were compared with the observed flows at the gauge station for the period 1993–2017. The peak discharges are not simulated perfectly using daily data through the HEC HMS model. The inputs taken up for the study need to be finer than the used ones to get higher correlation between observed and simulated flows. Comparisons to be needed with other hydrological models such as Soil and Water Assessment Tool (SWAT) using daily data for the same study area.

Acknowledgements This work was funded by the Department of Science and Technology (DST), Government of India, under the BRICS—DST project with grant no. DST/IMRCD/BRICS/PilotCall2/IWMMBIS/2018 (G).

References

1. Vijverberg J, Sibbing FA, Dejen E (2009) Lake Tana: source of the Blue Nile. In: Dumont J (ed) The Nile (H. Springer, Dordrecht, pp 163–192
2. Yamamoto EMS, Sayama T, Yamamoto K (2020) Comparison of runoff generation methods for land use impact assessment using the SWAT model in humid tropics. Hydrol Res Lett 14(2):81–88
3. Chow Te V (2010) Applied hydrology. Tata McGraw-Hill Education, New Delhi, India
4. Shaw EM, Beven KJ, Chappell NA Lamb R (2010) Hydrology in practice. CRC Press, Milton Park, Abingdon, UK
5. USACE-HEC (2010) Hydrologic modeling system, HEC-HMS v3.5. user's manual. US Army Corps of Engineers, Hydrologic Engineering Center, Davis, CA
6. Hamdan ANA, Almuktar S, Scholz M (2021) Rainfall-runoff modeling using the hec-hms model for the al-adhaim river catchment, northern iraq. Hydrology 8(2):58
7. Derdour A, Bouanani A, Babahamed K (2018) Modelling rainfall runoff relations using HEC-HMS in a semi-arid region: case study in Ain Sefra watershed, Ksour Mountains (SW Algeria). J Water Land Dev 36:45–55
8. Buri ES, Keesara VR, Loukika KN, Sridhar V (2022) Spatio-temporal analysis of climatic variables in the Munneru River Basin, India, using NEX-GDDP data and the REA approach. Sustainability 14(3):1715
9. Buri ES, Reddy KV, Loukika KN (2022) Hydrological appraisal of Munneru River basin using SWAT model innovative trends in hydrological and environmental systems. Lect Notes Civil Eng 234. https://doi.org/10.1007/978-981-19-0304-5_47

Spatio-Temporal Surface Urban Heat Island Effect Analysis Over Tiruchirappalli City, India, Using GIS Techniques

K. S. Arunab, Ajay Badugu, Aneesh Mathew, and Padala Raja Shekar

Abstract This study investigates the effects of urbanization on the climatic conditions of a region and the occurrence of urban heat islands (UHIs). The study area was analysed for variations in UHI effects during the summer and winter seasons over a 20-year period, and hot spots (HS) and cold spots (CS) were identified. Results indicate that the Land Surface Temperature (LST) is higher during summer nights than winter nights, and rural areas are cooler than urban areas. The study also found significant differences in LST within the study area, with the Cauvery delta experiencing a dip and the BHEL industrial area showing a significant spike in temperature. HS and CS maps were developed for Tiruchirappalli. The findings of this study can provide valuable insights for policymakers and urban planners to mitigate the UHI effects and improve the climate resilience of urban areas.

Keywords Urban climate · Land surface temperature · Urban heat island

1 Introduction

Urbanization has had significant impacts on various facets of human life, including the economy, society, and the environment. One of the consequences of rapid urbanization is the UHI effect [1, 2], which is caused by changes in land use that lead to the replacement of vegetation and soil with impervious and built-up surfaces [3]. This phenomenon causes urban areas to experience higher temperatures than surrounding suburban or rural areas, which leads to negative impacts on human health and the environment [4–7].

K. S. Arunab (✉) · A. Badugu · A. Mathew · P. R. Shekar
Department of Civil Engineering, National Institute of Technology, Tiruchirappalli 620015 Tamil Nadu,, India
e-mail: ksarunab@gmail.com

A. Mathew
e-mail: aneesh@nitt.edu

S. Mesapam et al. (eds.), *Developments and Applications of Geomatics*,
Lecture Notes in Civil Engineering 450, https://doi.org/10.1007/978-981-99-8568-5_33

The UHI effect is primarily caused by the thermal storage capacity of common materials such as asphalt, concrete, and brick, which absorb higher percentages of the sun's radiation than vegetation or soil [8]. Additionally, the urban geometry affects wind flow and direction, which can trap heat and pollutants [9–12]. Among other adverse effects, this effect causes rising energy demand [13], poor air quality [14, 15], and illnesses brought on by the heat [16–18].

One of the most effective ways to mitigate the UHI effect is by incorporating vegetation in urban areas [19]. Vegetation helps to reduce the amount of solar radiation absorbed by surfaces and provides shade, which reduces the temperature of the surrounding environment. By removing contaminants from the air and releasing oxygen, trees also aid in enhancing air quality. Moreover, vegetation helps to reduce the UHI effect by providing a cooling effect through the process of evapotranspiration, where plants release water vapour into the air, which cools the surrounding area.

Remote sensing techniques can be used to examine the geographical and temporal variations in UHI effects and the contribution of vegetation to reducing these effects [20]. One such technique is the Enhanced Vegetation Index (EVI), which uses the red and blue wavelengths of light to measure the density and health of vegetation. This index can be used to monitor the temporal and spatial changes in vegetation cover in urban areas, providing valuable information for urban planning and development [19].

The study of UHI effects and the importance of mitigating them is crucial for urban planning in the future and development, particularly in developing countries like India, which is rapidly urbanizing. One such city suffering the UHI impact is the Indian city of Tiruchirappalli in the Tamil Nadu state, a suitable location for studying the spatial and temporal fluctuations in the UHI impact, which has a dry summer tropical savanna environment.

In conclusion, rapid urbanization has led to changes in land use and subsequent increases in temperature, creating a significant challenge to the UHI effect in urban regions. Vegetation incorporation is an effective way to mitigate the UHI effect, but a comprehensive understanding of the thermal conditions of the research region is essential. The classification of the research region into thermal comfort zones and thermal HS allows for targeted interventions to address the UHI effect [21]. Remote sensing techniques, including the EVI, can be used to study the spatial and temporal changes in vegetation cover in urban areas and provide valuable information for urban planning and development. The study of UHI effects is crucial for upcoming urban development and planning, particularly in developing countries experiencing rapid urbanization.

This study aims to classify the study area of Tiruchirappalli into thermal comfort zones and thermal HS to understand the distribution and intensity of UHI effects in the city. The study will use Landsat 8 images and EVI data to analyse the spatial and temporal changes in vegetation cover and its relationship with the UHI effect. The study will provide valuable information for urban planning and development in Tiruchirappalli, which could potentially reduce the negative impacts of UHI effects on the city's residents and the environment.

2 Study Area

According to the 2021 census, Trichy, commonly known as Tiruchirappalli, is a tier-two Indian city in Tamil Nadu with a population of 9,28,772 people. It has a tropical savanna environment with dry summers and four seasons: summer (March, April, May); monsoon (June, July, August); wind season (September, October, November); and winter (December, January, and February). It is situated on the Deccan Plateau. The summer season is characterized by hot and dry days, while the winter season has cooler evenings with chilly winds from the southeast. With an average monthly temperature between 25 °C (77 °F) and 32 °C (90 °F) and an annual mean temperature of 28.9 °C (84.0 °F) (Tiruchirappalli (Trichy) City Municipal Corporation-E-Services Portal), the city is the warmest in the state. The 841.9 mm average annual rainfall is a little less than the 945 mm state average [22]. The city has a mild climate from June to September, with rain and thunderstorms, while rainfall is heaviest from October to December due to the northeast (N-E) monsoon winds. The climate is chilly and damp from December to February.

The MCD12Q1 MODIS Land Cover Type product and the Google Earth Engine are used to build the Tiruchirappalli city urban area polygon. The polygon is about 17.5 km in length from north to south and 16 km from east to west. The urban boundary has been extended by 7 kms to incorporate sufficient rural regions. Urban, rural, and suburban areas are all included in the 687.4 km^2 research region. The research region in Tiruchirappalli city is depicted in Fig. 1.

3 Methodology

The Aqua and Terra combined MODIS Land Cover Type product (MCD12Q1) is used to demarcate the study area. The MOD11A2 product from Terra was used to retrieve the LST [23] of the study area, this has an 8-day temporal resolution and a spatial resolution of 1 km, covering the period from 01-01-2001 to 31-12-2020.

Indicators like UHI intensity and UHI_{index} were employed to analyse the UHI effect. The UHI_{index} is a normalized index that may be used to compare the intensity of UHI across various seasons and times. The temperature differential between rural and urban regions is the UHI intensity. While the UHI_{index} was used to analyse the spatial variations of UHI, the UHI intensity was used to analyse its temporal variations.

Due to fluctuations in the temperature range that occur throughout the research area on a seasonal and diurnal basis, the intensity of UHI may vary. It is difficult to match UHI intensity over time because of the variable intensity of UHI and the change in positions corresponding to the minimum and maximum LST values. The UHI_{index} application was developed to make it simpler to investigate the UHI impact more thoroughly while taking seasonal and diurnal temperature variations into account. The UHI_{index} is derived using Eq. 1:

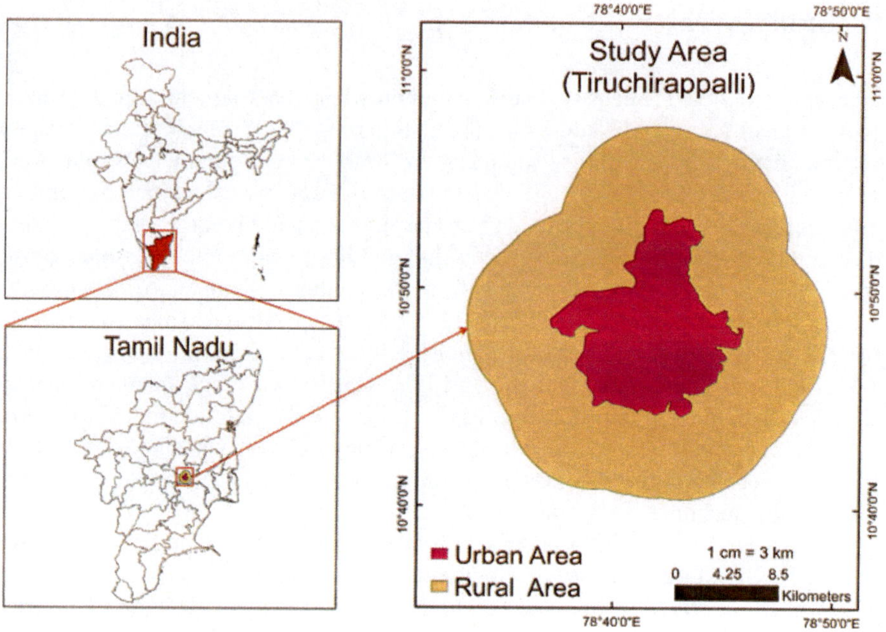

Fig. 1 Tiruchirappalli study region

$$\mathrm{UHI}_{\mathrm{index}} = \frac{\mathrm{LST_i} - \mathrm{LSR_{min}}}{\mathrm{LST_{max}} - \mathrm{LST_{min}}} \tag{1}$$

The LST of the ith location (pixel) in the same image is $\mathrm{LST_i}$. The research region's minimum and maximum LSTs are detected at the locations (pixels) known as $\mathrm{LST_{max}}$ and $\mathrm{LST_{min}}$. With this technique, the $\mathrm{UHI_{index}}$ is obtained for each research area pixel that corresponds to an image. The $\mathrm{UHI_{index}}$ can be used to compare the intensity of UHI across multiple times and seasons because it is a normalized index with only 0 and 1 values. The $\mathrm{UHI_{index}}$ was used to bring the variable LST range into a normal range via normalization.

The EVI is a remote sensing index that combines blue, red, and near-infrared bands to increase sensitivity to vegetation cover and reduce atmospheric influences. The MODIS data product MOD09GA on the Terra satellite provides high-quality vegetation indices, including EVI, which can be used to monitor vegetation growth and assess land cover changes. EVI values range from −1 to 1, where higher values indicate healthier vegetation cover. In urban areas, the vegetation cover is typically limited, resulting in lower EVI values than in rural areas with more abundant vegetation cover. Conversely, the LST in urban areas is typically higher due to the absence of vegetation cover and the presence of anthropogenic heat sources. By analysing the correlation between EVI and LST in urban areas, it is possible to identify areas vulnerable to the UHI effect and assess the effectiveness of mitigation strategies designed to reduce it. The correlation between EVI and LST is a useful metric for understanding the UHI effect in urban areas.

The identification of HS within urban areas is crucial in the context of mitigating the impact of UHI. This information can guide urban planners and policymakers to develop targeted strategies that address the most vulnerable areas. For instance, increasing vegetation cover, using high-reflectivity materials, and reducing anthropogenic heat sources can be implemented in these identified HS areas. Prioritizing resources and funds towards these mitigation strategies can lead to more efficient and effective solutions to combat the UHI effect. HS is calculated based on the maximum UHI_{index} pixels and the impact of the surrounding pixels. A group of pixels with a higher average temperature is known as an HS. However, only a considerable portion of those pixels are statistically significant as HS areas or zones. Spatial auto-correlation analysis is used to determine how HSs emerge and their relevance (ArcGIS Help—Global Moran's I). The thermal HS and CS map is generated using the Getis-Ord Gi* statistics (ArcGIS Help—Hot Spot Analysis).

The methodology used in the study allowed for a detailed analysis of the UHI effect in Tiruchirappalli, considering both temporal and spatial variations. While the HS analysis allowed for a detailed evaluation of the concentration of high temperatures in various zones of the research region, the UHI intensity and UHI_{index} were used to evaluate the intensity of UHI across different periods and seasons. The study provides useful insights into the UHI effect in Tiruchirappalli, which can be used to inform policies and strategies aimed at mitigating its impact on the urban environment and the health and well-being of its residents. Figure 2 provides the methodology's graphical abstract.

4 Results and Discussions

Annual and Seasonal Land Surface Temperature Analysis

Since the UHI effect is more common at night [16, 24], the study compared the winter and summer seasons while analysing the LST data for 20 years between 2001 and 2020. According to the research, Figs. 3 and 4 depict how the LST distribution in Tiruchirappalli follows a pattern that is comparable in the winter and summer. However, the summer months saw an average increase in temperature of 3.3772 K for a given location. The summer temperatures ranged from 296.653 to 301.853 K, while the winter temperatures ranged from 293.425 to 297.342 K. The yearly average temperatures ranged from 295.436 to 298.923 K. The south, southeast, and southwest portions of the research area saw the highest temperatures, as indicated in Figs. 3, 4, and 5. High-density urban structures, such as industrial as well as educational complexes, residential and commercial buildings, railway junctions, extensive road networks, and bridges, may be found throughout the city's centre area. The majority of Tiruchirappalli's UHI effect is attributed to this densely populated area.

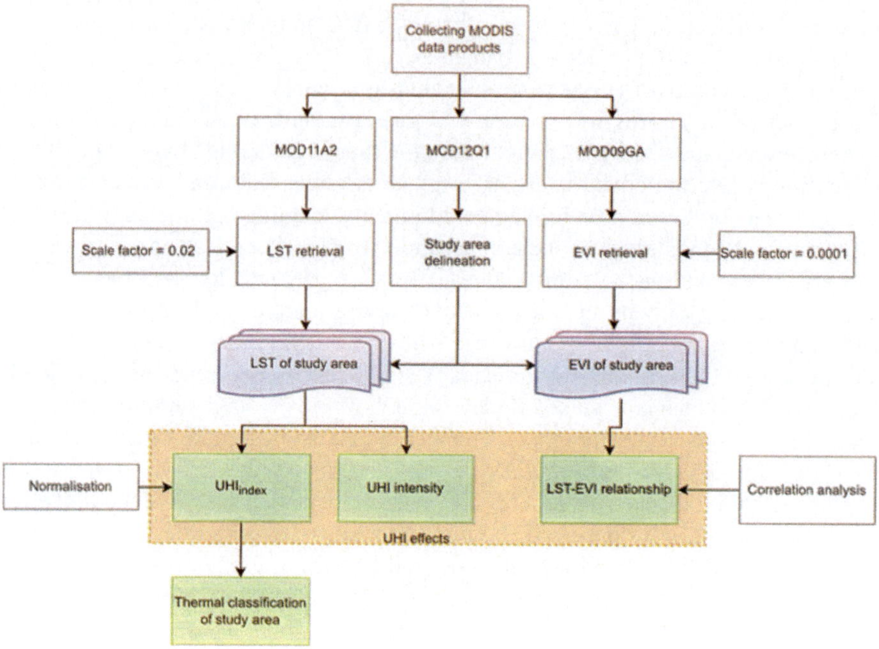

Fig. 2 Graphical abstract of methodology

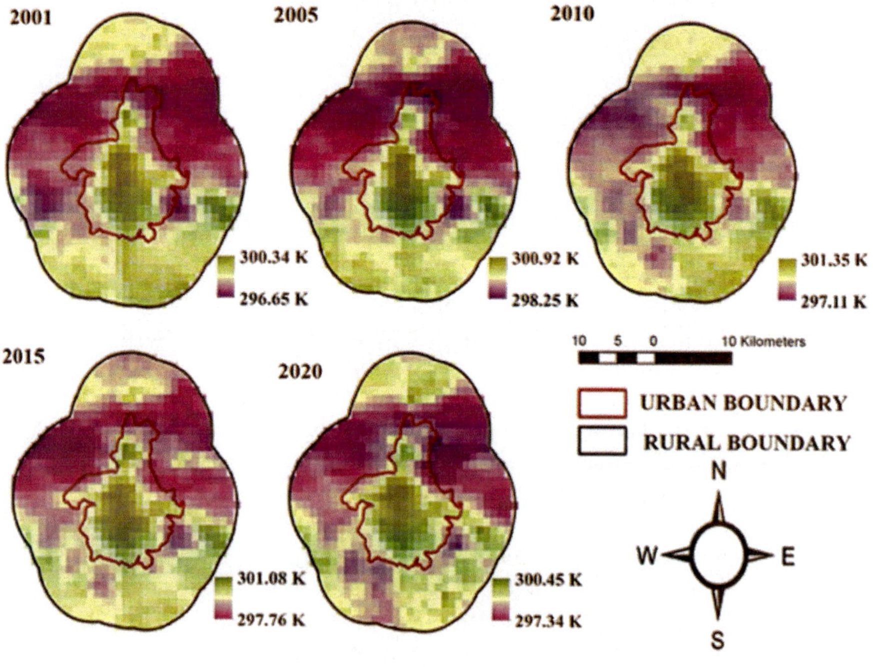

Fig. 3 Mean LST for summer

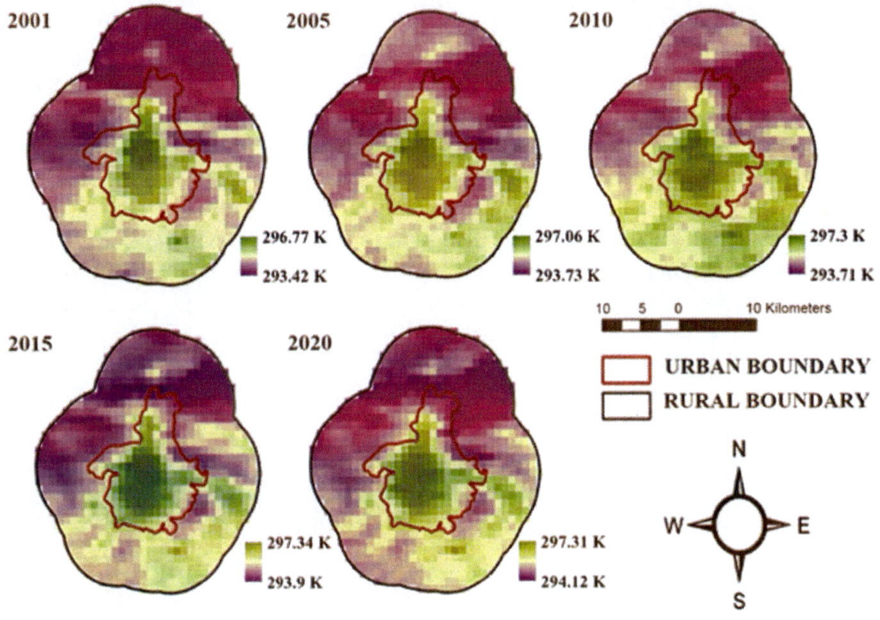

Fig. 4 Mean LST for winter

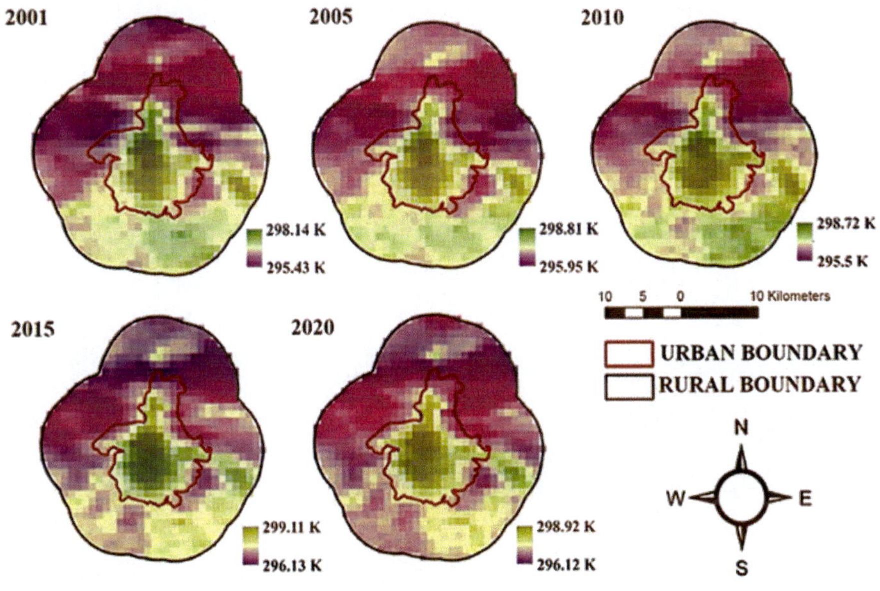

Fig. 5 Mean annual LST

The research area's rural areas in the southwest and south have a medium level of development and populated areas, which also add to the UHI effect. The Tiruchirappalli township and factory of Bharat Heavy Electronics Ltd (BHEL), which are in the southeast of the research region, are heavily developed and significantly warm the land surface. The area is also home to a few more towns, including Kumaramangalam and Navalpattu.

In comparison to all other regions, the northwest and northeast of the research region exhibit the lowest temperatures. The Cauvery River flows through the northwest and northeast parts of the study area, which are mostly covered with undeveloped land and vegetation. The northern region, however, exhibits a greater summer temperature than winter temperatures. The Cauvery River having less water and having less vegetation could be the causes of this.

Urban Heat Island Indicator

The mean UHI_{index} was calculated for the winter and summer seasons over a 20-year period, and the results are presented in Figs. 6, 7, and 8. In the urban area, the UHI_{index} was found to be highest in the centre and to drop towards the border. The average UHI_{index} of seventeen pixels within the city borders was discovered to be higher than 0.90, showing the predominance of high LST on these pixels and indicating that they are likely UHI HS. Tiruchirappalli UHI was focused on the urban core, with other high-temperature pixels grouped close by. LST dropped as the distance from the urban core grew.

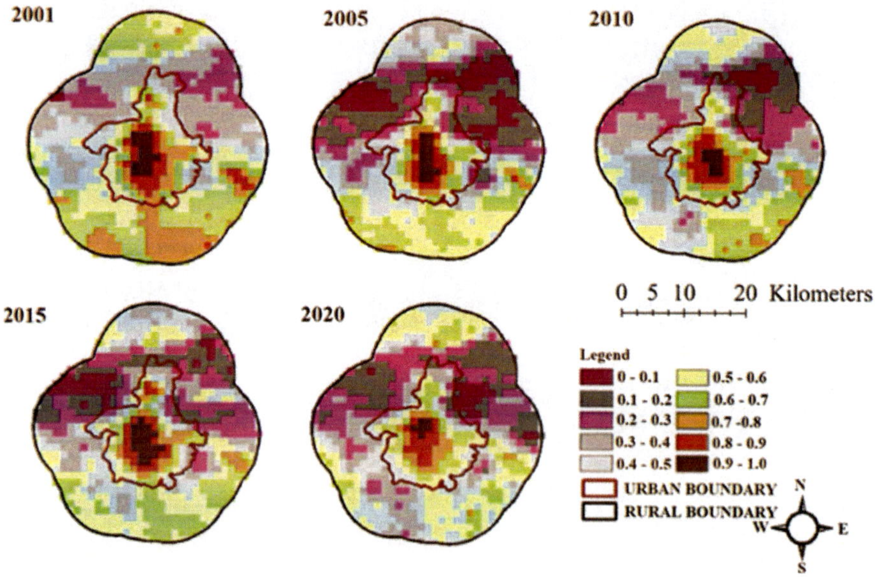

Fig. 6 Mean summer UHI_{index}

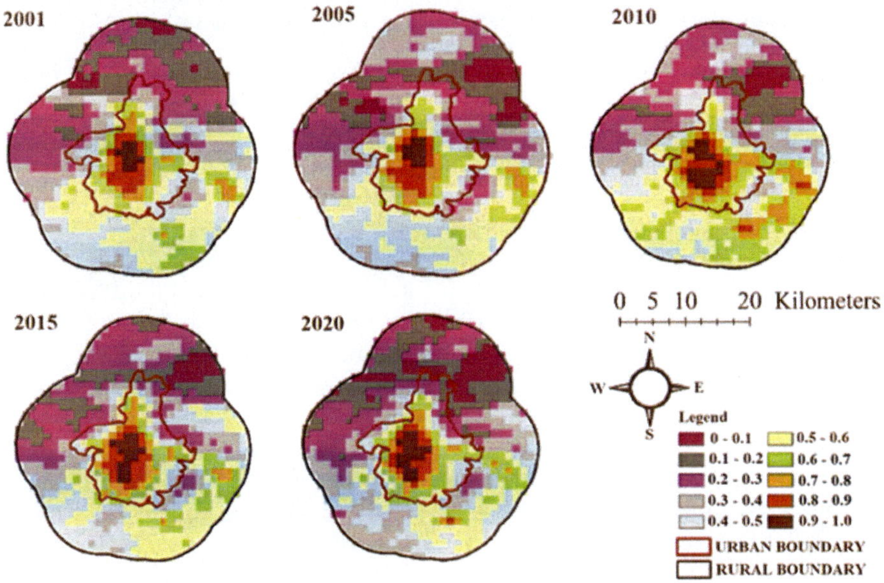

Fig. 7 Mean winter UHI_{index}

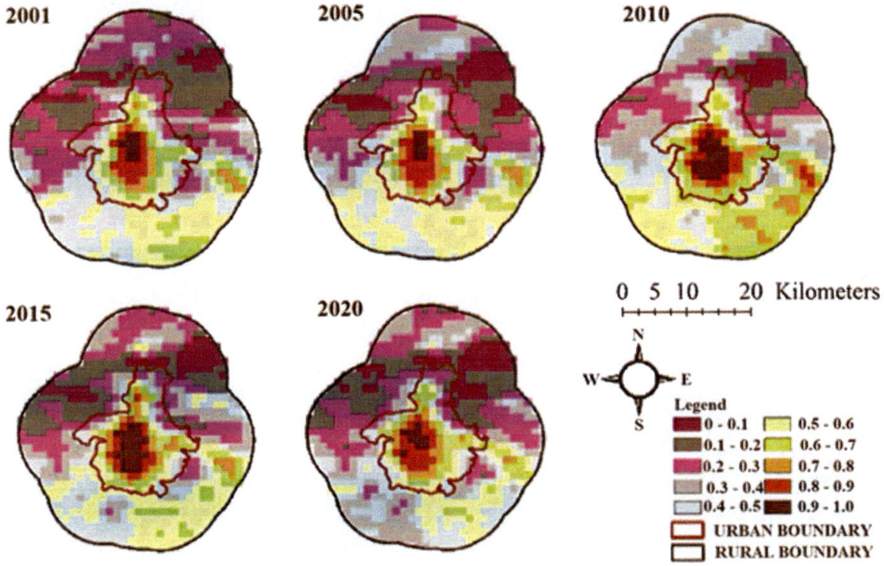

Fig. 8 Mean annual UHI_{index}

Relationship Between Land Surface Temperature and Vegetation

The amount of vegetation in a particular area is one of the most significant factors affecting LST. The study area, Tiruchirappalli, is a dry-summer tropical savanna environment that is primarily fed by rainfall and groundwater. The density of vegetation varies from year to year due to the monsoon rains, which are the primary source of irrigation for agriculture and vegetation. The present study examines the seasonal and spatial patterns of the EVI across the study area using 16-day EVI images representing two seasons, summer and winter, from 2001 to 2020.

The EVI images of different periods are presented in Figs. 9 and 10, depicting the variations in vegetation density across the study region. In areas with less vegetation, represented by the colour purple, EVI values are low, while in areas with more vegetation, represented by the colour green, EVI values are high. The rural areas of the study region show varying degrees of vegetation growth, with some areas being less vegetated than others. Agricultural areas, on the other hand, contribute to the EVI values. In comparison, the core of the city is significantly less vegetated, and its impervious built-up surface covers a large portion of the area, leading to lower EVI values.

The research area experiences high summer temperatures and a water shortage, which causes the vegetation density to decline. Because of this, summertime EVI levels are lower than wintertime EVI values. With a maximum value of 0.403 and a

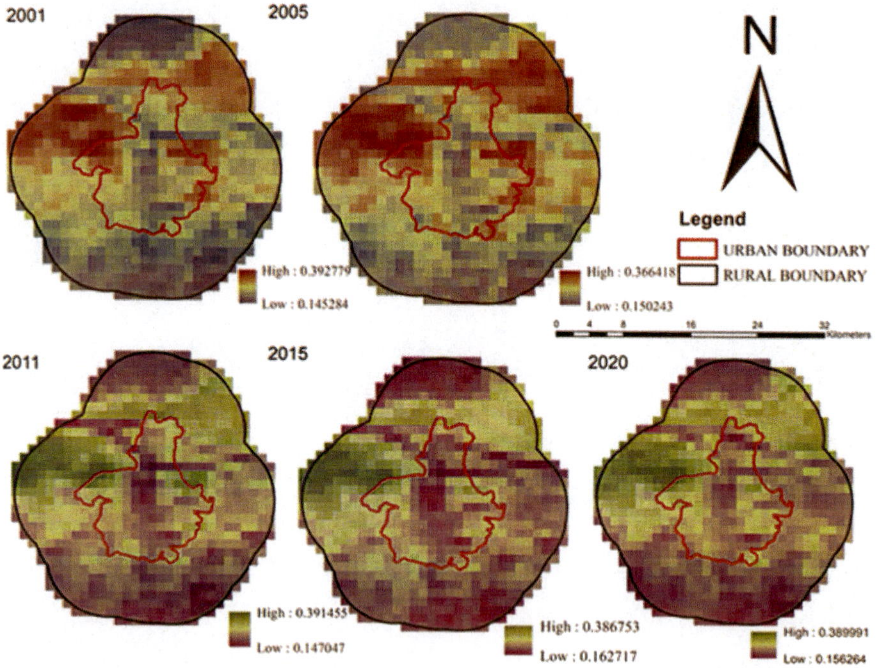

Fig. 9 EVI over the study area in summer

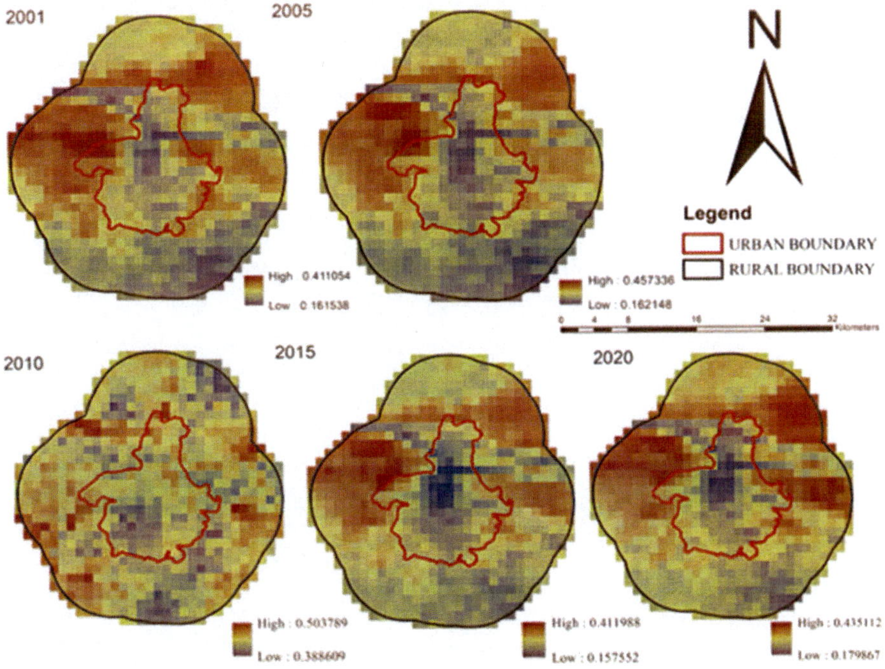

Fig. 10 EVI over the study area in winter

minimum value of 0.129, the summertime EVI average during the past 20 years has been 0.24. EVI, on the other hand, has an average value of 0.29 during the winter, with a maximum value of 0.462 and a minimum value of 0.143 during the course of 20 years.

Figures 11 and 12 display scatter plots and the combined LST-EVI relationship for the summer and winter seasons. In both seasons, the difference in temperature between low and high EVI values is higher. The relationship between LST and EVIs is season-dependent and nonlinear, with the climate of the research region influencing how relevant the EVI and LST relationship is in comparison to one another. The scatter plots' trend lines show a negative correlation between EVI and LST.

A second-order polynomial link between LST and EVI for several seasons is also shown in the study. The graph shows the correlation coefficient (R^2) for the polynomial relationship as well as the relation between the mean EVI and LSTs for both seasons. The winter EVI ranges are higher than those in the summer. By averaging the LST values of pixels with the same EVI index value, the mean LST values were determined. During the winter months of the study period, the R^2 value for the LST-EVI association ranges from 0.87 to 0.96, with an average value of 0.92. When compared, the summertime R^2 value has a mean of 0.90 and a range of 0.86 to 0.93. This shows that in the research location, the association between EVI and LST is more favourable during the winter than during the summer. The total mean

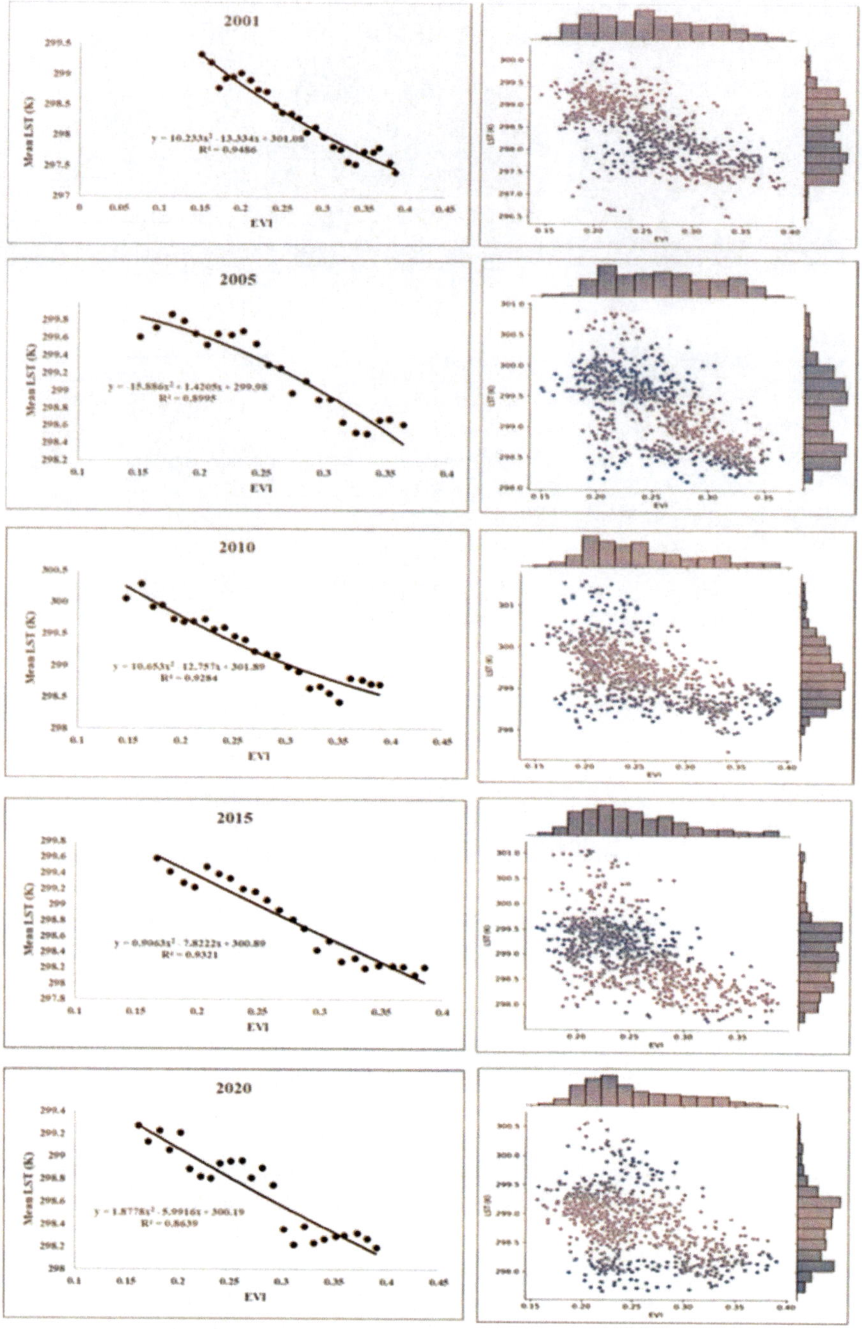

Fig. 11 Scatter plots and joint plots between LST and EVI for summer

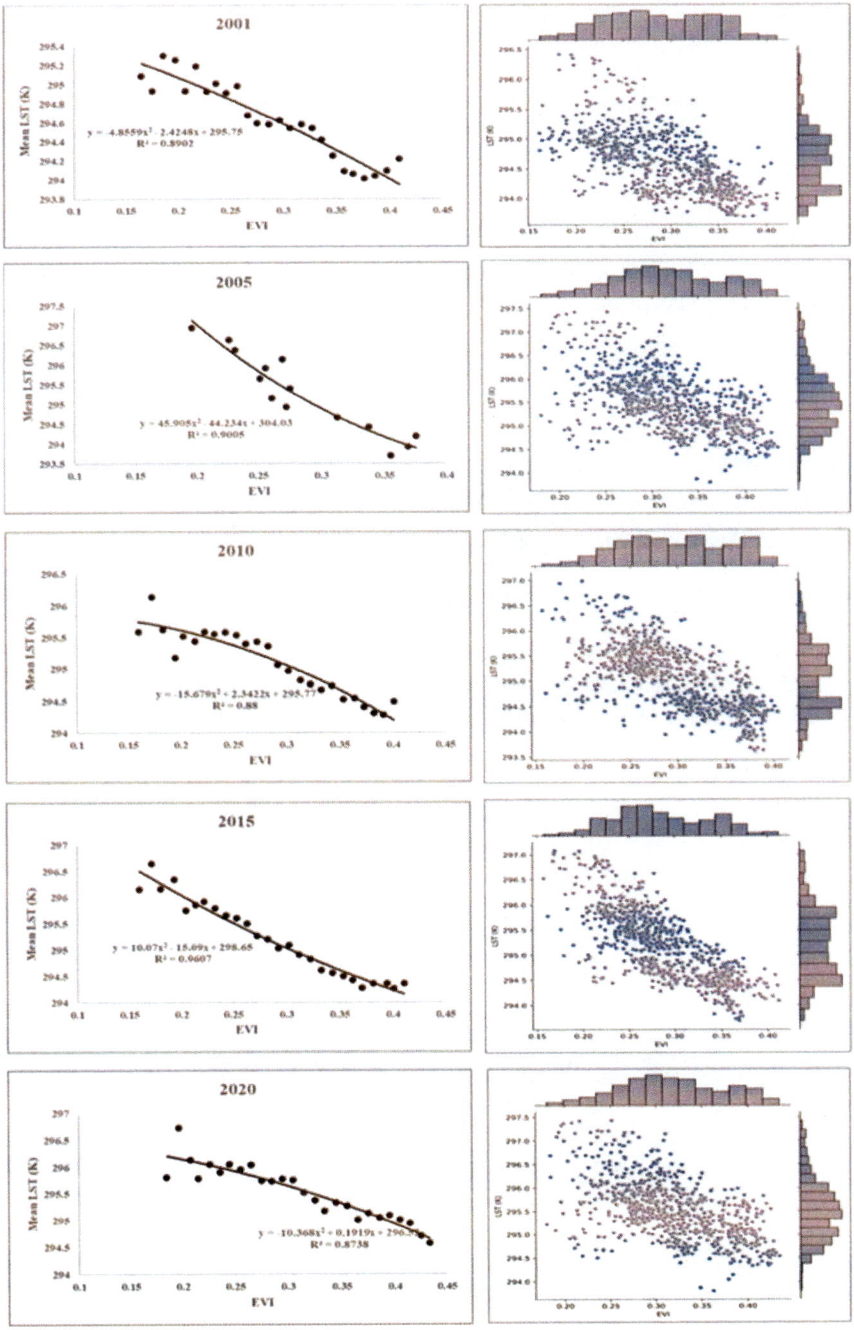

Fig. 12 Scatter plots and joint plots between LST and EVI for winter

R^2 value for the LST-EVI association is 0.88. With an average R^2 value of 0.92 for the winter season and 0.90 for the summer season, respectively, and a total mean R^2 value of 0.91, the association between mean EVI and LST is practically identical for both seasons.

As a key indicator of how the study area's surface temperature changes as a result of the various LULC types' varied behaviours, vegetation cover is an important parameter. In non-vegetated or low-vegetated areas, the behaviour of different LU types causes huge changes in surface temperatures, as the LULC cover in these locations is highly changeable. In contrast, temperature variations in thickly vegetated areas are much less pronounced.

In summary, the study shows a season-dependent and nonlinear relationship between LST and EVI, with the winter season having a better relationship than the summer season in the study area. The second-order polynomial relationship and R^2 values provide further insights into the LST-EVI relationship. The study highlights the importance of vegetation cover in determining the fluctuation in surface temperature and the different behaviour of LULC types.

Thermal Classification of Tiruchirappalli

Spatial autocorrelation analysis was performed on the UHI_{index} raster data set for yearly averages every five years for a period of twenty years, from 2001 to 2020, for the winter, summer, and annual data sets. Tables 1 and 2 present the findings. The summer average displayed a strong spatial correlation with a value of 0.9421 for Moran's index. The variance was 0.000649, while the predicted index value was −0.001250. The z-score was 37.234 and the p-value was 0, indicating a significant level of correlation. These results suggest that there is a high level of spatial clustering in the UHI_{index} values during the summer season.

The UHI_{index} raster dataset for the winter season was also examined for spatial autocorrelation. The final Moran's index value was 0.947, with an anticipated index of −0.001250 and a variance of 0.000648 for all images. The p-value was 0, and the z-score was 37.734, showing a significant correlation.

For the winter, summer, and annual averages, Table 3 displays the expected index, Moran's index, variance, p-value, and z-score. The estimated Moran's index values are positive, showing that there is a clustering tendency. The development of HS or CS is thus statistically significant, as evidenced by the enormous absolute values of the z-score and the evidently small p-value.

Table 1 Spatial correlation of the summer mean LST

Summer	Moran's index	Expected index	Variance	Z-score	P-value
2001	0.93	−0.0012	0.00064	36.81	0
2005	0.95	−0.0012	0.00064	37.17	0
2010	0.96	−0.0012	0.00064	37.26	0
2015	0.92	−0.0012	0.00064	36.48	0
2020	0.92	−0.0012	0.00064	36.42	0

Table 2 Spatial correlation of the winter mean LST

Winter	Moran's index	Expected index	Variance	Z-score	P-value
2001	0.94	−0.0012	0.00064	37.11	0
2005	0.95	−0.0012	0.00064	37.44	0
2010	0.95	−0.0012	0.00064	37.73	0
2015	0.94	−0.0012	0.00064	37.33	0
2020	0.95	−0.0012	0.00064	37.49	0

Table 3 Results of a spatial correlation analysis for winter, summer, and annual LST averages

For 20 years	Moran's index	Expected index	Variance	Z-score	P-value
Summer mean	0.95	−0.0012	0.00064	37.58	0
Winter mean	0.95	−0.0012	0.00064	37.73	0
Total mean	0.96	−0.0012	0.00064	37.79	0

Figure 13 shows the HS derived from the winter, summer, and annual average LST averages of the twenty-year average UHI_{index} maps. For each HS, the confidence levels are displayed, ranging from 90 to 99% for HS and from 90 to 99% for CS. The results indicate that pixels with higher UHI values have higher confidence levels. The Global Moran's Index (G.I.) value is also added as a lower layer on an Inverse Distance Weighted (IDW) map to show the HS zone. The IDW map has an orange area with lesser significance G.I. and a red area with higher significance G.I. The calculations revealed only a 90% confidence level CS; no 99% confidence level CS was found. A 90% confidence level CS for the northeastern portion of the area's G.I. is between −0.1 and −0.2.

The inhabitants of HS regions are exposed to a higher-than-average temperature that can adversely affect their health and well-being [25, 26]. The findings of this study can provide useful insights for reducing the UHI effects in Tiruchirappalli. By identifying the areas where the UHI effect is most severe, the study can inform policymakers and urban planners about the areas that require the most attention in terms of mitigation measures. Urban planning strategies that prioritize green spaces, vegetation cover, cool roofs, green roofs, shading devices, etc., incorporated into the city's development plans can help to reduce the UHI effect in the city [27]. Moreover, the study shows that undeveloped areas with vegetation experience lower temperatures, highlighting the importance of preserving and enhancing existing green spaces in the city. This finding can be used to inform policies related to the conservation of natural areas and the promotion of urban greening. Furthermore, the study identifies several locations with high UHI values, including the railway station, district collector's office, central bus stand, and Periyar College of Pharmaceuticals. These locations can be targeted for the implementation of mitigation measures to reduce the UHI effect and improve the thermal comfort of the residents.

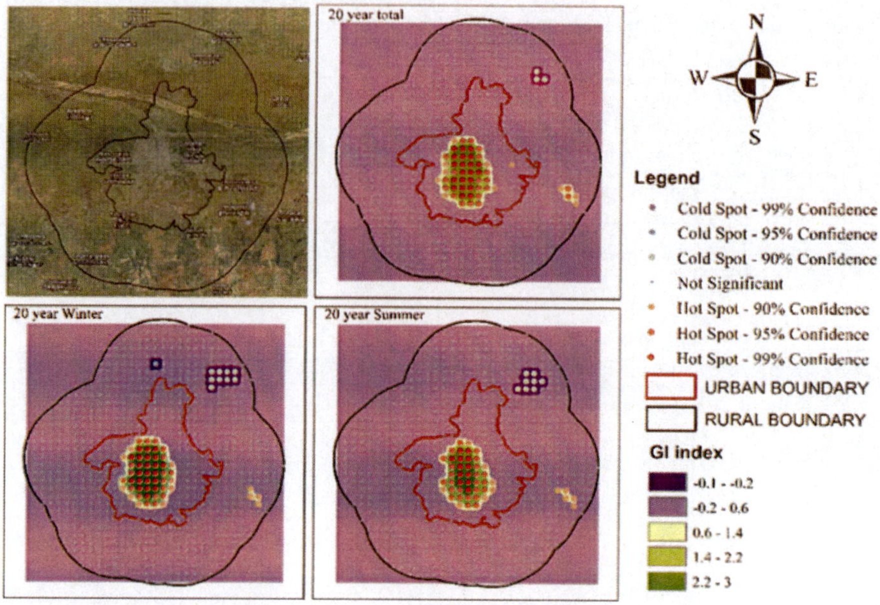

Fig. 13 Hot spot analysis for summer, winter, and annual mean LST

In summary, the study can guide sustainable development in Tiruchirappalli by informing urban planning and conservation policies that prioritize green infrastructure, promote urban greening, and target mitigation measures in areas with high UHI values.

This study focuses on investigating the spatio-temporal UHI effects in Tiruchirappalli, utilizing MODIS LST data for its daily temporal coverage. However, the use of coarser spatial resolution data may limit the identification of smaller scale UHI patterns within the cities. Also, the study is focused on a single city in India, so the results may not be generalizable to other cities or regions. To enhance the study's scope, future research could integrate higher spatial resolution data sources and consider a wider range of variables such as urban morphology and anthropogenic heat emissions. Longitudinal studies across different seasons and years, comparative analyses among various cities, and exploration of UHI mitigation strategies could provide a more comprehensive understanding of UHI dynamics and inform effective urban planning and climate resilience strategies.

5 Conclusions

The central urban area of Tiruchirappalli experiences the highest LST due to its high built-up area. However, the southeast portion of the rural area, which includes BHEL Tiruchirappalli, Navalpattu, and Kumaramangalam, also experiences similar high

LST values due to its development. The maximum UHI intensity in Tiruchirappalli is 4.16 K during the winter and 4.50 K during the summer, with an average of 4.30 K for the previous 20 years. The maximum UHI intensity outside the Tiruchirappalli city, in comparison, is 1.03 K in the summer and 1.05 K in the winter, with an average of 0.79 K. The EVI has a strong negative correlation with LST in both seasons, with the relationship being better in the winter. The R^2 value for the LST-EVI relationship is 0.88 on average. The district collector's office, central bus stop, railway station, and many educational and religious institutions are among the Tiruchirappalli areas recognized as HS with high temperatures. There is 90–99% confidence in these HS regions. With a 90 to 95% level of confidence, Thirumangalam, BHEL, and the Arulmiga Madhura Kaliyamman temple region are designated as HS outside the city.

Acknowledgements This study was funded by SEED grant—NITT/R&C/SEED GRANT/2021-22/P.14 from the National Institute of Technology, Tiruchirappalli. Furthermore, the authors would like to thank the United States Geological Survey (USGS) for making the satellite data available.

References

1. Gupta N, Mathew A, Khandelwal S (2020) Spatio-temporal impact assessment of land use/land cover (LU-LC) change on land surface temperatures over Jaipur city in India. Int J Urban Sustain Dev 12(3):283–299
2. Qiao Z, Tian G, Xiao L (2013) Diurnal and seasonal impacts of urbanization on the urban thermal environment: a case study of Beijing using MODIS data. ISPRS J Photogramm Remote Sens 85:93–101
3. Hassell JM, Begon M, Ward MJ, Fèvre EM (2017) Urbanization and disease emergence: dynamics at the wildlife–livestock–human interface. Trends Ecol Evol 32(1):55–67
4. Mackey TK, Liang BA (2012) Threats from emerging and re-emerging neglected tropical diseases (NTDs). Infect Ecol Epidemiol 2(1):18667
5. Mathew A, Khandelwal S, Kaul N (2016) Spatial and temporal variations of urban heat island effect and the effect of percentage impervious surface area and elevation on land surface temperature: Study of Chandigarh city, India. Sustain Cities Soc 26:264–277
6. Pandey A, Sadavarte P, Rao AB, Venkataraman C (2014) Trends in multi-pollutant emissions from a technology-linked inventory for India: II. Residential, agricultural and informal industry sectors. Atmos Environ 99:341–352
7. Radhi H, Assem E, Sharples S (2014) On the colours and properties of building surface materials to mitigate urban heat islands in highly productive solar regions. Build Environ 72:162–172
8. Levermore G, Cheung H (2012) A low-order canyon model to estimate the influence of canyon shape on the maximum urban heat island effect. Build Serv Eng Res Technol 33(4):371–385
9. Haizhu Z, Neng Z, Qingqin W (2021) Modelling and simulation of the urban heat island effect in a tropical seaside city considering multiple street canyons. Indoor Built Environ 30(8):1124–1141
10. Nakata-Osaki CM, Souza LCL, Rodrigues DS (2018) THIS-Tool for Heat Island simulation: a GIS extension model to calculate urban heat island intensity based on urban geometry. Comput Environ Urban Syst 67:157–168
11. Oke TR (1988) The urban energy balance. Prog Phys Geogr Earth Environ 12(4):471–508
12. Stewart ID, Oke TR (2012) Local climate zones for urban temperature studies. Bull Am Meteor Soc 93(12):1879–1900
13. Santamouris M, Ding L, Osmond P (2019) Urban heat island mitigation. In: Decarbonising the built environment. Palgrave Macmillan, Singapore, pp 337–355

14. US EPA (2022) Learn About Heat Islands. https://www.epa.gov/heatislands/heat-island-impacts. Accessed 28 June 2022
15. Simmons MT, Gardiner B, Windhager S, Tinsley J (2008) Green roofs are not created equal: the hydrologic and thermal performance of six different extensive green roofs and reflective and non-reflective roofs in a sub-tropical climate. Urban Ecosyst 11(4):339–348
16. Mathew A, Khandelwal S, Kaul N (2018) Analysis of diurnal surface temperature variations for the assessment of surface urban heat island effect over Indian cities. Energy Build 159:271–295
17. Townsend M, Henderson-Wilson C (2017) Greening the city: the health evidence of urban nature. In: de Leeuw E, Simos J (eds) Healthy cities. Springer, New York, NY
18. Voogt JA, Oke TR (2003) Thermal remote sensing of urban climates. Remote Sens Environ 86(3):370–384
19. Mathew A, Sarwesh P, Khandelwal S (2022) Investigating the contrast diurnal relationship of land surface temperatures with various surface parameters represent vegetation, soil, water, and urbanization over Ahmedabad city in India. Energy Nexus 5:100044
20. Zhang DL, Shou YX, Dickerson RR (2009) Upstream urbanization exacerbates urban heat island effects. Geophys Res Lett 36(24)
21. Mishra MK, Mathew A (2022) Investigating the spatio-temporal correlation between urban heat island and atmospheric pollution island interaction over Delhi, India using geospatial techniques. Arab J Geosci 15(20):1591
22. Government of India Ministry of Water Resources Central Ground Water Board South-eastern Coastal Region Chennai for official use Technical Report Series, 2008. https://cgwb.gov.in/District_Profile/TamilNadu/Trichy.pdf. Accessed 28 June 2022
23. Wan Z (2014) New refinements and validation of the collection-6 MODIS land-surface temperature/emissivity product. Remote Sens Environ 140:36–45
24. Badugu A, Arunab KS, Mathew A, Sarwesh P (2022) Spatial and temporal analysis of urban heat island effect over Tiruchirappalli city using geospatial techniques. Geodesy Geodyn
25. Ebi KL, Vanos J, Baldwin JW, Bell JE, Hondula DM, Errett NA, Hayes K, Reid CE, Saha S, Spector J, Berry P (2021) Extreme weather and climate change: population health and health system implications. Annu Rev Public Health 42:293
26. Zheng G, Li K, Wang Y (2019) The effects of high-temperature weather on human sleep quality and appetite. Int J Environ Res Public Health 16(2). https://doi.org/10.3390/ijerph16020270
27. Mathew A, Khandelwal S, Kaul N (2017) Investigating spatial and seasonal variations of urban heat island effect over Jaipur city and its relationship with vegetation, urbanization and elevation parameters. Sustain Cities Soc 35:157–177

Simulation of Streamflow and the Assessment of Nutrient Loadings for the Indravati River Basin of India using SWAT

Ch. Venkateswarlu, R. Manjula, P. Yuvaraja, and S. Hemavathi

Abstract Indravati River basin, a northern tributary of Godavari River, has a catchment area of 44,153 km^2 and is shared by the states of Maharashtra, Chhattisgarh, Telangana, and Odisha of India. It is one of the significant river basins for human and ecological needs and helps people through agriculture. However, due to rising nutrient toxicity from excessive agricultural and industrial runoff, a hydrological and water quality model is essential to assess streamflow and nutrient loadings and provide water management scenarios to reduce nutrient toxicity. The present study is focused on the development of a hydrological and water quality model to simulate streamflow, Nitrates and Ammonium ion loadings for the Indravati watershed for the period 2000–2012 using QGIS interface for SWAT (QSWAT) and for performing the Global Sensitivity analysis using SUFI-2 algorithm on SWAT-CUP for assessing the most sensitive parameters which affect the model's efficiency. SUFI-2 within SWAT-CUP was used to calibrate and validate the model over a monthly time step, with 8 parameters corresponding to streamflow and 6 parameters corresponding to nitrates and ammonium ion loadings in streamflow chosen based on the literature available. The simulation period 2000–2012 revealed a strong correlation between observed and simulated streamflow, nitrates and ammonium ion loadings. Statistical performance criteria using R^2 and Nash–Sutcliffe Efficiency (NSE) were 0.65, 0.70 and 0.67, and 0.64, 0.67 and 0.65 for streamflow, $\mathbf{NO_3^-}$ loading and $\mathbf{NH_4^+}$ loading, respectively. SWAT-CUP calibration and validation techniques provided the best-fitted values for all 14 parameters, 95PPU plots which show the best estimation curves for streamflow and nutrient loads considered, and global sensitivity analysis using t-Stat and P-value. This study has revealed that CN2, GW_DELAY, CH_N2, ESCO, GW_REVAP, and SOL_K are the most significant parameters affecting the streamflow and N_UPDIS, RCN, SOL_NO3, and CH_ONCO_BSN are the most significant parameters affecting the nutrient loadings of the Indravati River basin.

Ch. Venkateswarlu (✉) · R. Manjula · P. Yuvaraja
Department of Civil Engineering, National Institute of Technology, Tiruchirappalli, Tamil Nadu, India
e-mail: venkateswarluchalla100@gmail.com

S. Hemavathi
Department of Civil Engineering, K. Ramakrishnan College of Technology, Trichy, India

S. Mesapam et al. (eds.), *Developments and Applications of Geomatics*,
Lecture Notes in Civil Engineering 450, https://doi.org/10.1007/978-981-99-8568-5_34

There had not been any study of this type on this Indravati River basin. SWAT is used as it provides a flexible structure that allows finding solutions for different water resources, land management, and water pollution problems. The findings of this study would aid in managing land and water resources sustainably for future purposes.

Keywords SWAT · SWAT-CUP · 95PPU · Streamflow · Nutrient toxicity · Global sensitivity analysis · Significant parameters · Industrial runoff · Indravati River basin

1 Introduction

Rapid population growth, unplanned urbanization, and climate and land-use changes are significant challenges to water resources. The consequences of fertilizer and manure applications in irrigation processes, wastewater release, and industrial pollution are deteriorating river basins [1, 2]. In many areas of the world, surface water and groundwater nutrient toxicity have led to toxicity in the food chain, hurting aquatic and terrestrial living beings. Furthermore, temperature changes affect evapotranspiration, average soil water content, total water yield, streamflow, and nutrient loadings in river basins [3–6]. Furthermore, water scarcity is predicted to worsen as the world's population grows, resulting in increased demands for water for agricultural, industrial, and domestic usage [7]. In contrast, water availability has stayed stable or declined. Among all the models used, the Soil and Water Assessment Tool (SWAT) model is one of the finest watershed models on watershed and river basin scale used for a broad scale of hydrological and environmental problems [8–10].

SWAT is a comprehensive model that needs many model inputs for the parametrization of hydrology and basin description, which differ according to land use and soil characteristics. SWAT only accepts text-based inputs. In contrast, some software tools depend on scripting languages such as R or Python [11].

Disposal of pollutants into the river basins presents a novel global water quality challenge because of potentially severe threats to human health and ecosystems [12–14]. Indravati River basin is the primary source of water resources for the agricultural sector and people's livelihood but is being exploited due to industries in the nearby region. Agricultural runoff is more predominant, and continuous usage of fertilizers over decades is increasing nutrient pollution in the watershed. As a result, the dangers of eutrophication are increasing due to nutrient pollution of the water. An ongoing issue for environmental and water engineers is eutrophication [3, 15–17]. There is/are no or minimal literature available on the study of streamflow and nutrient loadings applied on the Indravati River basin. The prominent objectives of the present study are to develop a hydrological and nutrient loading model using QGIS interface for SWAT (QSWAT) for the Indravati River basin of India and to calibrate and validate the streamflow and nutrient loadings model using the SUFI-2 algorithm of SWAT-CUP. Finally, the most sensitive parameters influencing the changes in streamflow and nutrient loadings are to be found out. The findings of this present study may help in the sustainable management of land and water resources for future needs.

2 Materials and Methods

2.1 *Study Area*

Indravati is one of Godavari River's northern tributaries in its lower reaches. The Indravati catchment is located between latitudes 18°27′ and 20°41′ N, and longitudes 80°05′ and 83°07′ E. Indravati rises to a height of around 914 m near Thuamul Rampur village in Kalahandi district of Odisha state of India and enters the Godavari at roughly 125 m elevation. These locations are on the western slopes of the Eastern Ghats. The main river flows for approximately 477 kms. The total catchment area of 44,153.059 km^2 of the Indravati River basin is considered for this study, accounting for 13.32% of the whole Godavari basin. High hills, deep valleys, and enormous plateaus characterize the basin. In this region, there are no prominent irrigation projects. Forests are the most common land cover in the catchment (51.51%), followed by agriculture (17.73%). Agriculture is the main source of income in the study area. The location map of the Indravati River basin is shown in Fig. 1.

2.2 *Description of SWAT Model*

SWAT is a model designed to predict the impacts of land management practices on water, sediment, and agricultural chemical yields within extensive and intricate river basins or watersheds. These areas often exhibit diverse land uses, soil types, and management approaches, spanning extended timeframes [11]. SWAT operates as a continuous, physics-grounded, and partially distributed model, functioning at daily, monthly, and yearly intervals in its simulations. This modeling approach has been documented in various studies [10, 18–21].

Within the land phase of the hydrological cycle, SWAT has the capability to simulate a wide array of factors including climate, hydrology, land cover, plant growth, erosion, nutrient dynamics, pesticide behavior, and management practices within a watershed. To facilitate modeling accuracy, the watershed is divided into multiple sub-watersheds or sub-basins. Furthermore, each of these sub-basins is further subdivided into Hydrologic Response Units (HRUs) that are defined based on distinct characteristics such as land use, soil composition, and slope gradient. Input data organization for each sub-basin in SWAT encompasses climate information, hydrologic response units, groundwater dynamics, wetlands/ponds, and the primary channel or reach responsible for draining the respective sub-basin [1]. In the land phase of the hydrological cycle, the allocation of water, sediment, nutrients, and pesticides into

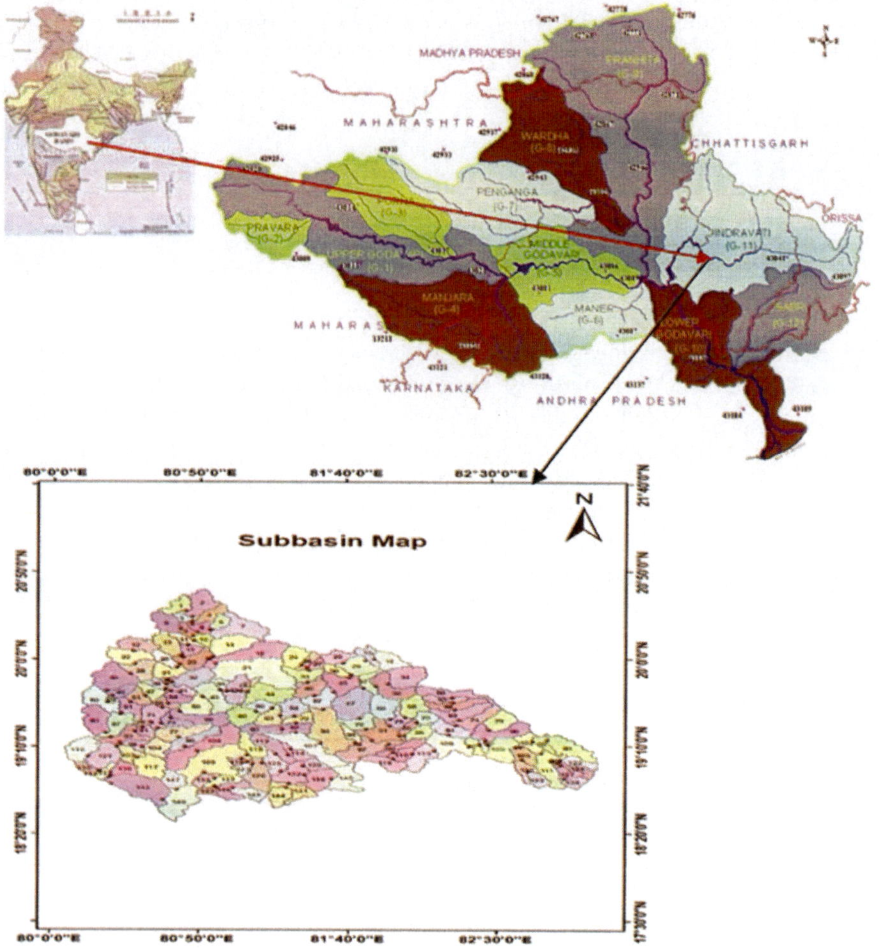

Fig. 1 Location map of Indravati River basin

the main channel of each sub-basin is managed. The subsequent movement of water, sediments, and other components through the network of channels within the basin, eventually reaching the outlet, constitutes the routing phase of the hydrological cycle [22].

2.3 *Data Sources*

Input data requirements for model simulation include a Land Use Land Cover (LULC) map of the basin, Digital Elevation Model (topographical properties), a Soil classification map which provides the complete soil classification and properties of the soil, and weather data which includes weather generator setup, solar radiation, precipitation, relative humidity, temperature, and wind speed. Observed precipitation and weather data (temperature, wind speed, solar radiation, and relative humidity) are processed, a calibration model is obtained for the period 2000–2008, and the model is validated for the period 2009–2012 using observed data of Streamflow and Nutrient Loadings obtained from the Ministry of Water Resources, India. Model spin-up time was calculated throughout each simulation's first three years (warm-up period). The observed flows and nutrient concentrations are used to calibrate sensitive parameters on a monthly time step. Then the model validation is carried out using field data obtained from hydrological observation stations. SWAT-CUP was used to calibrate streamflow, nitrates and ammonium ion parameters at the watershed scale.

2.4 *Digital Elevation Model*

NASA Earth data source was used to download digital elevation map polygons for Indravati region at a resolution of 30 $\times$ 30 m from the ASTER v3 Digital Elevation Model (DEM) dataset. For the study area, DEM has been of 25 parts, and with data management tools of QGIS interface, the 25 parts are mosaiced into a single raster dataset with WGS 1984 coordinate system. It will show unsigned 32-bit integer values for elevation. As an unsigned value, the correct UTM zone for Indravati is found, i.e., 44N, and the coordinate system is changed to WGS 1984 44N and projected. DEM has been clipped accurately for the study area with the help of the study area shapefile and is shown in Fig. 2. Highest elevation of the Indravati river basin is 1,367 m, the lowest elevation is 75 m, and the mean elevation is 490.57 m with the standard deviation of 183.11 m.

2.5 *LULC Map*

LULC maps for 2020 were obtained with a 10 m resolution from the Landsat ESRI land cover website. The maps were created using ESA Sentinel-2 images with a resolution of 10 m. It synthesizes LULC predictions for ten classes for the year, resulting in a representative glimpse of 2020. The raw imagery for this map comes from Sentinel-2 and the data projection is Universal Transverse Mercator (UTM) with WGS84 mosaic projection. This map was produced using a deep learning model that was trained using more than 5 billion manually annotated Sentinel-2 pixels gathered from more than 20,000 locations worldwide. The underlying deep learning model is trained using surface reflectance data from Sentinel-2, which comprises green, visible blue, red, near-infrared, and two shortwave infrared bands.

The model is run on several imaging dates throughout the year and results are combined to create a final map that is typical of 2020. The LULC classification

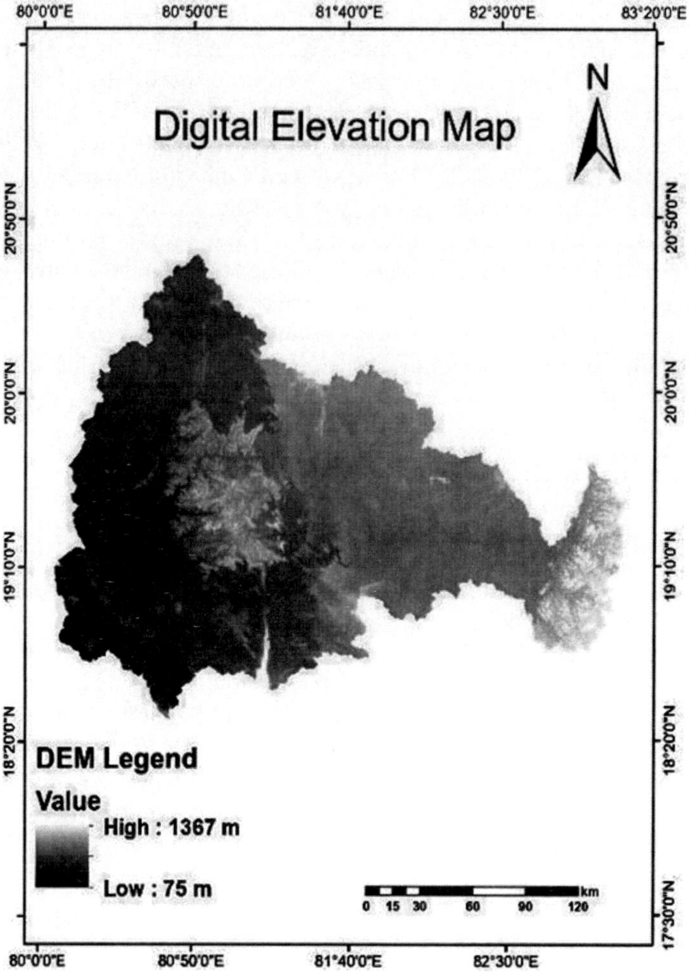

Fig. 2 Digital elevation model

of the Indravati River basin is shown in Fig. 3. Table 1 represents different LULC types and their corresponding percentages of area. Land use is classified into eight different classes: Water Body, Vegetation cover, Grass, Flooded vegetation, Agriculture, Scrub/Shrub, Built-up Area, and Bare ground with a maximum Vegetation cover and Agriculture at 51.51 percent and 17.73 percent, respectively.

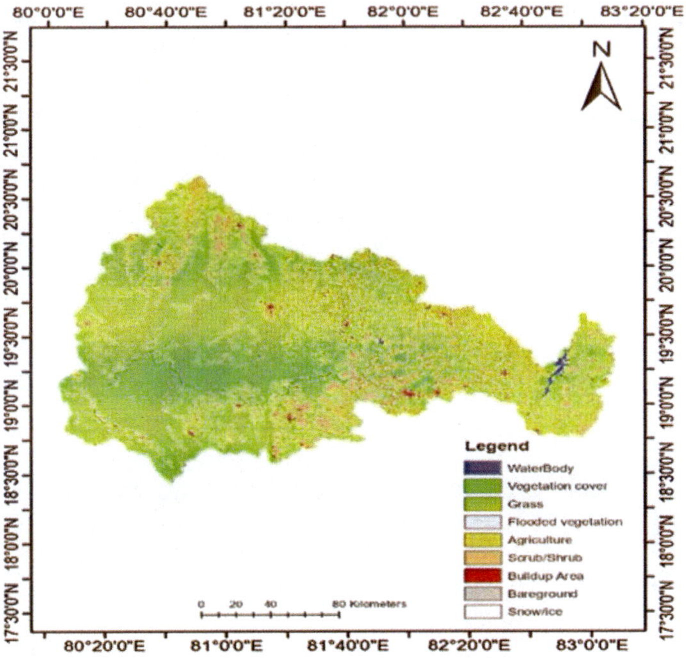

Fig. 3 LULC map for the study area

Table 1 LULC area percentages

Land cover type	Area (km^2)	Area (%)
Water body	432.507	0.98
Vegetation cover	22,742.43	51.51
Grass	35.159	0.08
Flooded vegetation	3.067	0.007
Agriculture	7827.327	17.73
Scrub/shrub	12,264.78	27.78
Built-up area	809.981	1.834
Bare ground	37.808	0.086

2.6 *Soil Map*

Soil cover classified map is extracted from the United Nations Food and Agriculture Organization (FAO) in a ESRI-2020 shapefile format. Soil text file for the Indian region is downloaded from the official SWAT website and imported to the SWAT database. With the help of the Indravati River basin shapefile, the soil cover is clipped on QGIS, and the polygon projection is converted to raster data with an equal number of cell sizes. The Soil Classification map of the Indravati River basin is shown in Fig. 4. Table 2 represents the different soil types and their corresponding area percentages. The soil cover is classified into eight classifications: Clay Loam1, Loam, Sandy Clay Loam1, Clay Loam2, Sandy Loam1, Sandy Loam2, Sandy Clay Loam2, and Clay. The Sandy Loam1 and Clay Loam1 covered most of the area at 48.3 and 21.2%, respectively. The Indian region soil text file is linked to SWAT with the FAO database. The attribute table for classification and delineation of soil cover provides the soil properties of the basin. It indicates that aquifers in the Indravati basin are divided into two layers, each with varying moist bulk density, saturated hydraulic conductivity, available water content, and organic carbon content.

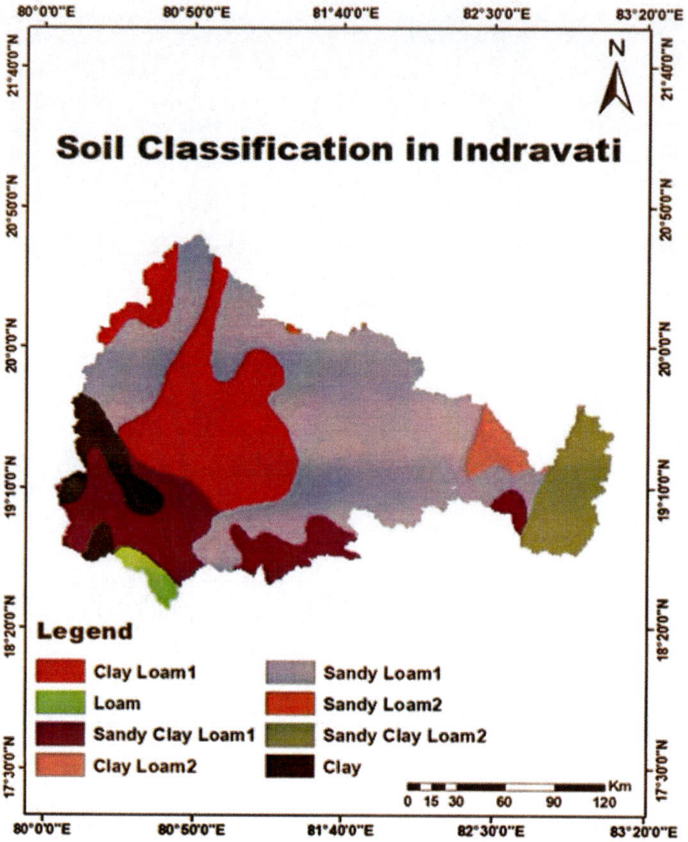

Fig. 4 Soil classification map

Table 2 Soil cover area percentages

Soil name	Area (km^2)	Area (%)
Clay loam1	9351	21.2
Loam	584	1.32
Sandy clay loam1	6330	14.3
Clay loam2	1090	2.47
Sandy loam1	21,308	48.3
Sandy loam2	58	0.13
Sandy clay loam2	3168	7.18
Clay	2264	5.13

3 Results and Discussion

3.1 Sub-basins and HRU Creation

The main outlet of the basin is given as Perur, the bottom-most elevated point in the Indravati River basin. The delineated sub-basins' map of study area is shown in Fig. 5. Indravati basin is classified into 148 sub-basins. HRUs are created by giving the threshold percentage as 15% over sub-basin for land use, soil type and slope class, respectively. The study area is divided into 837 HRUs. The Hydrologic Cycle of the watershed is shown in Fig. 6. It is inferred from the hydrologic cycle that the Average Curve Number, Precipitation, Evapotranspiration (ET), Potential Evapotranspiration (PET), Surface Runoff, Lateral Flow, and Return Flow are 75.53, 2,084.1, 558, 1,865.1, 668.94, 82.48, and 665.83 mm, respectively.

3.2 Hydrologic Cycle

Table 3 shows the complete hydrologic parameters affecting the model. It represents the average annual value of each variable in the watershed during a specific month. The parameters include rainfall, SUR Q (Surface Runoff), LAT Q (Lateral Flow), WATER YIELD, ET, SED YIELD (sediment yield), and PET. The hydrological cycle shows a large amount of spatio-temporal variation in the mid-months such as June, July, August, and September of each simulated year due to high precipitation in the south-west monsoon period of India. Due to high average annual rainfall in the watershed during June, July, August, and September such as 299.17, 547.41, 468.89, and 322.88 mm, respectively, the Surface Runoff, Lateral Flow, ET, and PET are increased to (74.83, 189.29, 149.20 and 74.81 mm), (6.09, 16.37, 17.57 and 14.75 mm), (66.64, 97.09, 88.02 and 94.79 mm), and (181.25, 111.93, 101.73 and 117.31 mm), respectively as inferred from Table 3. As the sediment yield concentration increases from 2.85 to 9.94 metric tons (MT) for the June to August period,

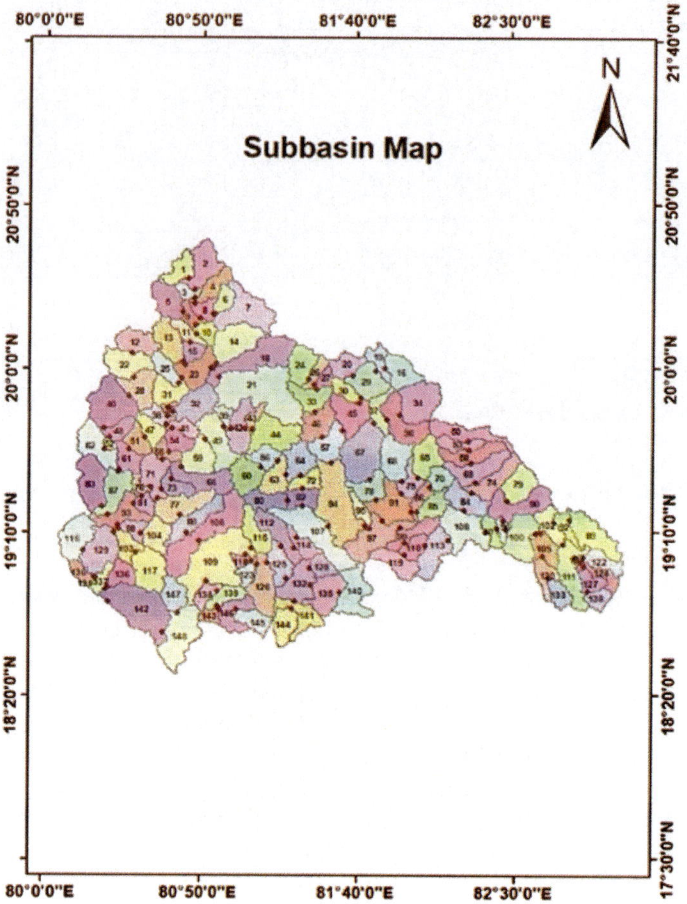

Fig. 5 Delineated sub-basins map

sediment yield retention occurs in September, October, and November are 8.5 MT, 2.55 MT, and 1.59 MT, respectively, whereas most of the sediments get out of the watershed in December (remained as 0.11 MT). Also, precipitation is mainly lost due to evapotranspiration and surface runoff.

Precipitation, evapotranspiration, lateral flow, surface runoff, groundwater, etc. are elements of a watershed's water balance. The study of long-term mean annual water balance components of the Indravati basin from 2000 to 2012 is divided into two parts: a warm-up period from 2000 to 2003 and an active period from 2004 to 2012. The SWAT-Check program determined the proportion of precipitation to each water balance component. Evapotranspiration of 558 mm, surface runoff of 668.94 mm, Percolation to shallow aquifer of 744.16 mm and followed by other components held in soils and/or lost from the channel by transmission were balanced with the incoming precipitation of 2,084.1 mm as shown in Fig. 6.

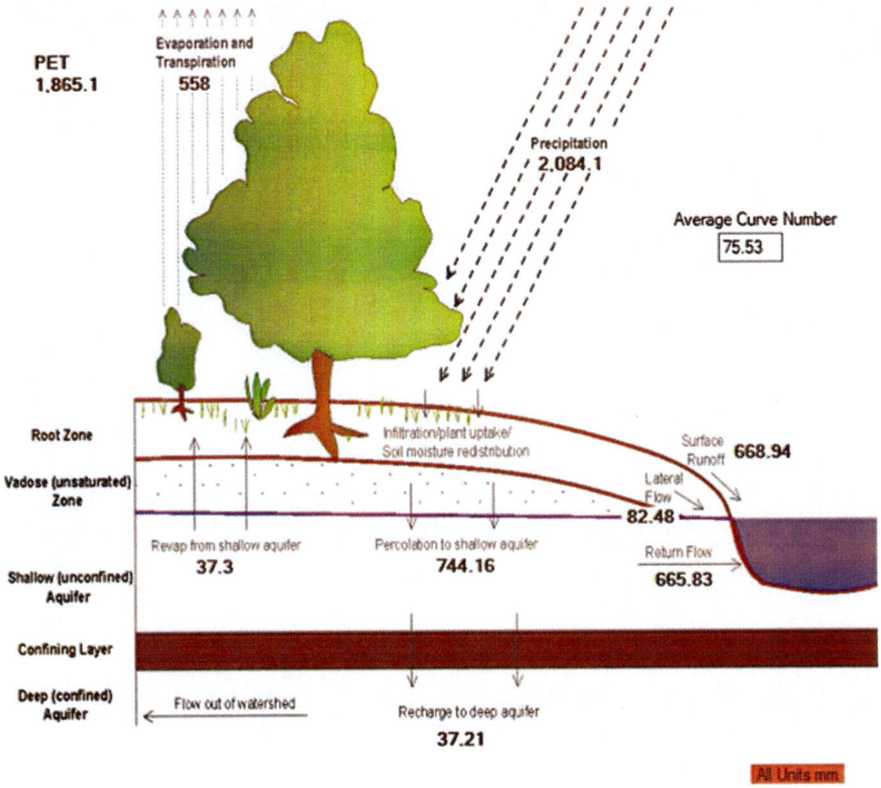

Fig. 6 Hydrologic cycle of the present study

3.3 Model Performance

SWAT outputs for monthly streamflow, NO_3^- and NH_4^+ loads, were calibrated using monthly observed data from the Ministry of Water Resources, India, for the Indravati watershed. The model was run for 13 years, from 2000 to 2012, with the first three years serving as a warm-up period, allowing the hydrologic cycle to be wholly represented. Three hundred simulations were evaluated for calibration and validation using SUFI-2 approach. In addition, land-use statistics from 2020 were used to achieve high land-use accuracy throughout the calibration process.

Table 3 Average annual values of simulated hydrological parameters in the watershed during a particular month

Month	Rainfall (mm)	SUR Q (mm)	LAT Q (mm)	Water yield (mm)	ET (mm)	SED yield (metric tons)	PET (mm)
Jan	15.98	0.43	1.24	15.08	15.92	0.11	134.66
Feb	6.90	0.71	0.94	4.77	16.77	0.19	159.37
Mar	17.69	1.20	0.85	3.89	44.88	0.36	214.31
Apr	15.14	1.01	0.71	2.97	46.86	0.09	239.29
May	14.69	0.26	0.57	1.78	18.5	0.02	255.36
Jun	299.17	74.83	6.09	81.13	66.64	2.85	181.25
Jul	547.41	189.29	16.37	239.53	97.09	6.95	111.93
Aug	468.89	149.20	17.57	290.19	88.02	9.94	101.73
Sep	322.88	74.81	14.75	250.71	94.79	8.5	117.31
Oct	112.50	15.85	7.35	169.79	73.43	2.55	132.39
Nov	51.43	7.83	3.97	101.34	44.83	1.59	120.57
Dec	9.44	0.41	1.81	48.89	22.93	0.11	121.01

3.4 Statistical Tests

The best-fitted values for all the parameters are found using the SUFI-2 calibration algorithm on SWAT-CUP and considering NSE as the objective function. The same is presented in Table 4.

Global Sensitivity Analysis is performed for all parameters and presented in the most to least sensitive order with the help of t-Stat and P-value, as available in Table 5. The most to least sensitive parameters for streamflow simulation are CN2, GW_DELAY, CH_N2, ESCO, GW_REVAP, SOL_K, ALPHA_BF, and SURLAG, respectively. On the other hand, the most to least sensitive parameters for simulating Nitrates and Ammonium ions are N_UPDIS, RCN, SOL_NO3, CH_ONCO_BSN, SOLN_CON, and SOL_ORGN.

The dotted plots derived from SUFI-2 considering NSE as the Objective Function at each simulation of SWAT-CUP by varying parameter values are shown in Fig. 7 and provide the variable range of parameters.

Scatter plots of the parameters considered

95PPU plots

The 95PPU plots for streamflow, nitrates and ammonium ions, are simulated using the SUFI-2 approach of SWAT-CUP and shown in Fig. 8. Table 6 gives the details about the performance of SWAT model in simulating streamflow, NO_3^- and NH_4^+, using Statistical Evaluators such as P-factor, R-factor, R^2 (Coefficient of Determination) and NSE. P-factor, which ranges from 0 to 1, is the percentage of measured data (plus

Table 4 Fitted parameters after calibration

Parameter name	Parameter	Fitted value	Min value	Max value
Moisture condition II curve number	r_CN2.mgt	0.1353	−0.2	0.2
Baseflow recession constant	v_ALPHA_BF.gw	0.3275	0	1
Delay time for aquifer recharge (days)	v_GW_DELAY.gw	442.2999	30	450
Groundwater Revap coefficient	v_GW_REVAP.gw	0.0515	0	0.2
Soil evaporation compensation coefficient	v_ESCO.hru	0.8165	0.8	1
Manning's "n" value for the main channel	v_CH_N2.rte	0.0232	0	0.3
Saturated hydraulic conductivity (mm/hr)	r_SOL_K.sol	0.588	−0.8	0.8
Surface runoff lag coefficient	v_SURLAG.bsn	0.0673	−0.2	0.2
Soluble nitrogen concentration in runoff	v_SOLN_CON.hru	0.2167	0	10
Initial organic N concentration in the soil layer (mg N/kg soil or ppm)	r_SOL_ORGN.chm	83.5	0	100
Initial NO_3^- concentration in the soil layer (mg N/kg soil or ppm)	r_SOL_NO3.chm	95.8333	0	100
Channel organic nitrogen concentration in basin (ppm)	v_CH_ONCO_BSN.bsn	91.8333	0	100
Concentration of nitrogen in rainfall (mg N/L)	v_RCN.bsn	2.125	0	15
Nitrogen uptake distribution parameter	v_N_UPDIS.bsn	76.8333	0	100

Table 5 Global sensitivity analysis of the parameters

Parameter	t-Stat	P-value	Rank
r_CN2.mgt	−7.403	0.00004	1
v_GW_DELAY.gw	−6.421	0.00006	2
v_CH_N2.rte	4.882	0.00012	3
v_N_UPDIS.bsn	2.725	0.00684	4
v_ESCO.hru	2.030	0.04327	5
v_GW_REVAP.gw	−1.909	0.05721	6
v_RCN.bsn	−1.699	0.09049	7
r_SOL_K.sol	1.541	0.12448	8
r_SOL_NO3.chm	−1.410	0.15962	9
v_ALPHA_BF.gw	1.089	0.27722	10
v_CH_ONCO_BSN.bsn	0.936	0.34987	11
v_SOLN_CON.hru	−0.692	0.48958	12
v_SURLAG.bsn	−0.666	0.50591	13
r_SOL_ORGN.chm	0.194	0.84623	14

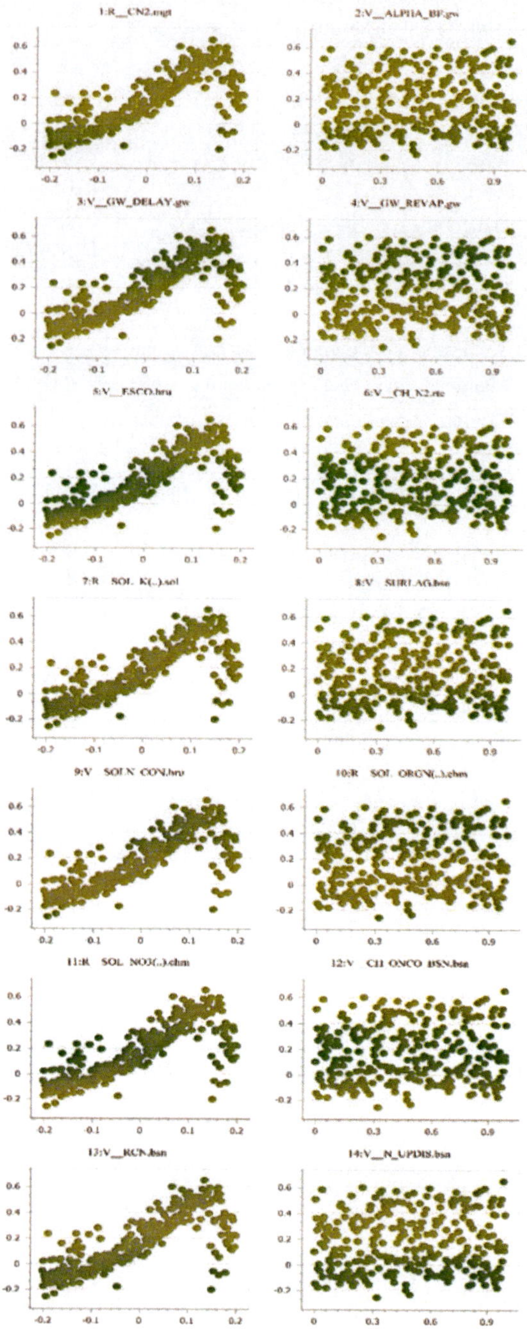

Fig. 7 Scatter plots for all the fourteen parameters varied concerning NS (Nash–Sutcliffe)

its error) bracketed by the 95PPU band. A P-factor of 1 means that all measured data is entirely bracketed by the 95PPU band, representing a perfect model simulation considering uncertainty. This means that value (1–(P-factor)) is inferred as an error fraction of the model. SWAT-CUP validation yielded P-factor of 0.64, 0.67, and 0.65 for streamflow, NO_3^- and NH_4^+, respectively. The average 95PPU band width ratio to measured variable's standard deviation is known as R-factor. R-factor of 0.29, 0.21, and 0.23 is obtained from validation for streamflow, NO_3^- and NH_4^+.

R^2 denotes the proportion of the variation in the simulated value from observed value and ranges 0–1, with the ideal model indicating 1 as the value. R^2 of 0.65,

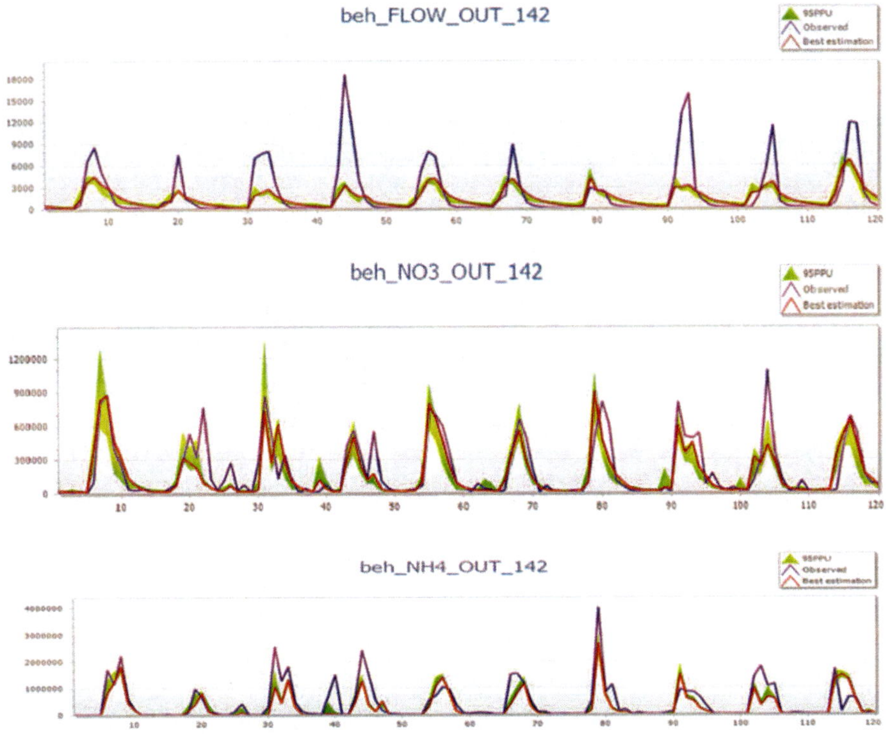

Fig. 8 95PPU plots of **a** Flow out, **b** NO_3^- Out, and **c** NH_4^+ Out

Table 6 Statistical evaluation of the model for the validation period (2009–2012)

Statistical evaluators	FLOW_OUT_142	**NO_3^-** _OUT_142	**NH_4^+** _OUT_142
P-factor	0.64	0.67	0.65
R-factor	0.29	0.21	0.23
R^2	0.65	0.70	0.67
NSE	0.64	0.67	0.65

0.70, and 0.67 are obtained for streamflow, NO_3^-, and NH_4^+ in the present study. The predicting skill of hydrological models is assessed using NSE, with a value of 1 being the perfect model. Nash–Sutcliffe factors of 0.64, 0.67, and 0.65 are obtained for streamflow, NO_3^-, and NH_4^+.

4 Conclusions

Monthly observed values obtained from the Ministry of Water Resources, India, were compared to simulated values from SWAT. The maximum streamflow was observed from June to September of every year of the simulation period due to high inputs of discharge from precipitation because of the south-west monsoon of India. There is an increase in Nitrates and Ammonium ion load due to considerable precipitation input, over-irrigation, and application of pesticides and fertilizers in those months of every year, as inferred from the simulated versus observed results plots. The most sensitive parameters for streamflow simulation are CN2, GW_DELAY, CH_N2, ESCO, GW_REVAP, and SOL_K, f whereas the most sensitive parameters related to the simulation of nutrient loads are N_UPDIS, RCN, SOL_NO3, and CH_ONCO_BSN. A high concentration of nutrient loads (NO_3^- and NH_4^+) is transported from intense farming as agricultural land cover and vegetation cover have a high percentage of land use at 51.51% and 17.73%, respectively. Therefore, results of this study would aid in managing water resources sustainably for future requirements.

References

1. Cheng J, Gong Y, Zhu DZ, Xiao M, Zhang Z, Bi J, Wang K (2021) Modeling the sources and retention of phosphorus nutrient in a coastal river system in China using SWAT. J Environ Manag 278. https://doi.org/10.1016/j.jenvman.2020.111556
2. Zettam A, Taleb A, Sauvage S, Boithias L, Belaidi N, Sanchez-Perez JM (2020) Applications of a SWAT model to evaluate the contribution of the Tafna catchment (north-west Africa) to the nitrate load entering the Mediterranean Sea. Environ Monitor Assess 192(8). https://doi.org/10.1007/s10661-020-08482-0
3. Bauwe A, Kahle P, Lennartz B (2019) Evaluating the SWAT model to predict streamflow, nitrate loadings and crop yields in a small agricultural catchment. Adv Geosci 48:1–9. https://doi.org/10.5194/adgeo-48-1-2019
4. Čerkasova N, Umgiesser G, Ertürk A (2018) Development of a hydrology and water quality model for a large transboundary river watershed to investigate the impacts of climate change—a SWAT application. Ecol Eng 124:99–115. https://doi.org/10.1016/j.ecoleng.2018.09.025
5. Hoghooghi N, Radcliffe DE, Habteselassie MY, Jeong J (2017) Modeling the effects of onsite wastewater treatment systems on nitrate loads using SWAT in an urban watershed of metropolitan Atlanta. J Environ Qual 46(3):632–640. https://doi.org/10.2134/jeq2016.08.0322
6. Neupane P, Bailey RT, Tavakoli-Kivi S (2020) Assessing controls on selenium fate and transport in watersheds using the SWAT model. Sci Total Environ 738. https://doi.org/10.1016/j.scitotenv.2020.140318

7. Rocha J, Carvalho-Santos C, Diogo P, Beça P, Keizer JJ, Nunes JP (2020) Impacts of climate change on reservoir water availability, quality and irrigation needs in a water scarce Mediterranean region (southern Portugal). Sci Total Environ 736. https://doi.org/10.1016/j.scitotenv.2020.139477
8. Bailey RT, Tavakoli-Kivi S, Wei X (2019) A salinity module for SWAT to simulate salt ion fate and transport at the watershed scale. Hydrol Earth Syst Sci 23(7):3155–3174. https://doi.org/10.5194/hess-23-3155-2019
9. Osei MA, Amekudzi LK, Wemegah DD, Preko K, Gyawu ES, Obiri-Danso K (2019) The impact of climate and land-use changes on the hydrological processes of Owabi catchment from SWAT analysis. J Hydrol Regional Stud 25. https://doi.org/10.1016/j.ejrh.2019.100620
10. Shi W, Huang M (2021) Predictions of soil and nutrient losses using a modified SWAT model in a large hilly-gully watershed of the Chinese Loess Plateau. Int Soil Water Conserv Res 9(2):291–304. https://doi.org/10.1016/j.iswcr.2020.12.002
11. Arnold JG, Moriasi DN, Gassman PW, Abbaspour KC, White MJ, Srinivasan R, Santhi C, Harmel RD, van Griensven A, Van Liew MW, Kannan N, Jha MK (2012) SWAT: Model Use, Calibration, And Validation. Trans ASABE 55(4):1491–1508. http://swatmodel.tamu.edu
12. Donmez C, Sari O, Berberoglu S, Cilek A, Satir O, Volk M (2020) Improving the applicability of the swat model to simulate flow and nitrate dynamics in a flat data-scarce agricultural region in the mediterranean. Water (Switzerland) 12(12):1–24. https://doi.org/10.3390/w12123479
13. Wang Y, Chen N (2021) Recent progress in coupled surface–ground water models and their potential in watershed hydro-biogeochemical studies: a review. Watershed Ecol Environ 3:17–29. https://doi.org/10.1016/j.wsee.2021.04.001
14. Zeiger SJ, Owen MR, Pavlowsky RT (2021) Simulating nonpoint source pollutant loading in a karst basin: a SWAT modeling application. Sci Total Environ 785. https://doi.org/10.1016/j.scitotenv.2021.147295
15. Moriasi DN, Gowda PH, Arnold JG, Mulla DJ, Ale S, Steiner JL (2013) Modeling the impact of nitrogen fertilizer application and tile drain configuration on nitrate leaching using SWAT. Agric Water Manag 130:36–43. https://doi.org/10.1016/j.agwat.2013.08.003
16. Peraza-Castro M, Ruiz-Romera E, Meaurio M, Sauvage S, Sánchez-Pérez JM (2018) Modelling the impact of climate and land cover change on hydrology and water quality in a forest watershed in the Basque Country (Northern Spain). Ecol Eng 122:315–326. https://doi.org/10.1016/j.ecoleng.2018.07.016
17. Woo SY, Kim SJ, Lee JW, Kim SH, Kim YW (2021) Evaluating the impact of interbasin water transfer on water quality in the recipient river basin with SWAT. Sci Total Environ 776. https://doi.org/10.1016/j.scitotenv.2021.145984
18. Narsimlu B, Gosain AK, Chahar BR, Singh SK, Srivastava PK (2015) SWAT model calibration and uncertainty analysis for streamflow prediction in the Kunwari River Basin, India, using sequential uncertainty fitting. Environ Proc 2(1):79–95. https://doi.org/10.1007/s40710-015-0064-8
19. Nilawar AP, Waikar ML (2018) Use of SWAT to determine the effects of climate and land use changes on streamflow and sediment concentration in the Purna River basin, India. Environ Earth Sci 77(23). https://doi.org/10.1007/s12665-018-7975-4
20. Uniyal B, Jha MK, Verma AK (2015) Assessing climate change impact on water balance components of a river basin using SWAT model. Water Resour Manage 29(13):4767–4785. https://doi.org/10.1007/s11269-015-1089-5
21. Vazquez-Amabile GG, Engel BA (2008) Fitting of time series models to forecast streamflow and groundwater using simulated data from SWAT. https://doi.org/10.1061/ASCE1084-0699200813:7554
22. Shrestha S, Bhatta B, Shrestha M, Shrestha PK (2018) Integrated assessment of the climate and land use change impact on hydrology and water quality in the Songkhram River Basin, Thailand. Sci Total Environ 643:1610–1622. https://doi.org/10.1016/j.scitotenv.2018.06.306

Spatiotemporal Analysis of Agricultural Drought in Krishna River Basin

Hussain Palagiri and Manali Pal

Abstract Agricultural drought is a period of time with decreasing Soil Moisture (SM) which is the main variable to define and identify agricultural drought. However, the actual water content in the soil is very rarely taken into consideration due to the unavailability of fine-resolution SM both temporally and spatially. Many agricultural drought studies for this reason are performed based on hydro meteorological variables like rainfall, temperature, evapotranspiration, runoff, etc. With the advance in microwave remote sensing, Soil Moisture Active Passive (SMAP) Level 3 data is able to provide daily SM estimates at 9 km resolution globally. This study uses SMAP L3 SM along with soil properties data from Harmonic World Soil Database (HWSD) to derive Soil Water Deficit Index (SWDI) for analyzing spatiotemporal agricultural drought in Krishna River Basin (KRB). Agricultural drought has been analyzed for both long-term (yearly) and short-term (monthly, seasonal) in KRB from 2016 to 2021. Temporal analysis of SWDI shows all the years and seasons from 2016 to 2021 in KRB are dry years with 2020 and 2021 having less drought than other years. Though all the seasons and years of the study period are dry, SWDI is able to capture inter-annual and inter-seasonal variation in the basin. SWDI when compared to the Aridity Index (AI) extracted from World Atlas of Desertification, dry/non dry lands captured by AI are justified by SWDI showing extreme dry/mild dry areas. Moreover, this study shows the potential of Remote Sensing SM in agricultural drought monitoring studies.

Keywords Agricultural drought · Soil water deficit index · Soil moisture · Soil moisture active passive

H. Palagiri · M. Pal (✉)
National Institute of Technology Warangal, Hanamkonda, Telangana 506004, India
e-mail: manalipal000@gmail.com

S. Mesapam et al. (eds.), *Developments and Applications of Geomatics*,
Lecture Notes in Civil Engineering 450, https://doi.org/10.1007/978-981-99-8568-5_35

1 Introduction

The agricultural industry is a critical element for ensuring stable livelihoods and economic growth in every state. The agricultural sector is an integral part of India's economy, with over half (54.6%) of the workforce involved in farming and related activities. For the fiscal year 2021–2022, it contributed 18.8% to the Gross Value Added of the country [1]. India has experienced a noteworthy rise in the frequency and spatial extent of agricultural droughts from 1980 to 2008 [2]. Therefore, identifying agricultural droughts in advance can help in mitigating the risk and implementing better agricultural policies and management. Agricultural drought is characterized by shortage of soil moisture (SM) that adversely affects plant growth or yields [3, 4].

Numerous indices were derived to identify agricultural droughts with SM as a basis. Examples include the Palmer Drought Severity Index (PDSI) [5], Crop Moisture Index (CMI) [6], Soil Moisture Index (SMI) [7], Soil Moisture Deficit Index (SMDI) [8], Empirical Standardized Soil Moisture Index (ESSMI) [9], Soil Water Deficit Index (SWDI) [10], and Soil Moisture Agricultural Drought Index (SMADI) [11]. Since meteorological drought and agricultural drought are closely related, indices such as Standardized Precipitation Index (SPI) [12] and Standardized Precipitation Evapotranspiration Index (SPEI) [13] which are based on meteorological variables are also frequently used to evaluate agricultural drought [14–17]. Additionally, vegetation indices, including Normalized Difference Vegetation Index (NDVI), Vegetation Condition Index (VCI) [18], and Vegetation Supply Water Index (VSWI) [19], are also frequently used to monitor agricultural drought [20–22].

In areas where the observed SM status is absent or limited, the use of indices may be difficult. In these cases, satellite SM data can serve as a suitable alternative [23, 24]. Satellite-based SM considers the interplay between meteorological inputs, vegetation, and evaporation, and can provide estimates that are comparable to actual observations [4]. Various remote sensing systems, including the European Remote Sensing Satellite Scatterometer (ERSCAT), Advanced Microwave Scanning Radiometer-2 (ASMR2), Advanced Scatterometer (ASCAT), Soil Moisture and Ocean Salinity (SMOS), and European Reanalysis Interim (ERAinterim) missions, are currently in operation. However, these systems have limited spatial and temporal resolutions. The Soil Moisture Active Passive (SMAP) mission, operated by the National Aeronautics and Space Administration (NASA), utilizes an L-band radiometer to produce high-resolution mapping of SM at intervals of 2 to 3 days. SMAP's recent addition of an accurate SM product now permits SM mapping closer to every single day. The accuracy, worldwide coverage, and frequent revisit of SMAP make it a valuable tool for monitoring SM and drought warning systems, especially in regions where ground measurement networks are limited. This study uses SMAP SM data to identify agricultural drought in KRB using SWDI.

The Soil Water Deficit Index (SWDI) developed by Martínez Fernández et al. [10] is used to identify agricultural drought conditions when long-term observations are unavailable. SWDI is very closely related to plant health as it is obtained by scaling SM data between the levels of Wilting Point (WP) and Field Capacity (FC). The WP and FC can be obtained by pedotransfer functions using soil characteristics [25, 26]. Several studies have shown the potential of SWDI in identifying agricultural drought conditions across different parts of the world [27–30].

It is clear that SMAP and SWDI were used successfully for agricultural drought identification in previous studies. Keeping in mind the agricultural dependency in India and its strong influence by SM, the objective of this study is to identify agricultural drought in Krishna River Basin (KRB) during the period of 2016 to 2021 at long term and short term using Soil Water Deficit Index (SWDI) using SMAP L3 data and compares it with Aridity Index (AI).

2 Study Area

Figure 1 shows the Krishna River Basin (KRB) which spans across four Indian states of Maharashtra (44%), Karnataka (26%), and Andhra Pradesh and Telangana (30%), draining a combined total of 258,948 km^2, near to 8% of the country's total geographic area. The basin's location is bounded by latitudes 13°10′ and 19°22′ North and longitudes 73°17′ to 81°9′ East. It has a maximum length of 701 km and a width of around 672 km. The basin experiences an average annual rainfall of 859 mm and an average annual mean temperature of 26.32 °C. The majority of the KRB is dedicated to agricultural land use with 75.86% utilized for agricultural purposes [31]. The KRB experiences tropical monsoons, with hot and dry summers and moderate to heavy rainfall during the monsoons. The region is prone to droughts, which lead to adverse effects on agriculture, human health, and the environment.

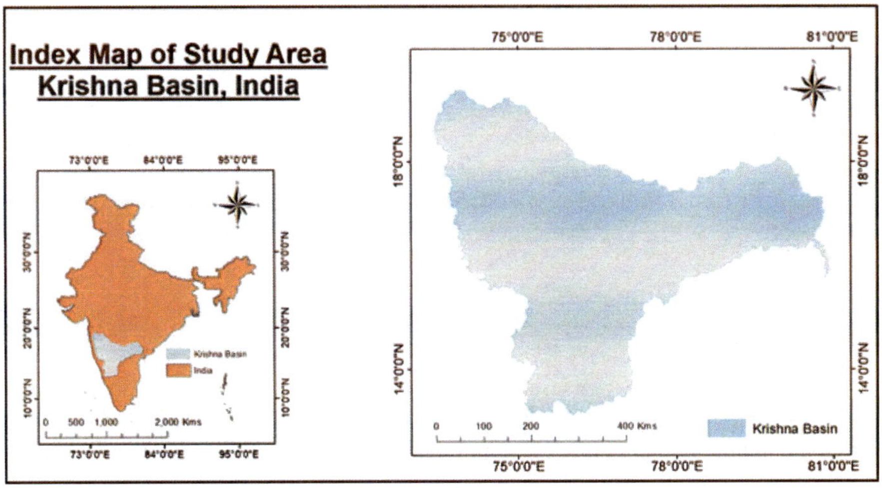

Fig. 1 Map of study area

3 Materials and Methods

3.1 Data

Soil Moisture Data. The Soil Moisture Active Passive (SMAP), which is a satellite-based mission launched by NASA in 2015, seeks to measure and map SM at a global level. The SMAP mission produces several products, including the Level 3 (L3) soil moisture data product, which is a gridded, global, daily dataset of surface SM and its uncertainty. The SMAP Level 3 data has a high spatial resolution of 9 km, which allows for detailed mapping of SM patterns at the regional scale, and a highly daily temporal resolution, which allows for the detection of rapid changes in SM over time. It also has a high accuracy, with an uncertainty of approximately 0.04 m^3/m^3 for SM and 0.5 cm for SM depth. In this study, SMAP L3 SM data is used in this study to calculate SWDI for the study basin from 2016 to 2021.

Soil Texture Data. Harmonized World Soil Database (HWSD) was created by compiling and harmonizing soil data from a variety of sources, including Food and Agriculture Organization (FAO), International Institute for Applied Systems Analysis (IIASA), European Soil Data Center (ESDAC), national soil survey organizations, research institutes, and other organizations. The Harmonic World Soil Database contains a wide range of soil information, including data on soil texture, pH, nutrient levels, organic matter content, soil erosion, salinity, and other soil properties. In this study, percentages of sand, clay, and organic matter from global HWSD are extracted for KRB to derive SWDI.

3.2 Methodology

Soil Water Deficit Index (SWDI). The SWDI for the KRB is derived by utilizing SMAP SM observations and HWSD soil characteristics. The SMAP-SM is re-gridded and brought to the resolution of HWSD by the nearest neighbor interpolation method, after which the SWDI is computed by the following equation:

$$SWDI = \left(\frac{SM - FC}{AWC}\right) \times 10 \tag{1}$$

In the above equation, SM refers to SMAP SM estimates, while FC and AWC refer to field capacity and available water capacity in the soil. The AWC is the difference between the FC and the wilting point (WP), and is calculated as follows:

$$AWC = FC - WP \tag{2}$$

Pedotransfer functions (PTFs) are used to determine FC and WP with the functions expressing the relations between FC, WP, and soil properties using the following equations:

$$WP = WP^* + \left(0.14 \times WP^* - 0.02\right) \quad (3)$$

$$WP^* = -0.024S + 0.487C + 0.006OM + 0.005(S \times OM) - 0.013(C \times OM) + 0.068(S \times C) + 0.031$$

$$FC = FC^* + [1.283\left(FC^*\right)^2 - 0.374FC^* - 0.015 \quad (4)$$

$$FC^* = -0.251S + 0.195C + 0.011OM + 0.006(S \times OM) - 0.027(C \times OM) + 0.452(S \times C) + 0.29$$

In the equations given above, S, C, and OM represent the % of sand, % of clay, and % of organic matter, respectively, which are obtained using improved pedotransfer functions (PTFs) [32]. To compute OM, Organic Carbon extracted from HWSD is divided by 0.58 [30]. Daily SM in KRB is averaged to monthly, seasonal, and yearly SM to derive seasonal and yearly SWDI from 2016 to 2021. The SWDI values are categorized into different drought classes as mild ($-2 \leq$ SWDI ≤ 0), moderate ($-5 \leq$ SWDI ≤ -2), severe ($-10 \leq$ SWDI ≤ -5), and extreme (SWDI ≤ -10) [10].

Aridity Index (AI). The AI of the KRB is extracted from the World Atlas of Desertification (WAD) data. Aridity is a climate phenomenon principally characterized by a shortage of water. Aridity is commonly computed by comparing the long-term averages of rainfall (R) and the potential evapotranspiration (PET). The AI which is an index of the average water available in the soil expressed as

$$AI = \frac{\sum_{i-1}^{30}\left[\frac{R_i}{PET_i}\right]}{30} \quad (5)$$

where R denotes Rainfall, PET denotes Potential Evapotranspiration, and i denotes the ith year from 1981 to 2010. WAD has used rainfall data from 1981 to 2010 from Global Precipitation Climatology Centre (GPCC) and PET from the Climate Research Unit (CRUTS v3.2) of the University of East Anglia to compute the AI.

The AI is a commonly utilized indicator for quantifying the level of dryness in a particular climate or geographic region. Land with AI < 0.65 are classified as drylands and AI $\geq$ 0.65 are classified as non-drylands.

Using the AI, different subtypes of lands are classified as shown in Table 1.

Table 1 Classification of land type based on AI [33]

Land type	Aridity index (AI)
Dry lands	AI < 0.65
Non-dry lands	AI ≥ 0.65

4 Results and Discussion

4.1 Temporal Analysis of SWDI

Figure 2 depicts the temporal distribution of spatially averaged monthly, seasonal, and yearly Soil Water Deficit Index (SWDI) in the Krishna River Basin (KRB) spanning from 2016 to 2021. In particular, Fig. 2a presents the annual distribution of spatially averaged SWDI values during the aforementioned period. Notably, this visualization underscores a consistent pattern of drought across all study years, as evidenced by negative SWDI values, signifying a shortfall in precipitation within the KRB. Examination of Fig. 2a reveals that the years 2016 through 2018 encountered extreme drought conditions, whereas the years 2019 to 2021 witnessed severe drought conditions. For the years characterized by extreme drought, namely 2016, 2017, and 2018, the corresponding SWDI values were −11.63, −10.75, and −10.80, respectively, underscoring the severity of water deficit within the KRB. Similarly, during the period of severe drought from 2019 to 2021, the SWDI values were recorded as −9.79, −8.34, and −8.12, respectively, further illustrating the sustained aridity. Noteworthy is the discernible trend in SWDI values, indicative of an escalating drought scenario over the study period. This trend is evident as SWDI progressively transitions from an extreme drought situation in 2016 to a severe drought scenario in 2021, reflecting the evolving hydroclimatic conditions within the KRB.

Figure 2b provides insight into the seasonal distribution of spatially averaged SWDI values from 2016 to 2021, with a focus on four distinct seasons: JFM (January–March), AMJ (April–June), JAS (July–August), and OND (October–December). Notably, all seasons exhibit negative SWDI values, indicative of prevailing drought conditions throughout the KRB. However, upon closer examination of Fig. 2b, it is apparent that the monsoon season (JAS) consistently registers higher SWDI values across all study years compared to the other seasons. Specifically, the monsoon season SWDI values for each year are −5.41 (2016), −5.49 (2017), −6.57 (2018), −3.83 (2019), −1.52 (2020), and −3.25 (2021), underscoring the enhanced precipitation capture during this critical period. Conversely, the initial two seasons display lower SWDI values, with respective values of −15.74 (2016 JFM), −14.81 (2017 JFM), −13.22 (2018 JFM), −14.21 (2019 AMJ), −12.72 (2020 JFM), and −11.15 (2021 JFM), thereby affirming SWDI's ability to delineate the monsoon's dominance in rainfall contribution.

Figure 2c illuminates the monthly distribution of spatially averaged SWDI values from 2016 to 2021 within the KRB. All months are characterized by negative SWDI values, attesting to the persistent drought conditions prevailing throughout the basin.

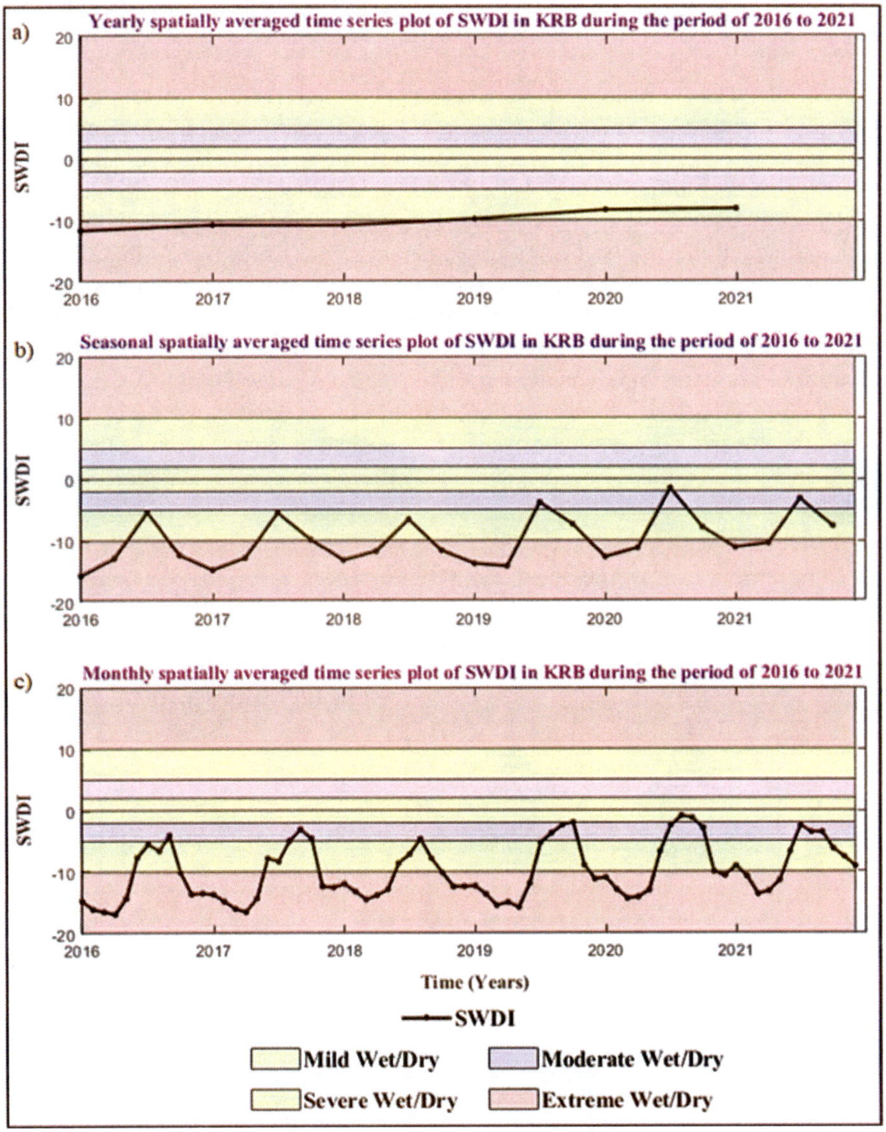

Fig. 2 Monthly/seasonal/yearly SWDI in KRB from 2016 to 2021

Notably, the intensity of drought during the monsoon months (July to August) is relatively subdued in comparison to the summer months (April to June). Evidently, the comprehensive analysis presented in Fig. 2 underscores the KRB's exposure to both short-term and long-term drought events during the entire study duration spanning from 2016 to 2021.

4.2 Aridity in KRB

Figure 3 depicts the Aridity Index (AI) analysis within the KRB, revealing distinct land classifications. The far-western regions of the study area are characterized as non-dry lands, denoted by AI values greater than or equal to 0.65, illustrated in a visually distinct blue hue. Conversely, the remaining expanse of the basin is identified as dry lands, distinguished by AI values below 0.65, and portrayed in a contrasting yellow shade. Remarkably, the non-dry lands situated in the far West exhibit a unique climatic profile marked by humidity and cold conditions, with a reference potential evapotranspiration (PET) of less than 400 mm. In contrast, the predominant yellow-colored dry lands, spanning a substantial portion of the basin, encompass a spectrum of aridity levels, encompassing hyper-arid ($AI < 0.05$), arid ($0.05 \leq AI < 0.2$), semi-arid ($0.2 \leq AI < 0.5$), and dry subhumid ($0.5 \leq AI < 0.65$) climates. Notably, the non-dry lands in the far west correspond to the Western Ghats region of India, characterized by dense forest cover, presenting an exceptional ecological facet within the broader KRB landscape.

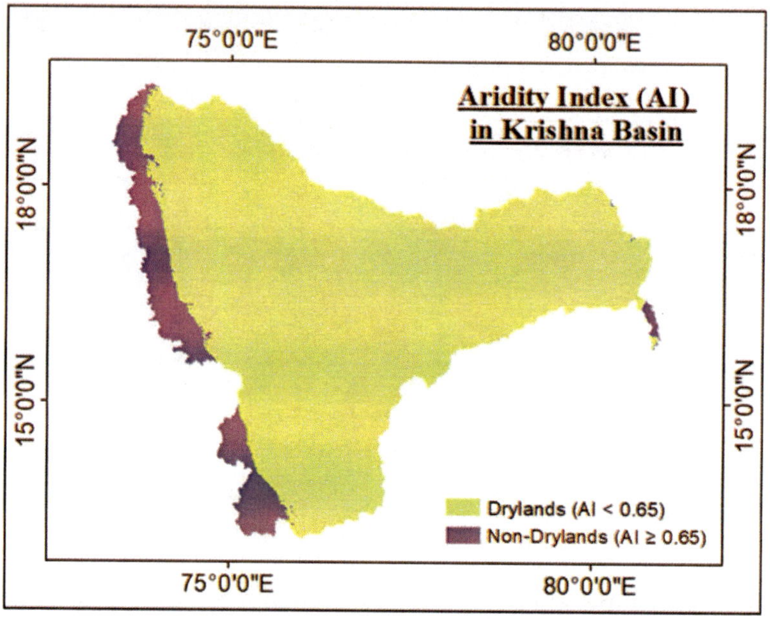

Fig. 3 Aridity index in KRB

4.3 Spatial Analysis of SWDI

Figure 4 provides a comprehensive overview of the annual spatially averaged SWDI distribution within the KRB across the years 2016 to 2021. Evidently, the far Eastern and far Western regions of the basin exhibit a distinct wetness pattern, while the predominant portion of the basin experiences dry conditions. This unique wetness trend in the far Eastern and Western regions can be attributed to the presence of the Eastern and Western Ghats, characterized by dense forest cover and contributing to higher SM content. The congruence of this wetness trend with the findings in Fig. 3 is striking, wherein the AI highlights the Western regions of the KRB as non-dry lands. The association between these two metrics reinforces the veracity of the depicted wetness phenomenon. The years 2020 and 2021 are notably characterized by decreased dryness and heightened wetness compared to the remaining study years. This discernible trend is mirrored in the temporal analysis presented in Fig. 2c, thereby corroborating the significance of these two years as periods of increased SM within the KRB. The spatial distribution of seasonal SWDI, as portrayed in Fig. 5, further underscores the wetter Eastern and Western regions juxtaposed against the drier expanse of the KRB. This pattern of wetness corresponds to the seasonal temporal analysis, reiterating the preeminence of the monsoon season (JAS) in terms of wetness and relative alleviation of drought conditions. Notably, while the plots for JFM, AMJ, and OND are characterized by darker yellow hues indicating elevated drought, the JAS plot exhibits lighter yellow tones signifying reduced drought severity, accompanied by darker blue shades depicting amplified wetness.

Figures 4 and 5 collectively establish the efficacy of the SWDI in capturing both inter-annual and inter-seasonal variations within the KRB. The consistent alignment between the spatial and temporal analyses solidifies SWDI's capacity to discern and depict these dynamic hydroclimatic fluctuations.

5 Conclusions

The comprehensive analysis presented in this study provides a profound understanding of the agricultural drought dynamics within the Krishna River Basin (KRB) from 2016 to 2021:

- The temporal distribution of spatially averaged Soil Water Deficit Index (SWDI) values, as depicted in Fig. 2, unveils a consistent and escalating pattern of drought across the study years. This pattern is clearly evident as the SWDI values transition from extreme drought in 2016 to severe drought in 2021. These findings emphasize the evolving aridity and the basin's vulnerability to sustained water deficit conditions.

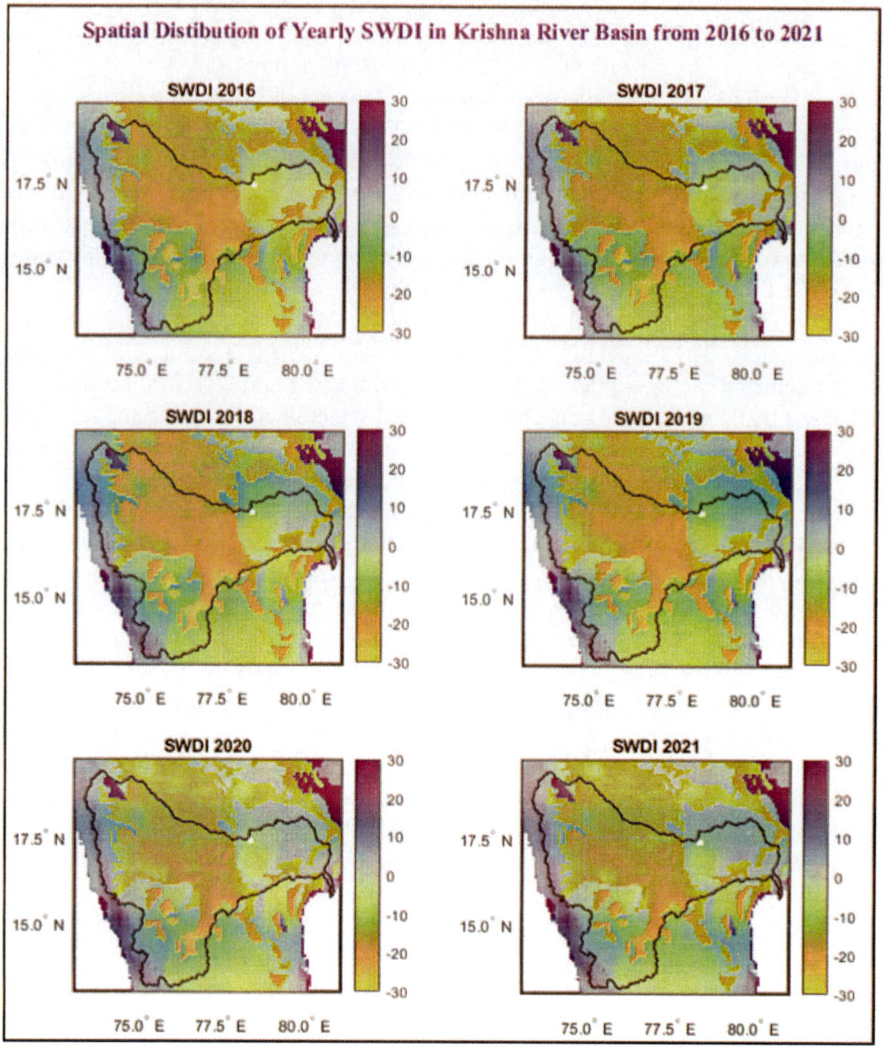

Fig. 4 Yearly SWDI in KRB from 2016 to 2021

- The seasonal distribution of SWDI values, illustrated in Fig. 2b, accentuates the crucial role of the monsoon season (JAS) in alleviating drought conditions. The monsoon season consistently registers higher SWDI values, underscoring its contribution to enhanced precipitation capture. Furthermore, the comparison of SWDI values with the Aridity Index (AI) in Fig. 3 corroborates the wetness trend observed in the far Western regions of the basin, attributed to their unique climatic profile.

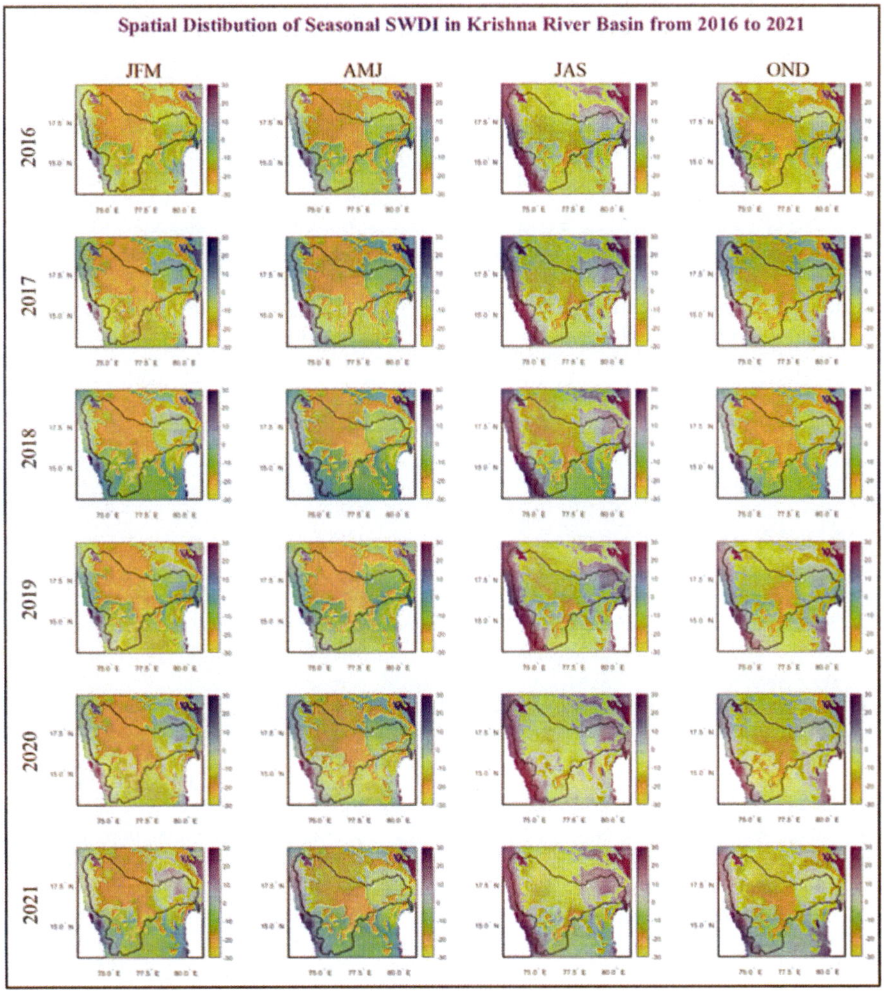

Fig. 5 Seasonal SWDI in KRB from 2016 to 2021

- The year-wise and spatial analyses presented in Figs. 4 and 5 highlight the distinctive wetness pattern in the far Eastern and Western regions of the KRB, reaffirming the role of dense forest cover in contributing to higher SM content. Additionally, the discernible decrease in dryness and heightened wetness in 2020 and 2021 aligns with both temporal and spatial analyses, underlining the significance of these years as periods of increased SM within the basin.
- The congruence between the spatial and temporal analyses firmly establishes the SWDI's effectiveness in capturing both inter-annual and inter-seasonal variations within the KRB. The SWDI's ability to delineate and depict dynamic hydroclimatic fluctuations further solidifies its utility as a valuable tool for assessing

drought conditions and guiding water resource management strategies in the Krishna River Basin.

References

1. Annual Report 2021–2022,.Department of Agriculture & Farmers Welfare Ministry of Agriculture & Farmers Welfare Government of India
2. Mishra V, Shah R, Thrasher B (2014) Soil moisture droughts under the retrospective and projected climate in India. J Hydrometeorol 15(6):2267–2292
3. Panu US, Sharma TC (2002) Challenges in drought research: some perspectives and future directions. Hydrol Sci J 47(S1):S19–S30. https://doi.org/10.1080/02626660209493019
4. Mishra AK, Singh VP (2010) A review of drought concepts. J Hydrol 391(1–2):202–216
5. Palmer WC (1965) Meteorological drought. In U.S. Weather Bureau Research Paper No. 45, (p 58)
6. Palmer WC (1968) Keeping track of crop moisture conditions, nationwide: the new crop moisture index. Weatherwise 21(4):156–161
7. Sridhar V, Hubbard KG, You J, Hunt ED (2008) Development of the soil moisture index to quantify agricultural drought and its "user friendliness" in severity-area-duration assessment. J Hydrometeorol 9(4):660–676
8. Narasimhan B, Srinivasan R (2005) Development and evaluation of soil moisture deficit index (SMDI) and evapotranspiration deficit index (ETDI) for agricultural drought monitoring. Agric For Meteorol 133(1–4):69–88
9. Carrão H, Russo S, Sepulcre-Canto G, Barbosa P (2016) An empirical standardized soil moisture index for agricultural drought assessment from remotely sensed data. Int J Appl Earth Obs Geoinf 48:74–84
10. Martínez-Fernández J, González-Zamora A, Sánchez N, Gumuzzio A (2015) A soil water based index as a suitable agricultural drought indicator. J Hydrol 522:265–273
11. Salvia MM, Sanchez N, Piles M, Gonzalez-Zamora A, Martínez-Fernández J (2020) Evaluation of the soil moisture agricultural drought index (SMADI) and precipitation-based drought indices in Argentina. In: 2020 IEEE Latin American GRSS & ISPRS remote sensing conference (LAGIRS). IEEE, pp 663–668
12. McKee TB, Doesken NJ, Kleist J (1993) The relationship of drought frequency and duration to time scales. In: Proceedings of the 8th conference on applied climatology, vol 17(22), pp 179–183
13. Vicente-Serrano SM, Beguería S, López-Moreno JI (2010) A multiscalar drought index sensitive to global warming: the standardized precipitation evapotranspiration index. J Clim 23(7):1696–1718
14. Sandeep P, Reddy GO, Jegankumar R, Kumar KA (2021) Monitoring of agricultural drought in semi-arid ecosystem of Peninsular India through indices derived from time-series CHIRPS and MODIS datasets. Ecol Ind 121:107033
15. Kumar U, Singh S, Bisht JK, Kant L (2021) Use of meteorological data for identification of agricultural drought in Kumaon region of Uttarakhand. J Earth Syst Sci 130(3):1–13
16. Roy S, Hazra S, Chanda A (2022) Assessment of wet season agricultural droughts using monthly MODIS and SAR data in the red and lateritic zone of West Bengal, India. Spat Inform Res 1–16
17. Palagiri H, Pal M (2022) Agricultural drought assessment using satellite-based surface soil moisture estimate. In: river dynamics and flood hazards: studies on risk and mitigation. Springer Nature, Singapore, pp 411–431
18. Kogan F (2002) World droughts in the new millennium from AVHRR-based vegetation health indices. EOS Trans Am Geophys Union 83(48):557–563

19. Carlson TN, Gillies RR, Perry EM (1994) A method to make use of thermal infrared temperature and NDVI measurements to infer surface soil water content and fractional vegetation cover. Remote Sens Rev 9(1–2):161–173
20. Dutta D, Kundu A, Patel NR, Saha SK, Siddiqui AR (2015) Assessment of agricultural drought in Rajasthan (India) using remote sensing derived Vegetation Condition Index (VCI) and Standardized Precipitation Index (SPI). Egypt J Remote Sens Space Sci 18(1):53–63
21. Vaani N, Porchelvan P (2018) Monitoring of agricultural drought using fortnightly variation of vegetation condition index (VCI) for the state of Tamil Nadu, India. Int Arch Photogram Remote Sens Spat Inform Sci 42:159–164
22. Kamble DB, Gautam S, Bisht H, Rawat S, Kundu A (2019) Drought assessment for kharif rice using standardized precipitation index (SPI) and vegetation condition index (VCI). J Agrometeorol 21(2):182–187
23. McNally A, Arsenault K, Kumar S, Shukla S, Peterson P, Wang S, Verdin JP (2017) A land data assimilation system for sub-Saharan Africa food and water security applications. Sci Daa 4(1):1–19
24. Yuan S, Quiring SM, Zhao C (2020) Evaluating the utility of drought indices as soil moisture proxies for drought monitoring and land–atmosphere interactions. J Hydrometeorol 21(9):2157–2175
25. Martínez-Fernández J, Sánchez N, González-Zamora Á (2017) Agricultural drought monitoring using satellite soil moisture and other remote sensing data over the Iberian Peninsula. In: Remote sensing of hydrometeorological hazards. CRC Press, pp 33–52
26. Pablos M, Martínez-Fernández J, Sanchez N, Gonzalez-Zamora A (2017) Temporal and spatial comparison of agricultural drought indices from moderate resolution satellite soil moisture data over Northwest Spain. Remote Sens 9(11):1168
27. Martínez-Fernández J, González-Zamora A, Sánchez N, Gumuzzio A, Herrero-Jiménez CM (2016) Satellite soil moisture for agricultural drought monitoring: assessment of the SMOS derived soil water deficit index. Remote Sens Environ 177:277–286
28. Paredes-Trejo F, Barbosa H (2017) Evaluation of the SMOS-derived soil water deficit index as agricultural drought index in Northeast of Brazil. Water 9(6):377
29. Mishra A, Vu T, Veettil AV, Entekhabi D (2017) Drought monitoring with soil moisture active passive (SMAP) measurements. J Hydrol 552:620–632
30. Fang B, Kansara P, Dandridge C, Lakshmi V (2021) Drought monitoring using high spatial resolution soil moisture data over Australia in 2015–2019. J Hydrol 594:125960
31. Krishna Basin Report, by Ministry of Water Resources, Government of India (2014) https://indiawris.gov.in/downloads/Krishna%20Basin.pdf
32. Saxton KE, Rawls WJ (2006) Soil water characteristic estimates by texture and organic matter for hydrologic solutions. Soil Sci Soc Am J 70(5):1569–1578
33. Cherlet M, Hutchinson C, Reynolds J, Hill J, Sommer S, von Maltitz G (eds) (2018) World atlas of desertification. Publication Office of the European Union, Luxembourg

Towards Imaging-based Quantification of Deterioration Using Colour Space Study

V. Guru Prathap Reddy, K. Bhanu, T. Tadepalli, and Rathish Kumar Pancharathi

Abstract This work presents an approach for imaging-based measurement of structure deterioration for M30 grade concrete. The samples are subjected to a variety of concrete deterioration chemicals including 5% concentrated solutions of H_2SO_4, HCl, NaCl, and $MgSO_4$ for 3, 7, 14, and 28 days, respectively. Red, Green, and Blue (RGB) colour space is not advisable for colour-based detection or colour analysis because of the mixing of colour and intensity information as well as non-uniform features. Hue, Saturation, and Value (HSV) colour space can quantify the surface colour detection of deterioration by the various chemical agents with respect to the duration of the attack. The surface colour change of the concrete cube specimen can be identified with respect to the duration of the attack and the type of chemical involved; meanwhile, it gives the residual strength of the specimen. This method enables the construction of a visual spectrum-based quantification tool for the degree and duration of deterioration by establishing empirical equations.

Keywords HSV colour space · Concrete · Chemical attack · Colour space · Image processing

1 Introduction

Deterioration is a major problem for existing infrastructure in India like buildings, dams, and bridges. Concrete is vulnerable to chemical attacks, because of the acidic, basic, and caustic nature of the environment. Cement paste-based materials are affected by the deterioration process and the original properties of the concrete. The phenomenon is that the chloride ions penetrate the concrete surface by its capillarity nature by absorption. Durability is an important characteristic because performance

V. Guru Prathap Reddy (✉)
Department of Civil Engineering, National Institute of Technology Nagaland, Chumukedima 797103, Nagaland, India
e-mail: guruv@nitnagaland.ac.in

K. Bhanu · T. Tadepalli · R. K. Pancharathi
Department of Civil Engineering, National Institute of Technology Warangal, Warangal, India

S. Mesapam et al. (eds.), *Developments and Applications of Geomatics*, Lecture Notes in Civil Engineering 450, https://doi.org/10.1007/978-981-99-8568-5_36

can be affected by severe pollution. Visual inspection can be used to evaluate the strength and weight losses of the cast specimen by proper quality check analysis [1]. One needs to take the required precautionary measures for the structure with the help of structural health monitoring system guidelines.

Normally, the image processing technique is used for object recognition by integrating the discrete objects in the pixels. Nowadays, these technologies have a variety of applications in the field of the medical industry, mineral identification, agriculture, mobile robots, etc. Red, green, and blue (RGB), hue, saturation, and value (HSV), and L*a*b* are the most often used colour categorization schemes. These colour spaces have been compared in earlier research for mineral identification [2].

A colour detection algorithm explores an image for pixels that have a specific colour in nature. The HSV colour space has been used in the present study, instead of the RGB space. According to this particular model, H (hue) is a dimension which represents the 'colour' of an image, S (saturation) is a dimension that represents the dominance of a colour, and V (value) dimension represents the 'brightness'.

2 HSV Colour Space

HSV stands for Hue, Saturation, and Value. HSV states or describes exactly the nature and type of colour space in an image. This particular colour space has three basic parts called hue, saturation, and value. 'Value' is in some cases replaced with 'brightness', and then it is known as HSB. In 1987, Alvy Ray Smith created the HSV model. It is in the shape of a Hexa cone, which is why it is also called the hex-cone colour model. Developing high-quality computer graphics in the current scenario, HSV is the most economical and accurate technique in the image processing techniques. Manmade colour can be assigned to the image after image processing. In this model, Fig. 1, hue can be represented circumferentially from 0° to 360°. Based on the hue angles of an image, the colours can be identified accurately. Normally, saturation can be expressed in the form of grey colour in the image strata with 0–100%. The numerical order represents the type of grey colour in the colour space band such as full grey, faded grey, and dull grey. The value of an image represents the brightness part of an image from the range of 0–100%, respectively. If the value of the colour space is '0', then it shows the black colour in the band.

This particular HSV colour space is used to create images with superior-rated computer graphics with desired colour selection. Generally the HSV model, is started with a conical shape object, Due to the conical shape of the HSV model, one can select a unique combination of with HSV colour. The following are the RGB values used to calculate the HSV values for the deteriorated concrete images, in the respective calculations:

$$V = \mathrm{MAX}\begin{cases} \mathrm{MAX} = \max(R, G, B) \\ \mathrm{MIN} = \min(R, G, B) \\ \delta = \mathrm{MAX} - \mathrm{MIN} \end{cases} \quad (1)$$

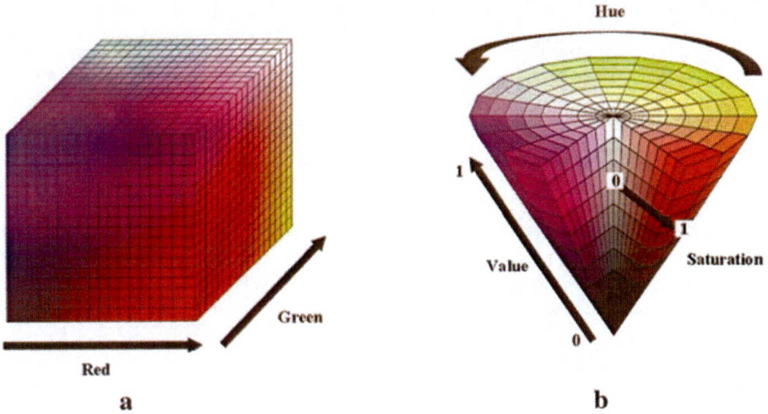

Fig. 1 HSV colour space

$$S = \begin{cases} \delta/\mathrm{MAX}, & \text{if MAX} \neq 0 \\ 0, & \text{if MAX} = 0 \end{cases} \tag{2}$$

$$H = \begin{cases} \frac{60(G-B)}{\delta}, & \text{if Max} = R \\ \frac{60(B-R)}{(\delta+2)}, & \text{if Max} = G \\ \frac{60(R-G)}{(\delta+4)}, & \text{if Max} = B \end{cases} \tag{3}$$

When MAX = 0, H is not defined. And 'H' should be confined to the range of 0°–360° only.

3 Materials

The following materials are employed in the experimental investigation.

3.1 Cement

In the present study, Ordinary Portland cement of 53 grade is used as per IS 12269-2013, ASTM C150-07 [3, 4]. The specific gravity is 3.12, while the initial and final setting times are 93 and 255 min, respectively, as per IS 4031 [5].

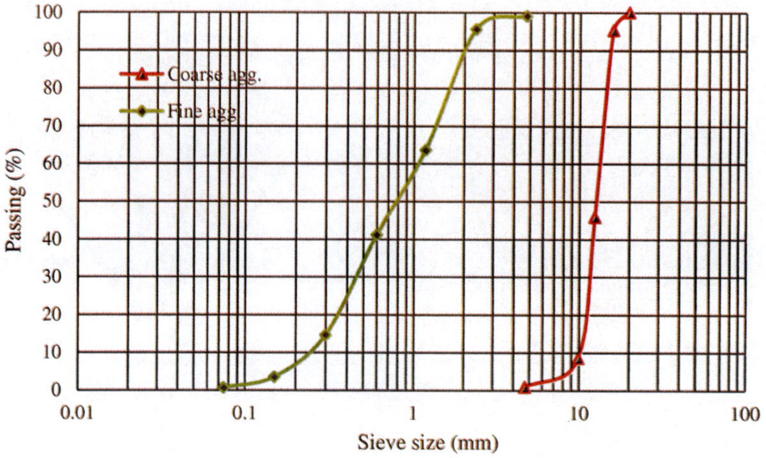

Fig. 2 Particle size distribution of coarse and fine aggregates

3.2 Fine Aggregate

River sand is used as fine aggregate and conforms to Zone-II of IS 383-2016, ASTM C128 [6, 7]. The grading curve is shown in Fig. 2, while bulk density (kg/m^3), specific gravity, void content (%), and water absorption are 1485, 2.65, 33.6, and 0.9, respectively.

3.3 Coarse Aggregate

Machine-crushed granite aggregate is utilized as coarse aggregate, and its properties have been evaluated in accordance with IS 2386 part IV-1963(R2016), ASTM C33 [8, 9]. The grading curve is shown in Fig. 2, while the bulk density (kg/m^3), specific gravity, void content (%), and water absorption are 1455, 2.7, 40.7, and 0.77, respectively.

3.4 Water

Potable water [10] is used in the concrete mix for preparing the specimen. The grade of concrete designed in this study is M30 (meaning the compressive strength is not less than 30 MPa strength at end of 28 days) designed as per IS 10262:2019 [11]. The mix proportions per m^3 of concrete are water—184 lts, cement—412 kgs, w/c ratio—0.45, fine aggregates—645 kgs, and coarse aggregates—1195 kgs.

4 Experimental Programme

This experimental setup consists of a high-resolution DSLR camera with a colour board checker, deteriorated concrete cube specimens on the front side, and a tripod stand with camera on the opposite side as shown in Fig. 3. This particular spectrum of images of deteriorated concrete cubes is taken at 3 levels of lighting, which are bright lighting conditions outdoor, indoor medium lighting in the second level, and ambient conditions indoor [12–15].

The image capturing area should consist of good lighting, and a room temperature of 27 °C can be maintained throughout the image capturing process. The deteriorated concrete cube specimen is placed on a 0.75–0.9 m high table with a white coloured background. The experimental setup consists of a tripod stand with a commercial DSLR high-resolution camera with 6000 × 4000 pixels, which has the ability to autofocus. The colour checker is kept beside the cube specimen, for photographic records of standard colour wavelengths. This chart includes 24 colours for colour characterization and analysis of an image. The illumination intensity of the surface specimen as well as the ambient light intensity or brightness values are measured using an HTC digital lux metre. This particular test is conducted for the measurement of light intensity conditions under the tube lights indoors. Generally, the average illumination is maintained at 210 lx next to the specimen. The captured images are processed as shown in Fig. 4. The HSV values are calculated using Equations [1–3].

Fig. 3 Experimental setup

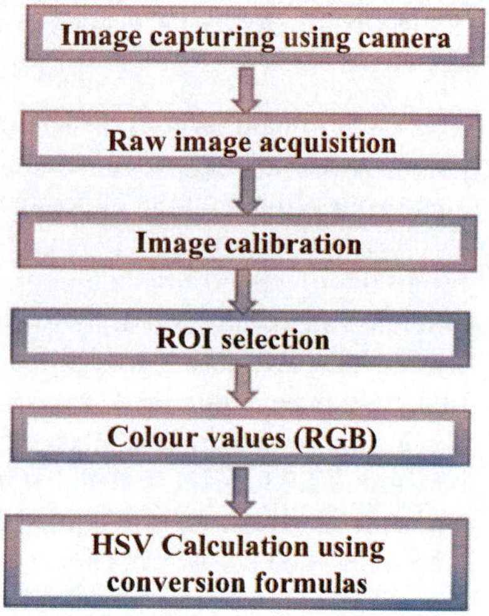

Fig. 4 Workflow of image analysis

5 Results and Discussion

The concrete cube specimens are exposed to aggressive environments like acidic and salt-based conditions with respect to the duration periods such as 3, 7, 14, and 28 days. Due to the acidic environmental conditions, the surface colour of the concrete cube changes drastically with respect to the time period. Based on the plot area, the exposed surface conditions can be plotted through the image enhancement tool. This surface plotting technique consists of some elements of good surface texture, which are mainly grid size, smoothening, the perspective of an image, lighting, and scale (X, Y, Z-axes). The surface colour plots for different chemicals vary; for example, salts such as sodium chloride are visible as white layer. The surface plots for H_2SO_4 and $MgSO_4$ are wine red and at 14 days, HCl exhibits a red colour along with exposed aggregate; thereafter, the cement becomes white. For different chemicals and exposure times, the surface plots vary as shown in Fig. 5.

These surface colouring features such as RGB values of the deteriorated cube specimen are evaluated by the image processing technique. With the help of conversion formulas, for a particular cube specimen, the HSV values are determined. Figures 6, 7 and 8 show how the H, S, and V colour spaces change after exposure to chemicals for 3, 7, 14, and 28 days. The mean value of three specimens is reported in each case.

Figure 6 depicts the change in the H value of a concrete surface after being exposed to HCl, H_2SO_4, NaCl, and $MgSO_4$ for 3, 7, 14, and 28 days, respectively. Figure 6 shows that the control specimen initially has a H of 59.5, which decreases to 45, 48, 41, and 55 after 3, 7, 14, and 28 days, respectively, of exposure.

Dark red colour causes more variance to be seen at 14 days, and photos also exhibit variation at this exposure time compared to the other exposure durations, while a white surface forms after 28 days. Figure 6 shows that exposure to H_2SO_4 for up to 14 days results in a white surface colouration with minimal variation owing

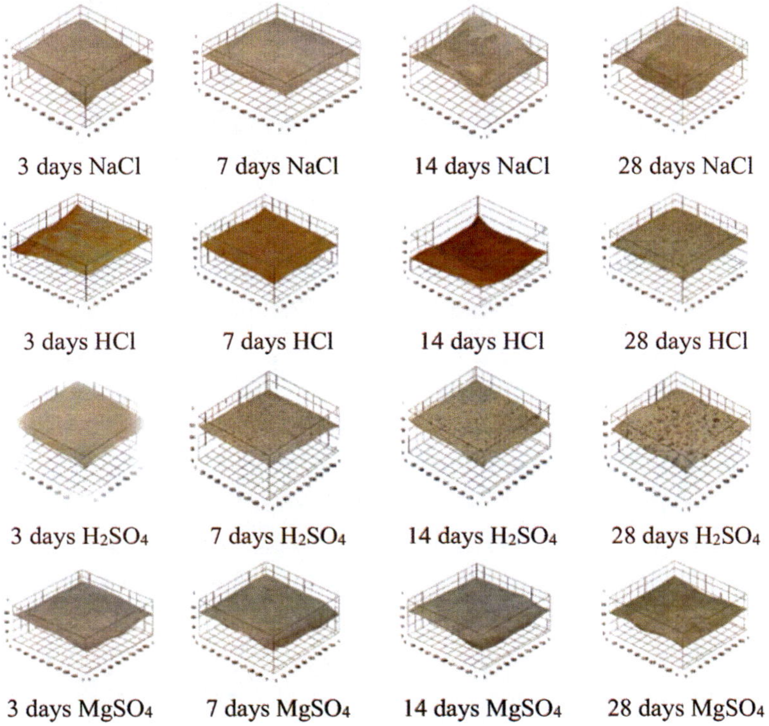

Fig. 5 Surface variations of concrete due to different exposures

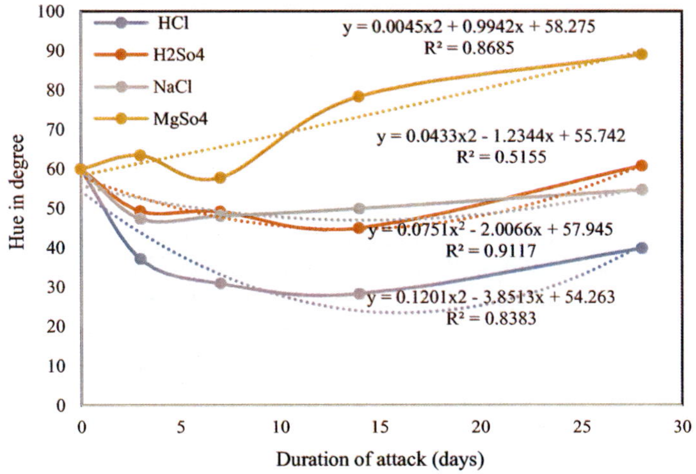

Fig. 6 Variation of Hue due to chemical attack

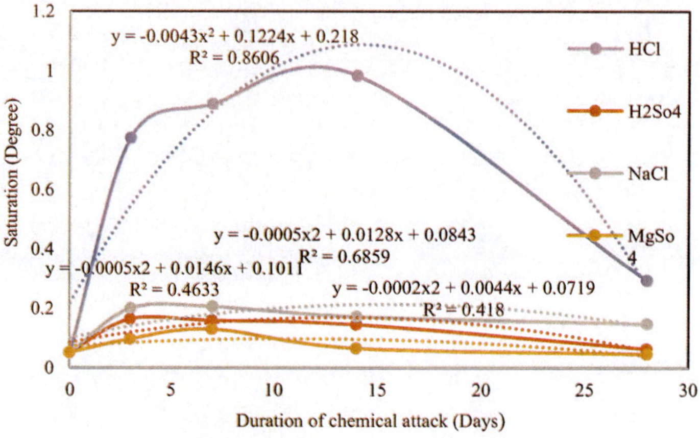

Fig. 7 Variation of saturation due to chemical attack

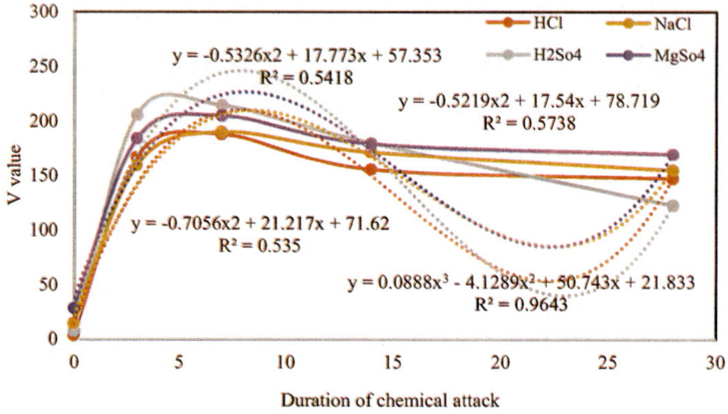

Fig. 8 Variation of value due to chemical attack

to the formation of Ettringite. A greater H value is shown at 28 days as a result of the aggregate phase's visibility following cement paste erosion.

The specimens subjected to NaCl generate white patches with just a minor fluctuation in H over time (0–28 days). Due to the Sulphate reaction, the specimens exposed to $MgSO_4$ show a fluctuation in hue (at first) from 59.5 (light grey) to 55 (medium grey) after 28 days. Saturation refers to how much colour is diluted by adding white colour. The S value varies from 0 to 100% and rises as one moves away from the central axis. Figure 7 represents the change in S value for increasing exposure times to chemicals. Figure 7 shows an increase in the S value after exposure to HCl for 14 days, followed by a decrease at 28 days, and shows a drastic change in the S value.

At 0 days, concrete appears greyish with a S value of 3%; after exposure to HCl for 14 days, the surface turns bright red colour with a 92% S value and after 28 days it decreases to an 14% S value due to erosion of cement from its surface. After three days of exposure, the sulphuric acid-exposed surface exhibits the greatest S value of 10%, which thereafter gradually declines. After exposure to NaCl and $MgSO_4$, the specimens initially showed similar trends of high S-values of 8% after 3 days and 5% after 7 days, respectively. The surface colour does not significantly alter throughout the course of continued exposure (up to 28 days) owing to either $MgSO_4$ or NaCl (little white colour development) (dark grey colour formation).

The maintenance of continuous lighting ensures that the light intensity is also consistent. Yet, the concrete surface changes with time as a result of surface deterioration. Figure 8 depicts the fluctuations in H values of exposed concrete surfaces caused by HCl, H_2SO_4, NaCl, and $MgSO_4$ over varied periods of 3, 7, 14, and 28 days, respectively.

6 Conclusions

Analysis of digital images using the HSV colour space enables the identification of chemical attacks at inaccessible locations. The study shows that the variation in the Hue value is very small for the deterioration induced in this study. However, the additional combination of saturation 'S' and 'V' values provide sufficient uniqueness in the colour space to enable the assessment of deterioration. Based on the results, we can assess the history of deterioration on concrete surface using developed empirical equations due to individual chemical attack.

References

1. Khan AR (2009) Performance of different types of Pakistani cements exposed to aggressive environments. In: Concrete repair, rehabilitation and retrofitting II. CRC Press, pp 121–122
2. Swenson RL, Dimond KR (1998) A universal colour transformation architecture. Pattern Recogn Lett 19(9):805–813
3. Indian Standards:12269-2013, Specifications for 53 grade ordinary Portland cement, Bureau of Indian Standards, New Delhi, India
4. ASTM C 150-07 (2009) Standard specification for Portland cement. Annual Book of ASTM Standards, Vol 4.01, ASTM International
5. IS 4031 (2005) Indian standard specification for method of physical tests for hydraulic cement, determination of fineness by dry sieving, Part 1
6. Indian Standards: 383-2016, Coarse and fine aggregate for concrete-specification. Bureau of Indian Standards, New Delhi, India
7. Standard test method for specific gravity and absorption of fine aggregate, ASTM C 128, American Society for Testing and Materials, ASTM specification, Philadelphia (1993)
8. Indian Standards: 2386-4:1963 (Reaffirmed 2016), Method of test for aggregates for concrete-mechanical properties. Bureau of Indian Standards, New Delhi, India
9. Standard specification for concrete aggregates, ASTM C 33, American Society for Testing and Materials, ASTM specification, Philadelphia (1999)
10. Indian Standards: 10500-2012, Drinking water-specification, Bureau of Indian Standards, New Delhi, India

11. Indian Standards: 10262-2019, Concrete mix proportioning-guidelines, Bureau of Indian Standards, New Delhi, India
12. Reddy VGP, Krishna BM, Tadepalli T, Kumar PR (2020) Image-based deterioration assessment of concrete. Mater Today Proc 32
13. Reddy VGP, Tadepalli T, Pancharathi RK (2022) Surface imaging based non-destructive assessment of concrete deterioration using hue-saturation-intensity colour space. Measurement 197:111311
14. Alamandala S, Sai Prasad RLN, Pancharathi RK, Guru Prathap Reddy V (2022) A comparative study of digital image correlation and fiber Bragg grating sensors based on strain measurements. Innov Infrastruct Solutions 7(2):167
15. Priyanka GTL, Kumar K, Reddy VGP, Tadepalli T (2021) Mechanical reliability of extruded PLA filaments. Materialia 16:101075

Trend Analysis of Climate Variables and Extremes Over Maner River Basin, India

Koppuravuri Ramabrahmam and Keesara Venkatareddy

Abstract Long-term trend analysis of meteorological variables is required for implementing any hydrological model or a basin. Spatio-temporal variations in precipitation and temperature of a basin are helpful for meteorologists, agriculturists, and policymakers to make appropriate decisions. It is necessary to understand the changes in rainfall characteristics using gridded precipitation data and robust statistical analysis for making decisions. In this study, the long-term trend of climate variables, i.e., precipitation, temperature over the Maner basin is studied during 1951–2020 at seasonal and annual timesteps. The non-parametric statistical methods of Mann-Kendall (MK) tests was performed using IMD daily gridded (0.25×0.25) data for 70 years. The magnitudes in the rainfall, temperature, and their extremes (CDD, CWD, PRCPTOT, R10, R20, R95, RX1DAY, RX5DAY, TNN, TNX, TXN, and TXX) are analyzed using Sen's slope method. The increase trend was noticed at the upper and lower Maner basins successively during the winter and pre-monsoon seasons. At the upper Maner basin, a decline in monsoon precipitation is visible. In terms of annual precipitation and post-monsoon, no significant trend was found. In the Maner basin, there was an upward trend in both the maximum and minimum annual temperature trends. According to an analysis of extreme indices, the annual total rainfall days (wet days) (R95PTOT) were on the rise across the basin, despite the fact that the consecutive wet days (CWD) and dry days (CDD) showed a decreasing trend. While the extreme indices for minimum and maximum temperatures (TNX and TNN) showed no discernible trend, respectively. The TXX and TXN of maximum temperature indices showed an increasing trend. These increasing trends in wet days and extreme temperatures in the Maner basin may be majorly attributed to changes in urbanization in the city of Karimnagar and land use changes in the basin. These all changes experience changes in precipitation patterns, which could have implications for water availability and management in the region.

Keywords Trend analysis · Precipitation · Maximum temperature · Minimum temperature · Extreme indices

K. Ramabrahmam (✉) · K. Venkatareddy
National Institute of Technology, Warangal 506004, Telangana, India
e-mail: rambrahmam@student.nitw.ac.in

S. Mesapam et al. (eds.), *Developments and Applications of Geomatics*,
Lecture Notes in Civil Engineering 450, https://doi.org/10.1007/978-981-99-8568-5_37

1 Introduction

India is a country that is heavily dependent on rainfall for its agriculture and economy. Although global precipitation was expected to increase, nonuniformity in its pattern was observed at regional scales [1]. Also, the globally averaged land and ocean surface temperatures were found to increase during the period 1880–2012 [2]. Surface warming is expected to increase water vapor concentration by about 6–7% per °C change in temperature, according to the Clausius-Clapeyron relation [3]. As a result, it was discovered that the severity and frequency of precipitation extremes increased under the condition of global warming [2]. India has experienced both positive and negative trends in rainfall over the last few decades, as well as an increase in extreme rainfall events. Therefore, the detection of trends in hydrologic and climate variables at the basin scale has gained importance in recent studies for climate impact assessments [4–6]. Analyzing precipitation, temperature trends and their extremes over the Basin is essential because it can provide insights into changes in the hydrological cycle in the region.

Numerous studies have been conducted on the long-term trends and variability of various climatic variables. Some of the more significant worldwide studies [7, 8] and at the Indian sub-content basin scale [5, 6, 9–14] give a brief analysis of trend detection analysis using parametric and non-parametric tests (Majorly MK test). The trend analysis of extremes and extreme indices were addressed [5, 6, 15–18] over the river basins of India, explaining the influence of extremes over the basin and its trend behavior. The studies [19–21] majorly focused on the trend analysis of climate variables over the Godavari River sub-basins. The long-term historical trend analysis of precipitation, temperature, and extreme indices were important at the basin scale for sustainable watershed management. In the recent decade, the Telangana state has shown high water availability majorly due to the Kaleshwaram project and mission Kakatiya programs; therefore, the hydrology and climate impact parameters may also change within the basins of the state. Hence, the Maner basin (sub-basin of Godavari basin) has been taken for the study, and this has also not yet been addressed for the long-term historical trend analysis. In the present study, the non-parametric trend detection analysis has been performed for climatic parameters including precipitation, maximum and minimum temperature, and extreme indices over the Maner basin.

2 Materials and Methods

2.1 *Study Area*

The study performed on the Maner watershed is the sub-basin of Godavari basin, India (Fig. 1a). The basin is located between latitudes 17° 41′ 20 N to 18° 40′ N and longitudes 78° 13′ E to 79° 59′ 40 E. The Maner river is a right bank tributary of

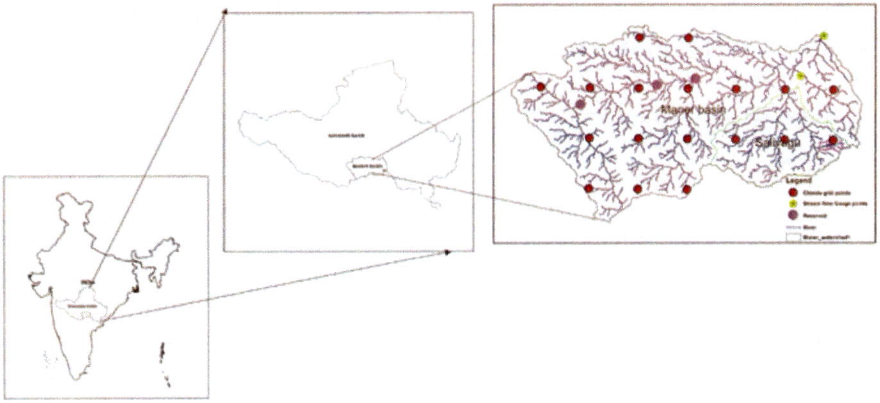

Fig. 1 Study area of Maner basin

the Godavari River that flows in a W–E direction from Narmala village in Rajanna Sirisilla district, Telangana, to Karimnagar district, covering an area of 13,106 km^2. The average annual rainfall of the catchment is 932 mm (1951–2005) and the south-west monsoon (June–October) is the only primary rainy season. The Maner basin Agro-climatic zone is divided into two regions: the southern plateau and hills region and the hot semi-arid ecoregion with red and black soils.

2.2 *Data Used*

The daily rainfall and maximum and minimum temperature data of 0.25 * 0.25 grids are extracted from the Indian metrological department in Pune from 1951 to 2020 [22]. The ETCCDI climate change indices are CDD, CWD, R10MM, R20MM, R95PTOT, PRCPTOT, RX1DAY, RX5DAY, TXX, TXN, TNX, and TNN [22].

2.3 *Methodology*

The methodology used for the study includes calculating average monthly values and applying trend detection for various climatic variables using historical Maner basin data. In the analysis, non-parametric MK trend tests and Sens slope statistics were applied to precipitation, maximum and minimum temperature, and extreme indices. Calculating the MK statistic and the corresponding z-value indicates the significance of the trend. In this study, MK1, MK2, and MK3 [23] (Kendall 1962) tests and Sen's slope test [24] were carried out using the R programming language with the "modifiedmk" package. The statistics 'S' and standardized test statistics (z) in the MK trend test were referred from the (i) Mann-Kendall test (MK1) [25],

(ii) Mann-Kendall test with trend-free pre-whitening (MK2) [26], and (iii) Modified Mann-Kendall test 3 (MK3) [27]. A decreasing trend is indicated by negative S values and vice versa.

The test statistic Z gives confidence levels (CL = 1 − SL) (i.e., 99, 95, and 90%) for rejecting the null hypothesis as well as significance levels (SL) (1, 5, and 10%). The standard value of Z at α % level of significance is obtained from the standard normal cumulative distribution tables. In the present study, significance levels at $\alpha = 0.1$ for which Z = 1.96 were taken into consideration. The magnitude of trends is determined by the Theil-Sen Approach (TSA) with a slope equation [24]. The trend will be accepted with a significant level of α % if the condition in Eq. 1 is satisfied:

$$-Z_{k(1-(\alpha/2))} < Z_k < Z_{k(1+(\alpha/2))} \quad (1)$$

The stepwise procedure adopted to determine the long-term temporal and spatial variability in climatic variables and extreme climatic indices is included in Fig. 2.

Extreme Climate Indices

In the present study, to analyze and describe the spatiotemporal variations in extreme rainfall and temperature, a total of 8 rainfall and 4 temperature indices given in Table 1 have been used to analyze the trends in extreme climatic conditions in the basin. These indices were established by Expert Team on Climate Change Detection Monitoring Indices (ETCCDMI) and have been used in several studies to understand the behavior of climate extremes and their trends [4, 29]. The indices were calculated using R programming with the ClimDex.pcic [28] library for identifying the anomalies in precipitation and temperature for observed and future periods.

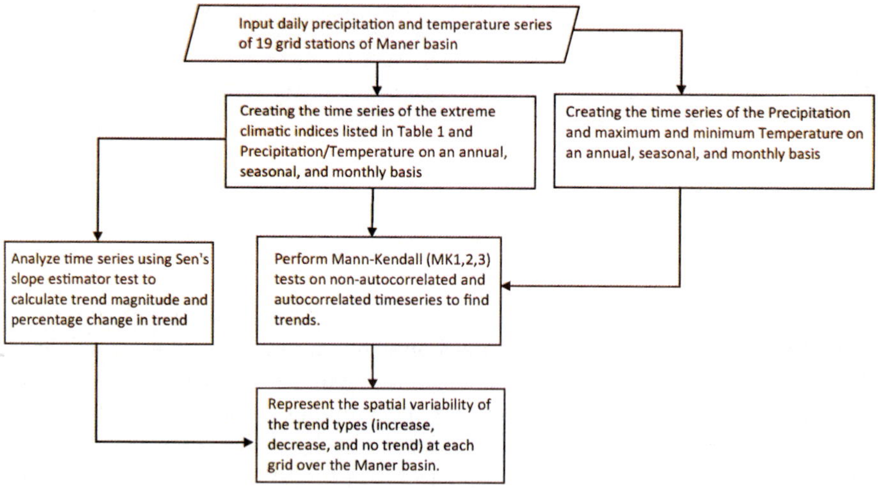

Fig. 2 Methodology for trend analysis of metrological parameters and extreme indices used in the study

Table 1 List of extreme indices used in the study

Extreme indices	Definition	Units
CDD	Maximum number of consecutive days when daily rainfall < 1 mm	Day
CWD	Maximum number of consecutive days when rainfall ≥ 1 mm	Day
PRCPTOT	Total amount of rainfall in wet days	mm
R10MM	Number of days when RR ≥ 10 mm. Count the number of days where RR_{ij} ≥ 10 mm	Day
R20MM	Number of days when RR ≥ 20 mm. Count the number of days where RR_{ij} ≥ 20 mm	Day
R95PTOT	Total rainfall when RR > 95p	mm
RX1DAY	Monthly maximum 1-day precipitation	Day
RX5DAY	Monthly maximum consecutive 5-day precipitation	Day
TXx	Monthly maximum value of daily maximum temperature	°C
TXn	Monthly minimum value of daily maximum temperature	°C
TNx	Monthly maximum value of daily minimum temperature	°C
TNn	Monthly minimum value of daily minimum temperature	°C

RR_{ij} is the rainfall amount on day i in the period j

3 Results

Three different MK tests (i.e., MK1, MK2, and MK3) with a 90% confidence interval were used to analyze the trend pattern of precipitation, temperature, and their climate extreme indices at monthly, seasonal, and annual scales in the Maner basin. It is regarded as a threshold value, and the trends of those grids are examined if any grid exhibits either positive or negative trends in at least two tests. The spatial maps were prepared with Z values using kriging interpolation for the whole basin of precipitation, maximum and minimum temperature, and climate extreme indices (Figs. 3, 4 and 5). The detailed analysis is explained below.

3.1 Observed Trends in the Seasonal and Annual Precipitation and Temperature

There was no significant trend in 16 grids and 3 grids that showed a negative trend in the upper Maner basin for monsoon precipitation (Fig. 3a). In the post-monsoon precipitation, no trend scenario was seen. During the winter season, 11 grids showed no significant trend and other 5 grids showed increasing trend at upper Maner basin. Increasing trends showed in 8 grids of the upper and lower Maner basin during pre-monsoon (Fig. 3b). In the annual scale analysis, the precipitation showed no trend

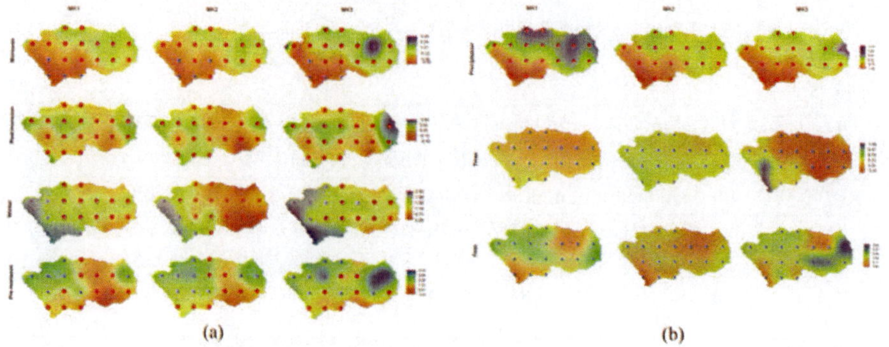

Fig. 3 The spatial plot and trends of precipitation and temperature **a** at seasonal scale and **b** at annual scale

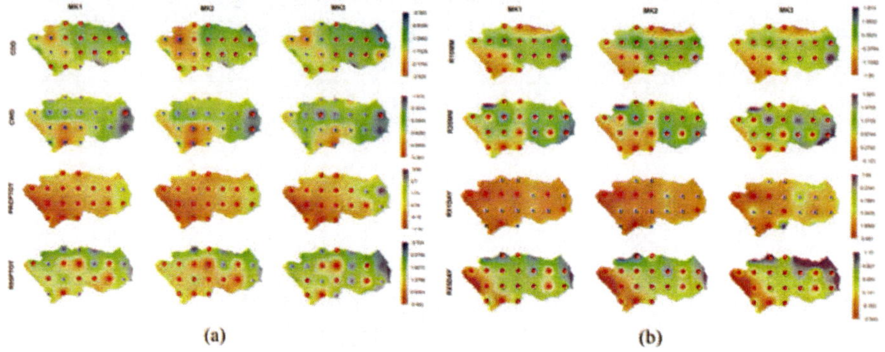

Fig. 4 The spatial plot and trends of precipitation extreme indices

scenario in the entire basin. The maximum and minimum annual temperature trends in the Maner basin showed an increasing trend.

3.2 *Observed Trends in the Climate Extreme Indices*

For the trend analysis in this study, four temperature and eight precipitation climate extreme indices were taken into account. Seven extreme indices were shown a significant trend out of twelve. The precipitation indices of CDD showed no trend in 13 grids, and six grids at the upper and lower Maner basins showed a decreasing trend (Fig. 4a). CWD showed a decreasing trend and the PRCPTOT indices showed no significant trend. In 12 grids, the R95PTOT showed no trend, and in 7 grids dispersed throughout the basin, it showed an increasing trend. The R10 MM, R20MM, and RX5DAY indices showed no significant trend. In 13 grids, an increasing trend was seen for RX1DAY, while 6 grids showed no trend at all (Fig. 4b).

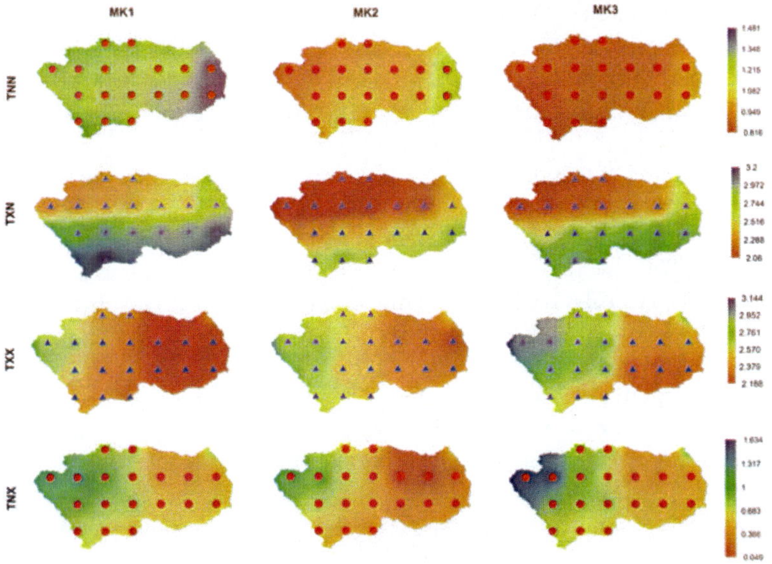

Fig. 5 The spatial plot and trends of temperature extreme indices

Analyzing extreme indices Out of the twelve extreme indices, seven showed a significant trend. Although the number of consecutive wet days and dry days (CWD and CDD) showed a decreasing trend, the number of annual total rainfall days (wet days) (R95pTOT) increased across the basin. The upper and lower Maner basins showed increasing maximum 1-day rainfall amounts (RX1 DAY), while other 5-day precipitations (RX5 DAY) showed no significant trend. There was no significant trend in the monsoon or the annual precipitation, so the increase in the number of days with more than 10 mm (RX10 Day) or 20 mm (RX20 Day) of precipitation is also not significant. The extreme indices of maximum temperature (TXX and TXN) showed increasing trend, while minimum temperature (TNX and TNN) shown no significant trend. The maximum temperature indices of TXN and TXX showed an increasing trend in all the grids, while the minimum temperature indices of TNN and TNX showed no significant changes (Fig. 5). These results were similarity to the studies done by Praveen Kumar and Jothi Prakash [21] and Das et al. [17] performed on Godavari basin.

4 Conclusions

This study considered Sen's slope approach and the Mann-Kendall test to analyze trends in temperature and precipitation. In the trend analysis, the upper and lower Maner basins showed an increasing trend over the winter and pre-monsoon. However, the monsoon and annual precipitation have been declining in the upper Maner basin. These short-term changes may be attributable to human activities, which dominate natural climate variability.

The increase in the one-day precipitation (RX1 DAY) at the upper and lower Maner basin is due to the presence of reservoirs in this region. As the study area comes under a semi-arid region, this is the reason the maximum and minimum annual temperature trends showed a rising trend.

According to an analysis of extreme indices, the annual total rainfall days (wet days) (R95PTOT) were on the rise across the basin, despite a decreasing trend in wet and dry days (CWD, CDD) but there was increased trend in the maximum and minimum temperatures (TXX, TXN). These conditions might be attributed to changes in atmospheric circulation.

It's important to note that no significant trend was observed in the annual precipitation and post-monsoon, which means that there has been no overall increase or decrease in rainfall throughout the year or after the monsoon season. These trends suggest that the Maner basin may be experiencing changes in precipitation patterns, which could have implications for water availability and management in the region.

Acknowledgements The Science and Engineering Research Board (SERB) provided funds to carry out the research that is described in this paper.

References

1. Stocker TF, Qin D, Plattner GK, Alexander LV, Allen SK, Bindoff NL, Bréon FM, Church JA, Cubasch U, Emori S, Forster P (2013) Technical summary. In: Climate change 2013: the physical science basis. Contribution of working group I to the fifth assessment report of the intergovernmental panel on climate change. Cambridge University Press, pp 33–115
2. IPCC (2014) Climate change 2014: synthesis report contribution of working groups I, II and III to the fifth assessment report of the intergovernmental panel on climate change [Core Writing Team, RK Pachauri and LA Meyer (eds)]. IPCC, Geneva, 151 pp
3. Trenberth KE (2011) Changes in precipitation with climate change. ClimRes 47(1/2):123–138. https://doi.org/10.3354/cr00953
4. Dubey SK, Sharma D (2018) Spatio-temporal trends and projections of climate indices in the Banas River Basin, India. Environ Process 5(4):743–768
5. Rao GV, Reddy KV, Srinivasan R, Sridhar V, Umamahesh NV, Pratap D (2020) Spatio-temporal analysis of rainfall extremes in the flood-prone Nagavali and Vamsadhara Basins in eastern India. Weather Clim Extremes 29:100265
6. Buri ES, Keesara VR, Loukika KN, Sridhar V (2022) Spatio-temporal analysis of climatic variables in the Munneru River Basin, India, using NEX-GDDP data and the REA approach. Sustainability 14(3):1715
7. Hamed KH (2008) Trend detection in hydrologic data: the Mann-Kendall trend test under the scaling hypothesis. J Hydrol 349(3–4):350–363
8. Alemu ZA, Dioha MO (2020) Climate change and trend analysis of temperature: the case of Addis Ababa, Ethiopia. Environ Syst Res 9(1):1–15
9. Hingane LS, Rupa Kumar K, Ramana Murty BV (1985) Long-term trends of surface air temperature in India. J Climatol 5(5):521–528

10. Kumar V, Jain SK, Singh Y (2010) Analysis of long-term rainfall trends in India. Hydrol Sci J-Journal des Sciences Hydrologiques 55(4):484–496
11. Jain SK, Kumar V (2012) Trend analysis of rainfall and temperature data for India. Curr Sci 37–49
12. Jaiswal RK, Lohani AK, Tiwari HL (2015) Statistical analysis for change detection and trend assessment in climatological parameters. Environ Process 2:729–749
13. Bisht DS, Chatterjee C, Raghuwanshi NS, Sridhar V (2018) Spatio-temporal trends of rainfall across Indian river basins. Theoret Appl Climatol 132:419–436
14. Biswas B, Jadhav RS, Tikone N (2019) Rainfall distribution and trend analysis for upper Godavari basin, India, from 100 years record (1911–2010). J Indian Soc Remote Sens 47:1781–1792
15. Deshpande NR, Kothawale DR, Kulkarni A (2016) Changes in climate extremes over major river basins of India. Int J Climatol 36(14):4548–4559
16. Bothale RV, Katpatal YB (2016) Trends and anomalies in extreme climate indices and influence of El Niño and La Niña over Pranhita catchment in Godavari Basin, India. J Hydrol Eng 21(2):05015023
17. Das J, Nanduri UV (2018) Future projection of precipitation and temperature extremes using change factor method over a river basin: case study. J Hazard Toxic Radioact Waste 22(3):04018006
18. Sharma PJ, Loliyana VD, Timbadiya PV, Patel PL (2018) Spatiotemporal trends in extreme rainfall and temperature indices over Upper Tapi Basin, India. Theoret Appl Climatol 134:1329–1354
19. Jhajharia D, Dinpashoh Y, Kahya E, Choudhary RR, Singh VP (2014) Trends in temperature over Godavari river basin in southern peninsular India. Int J Climatol 34(5):1369–1384
20. Aher MC, Yadav SM (2021) Assessment of rainfall trend and variability of semi-arid regions of Upper and Middle Godavari basin, India. J Water Clim Change 12(8):3992–4006
21. Praveenkumar CH, Jothiprakash V (2021) Spatio-temporal trend analysis of long-term IMD-gridded precipitation in Godavari river basin, India. In: Climate change impacts on water resources: hydraulics, water resources and coastal engineering, pp 59–66
22. ETCCDI home page. etccdi.pacificclimate.org/list_27_indices.shtml. Accessed 05 Oct 202
23. Mann HB (1945) Nonparametric tests against trend. Econometrica 13:245–259
24. Sen PK (1968) Estimates of the regression coefficient based on Kendall's tau. J Am Stat Assoc 63:1379–1389
25. Kumar S, Merwade V, Kam J, Thurner K (2009) Streamflow trends in Indiana: effects of long term persistence, precipitation and subsurface drains. J Hydrol 374:171–183. https://doi.org/10.1016/j.jhydrol.2009.06.012
26. Yue S, Pilon P, Phinney B (2003) Canadian streamflow trend detection: impacts of serial and cross-correlation. Hydrol Sci J 48:51–64. https://doi.org/10.1623/hysj.48.1.51.43478
27. Hamed KH, Rao AR (1998) A modified Mann-Kendall trend test for autocorrelated data. J Hydrol 204(1–4):182–196
28. Zhang X, Alexander L, Hegerl GC, Jones P, Klein Tank A, Peterson TC, Trewin B, Zwiers FW (2011) Indices for monitoring changes in extremes based on daily temperature and precipitation data. WIREs Clim Change. https://doi.org/10.1002/wcc.147
29. Easterling DR, Kunkel KE, Wehner MF, Sun L (2016) Detection and attribution of climate extremes in the observed record. Weather Clim Extremes 11:17–27

Urban Dynamics and Impact Assessment of Bengaluru–Mysuru Expressway Corridor

S. Suhas, V. Bhavani, B. M. Vishwanath, Ruthvik Krishna, and M. C. Chandan

Abstract The focus on the provision of basic amenities and infrastructure has been a major issue in urban planning because of urban growth and sprawl. It is one of the reasons that adversely contributes to the loss of natural resources and encourages unbalanced and uneven urbanization in developing nations like India. The government and planning authorities are increasingly considering regional development to accommodate the incursion of people into downtown urban areas therefore encouraging comprehensive, sustainable development. By a methodological approach and the use of multi-temporal satellite data, this study aims to explain the urban growth along the Bangalore–Mysuru motorway as influenced by the transportation corridor. The objective of the transport corridor development is to increase the efficiency of the corridor's transportation and logistics systems and to spur regional economic growth by taking advantage of better connectivity and transportation infrastructure. A transport corridor can help the region along the corridor achieve more sustainable and spatially balanced economic growth. To analyse the existing development pattern and provide a framework for predicting how land will be used in the future, land use mapping and related data are developed. Later for the analysis, a cellular automata-based model derives the name from the input layers utilized, namely Slope Land use Excluded Urban Transport Hillshade (SLEUTH) was performed to meticulously examine subdivision-level specifics of the area affected by the corridor. The results of this study will aid policymakers and planners in making choices about upcoming urban trends that support a more secure, healthy, efficient, sustainable, and livable urban ecosystem.

Keywords Urban growth · Land use · Cellular automata · SLEUTH model · Bangalore–Mysuru expressway

S. Suhas · V. Bhavani · B. M. Vishwanath · R. Krishna · M. C. Chandan (✉)
Department of Civil Engineering, The National Institute of Engineering, Mysuru 570008, India
e-mail: chandanmc@nie.ac.in

S. Mesapam et al. (eds.), *Developments and Applications of Geomatics*,
Lecture Notes in Civil Engineering 450, https://doi.org/10.1007/978-981-99-8568-5_38

1 Introduction

1.1 Introduction/Overview

Urbanization is the term used to describe the movement of people into a certain area in search of employment and other means of subsistence. As a result, a region's population grows gradually. Urban sprawl is the movement of people out of central metropolitan districts and into low-density, mobility-dependent communities because of unplanned urbanization. Since the planners were unable to anticipate the inevitable growth, the sprawl regions typically lack basic infrastructure and facilities. Thus, urbanization is caused by the redistribution and redirection of resources into urban regions, extensive development, and population movement in a quest of employment from rural to urban areas, and many other factors that result in changes in land use patterns, unsustainable development, and socioeconomic, water, and environmental problems [1]. According to World Urbanization Prospects, urban population expanded from 30% in 1950 to 54% in 2014 due to enhanced connectivity and is expected to grow to 66% by 2050 [2]. Modern concepts of urban expansion like urban corridors, on the other hand, share the characteristic of extending the constraints of urban growth outside of administrative borders, creating enormous territories around the central metropolis. An urban corridor is one that has been constructed with several polycentric city clusters connected by transportation lines. Since it is essential to improve both domestic and international trade, the idea of urban corridors is intended to encourage economic growth.

Globalization and the subsequent opening of Indian markets in the early 1990s resulted in rapid urban development of the landscape. A variety of chances for increasing basic necessities have been created by economic changes and infrastructure upgrades over the past ten years, with implementation anticipated in many Indian cities. Although rural development initiatives remain the core of our anti-poverty strategy, two proactive urban-industry policies were introduced in the last decade pertaining to (a) Special Economic Zones (SEZs) and (b) Industrial and Economic Corridors [1]. These two models are highly dependent on one another. The corridor policy represents the Indian national government's attempt to explicitly link economic and industrial development to urbanization. Corridors are transportation arteries—railways, roads, canals, and rivers that necessitate long swaths of linear physical space that pass through multiple administrative jurisdictions. Geospatial industrial clusters, on the other hand, are compact spaces that are typically located within a state.

Infrastructure development is crucial to the general development of human necessities, determines quality of life, and helps identify new elements that may be used to enhance things and promote urban expansion. India has experienced widespread industrialization, and political, cultural, and other socioeconomic elements that have resulted in urbanization in cities and towns during the 1990s have neglected the need for infrastructure improvements and the supply of essential facilities to the population in a planned, staged manner. India's infrastructure development has recently placed significant emphasis on the establishment of industrial corridors, connecting several hundred towns with two or more major cities. These corridors have been essential for connectivity throughout the previous 10 years. The economic exchange between cities is anticipated to increase as a result.

In India, the conception of urban corridors has been around for a decade, with the primary goal of connecting megacities, industrial centres, and key ports throughout the country. The most significant Indian urban corridors are Mumbai–Jalgaon, Hyderabad–Vijayawada, Bangalore–Mysore Infrastructure Corridor, Delhi National Capital Region, Chennai–Bengaluru–Mumbai Industrial Corridor, Delhi–Mumbai Industrial Corridor, Bangalore–Pune Infrastructure Corridor, and Bengaluru–Mysuru Expressway, also known as NH275 and maintained by the National Highway Authority of India (NHAI) [3].

The Indian government has recently started ambitious, ground-breaking initiatives like Made in India, Skill India, Smart Cities, Industrial Corridors, Digital India, etc. These initiatives aimed to increase employment possibilities and strengthen the economies of the countries through planned capacities and tactics of international competition [2].

An economic corridor is a more extensive approach that blends transportation network expansion, manufacturing, trading, commerce, and real estate investment, as well as the proper management of environmental safeguards. The corridors offer crucial connections between metropolitan areas, ensuring that each city is a part of a wider regional system rather than a standalone development hotspot. It is vital to understand a city's relationships and patterns with other cities in order to properly understand its significance in a system. A social network is an example of this type of study, which consists of a collection of nodes (or cities) connected by different links. Industrial corridors connect certain geographic regions and enable goods to move more swiftly from the point of origin to the point of destination, improving supply chain management.

In order to promote industrial development by lowering transit time and expense, the corridors are usually constructed along existing train and road networks, allowing for faster supply, and so on. To prevent the environmental and ecological implications of land use changes, this demands earlier visualization of spatial growth. Mapping, tracking, and quantifying changes in the traditional way is time-consuming. Remote sensing data and Geographic Information Systems (GIS) have been vital for analysing past land use change and predicting future development. Human resources are being used to boost socioeconomic development and education.

To predict land change, assessing the state of spatial–temporal dynamics is the first step. A Markov chain paired with CA can control a region's temporal dynamics, whereas CA would capture and govern spatial dynamics. SLEUTH, on the other hand, can anticipate urban growth by taking into account both geographical and temporal factors.

The corridor strategies that aim to preserve or improve transportation capacity while also supporting development are investigated. To encourage ride sharing and limit automotive journeys to a major development location, land use constraints such as zoning and parking laws are proposed. The objectives of this research are to

(a) understand spatio-temporal LU change along the urban corridor connecting two important cities of India, (b) evaluate urban transitions and study causal effects, (c) analyse the influence of corridor on sub-regions on a minute scale, and (d) simulate the urban growth influence along major cities of a corridor using SLEUTH.

1.2 Literature Review

Chandan and Aithal [1] In this paper, SLEUTH-UGM is used to predict land change acknowledging the status of spatial–temporal dynamics. SLEUTH has the ability to assess urban growth by taking into account the spatial and temporal characteristics of urban expansion and development. The standard GNU C compiler is used to operate the C language-based SLEUTH model (GCC). As a result of the ranking research, it was possible to determine the city's supremacy and role. In order to prevent the environmental effects of urbanization brought on by the development and building of industrial corridors, regional decision-makers need to have access to vital information about the spatial dynamics of the landscape.

Ramachandra et al. [4] This study can be used as the technical basis such that due to its simplicity, adaptability, and intuitiveness, a cellular automata (CA) driven modelling framework has become one of the most extensively used methods for urban simulation and modelling. This opens up the possibility of merging CA and Markov Chains (MC) to produce a potent modelling framework that might be applied to spatial–temporal simulation. The project used temporal remote sensing data to examine landscape dynamics, identify local growth pole agents, analyse spatial patterns using segment (zones) techniques, and anticipate future land uses for two main industrial corridors in India.

Li et al. [5] This study used a combination of multi-remote sensing images, landscape measurements, and detailed structured investigation focusing on urban growth and regional level changes. To examine the spatial distribution and rate of urban growth, they employed the urbanization rate (UR) and the urban expansion rate as their two indicators (UER). UR can be used to visualize the proportion of urban areas within the research area buffer zones. A UER can be used to demonstrate how the urban area evolves over time as different construction phases are carried out. These metrics are crucial for evaluating the geographical and temporal shifts brought on by urban expansion along a metro line.

Liu and Feng [6] By the use of logistic-regression-based cellular automata (also known as LogCA) modelling, this study investigated the implications of competing factors such as economy, environment and ecosystem, and conservation policies on hypothetical potential urban scenarios in Ningbo, China. For modelling the spatio–temporal process of urban growth in Ningbo, a LogCA model was used to forecast future scenarios after being validated using urban patterns. Based on the urban patterns that have been observed, the LogCA model is calibrated. By simulating future scenarios and evaluating the influence of economic development and ecological protection methods on urban growth, CA modelling provides a diagnostic for China's urban development.

Chandan et al. [7] In this study, temporal remote sensing data were utilized to assess the dynamics of land cover and land use. Urbanization patterns at the local level are analysed using land use with gradients and zonal borders, together with spatial metrics, to track patterns of land use change. The measures that come from a digital analysis of theme category maps that show spatial heterogeneity of patches and classes of patches are called spatial metrics. The research shows that rapid urban growth is highly fragmented in the city's periphery and outskirts, and the spatial metrics produced by this method can really be used to visualize and measure the growing footprints of developed areas.

Mallouk et al. [8] This study shows that SLEUTH is the most used cellular automaton model for simulating and forecasting urban growth. The SLEUTH model is open-source software that uses the LINUX or UNIX operating system and was created in the C programming language. In order to investigate and model the expansion of urban areas in relation to neighbouring territories and the potential effects of this expansion on the environment of the region in question, this model is primarily created to simulate urban growth. SLEUTH's popularity and growing user base worldwide are ascribed to its open-source nature, source code accessibility, regional modelling capability, and relative simplicity of computation and application.

Quium [9] This paper examines a few of the difficulties, lessons, and consequences related to the development of transportation corridors that may be relevant to future designs of corridors. According to this, the development of transportation infrastructure often has a positive impact on the economy, income, poverty, employment, equity, and inclusion. Economic growth and social well-being can, however, be seriously compromised, and there may also be negative environmental effects. Research is necessary to comprehend the uneven effects of various infrastructure types across geographic regions and population groups, as well as complementary policies, interventions, and institutions that may be taken into consideration to lessen the negative effects of uneven impacts. Although vital, transportation development might not be enough to produce broader economic advantages. There is also a need for further complementary interventions.

Manjunatha and Basavarajappa [10] In their study, they investigated the LU/LC aspects of Mysore and surroundings. They divided the region into three categories, and the information gained from the classifications aided in attempting to address spatial alterations, environmental and socioeconomic problems, rising demand for economic natural resources, farming habits, and sensitivity to particular management methods.

Reis et al. [11] In this study on spatial metrics, a repository of measures that address the geographical patterns of expanding and contracting cities are presented, along with discussions of their advantages and disadvantages. While the majority of the documented geographical patterns of urban growth are addressed by these measures, those used in studies of urban decline are significantly more limited and far from being enough to provide a thorough evaluation of their spatial patterns. The study shows that there is a great potential for new geographical measures or mixed indicators, especially under shrinkage scenarios. Despite taking into account a very large and diverse collection of metrics and concentrating on both the effects of urban growth and those of the increasingly prevalent shrinking phenomena, it encompasses reviewing and constructing measures for specific spatial patterns.

Roy and Roy [12] Among the most significant visible changes occurring in human environment are changes to land use and land cover. Although noticeable, the size, diversity, and regional unpredictability of the changes have made it difficult for scientists to quantify and evaluate changes in land use and land cover. Furthermore, because human activities influence the majority of land use and land cover changes, conventional ecological ideas are rarely followed. Areas of change in a region can be determined using the temporal data on land use and land cover. With the help of geoinformatics, we have been able to link the factors driving land use and land cover changes such as population pressure, climate, and terrain—to specific geographic locations. Scientists have been able to quantify these techniques and forecast various eventualities, thanks to this. Numerical modelling of land use and land cover change makes it feasible to comprehend and quantify the processes involved in this transition, allowing us to take the required steps to halt the various land use and land cover changes that are already under progress.

The literature review suggests that the management of urban land necessitates a variety of data on ecological, environmental, and urban planning and development processes being combined. This research develops a spatial analytic system for managing urban expansion land using remote sensing (RS), landscape ecological analysis, and geographic information systems (GIS). For the suitability assessment of land for urban expansion, the spatial analysis system uses a multi-criteria mechanism in GIS. This strategy was used to specifically identify barriers and openings for the conservation and development of land in the future.

2 Materials and Methods

2.1 Study Area

The Bengaluru–Mysuru Infrastructure Corridor (BMIC) connects the two important cities Bengaluru and Mysuru in the Indian state of Karnataka. The geographical extents of the study area are latitude of 12° 20′ N to 13° 00′ N and longitude 76° 40′ E to 77° 30′ E which consist an area of 2582 km^2. Figure 1 shows the study area. To better comprehend future urban dynamics, our analysis took into account administrative borders with a 10 km buffer around the corridor. A rural area with tiny towns and villages exhibits a very sensitive trend towards rapid urbanization; the rapid change in land use is anticipated as a result of significant investments in industries and economic zones. It is reasonable to contemplate a 10 km buffer in order to take into account the geographic changes happening in these peri-urban or hinterland

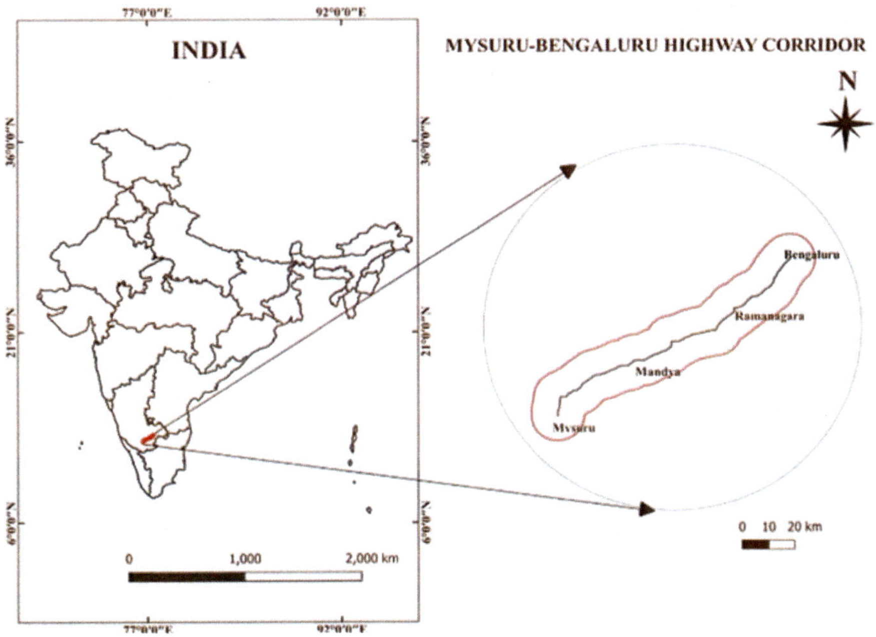

Fig. 1 Mysuru–Bengaluru expressway study area with 10 km buffer on either side

areas and to see how these areas will look in a few years. Understanding the current urban extents would help in modelling and visualizing these changes. Due to internal migration, population growth, and the establishment of new urban centres, the two urban ends of the region are rapidly urbanizing.

The BMIC was a 4–6 lane expressway which is now being expanded to a 6–10 lane expressway. It has a length of 119 km. The construction is expected to be completed at a deadline of October 2022 with an estimated cost of Rs. 8066 crore. There are two sections: a 58 km Phase 1 between Bengaluru and Nidaghatta, and a 61 km Phase 2 connecting Nidaghatta and Mysore. It will have a total of 50 underpasses for automobiles and pedestrians, 19 big bridges, 44 small bridges, 4 ROBs, and 50 underpasses for trains.

Mysuru city is in the southern region of the Indian state of Karnataka. It occupies a 155 km^2 area and is located near the Chamundi Hills' foothills, approximately 145.2 km (90 mi) southwest of Bengaluru (60 sq mi). Mysore is well-known for its historical virtue, buildings, and many palaces, such as the Mysore Palace, as well as for the celebrations that take place during the Dasara festival, during which millions of visitors come from all over the world. It gives its name to many different forms of art and culture, and Mysore's cultural achievements have earned it the title of Cultural Capital of Karnataka. The city's civil administration is under the control of Mysore City Corporation. It is a popular tourist site and acts as a hub for nearby tourist attractions.

Table 1 Satellite data sources

Layers	Purpose/Source
Land use	Landsat—Land use analysis and urban data extraction Sentinel data from ArcGIS home page, i.e., 10 m land use/land cover time series
Hill shade and Slope	Processed from SRTM DEM (raster)
Road data	Classified street and road data (OSM and Google Maps)
Google earth	For collection of point, line, polygon data, and validation
Boundary maps and raster layers	Used to create excluded maps for restricted regions from city development plans

Bengaluru is the capital and the largest city of the Indian state of Karnataka. It is India's third most populated city and fifth most populous urban agglomeration, with a metro population of over 8 million and a total population of almost 11 million. The city is in the Deccan Plateau that has a total area of 741 km^2 (286 sq mi). Bangalore is renowned for having a comfortable climate all year round and is located at a height of over 900 m (3000 ft). Bangalore is known as the "Silicon Valley of India" due to its status as the top services offered through information technology (IT). Bangalore is India's second fastest growing major metropolis and a city with a diversified population. The Bruhat Bengaluru Mahanagara Palike (BBMP) is in charge of running the city's civic affairs. Bangalore's rapid expansion has led to a number of issues with infrastructure obsolescence and traffic congestion that the BBMP has found difficult to resolve. It has the second fastest population growth in India. It is a good location for industrial growth and a hub for automobiles.

2.2 Data

Data, software used for the study are listed in Tables 1, 2 and number of classes considered is described in Table 3.

2.3 Methodology

A three-step framework is included in this method. Initially, it includes data collection, preprocessing, and analysis of land use change. The collection of primary and secondary data is referred to as data acquisition. Remote sensing data from 2017 to 2021 is included in the primary data. Secondary data from sources like historical land use maps, information from the virtual earth database, and Google Earth were used to supplement the source data for geometric adjustments and land use classification.

Table 2 Software and analysis

Software	Authority	Description
Google earth	Google Inc.	Virtual earth system formed by satellite imagery, maps, terrain, 3D buildings, etc.
Grass GIS	OSGEO	Image processing, raster and vector data management, visualization, spatial modelling, etc. https://grass.osgeo.org/
Quantum GIS	OSGEO	Helps in visual interpretation, manages, edits, and analyses data, and composes geospatial maps https://qgis.org/en/site/
Cygwin	Cygnus solutions (Now Red Hat, Inc.)	Its extensive array of GNU and Open-Source utilities offers Windows users capability akin to a Linux installation. Acts as the framework for processing SLEUTH codes https://www.cygwin.com/
Project Gigalopolis	UCSB	It's an open-source project that extends and refines the SLEUTH model to predict urban growth at local, regional, and global scales http://www.ncgia.ucsb.edu/projects/gig/index.html

Table 3 Various classes of land use and features falling under the category

Sl. no	LU class	Features included
1	Urban	Urban area including residential, industrial, commercial, paved materials, hard concrete roof-tops, bituminous material, etc.
2	Vegetation	Forest, plantations, flooded vegetations, etc.
3	Water	Lakes, reservoirs, rivers, etc.
4	Others	Quarry pits, open areas, unpaved roads, current fallow farmlands, bare lands, wastelands, etc.

Geometric and positional errors were removed from the remote sensing data during preprocessing, and for faulty datasets, geometric correction was done using GCPs. To preserve consistency, temporal datasets were resampled to a standard spatial resolution of 30 * 30 m. The study zone (119 km long highway with 10 km of buffer space on either side of the corridor) was cropped out of the main satellite imagery (Fig. 2).

Land Use Classification

In order to visualize urban enclaves, base layer satellite data land use maps are created and used. When geographical limitations or security and privacy laws made it difficult to collect data, Google Earth was employed as a backup source of data to generate training polygons for heterogeneous patches utilizing ground truth points gathered using GPS during the field data collection. Urban, vegetation, water, and other land use categories were examined, and related training polygons were defined

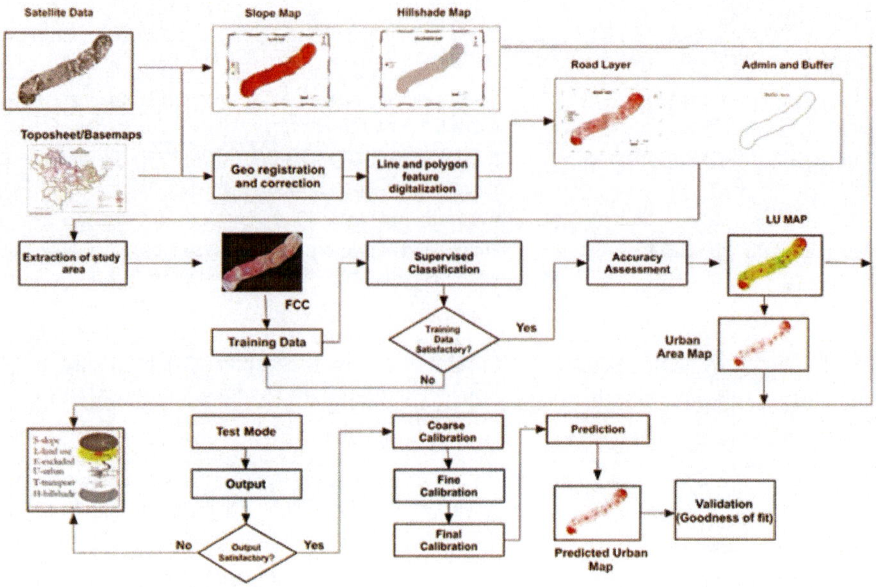

Fig. 2 Method approach to assess the impact on land use and modelling

in such a way that they were evenly distributed over the discrete region. Land use was classified using a maximum likelihood classifier, a guided technique based on the Gaussian distribution function. Remote sensing data classification accuracy is evaluated using kappa statistics and overall accuracy.

Estimation of Future Land Use Transition

Only two urban models have been devised and used to evaluate theories about spatial placement and interactions across land use categories, as well as related activities. Population, land use, transportation, and other socioeconomic issues are all considered and represented in urban models. In order to identify and forecast urban development, it is crucial to gather and prepare data. As resource layers for the SLEUTH model, slope and hill-shade images were generated from a digital elevation model (DEM). Urban and LU maps are categorized as previously mentioned. A restricted layer or an excluded map were included in the city development plan. The transportation layer, often known as the road layer, was produced using Google Earth and OpenStreetMap virtual repositories.

3 Results and Discussions

3.1 *Land Use Classification*

For Mysore–Bangalore Expressway Corridor, substantial alteration in the water, vegetation, as well as built-up area has been observed from 2018 to 2021. Urban setting has been gradually increased from 20.27 to 22.81%. Class of vegetation has been noted to slightly decline from 4.73% in 2018 to 3.75% in 2019. Later, there was an abrupt rise in the vegetation in 2020 at 6.22% mainly due to the year's abundant rainfall and then it increased steadily in 2021 at 7.02%. Rapid transition in adjoining districts such as Mandya, Ramanagara, and Channapatna was observed mainly due to industrialization, with a push in the IT and BT sectors and development in infrastructure: for instance, areas such as KIADB Bidadi, industrial layout near Ramanagara, and Eagleton Golf Vourse near Bidadi. The category of water, including open water bodies, rivers, lakes, etc. has also recorded a drop in 2019 (1.86%) and increased slowly from 2020 (2.12%), delivering a dire warning that water must be saved in order to maintain a viable urban ecosystem. Other land use groups, like bare land, open fields, and agricultural fields have decreased and changed over time into built-up areas, according to the data. Overall accuracy and kappa statistics were within the satisfactory range.

3.2 *Future Land Use Transition*

The input resource layers for the SLEUTH model were 8-bit unsigned GIF and common greyscale data types. The number of Monte Carlo iterations used for the coarse, fine, final, and prediction modes, respectively, for the corridor region was 3125, 25,920, 49,896, and 100. The final calibration mode's results were used in the prediction stage. The prediction coefficient values for the study area are listed in Table 4. Land use prediction for the corridor region was performed for the year 2030, depicted in Fig. 3. Urban growth in the area is projected to increase from 22.81% (2021) to 26.57% (2030).

Table 4 Land use classes and associated features

Parameters	Prediction values
Diffusion	60
Breed	35
Spread	35
Slope	98
Road gravity	57
Monte Carlo iterations	100

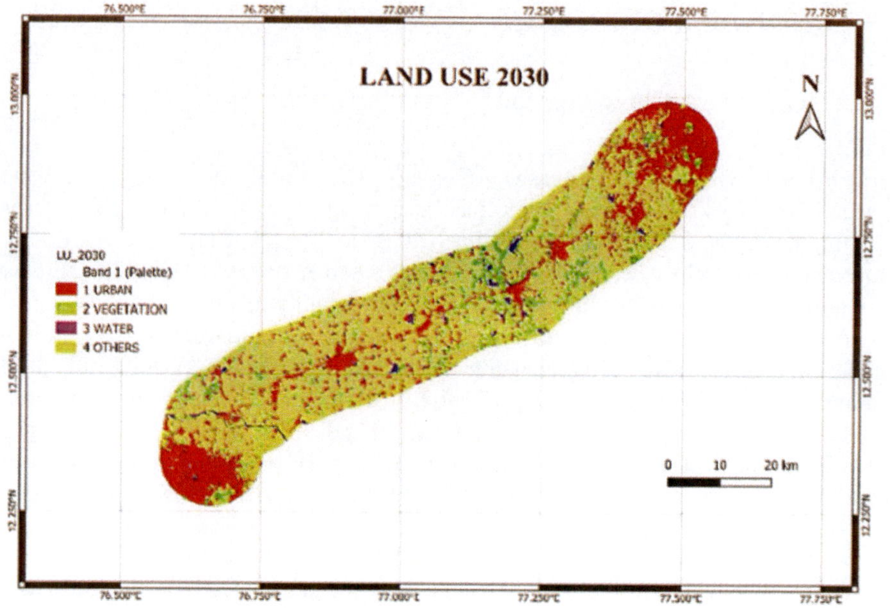

Fig. 3 Predicted output map

The diversion of the highway in Srirangpatna, Mandya, Maddur, Channapatna, and Ramanagara has led to an increase in the value of diffusion (60), suggesting that there is a chance of dispersed urban settlements at the outskirts of the corridor region limits. Whereas the medium range of spread (35) and breed (35) coefficient specifies that there is decreased potential for edge growth and the formation of new urban pockets. An extremely high slope resistance value (98) that allows for constrained expansion outside of the bounds controls and justifies the topography in and around the corridor zone. For instance, greenfield sections will form bypasses around Bidadi (7 km), Ramanagara-Channapatna (22 km), Maddur (7 km including 3.5 km of elevated highway), Mandya (10 km), and Srirangapatna (7 km) depicts sudden terrain change. The medium value of the road gravity (57) coefficient reflects the road as the chief urban attraction factor and shows the significance of urban growth along major roads and highways. NH275 connects major cities in Karnataka like Bengaluru, Mysuru, Mandya, Madikeri, Hassan, etc. These new developments are of the spreading variety, posing a danger to other land use categories and resulting in a discontinuous pattern of landscape development as the central city continues to grow compactly (Table 5).

A substantial alteration in the parameters has been observed from 2021 to 2030 for the study area. Urban settlements have increased from 22.81 to 26.57%, while vegetation has declined from 7.02 to 6.45%. The reasons for the spontaneous urban growth are mainly new job opportunities, better livelihood conditions, and advancements in infrastructure. Significant urbanization has been observed in the Channapatna region due to future development in industries and service sectors and also due to the effect

Table 5 Predicted land use dynamics

Class	Urban		Vegetation		Water		Others	
Year	Km^2	%	Km^2	%	Km^2	%	Km^2	%
2018	523.83	20.27	122.35	4.73	51.99	2.01	1885.1	72.97
2019	544.61	21.08	96.95	3.75	48.27	1.86	1893.42	73.29
2020	579.68	22.43	160.51	6.22	54.82	2.12	1788.25	69.23
2021	589.55	22.81	181.22	7.02	58.15	2.25	1754.3	67.92
Predicted data								
2030	686.46	26.57	166.55	6.45	56.73	2.20	1673.57	64.78

of the newly built expressway corridor, which also exhibits a loss in vegetation and water categories. There is a slight drop in the water category from 2.25 to 2.20%, which includes open water bodies like rivers, lakes, etc., due to the growth of cities in the Channapatna and Bidadi regions influenced by the development of industries and infrastructure like agro-industries, toy industries, and other commercial buildings.

3.3 *Sub-regional Analysis*

The expressway connects some of the major cities and towns in the state of Karnataka; when analysed the type of growth was different in different cities; this was mainly due to their geographical attributions and economic aspects of the region (Fig. 4).

The type of urban growth is "spread" in the major capital cities such as Mysuru and Bengaluru, where its outskirts are expected to expand rapidly. In Mandya and Ramnagara, the growth was due to "slope" and "road gravity," as new roads, bypasses, and elevated highways were constructed around the cities, leading to changes in the landscape. Also, there are high chances of the construction of commercial buildings around these new roads. In Bidadi and Channapatna cities, the growth is expected to be in the form of "breed" and "diffusion," mainly due to the presence of factories and industries that are currently present far from the city centre but predicted to eventually fuse due to the construction of residential buildings nearer to it.

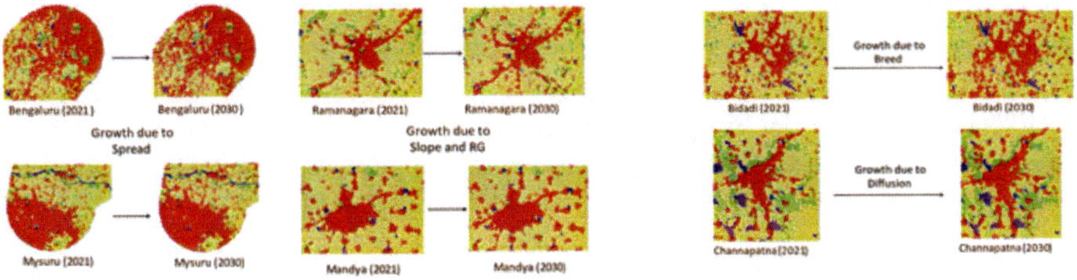

Fig. 4 Type of urban growth in different cities along the expressway

3.4 Conclusion

Extensive research on urban expansion and the urbanization process has been done in the last ten years by researchers. Urban corridor studies are still in their early phases, especially in emerging countries like India where infrastructure planning for the next 10 years is already in great progress. These studies also examine associated development and how it affects land usage. A better visualization is required when planning for cities, corridors, and land use-related changes in present policies, integration of various driving factors in both spatial and temporal domains, and understanding the effects of development on the urban ecology. Numerous open questions are being addressed using advanced geospatial technology to understand the shifting characteristics of landscape changes and future urban growth. In this direction, the present study was designed to assess the impact and urban dynamics of Bengaluru–Mysuru Expressway Corridor; the following key observations provide an insight to varying urban dynamics along the corridor.

The Mysore–Bangalore Expressway route underwent a thorough land use analysis with a 10 km buffer barrier. Validation was performed and the accuracy assessment was within satisfactory range. In order to forecast future land use change for the year 2030, the SLEUTH model was also used. To find the best fit and ideal prediction parameters, test phase and a rigorous three-step calibration phase were effectively carried out. Based on a literature review, various model fit statistics were selected. The results were satisfactory, indicating that the simulated urban growth is exceptionally consistent with the actual urban expansion during the control years [1].

Predicted maps of the corridor region depict areas of expansion with urban sprawl taking place on the fringes and in close proximity to existing urban areas for upcoming government projects. The Mysore–Bangalore Expressway Corridor region demonstrated a rise in the built-up category from 22.81 to 25.47% during 2021 and 2030.

The increased rate of urbanization along the corridors is a result of the industry and better infrastructure networks. The current trend in land use is anticipated to influence the natural vegetation cover and agricultural lands alongside highways and close to CBDs. The forecasting and depiction of geographical patterns of expected land use changes related to the construction of corridors would be beneficial for informed decision-making. The suggested modelling approach can be used to predict transition for any key Indian corridor, making it a crucial input for policymakers and administrators as they plan India's future cities methodically. It is possible to gain knowledge of the insights and practical steps that need to be taken to address the clustered cities' sustainable urban growth issues in the near future. Results from the SLEUTH study can significantly aid in reducing the negative effects on the natural environment nearby, and hence in reaching and comprehending SDGs.

Similar recent international programs have used CA-based predictive models to forecast future land use change and urban growth, giving a real-time picture of dynamically developing cities. The study's future focus could be on merging different improvements with SLEUTH to see how urban growth might alter over the next two decades.

References

1. Chandan MC, Aithal BH (2021) Impact assessment of corridor oriented development a case of urban agglomerations of India. Int Rev Spat Plan Sustain Dev 9(2):172–194
2. Ramachandran H (2019) Review of industrial and development corridors in India. Working paper 217. Institute for Studies in Industrial Development, pp 1–23
3. Jain S, Panda J, Kant S (2014) Possible socio-scientific issues of land-use and land- cover change impact and associated tools of study with a special reference to Delhi–Mumbai industrial corridor region. Int J Earth Atmos Sci 1(2):58–70
4. Ramachandra TV, Sellers JM, Bharath HA, Vinay S (2019) Modeling urban dynamics along two major industrial corridors in India. Spat Inf Res 27(1):37–48
5. Li S, Liu X, Li Z, Wu Z, Yan Z, Chen Y, Gao F (2018) Spatial and temporal dynamics of urban expansion along the Guangzhou-Foshan inter-city rail transit corridor, China. Sustainability 10(3):593
6. Liu Y, Feng Y (2016) Simulating the impact of economic and environmental strategies on future urban growth scenarios in Ningbo, China. Sustainability 8(10):1045
7. Chandan MC, Bharath HA, Ramachandra TV (2014). Quantifying urbanisation using geospatial data and spatial metrics-a case study of madras. In: Proceedings of lake 2014: conference on conservation and sustainable management of wetland ecosystems in Western Ghats, 13th-15th Nov., Uttara Kannada, India
8. Mallouk A, Elhadrachi H, Malaainine MEI, Rhinane H (2019) Using the SLEUTH urban growth model coupled with a GIS to simulate and predict the future urban expansion of Casablanca region, Morocco. Int Arch Photogram Remote Sens Spat Inf Sci 42:139–145
9. Quium ASM (2019) Transport corridors for wider socio–economic development. Sustainability 11(19):5248
10. Manjunatha MC, Basavarajappa HT (2020) Assessment of land use land cover classification through geospatial approach: a case study of Mysuru Taluk of Karnataka State, India. J Environ Waste Manag 7(1):326–338
11. Reis JP, Silva EA, Pinho P (2016) Spatial metrics to study urban patterns in growing and shrinking cities. Urban Geogr 37(2):246–271
12. Roy PS, Roy A (2010) Land use and land cover change in India: a remote sensing & GIS perspective. J Indian Inst Sci 90(4):489–502